Praise for
Life in Our Phage World

"...spectacular, unique, trailblazing...I have never seen such a display of scholarship and artistry. You have freed scientific writing from its conventional shackles."

—Moselio Schaechter, Distinguished Professor, emeritus, Tufts University, and author of *Microbe and In the Company of Mushrooms*

"...an excellent piece of written and visual art for newcomers to phage research and seasoned phage biologists alike. If you think phage are not relevant to your life or research, reading this is sure to change your mind!"

—Mya Breitbart, Associate Professor, University of South Florida

"Beautiful art, fascinating book and a wonderful historical perspective on the field."

—Lita M. Proctor, Project Coordinator, NIH Human Microbiome Project

"The illustrations, the stories, and the vignettes are just delightful. It is very difficult to create such a perfect combination of science, art, and human warmth, but the authors have managed this superbly."

—Eugene Koonin, Senior Investigator, National Center for Biotechnology Information

"the 21st-century hitchhiker's guide to the (phage) universe...a welcome refresher on phage complexity and diversity that would serve as an amazing resource for biology instructors...even accessible enough for the casual science aficionado to browse..."

—Michael Koeris, *Science*

"a treasure trove, not only of phage information presented in a scholarly fashion, but also of amusing tales of their various roles in the vast living world, including personal stories and biographical sketches of scientists that make it fun to read."

—Abraham Eisenstark, *Microbe*

"a field guide to the tiny portion of phagedom that has so far been explored...handsome illustrations...full of astonishing phage statistics"

—Nicola Twilley, *The New Yorker* online

"drawings show the strange beauty of phages"

—Brian Handwerk, Smithsonian.com

Life in Our Phage World

*A centennial field guide to the Earth's
most diverse inhabitants*

by

Forest Rohwer

Merry Youle

Heather Maughan

Nao Hisakawa

Illustrations by Leah L Pantéa and Benjamin Darby

Wholon

San Diego, CA

Library of Congress Cataloging-in-Publication Data
Rohwer, Forest
Life in our phage world: a centennial field guide to the Earth's most diverse inhabitants. 1ˢᵗ ed.
Includes bibliographical references.
Library of Congress Control Number: 2014945859

ISBN: 978-0-9904943-0-0

Cover art by Leah L Pantéa
Book design by Alexis Morrison, Set Right Typography

Published by Wholon, San Diego, CA
http://www.wholon.org/

Printed in the United States of America

Dedication

To the second century of phage explorers

Table of Contents

Featured Phages

Note: Bacteriophages and archaeal viruses are grouped by their PPT-based family assignments, the eukaryotic viruses by their ICTV family.

Preface

A field guide helps everyone from novice to expert to identify plants, animals, sea shells, and other natural objects. It seemed unfair that the world's most numerous and dynamic life forms, the phage, did not yet have a field guide of their own. The desire to set this right was my motivation for creating *Life in Our Phage World: A centennial field guide to the Earth's most diverse inhabitants*.

Even though no one can *see* phages or other viruses without specialized tools, current imaging methods have unveiled the beauty of their typically icosahedral virions—embellished versions of one of Plato's ideal forms. This is one reason why artistic renditions of phage virions warrant widespread recognition and appreciation, much as birds are known from Peterson's photographs and Audubon's paintings, and as plankton, such as diatoms, were captured in Haeckel's lithographs. Such renditions are to be found in this first phage phield guide.

Why now? Appropriately, we are celebrating 2015 as The Year of the Phage in recognition of mankind's first documentation of their existence in 1915 by Frederick W. Twort. At that time, not one person realized that over the following century and beyond these elemental creatures would radically change our understanding of life. Exploration of their adroit maneuvers has produced our most fundamental understandings of how life works; attempts to identify or count them have revealed the glory of their diversity.

Even so, phages remain all too often ignored, overlooked, discounted. To omit them from the picture is to leave a gaping hole in biology; including them is becoming within reach for many researchers and students. We are entering an age when affordable DNA sequencing will make it relatively common for novices and experts alike to get information about the phages in their samples. Opportunity alone is not sufficient. What is needed is some reference that brings these creatures to life in an easy, quick format with sufficient detail to whet the appetite to learn more. We sought to create just such a phield guide to inform, but more importantly to excite, intrigue, and inspire.

Phages are the winners in the game of life. Let's give them their due.

Forest Rohwer
San Diego, CA 2014

Acknowledgements

I can trace my entry into the phage world back to the influence of one person, Dr. Anca Segall. It was Anca who explained to this recovering immunologist the difference between bacteriophages and macrophages. It was also she who convinced me to attend a seminar by Dr. Farooq Azam. In his famously rambling way, Farooq introduced me to the world of marine microbiology and the ten million phages to be found in every marine milliliter. Also about this time, Anca brought me together with Dr. Rick Bushman, one of the best virologists in the world. During what started as a recreational SCUBA diving trip to Borneo, Rick, Anca, and Dr. Kathie McGuire (my PhD advisor) convinced me to forget law school and go study marine phage with Farooq (at Scripps Institution of Oceanography, UCSD). It was Anca who also introduced me to Drs. Stanley Maloy and Rob Edwards. Rob eventually developed most of the bioinformatics approaches used by our SDSU Phage Group and that then spread throughout the phage field. Stanley has been the long suffering colleague, as well as The Dean that mostly gets blamed for everything. He is an eternal optimist who has created amazing opportunities for myself and thousands of others. While still in Farooq's lab, I had met an extremely talkative undergraduate named Mya Breitbart, now a professor at University of South Florida. Mya later joined my lab at SDSU as my first PhD student, and together we developed the methods for shotgun genomics, which eventually became metagenomics and viromics (a term coined by Dr. John Paul). Again, it was Anca that made this step possible. Even though we had the methods, we had no money for sequencing. Anca sponsored us with 2000 sequencing reactions—a most generous gift at that time. At her suggestion, we started working with Drs. Peter Salamon, Bjarne Andresen, and Joe Mahaffey. Eventually, with the addition of Ben Felts and Jim Nulton, this kernel became the SDSU BioMath Group. This group came up with methods to determine phage diversity from assembly data and over the years cracked many other biology-meets-math problems. This book would have never happened without the mentorship and support of this incredible group of people.

Of course, these developments would not have been possible without a lab to provide the data. Dr. Linda Wegley Kelly has run the lab and been my right hand for over a decade. I would truly be lost without her. Dr. Matt Haynes, Mike Furlan and Yanwei Lim optimized the protocols to study phage in everything from deep sediments to sputum to oxygen minimum zones. Drs. Beltran Rodriguez-Brito, Dana Willner, and Barb

Bailey taught me and the rest of the lab statistics and applied them to the phage world. Over the years, more than a hundred students, post-docs, visiting scientists, technicians, etc., have populated my lab. I learned much more from them than vice versa, and I want to acknowledge them here for all their efforts.

One group of people that rarely get mentioned, but are essential, are the funding and other administrators. Nothing could be done without them. Dr. Lita Proctor has been amazing. She helped design and implement the first Marine Microbiology Initiative at the Gordon and Betty Moore Foundation (GBMF) and then became a leader in microbial matters at the National Science Foundation (NSF) and National Institutes of Health (NIH). Also invaluable are Drs. Matt Kane, Dave Garrison, Phil Taylor, as well as dozens of other program officers and others that make the NSF an awesome organization that really warrants more generous funding from Congress. This book grew out of the first NSF Dimensions program with additional support by the GBMF. For their support of our work, thank you to the current GBMF administrators and, at SDSU, the numerous people in the College of Sciences, Research Affairs, the President's Office, and the Research Foundation. A personal thank you goes to Leslie Rodelander and Gina Spidel who administer the money, create sanity where there often is none, and have gotten my butt of out more than one tight spot.

To the phage phield in general—Drs. Bamford, Beja, Campbell, Dinsdale, Fuhrman, Hatfull, Hendrix, Kutter, Lawrence, Lucigen, Molineux, Nash, Paul, Prangishvili, Rainey, Raoult, Riemann, Shub, Stedman, Stewart, Sullivan, Suttle, Wommack, Young, and so many others—thank you for your criticisms and support, and for teaching a wayward immunologist how to really understand biology.

Many hands helped to assemble this tome: the NSF Viral Dark Matter PIs and people that actually did the work, including Drs. Alex Burgin, Don Lorimer, Anca Segall, and Rob Edwards, as well as Savannah Sanchez, Jason Rostron, Daniel Cuevas, and Katelyn McNair; the guest authors who shared their personal perspectives on phage; and Dr. Matt Sullivan and all of the members of the Tara Oceans expedition who collected phage all over the world.

And last, but truly not least, I say thank you to Anca's "best-worst idea," Willow Segall, for humoring two phage-obsessed parents.

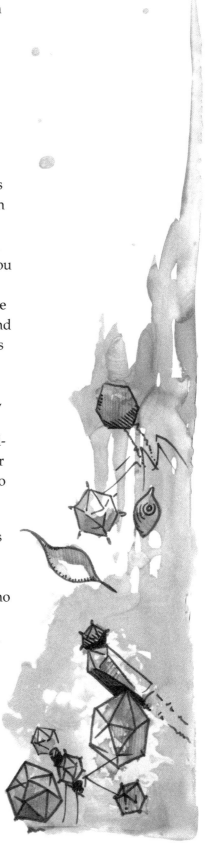

Introduction

Two in the morning in a roadside hotel in the middle of California's Eastern Sierra Mountains. Over the last 48 hours, Mya Breitbart, Tom Schoenfeld, and I had driven over a thousand miles, then carried several hundred pounds of filters, pumps, car batteries, and water up and down steep slopes in the 95° F plus temperatures, all so that we could sit next to much hotter springs for several hours watching the pumps run. Now Tom has fallen asleep with Cheaters playing on the TV and Mya is in the bathroom, finishing the filtering for the day. When she is almost done, I jokingly say, "Just one more thing." She throws a pipetter at me and collapses on the floor. Tom doesn't stir. We'll grab some sleep and then get up at 5 am, drink a lot of coffee, and head back out to hunt the most voracious predators on the planet.

Much of biology is about feeding the phages[1]. By killing nonillions of Bacteria, they have major effects on global energy and nutrient cycles. Phages are the friend of the underdog. When a bacterial strain prospers and threatens to take over the local community, their phages feast and decimate that strain, thereby successfully maintaining microbial diversity in the face of a winner-take-all threat. This behavior can be a nuisance. When we

populate a million dollar lysine fermenter with our bacterial workers of choice, one phage invader can multiply and crash the worker population in a couple of hours. But the phages must be forgiven for such pranks as so many of the major breakthroughs in biology over the past century emerged from the study of phage. Trace most any aspect of molecular biology back to its roots, and there you'll find a phage. Phages were there early on to provide experimental proof that nucleic acid, not protein, was the genetic material and to assist in the recognition of the triplet genetic code. Later they were used to uncover mechanisms of gene regulation, protein binding to DNA, protein folding, assembly of macromolecular structures, and genetic recombination. They have demonstrated evolution by flagrant horizontal gene transfer and provided proof that mutations arise independent of—not as a response to—the pressure of natural selection.

Enzymes from phage launched the molecular biology revolution and remain essential tools for genetic engineering. Phage genomes were the test subjects used for the first genomic and shotgun genomic/metagenomic sequencing, the first fully synthetic life forms, etc. Phage

---

[1] We use the term 'phage' sensu lato to encompass all microbial viruses, i.e., the bacteriophages (Bacteria-eaters), viruses of the Archaea, and viruses of single-celled eukaryotes.

biologists were at the forefront of advances in cancer biology. Most of the stuff of life itself—the global pool of genetic diversity—is encoded by phage. Closer to our individual homes, in the last ten years we finally came to realize that of all the varied genes we carry in our own bodies, the majority reside within our phages. Phages are essential bionts within the human—and every other—holobiont.

Despite their paramount importance to human health, to science, and to all life on the planet, the phage field remains a niche area of study. One reason that phages (as well as most viruses that don't make us or our domesticates sick) remain overlooked is that you can't just go out or look inside and observe them. When outside a host cell, they travel as virions so small that seeing them requires an electron microscope or other sophisticated and costly equipment. Most can't be cultured and interrogated in the lab because their hosts are not known or not yet culturable. Community metagenomics, likewise, is still relatively difficult and costly. This inability to 'see' phages leads to a disconnect between them and all other life forms. Most scientists and others just don't think of them as alive. So this major component of life is reduced to its inert intercellular transport form that is then subjected to biochemical analysis and described in lifeless terms, leaving us blind to their nature as active agents. This is somewhat of a travesty, as these bits of biochemistry are the most successful predators on the planet. They are promiscuous and engage in kinky sex games (e.g., homologous and illegitimate recombination with related and completely alien genomes, orgies of hundreds of genomes). Humans observing the virions perceive them to be inert. But these 'inert' particles, given contact with a potential host, reveal their true nature as complicated nanomachines primed for action. Their performance is precise; milliseconds or nanometers mean the difference between life and death.

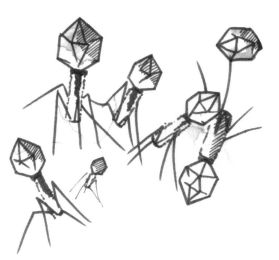

It is not possible to understand the biological world without 'seeing' the phages. This book provides a glimpse of the rich and diverse phage life that has been sampled over the past one hundred years. The overall organization of the book parallels the phage life cycle. It arbitrarily starts with their virions on the prowl, observes them as their genome enters and takes over a host cell, describes their replication, then applauds as the progeny virions assemble and make their escape into the world. For each stage we chose a few diverse phages to feature.

Field guide pages provide basic information for each of these phages, the kinds of information a naturalist would have at hand for any life form they wanted to study. For each we also relate a lively, thoroughly researched story revealing some of this phage's secrets for success. Terms in boldface within the stories are defined in our glossary. Each story plays out visually in an illustration by San Diego fine artist Leah L. Pantéa. These illustrations are rich in detailed information intended to complement your reading of the text. The 30 featured phages were selected to illustrate the great diversity that exists in even the small fraction of the phage world that has been characterized. Although you will find well-studied phages such as λ, T4, and T7 in these pages, we made no attempt to include the wealth of information available for them; there are many good books that already do this. Each chapter ends with one or two longer, personalized perspectives. Each informs about a particular aspect of The Big Picture and relates part of the recent history of phage research. What makes them so delightful to read is that each is infused with the excitement and humor that has characterized phage research and phage researchers.

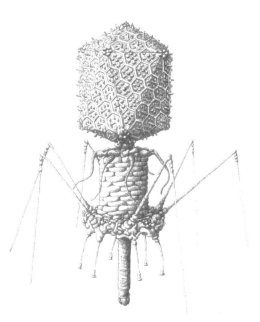

Since we envisioned an Audubon-like field guide to the phages of the world, the portraits of the 30 phages were rendered in pen and ink by Benjamin Darby, an imaginative San Diego artist. As typical of a field guide, he emphasized important or identifying characters of each specimen and added a touch of elegance. When no photo or virion structure was available for that particular phage, we turned to its close relatives for a stand-in. Such a field guide would also group the objects of study into related groups. This is not so easy to do for the phages. Observable virion morphology is not an adequate basis for such classification as great phage diversity lurks within each virion type.

The recent accumulation of genome data provides another handle on phage taxonomy, but application of this approach remains challenging. The now familiar Tree of Life portrays the evolutionary relationships among all members of the three domains based on the rRNA genes that they all carry. A similar tree could in theory be constructed for the phages if any single gene were present in all phage genomes, but there is no such gene, thus there can be no such tree. At most, such 'signature genes' can serve to elucidate relationships within closely-related groups. Moreover, the evolution of viruses has not followed the same strict pattern of vertical descent from a common ancestor as predominates in many cellular organisms. Phages may not all share a common ancestor, and moreover they have exchanged genes horizontally. This argues for a different approach to their classification. To that end, we compared the genomes of 1220 phages and built a taxonomic tree based on their

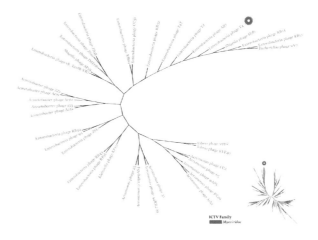

similarities (*see page 8-8*). For each featured phage in the field guide, we show its relationship to all the other 1219 phages on that tree and also zoom in on its local tree neighborhood.

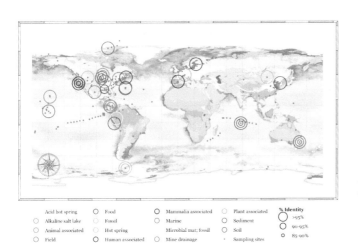

In the tradition of other field guides, we have included a global map showing the known geographic range of each featured phage as well as the habitats where it has been found so far. These 'sightings' (*see page 8-20*) are based on BLAST hits between that phage genome and publicly-available metagenomes from around the world. For a guide to interpreting these maps, see Appendix A4 (*page 8-20*). While the data displayed here is interesting, more important is what is missing. Most of the globe and many ecosystems have not been sampled nor have their phage communities been characterized. Microbes have been found everywhere people have looked on Earth—on the land, in the sea, in the air, inside rocks and inside host cells—even under extreme conditions previously thought to be unable to support life. Wherever there are microbes, there are phages. For phage explorers, most of the Earth remains a terra incognita. It is time to get to work and put phages on the map.

To portray phage genomes as the lively, evolving molecules that they are, we present two versions of each phage's genome. First, an artist-created overview shows the variety of genome structures used by our featured

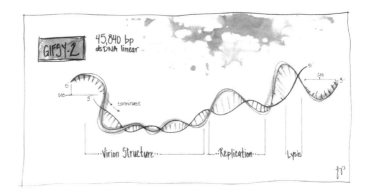

phages when traveling by virion between hosts (e.g., linear or circular, single-stranded or double-stranded, sticky ends, direct or inverted terminal repeats). Here we have also delineated functional modules and highlighted landmarks that are featured in the stories or are well appreciated among phageophiles. Each overview is followed by a detailed genome map that allows for admiration of each gene including information (if available) about its function, its homology with other phage genes, and/or (if applicable) the localization of its protein product in the virocell. Genes are counted as open reading frames (ORFs) if they encode a protein and as RNAs if their transcripts are not translated (e.g., tRNAs).

To emphasize the dynamic nature of phage in the writings, we have developed a lexicon based on ethology (*see page 8-27*) and used its terms in our writing. The goal of this writing style is to bring each phage to life, without seriously compromising scientific accuracy. It is also to remind us that there are many phage behaviors that we expect to observe, but haven't studied yet. In some cases, we can link a particular behavior to one or more genes, but the genetic basis for many remains to be discovered. No doubt clues are hiding in the ~80% of phage genes that are completely novel.

The first 100 years of phage research have fundamentally changed our lives and our understanding of the natural world. In the near future we expect to see a new synthesis in biology that puts phage at the center of the field, no longer to languish in a dimly lit corner as a biological novelty, an after-thought. But that will occur only when many people, such as yourself, include the phage in your research, in your study, in your teaching, and in your understanding of life on Earth. The second century of phage study is beginning. Be there.

Fast-forward to a decade after the Sierra Phage Hunting expedition and I'm walking around in a Wisconsin winter in shorts; −20° F is not a great place to make a San Diego fashion statement. My latest phage hunting had taken me to the Arctic and I am still waiting for my winter clothing to be shipped back from Russia. Mya has gone on to become a leader in the field of phage ecology, despite a history of throwing things at her PhD advisor. I am crunching through the snow with Tom and his ever-excitable business partner David Mead. Together they had built Lucigen into a leading company in the realm of enzymes and cloning, now expanding into diagnostics. Many of their products are based on enzymes originally found in phages isolated from hot springs, enzymes such as DNA polymerases that are also primases and reverse transcriptases, incredibly efficient ligases, and many others. But neither one of them is talking about their business successes. They are both happily planning yet another sampling trip to find yet more weird and wonderful phages. They know the phages are out there, waiting for someone to notice.

Chapter 1:
Welcome to the Phage World

STORIES

PERSPECTIVE

The Phage Life Cycle: Why So Many Genes?

Merry Youle

Note: This story is an introduction to the ways of the phages. If you are already acquainted, consider skipping this one.

As with cellular organisms, phage replication proceeds via a precisely orchestrated life cycle. However, for phages the pace is rapid and the numbers astronomical. Given plush culture conditions, one **virion** can produce more than one hundred infectious progeny in less than one hour. In the environment, phages launch approximately 10^{24} productive infections every second to maintain their estimated 10^{31} global population (Hendrix 2010). Each turn of the life cycle, a virion becomes part of a **virocell** that produces and releases more virions to repeat the cycle yet again. What are the tasks that a phage must accomplish to keep their lytic life cycle turning?

One turn of the cycle

Being a cycle, there is no beginning, but let's start with a virion adrift. Each virion contains one copy of the phage genome, as DNA or RNA, encased in a protective protein shell, or capsid. Virions are the beautiful, intricate structures visualized with an electron microscope that were used earlier to classify phages into a few major families. But the virion is not the phage. Virions are the inert dispersal form, sometimes likened to a spore, that transports the phage genome between hosts. Even this step is not a simple task. The capsid must protect the genome from environmental dangers such as UV irradiation and **nucleases**. It must be quick to recognize a host when it collides with one. Although it may explore the cell surface for a while, when it detects its specific **receptor**, it must irreversibly bind to it (**adsorption**) and deliver the phage genome into the host cell. The capsid itself remains outside, empty, its job completed. Once the phage adsorbs to the host cell, the phage and cell together are referred to as a virocell.

If that infecting genome evades host defenses, it then redirects the cell's labors to the production of many progeny virions. In this finely-tuned takeover, the transcription of host genes may be shut down and the host's DNA destroyed, while energy produced by host metabolism is expropriated to fuel phage reproduction. The phage genome replicates repeatedly to yield 25, 50, or even hundreds of copies. These copies engage in promiscuous sex by exchanging genome segments with one another or with the host chromosome through **recombination**. Meanwhile, all of the structural proteins comprising the capsid are synthesized in the correct relative numbers. Typically the proteins are assembled into **procapsids** and then a genome is **packaged** into each. Mature virions accumulate until the infection is abruptly terminated by lysis of the host cell. This cell lysis, like the other intracellular steps in the life cycle, is deliberately timed and executed by the phage. The escaping virions then set out in search of new hosts, therein to repeat the cycle yet once again.

Minimal genomes

How many genes are required to carry out these life cycle steps? Consider phage Qβ, a minimalist Leviphage with a single-stranded RNA (ssRNA) genome that encodes only four genes. Using RNA for its genome is efficient in that copies of the **positive-sense** RNA genome function also as mRNA for translation of phage proteins. For Qβ, DNA synthesis is expendable, but there is a price to be paid. Since its bacterial hosts do not have enzymes for replicating RNA, Qβ must dedicate one gene to encoding its own replicase. Two more genes are used for structural proteins. Qβ's simple icosahedral capsid (T=3) is built from 180 copies of the major capsid protein and 12 copies of the minor capsid protein. The fourth gene encodes a multifunctional protein termed the maturation protein, one copy of which is found in each capsid. Its essential tasks are: (1) to lyse the host cell so the assembled virions can exit; (2) to protect the encapsidated genome from RNases while in transit between hosts; and (3) to recognize and adsorb to a **pilus** on a potential host, the first step in launching a new infection.

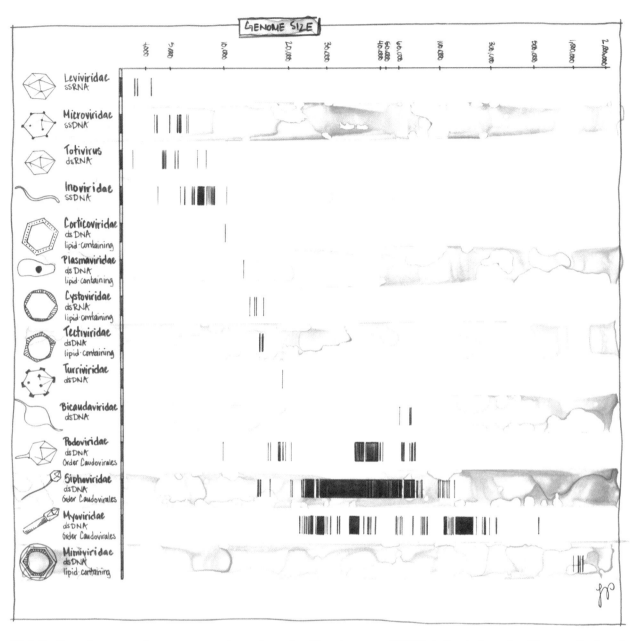

Distribution of genome sizes among various phage families. A range of ICTV-defined phage families are displayed along the y-axis along with their virion morphologies. The x-axis corresponds to genome size. Each 'lane' highlights the range of genome sizes for a particular ICTV family. Varying bar thicknesses result from adding one thin line for each phage with that genome size. These ICTV families correspond closely to the PPT-based families used throughout this book (*see page page 8-8*). This illustration was inspired by Figure 2 from Hyman, Abedon 2012. Genome sizes for families were drawn from that source and from GenBank (http://www.ncbi.nlm.nih.gov/genomes/GenomesHome.cgi?taxid=10239).

These four genes are all encoded within a 4217 nt single-stranded RNA (ssRNA) genome. Extra coding economy is provided by the overlapping of the genes for the two capsid proteins (*see page 7-10*). When translating the major capsid protein, ~5% of the time the ribosome 'reads through' the 'leaky' stop codon and continues translating, thus yielding the longer minor capsid protein.

The smallest DNA phage genomes are found among the Microphage. The small circular, single-stranded DNA (ssDNA) genome of Microphage ϕX174 was the first DNA genome sequenced (Sanger et al. 1977). With only 5,386 nt, its genome appeared to be too small to encode all of its proteins. It can, in fact, accommodate those eleven genes because several of them overlap by using different reading frames. The jump from four genes to eleven reflects their more complex, although still tailless, icosahedral capsid whose assembly requires the products of six genes. Even in this extremely small genome of a well-studied phage, two genes are not essential for phage replication in the lab, and thus their function has not been determined.

More genes, more capabilities

Most phages have much larger genomes, with 50-100 kbp being typical in various environments (Angly et al. 2009) The well-studied myophage T4 uses a 169 kbp genome that encodes about 300 genes to carry out the same life cycle steps (*see page 4-39*; [Miller et al. 2003]). Why so many genes? For what purposes? The functions of most phage genes are still unknown. We do know that phages dedicate a substantial number of genes to encoding virion structural proteins. At the upper end of the known range is the phage *B. thuringiensis* 0305φ8-36 that uses 42% of its genome to encode 55 structural proteins (Thomas et al. 2007). Of T4's 300 genes, only 62 are essential under standard lab conditions, and 36 of those encode structural proteins. Increasing the number of genes above the minimum can allow a phage to carry out the basic life cycle steps with greater finesse. For instance, Qβ employs only a single multi-purpose gene to effect host lysis (*see page 7-11*), while many other phages use a holin-**endolysin** system (*see page 7-5*). The latter method provides more versatile lysis timing, but it costs two or more genes (Zheng et al. 2008).

Apparently the vast majority of the 'non-essential' genes in any phage genome are essential for the phage to compete successfully in the world outside the lab. Such genes may function to counter host defenses, to compete with other phages wanting the same host, or to do battle with other mobile genetic elements. Others are used to obtain the extracellular resources needed for phage replication, such as phosphate, or to precisely manipulate the host's metabolism to provide for the needs of the phage. Still others may enable the phage to expand its host range or to thrive under other environmental conditions. In addition, temperate phages (*see page 1-5*) that co-exist with their host as a virocell for extended periods often carry metabolic genes that benefit their host, thereby serving the phage's interests, as well. Considering the great number of currently uncharacterized phage genes, undoubtedly many novel protein structures and functions await discovery within the vast dark matter of phage genetic diversity.

Cited references

Angly, FE, D Willner, A Prieto-Davó, RA Edwards, R Schmieder, R Vega-Thurber, DA Antonopoulos, K Barott, MT Cottrell, C Desnues. 2009. The GAAS metagenomic tool and its estimations of viral and microbial average genome size in four major biomes. PLoS Comput Biol 5:e1000593.

Hendrix, RW. 2010. Recoding in bacteriophages. In: JF Atkins, RF Gesteland, editors. *Recoding: Expansion of Decoding Rules Enriches Gene Expression*. Springer. p. 249-258.

Hyman P, ST Abedon. 2012. Smaller fleas: Viruses of microorganisms. *Scientifica* 2012:734023.

Miller, ES, E Kutter, G Mosig, F Arisaka, T Kunisawa, W Ruger. 2003. Bacteriophage T4 genome. Microbiol Mol Biol Rev 67:86-156.

Sanger, F, G Air, B Barrell, N Brown, A Coulson, J Fiddes, C Hutchison, P Slocombe, M Smith. 1977. Nucleotide sequence of bacteriophage φX174 DNA. Nature 265:687-695.

Thomas, JA, SC Hardies, M Rolando, SJ Hayes, K Lieman, CA Carroll, ST Weintraub, P Serwer. 2007. Complete genomic sequence and mass spectrometric analysis of highly diverse, atypical *Bacillus thuringiensis* phage 0305φ8–36. Virology 368:405-421.

Zheng, Y, DK Struck, CA Dankenbring, R Young. 2008. Evolutionary dominance of holin lysis systems derives from superior genetic malleability. Microbiology 154:1710-1718.

The Phage Life Cycle: Why Be Temperate?

Merry Youle

Note: This story is an introduction to phage lifestyles. If you can already define *lysogen* and are familiar with the ins and outs of prophages, consider skipping this one.

Lytic replication can lead to a dead end. Our usual view of explosive phage replication is biased by culture-based studies in which abundant, well-fed, rapidly-growing hosts are provided and competing phages excluded. Thus provisioned, one **virion** often produces more than a hundred progeny in less than an hour. Conditions in the world outside the lab do not foster such exuberant proliferation. Susceptible hosts may be so scarce that virions perish before finding one. And when the right bacterial strain is encountered, the cell is apt to be starving and unable to support virion production. Such hurdles are the norm for phages in many environments.

An alternate strategy

Some phages, including a large majority of the tailed phages (**Caudovirales**), have a second strategy at hand: they establish a temporary partnership with the host cell. In this case the phage postpones virion production in exchange for interim preservation along with slow, host-paced replication. Phages that are able to abstain from immediate rapid replication culminating in host lysis are termed **temperate**. Soon after arrival in the host cell, they can opt to synthesize an **integrase** that will insert their genome into the host's chromosome at a specific location by **site-specific recombination**. An inserted genome is known as a **prophage**; a bacterium with a prophage is a **lysogen**; the process is referred to as establishing **lysogeny**.

Although not actively replicating, a prophage still transcribes one or more of its genes. At least one phage-encoded protein is necessary to repress transcription of the genes that would otherwise trigger the lytic pathway. Often that same protein also protects the **virocell** from infection by related phages, a defense known as **superinfection immunity**. Other active genes may increase host fitness in some manner. By thus favoring growth and

survival of its partner, the prophage also prospers. Each time the host replicates its chromosome, the prophage is replicated along with the rest of the host chromosome; each time the host divides, one lysogen becomes two and one prophage becomes two.

When conditions deteriorate, the prophage can mutiny and turn on its partner. It then excises from the chromosome and converts the cell into a virion factory, with lysis following soon thereafter. This **induction** of the prophage can be triggered by damage to host DNA. It also occurs spontaneously at a low frequency, thus ensuring that there are always some virions drifting about in the environment and available to launch lytic infections when environmental conditions improve.

Lysis or lysogeny?

How does a temperate phage genome decide what to do when it arrives in a host cell? The optimal choice is the one that will yield the most progeny *over time*. Choose lysogeny, and at best you duplicate at a slow pace; at worst you, along with your host, are consumed by a hungry protist. Lytic replication can potentially produce many progeny, quickly, but only if conditions are favorable. Furthermore, the phage that seeks victory by rapid lytic replication will win only a pyrrhic victory if those progeny cannot launch future infections. If the phage were to hire a consultant to advise it, said consultant would formulate a mathematical model and assign values to parameters such as the concentration of potential hosts, the concentration of competing phages, the percent of potential hosts that are immune lysogens, virion half-life, host metabolic state, and predicted **burst size**. Phage λ, a model temperate coliphage, likewise takes those parameters into account and makes its own decision efficaciously. The two key factors are the ability of the cell at hand to support virion production and the likelihood that the progeny produced will find hosts. λ assesses one parameter that integrates both of those factors: the concentration of phages in the infected cell.

It had long been observed in lab experiments that, at a low **MOI** (multiplicity of infection), λ almost always goes lytic; the probability of lysogeny increases with increasing MOI, eventually reaching one. More recent work uncovered some of the molecular mechanisms underlying this behavior. Al-though the **porins** used by λ as its **receptor** (LamB) are numerous and scattered over the entire outer membrane of its host, DNA entry requires an inner membrane protein (ManY) that is concentrated around the cell poles (Edgar et al. 2008). Thus λ usually delivers its genome into a polar region. There it quickly initiates synthesis of its early proteins including a key regulatory protein, CII. A high concentration of CII establishes lysogeny. The higher the MOI, the more CII synthesized, the higher the probability of lysogeny. By sensing the

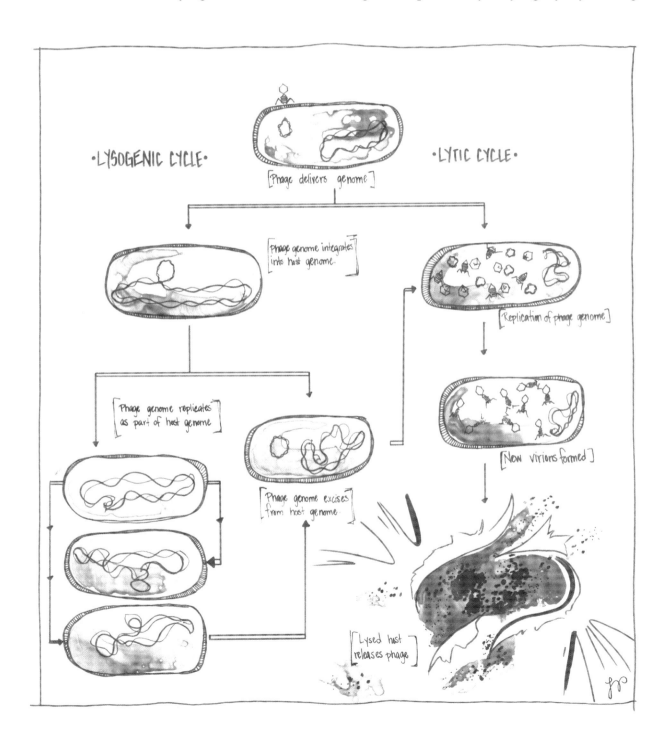

MOI in this fashion, λ assesses the relative concentration of hosts and competing phages.

λ also factors in the metabolic state of the host cell. The concentration of CII depends not only on the amount synthesized, but also on the cell volume. Thus, given the same MOI and the same amount of CII synthesized, CII concentration will be higher in smaller cells, thus favoring lysogeny when the host is starved. The concentration of CII is also affected by an essential host protease, FtsH, that is localized at the poles and that specifically degrades CII (Hendrix 2008). Since the cellular FtsH level is regulated by other aspects of host metabolism, this provides λ with yet another indication of the well-being of this host.

Variations on the theme

While lysogeny as practiced by λ has become the paradigm, many other temperate phages add their own distinctive twists (*see page 5-5 and page 5-37*)

to the story. Some prophages do not integrate, but persist stably as **plasmids**, replicating with the cell cycle and partitioning to both daughter cells when their host divides (*see page 5-25*; Ravin 2011). Other phages, when stalled by unfavorable conditions, settle for **pseudolysogeny** (Łoś, Węgrzyn 2012). In this state they do not replicate with the host, but simply persist, ready to resume activity when conditions improve. Many archaeal viruses that infect the **hyperthermophilic** Crenarchaeota also display temperance, but the underlying mechanisms are still unknown (Prangishvili, Garrett 2005).

Nowhere else is the association between phage and host more intimately intertwined. Exploration of this interplay can reveal much about cellular biology, but unraveling those same interactions poses challenges to researchers. Thus although lysogeny is known to be important in diverse environments ranging from marine waters to the human gut, much remains to be investigated.

Cited references

Edgar, R, A Rokney, M Feeney, S Semsey, M Kessel, MB Goldberg, S Adhya, AB Oppenheim. 2008. Bacteriophage infection is targeted to cellular poles. Mol Microbiol 68:1107-1116.

Hendrix, R. 2008. Cell architecture comes to phage biology. Mol Microbiol 68:1077-1078.

Łoś, M, G Węgrzyn. 2012. Pseudolysogeny. Adv Virus Res 82:339-349.

Prangishvili, D, RA Garrett. 2005. Viruses of hyperthermophilic Crenarchaea. Trends Microbiol 13:535-542.

Ravin, NV. 2011. N15: The linear phage-plasmid. Plasmid 65:102-109.

Recommended reviews

Canchaya, C, G Fournous, H Brüssow. 2004. The impact of prophages on bacterial chromosomes. Mol Microbiol 53:9-18.

Paul, JH. 2008. Prophages in marine bacteria: Dangerous molecular time bombs or the key to survival in the seas? ISME J 2:579-589.

Phage Classification for the 21st Century

Daniel C. Nelson[†]

Abstract: *Taxonomic classification of bacteriophages, and all viruses, has been hindered by the lack of a common protein or genetic locus similar to the 16S rRNA in Bacteria on which to base a phylogenetic tree. Traditional taxonomy schemes for phage have been based on morphology (i.e., electron microscope analysis) and biochemical evidence (i.e., type of nucleic acid, strandedness, etc.). However, these approaches do not utilize the significant accumulation of genomic data generated during the past 20 years, nor do they consider the mosaic nature of phage evolution. In response to these shortcomings, several competing genomic-based alternative classification schemes have been proposed and are being passionately debated. Whether one or several methods are eventually adopted, it is clear that the current paradigm must evolve to keep pace with the discovery of new phages. Meanwhile, the current indexing by the International Committee on the Taxonomy of Viruses (ICTV) continues to fall farther behind. A shift in emphasis from higher order taxa designations to more detailed descriptions of monophyletic groups or even individual viruses in an open source, user-curated format may be a trend for the future of phage taxonomy.*

[†] Institute for Bioscience and Biotechnology Research, University of Maryland, Rockville, MD
Email: nelsond@umd.edu
Website: https://www.ibbr.umd.edu/profiles/daniel-nelson

My introduction to phage taxonomy began in the fall of 2004. I had spent five years in Vince Fischetti's lab at The Rockefeller University, first as a postdoctoral fellow, then as a research associate (i.e., a glorified postdoc), exploring the potential therapeutic use of phage-encoded endolysins. Having trained as a protein biochemist and now being focused solely on biochemically characterizing these enzymes, I was not particularly interested in phage biology, phage genomics, or phage-host interactions. Phage taxonomy was probably the furthest thing from my mind. In fact, at large meetings such as the American Society for Microbiology General Meeting, I actively avoided the "phage group" by entering my posters in the biochemistry sessions rather than the phage sessions because I felt more connected to my peers on the enzyme level than the phage level.

I awakened to phage taxonomy abruptly that fall when an email from an editor at the *Journal of Bacteriology* asked me to review a manuscript on that very subject. That paper classified phages based on conserved features in their structural proteins as discovered through comparative phage genomics (Chibani-Chennoufi et al. 2004). I was flattered and excited—it was the first time I had been asked to serve as a reviewer—but scared at the same time since I felt I had no real expertise in comparative phage genomics. I accepted the assignment and began earnestly reading the manuscript, all the while scrambling to read recent articles on phage classification and taxonomy to supplement my limited knowledge. This reading showed immediately that there was no universally accepted taxonomic method; animated controversies swirled around the various taxonomic methods. Since this was my first critique, I wanted to show the editor that I took the assignment seriously and was current in the literature of the field, so with the critique I included a comprehensive background summarizing the debate over phage taxonomy, listed several recently described phage taxonomy methods, and explained how the manuscript I was reviewing fit with one of those methods. Much to my surprise, the editor wrote back the next day saying he was impressed with the depth of my review and wanted me to write a guest commentary expanding on the taxonomy controversies I cited in the critique. Now I was truly petrified. I, an 'outsider' and a junior scientist to boot, was about to publish my views on a subject that was currently hotly debated by real experts who had devoted their entire career to this field. While writing the commentary (Nelson 2004), I had nightmares of being confronted at phage meetings by scien-

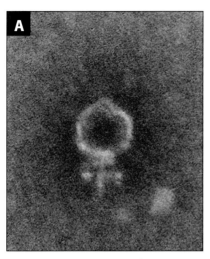
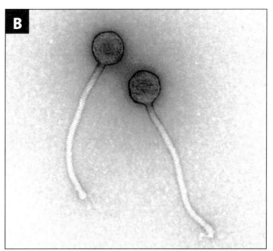
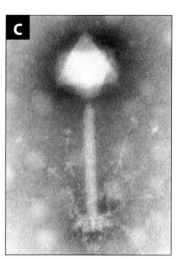

Figure 1. Transmission electron micrographs of representatives of the three families within the ICTV order *Caudovirales*. (A) the *Podoviridae*: *Staphylococcus aureus* phage GRCS courtesy of Daniel C. Nelson. (B) the *Siphoviridae*: Mycobacteriophage Badfish courtesy of Matt Olm, Deborah Jacobs-Sera, and Graham Hatfull. (C) the *Myoviridae*: *Synechococcus* phage S-PM2 (Mann 2005).

tists whose classification schemes I had disagreed with, or worse still, by those whose theories I had left out of my commentary altogether. Although I did not immediately know what I would write, I knew instantly what the title would be—*Phage Taxonomy: We Agree To Disagree.*

Now, a decade later, the editors of this book have asked me to share my current views, refreshed to reflect the many significant changes that have accompanied the explosion of sequence data and other tools now available to phage taxonomists. While I continue to have close ties to the phage community and to study phage-derived enzymes, I have remained on the fringes of the phage classification conversation. Perhaps I am once again a questionable choice to author such a tome, but, at the same time, my outsider status allows an unbiased perspective on the various alternative taxonomic strategies.

The evolution of phage taxonomy

Since the discovery of bacteriophage by Frederick Twort in 1915 (Twort 1915) and Félix d'Hérelle in 1917 (d'Hérelle 1917), taxonomic classification of phage has been a continually evolving process. In the 1920s and 1930s, phages were known to differ in their bacterial host specificity. In a hallmark 1934 paper, Alice Evans used different 'races' of streptococcal phage to discern streptococci that caused

human infection (e.g., *Streptococcus pyogenes*) from those that caused bovine infection (e.g., *Streptococcus dysgalactiae*) (Evans 1934). This approach was instrumental in establishing the field of phage typing, but in so doing it also developed a rudimentary phage classification system.

During the 1940s and 1950s, the electron microscope afforded biologists the ability for the first time to not only see phage virions, but to make observations about their physical size, tail fibers, and capsid symmetry (Luria, Delbruck, Anderson 1943). This resulted in a simple classification system based on phage morphology—a concept that would have a profound influence on future taxonomic schemes (Fig. 1). In 2011, it was reported that at least 6,000 phages had been investigated by electron microscopy (Ackermann 2011).

With advances in biochemical methods in the 1960s came the isolation of nucleic acids from phage virions and the consequent ability to resolve both genome size and type (DNA, RNA, single-stranded [ss], double-stranded [ds], linear, circular), adding further information to complement the morphological taxonomic schemes (Thomas Jr., Abelson 1966). By 1967, Bradley had created the first unified phage classification system based on all the data available at the time (Bradley 1967). His proposal included six divisions of phages based on both morphology and nucleic acid, including three with dsDNA (with long contractile tails, long non-contractile tails, or short tails); two with ssDNA (filamentous and small tailless), and small tailless ssRNA phages. The late 1960s also saw the formation of various committees concerned with viral taxonomy and nomenclature. Eventually, the International Committee on the Taxonomy of Viruses (ICTV) was established to develop a universal taxonomic system for viruses infecting Bacteria, fungi, plants, animals, and, later, Archaea. In 1971, the ICTV published its first report on virus taxonomy, and in so doing, incorporated many of Bradley's ideas for grouping bacterial viruses by their tail morphology (Wildy 1971).

A competing viral classification scheme, developed primarily for animal viruses, was suggested by David Baltimore in 1971 (Baltimore 1971). This system placed viruses into one of six, and eventually seven, groups based on their nucleic acid: whether RNA or DNA, its strandedness, its sense, and its mode of information transfer from genome to protein (Fig. 2). While popular with many virologists, the Baltimore system does not adequately describe the diverse morphology of phage at the lower taxonomic levels, particularly for dsDNA tailed phage, and therefore has fallen out of favor with most phage taxonomists.

In contrast, the ICTV taxonomy is based on the hierarchical system originally devised by Linnaeus in 1758 to classify plants and animals (Linnaeus 1758). The recognized viral taxa, progressing from the highest to the lowest rank, are: order, family, (sub-family), genus, and species. The term 'species,' which has generated the most discussion, was until recently defined as "a polythetic class of viruses that constitute a replicating lineage and occupy a particular ecological niche" (http://ictvonline.org/). A polythetic class is, in turn, loosely defined by the possession of a consensus group of properties, although no single property is necessarily shared by all members. Accordingly, phage have been historically classified by the ICTV based largely on their nucleic acid composition (i.e., dsDNA, ssDNA, dsRNA, ssRNA), morphology (i.e., tail presence,

type, and length), and other virion structural properties (e.g., the presence of a lipid membrane). That there has been little to no input from genome data has caused considerable lively debate in the phage community, escalating with the rapid expansion of genomic data available (see below).

Since their first report in 1971, the ICTV has published new reports every three to six years, with annual online updates in recent years. The most recent full report of the ICTV, the ninth, published in 2011 (King et al. 2011) has been supplemented by three subsequent updates, the most recent being in July, 2013. The current taxonomy release includes seven orders, 103 families, 22 subfamilies, and 455 genera for all known viruses (http://ictvonline.org/). Of these, bacteriophage presently encompass one order and ten families (Table 1).

The archaeal viruses pose similar classification challenges for the ICTV. The diverse hosts of these viruses include members of the phyla Euryarchaeota (encompassing numerous methanogens and halophiles) and the Crenarchaeota (notably many thermophiles and hyperthermophiles). Although the virions of some archaeal viruses have the icosahedral morphology common among viruses of Bacteria and Eukaryota, others embody shapes that are unique to these viruses such as extremely long filaments with terminal 'claws,' bottle-shapes, and spindles with or without slender tails (Ortmann et al. 2006).

Challenges facing the ICTV

Despite its 40+ year history, critics are quick to point out obvious shortcomings in the ICTV paradigm. Specifically, the model puts little, if any, weight on genomic information, with the single order and three main families that contain the majority of phage being based strictly on morphological characteristics observed by electron microscopy. In this respect, there has been little advancement since Bradley's 1967 classification. The importance placed on morphology and the concomitant disregard for functional genomics has led many phage biologists to question the ICTV rationale. It has also led to some conflicting taxonomic assign-

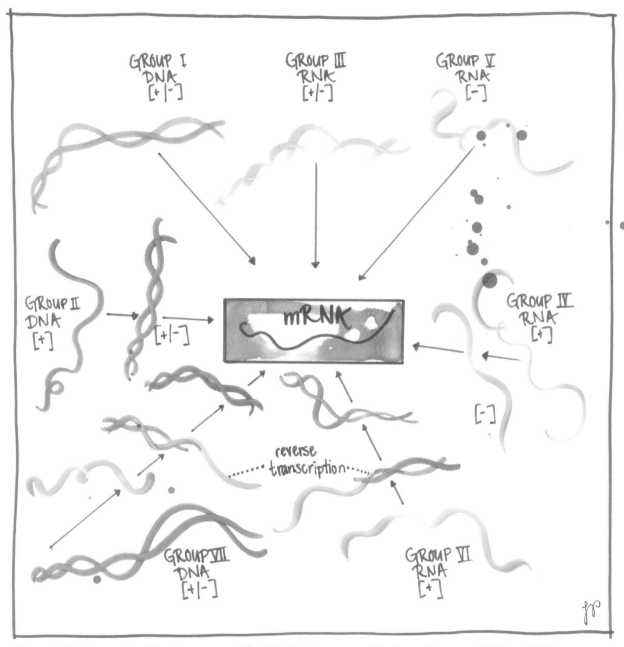

Figure 2. Baltimore classification system. The [+/-] designates **positive/negative sense** RNA or DNA.

ments. For example, *Salmonella* phage P22 and Enterobacteria phage T7 are, according to the ICTV, both members of the *Podoviridae* family based on the presence of short tails in both. However, on the genomic level, P22 is so closely related to Enterobacteria phage λ of the *Siphoviridae* that, as has been known for 40 years, recombination between their genomes forms functional hybrids (Botstein, Herskowitz 1974). Given the current bourgeoning

of metagenomic approaches for sequencing whole viral communities (Casas, Rohwer 2007; Mokili, Rohwer, Dutilh 2012), we now have vast amounts of data about a great diversity of environmental phages, none of which have been isolated for EM studies. Likewise, prophage or prophage-like elements are abundant in streptococcal, mycobacterial, and other bacterial genomes (Ferretti et al. 2001; Fan et al. 2014), yet few are recognized by ICTV or

Table 1. ICTV classification of bacteriophages and archaeal viruses. Many remain unclassified even at the family level. Bolded text denotes a phage featured in this book. The approval of families *"Pleolipoviridae," "Sphaerolipoviridae," "Spiraviridae," and "Turriviridae"* and their constituent species is pending at the ICTV. Sources: Pietilä et al. 2009; Pietilä et al. 2013a; Pietilä et al. 2013b; Prangishvili 2013; Pawlowski et al. 2014; N. Atanasova, personal communication; ICTV (http://ictvonline.org/).

ICTV Order	ICTV Family	Genome	Virion Morphology	Examples
Bacteriophages				
Caudovirales	*Myoviridae*	dsDNA	Icosahedral head; long, contractile tail	*Bacillus* phage SPO1, *Staphylococcus* phage Twort, **Enterobacteria phage T4**
Caudovirales	*Podoviridae*	dsDNA	Icosahedral head; short tail	**Enterobacteria phage T7**, *Streptococcus* phage C1, ***Bacillus* phage φ29**
Caudovirales	*Siphoviridae*	dsDNA	Icosahedral head; long, non-contractile tail	**Enterobacteria phage λ**, *Bacillus* phage SPβ, **Mycobacteriophage Brujita**
—	*Corticoviridae*	dsDNA	Complex icosahedral capsid with internal lipid membrane	*Alteromonas* phage PM2
—	*Cystoviridae*	Segmented dsRNA	Complex icosahedral capsid, lipid envelope	***Pseudomonas* phage φ6, *Pseudomonas* phage φ8**
—	*Inoviridae*	ssDNA	Filamentous	**Enterobacteria phage f1**, Enterobacteria phage M13, *Vibrio* phage CTX
—	*Leviviridae*	ssRNA	Icosahedral capsid	**Enterobacteria phage Qβ**, Enterobacteria phage MS2
—	*Microviridae*	ssDNA	Icosahedral capsid	Enterobacteria phage φX174, *Spiroplasma* phage 4, *Chlamydia* phage 1, *Bdellovibrio* phage MAC 1
—	*Plasmaviridae*	dsDNA	No capsid, lipid envelope	***Acholeplasma* phage L2**
—	*Tectiviridae*	dsDNA	Complex icosahedral capsid with internal lipid membrane	*Bacillus* phage AP50, **Enterobacteria phage PRD1**, *Thermus* phage P37-14
Archaeal viruses				
Ligamenvirales	*Lipothrixviridae*	dsDNA	Flexible, rod-shaped, lipid envelope	*Sulfolobus islandicus* filamentous virus (SIFV), *Thermoproteus tenax* virus 1 (TTV1)
Ligamenvirales	*Rudiviridae*	dsDNA	Stiff, rod-shaped	*Sulfolobus islandicus* rod-shaped virus 2 (SIRV2), *Acidianus* rod-shaped virus 1 (ARV1)
Caudovirales	*Myoviridae*	dsDNA	Icosahedral head; long, contractile tail	*Halorubrum sodomense* tailed virus 2 (HSTV-2), φH
Caudovirales	*Podoviridae*	dsDNA	Icosahedral head; short tail	*Haloarcula sinaiiensis* tailed virus 1 (HSTV-1)
Caudovirales	*Siphoviridae*	dsDNA	Icosahedral head; long, non-contractile tail	*Haloarcula vallismortis* tailed virus 1 (HVTV-1), ψM1

—	*Ampullaviridae*	dsDNA	Bottle-shaped, lipid envelope	*Acidianus* bottle-shaped virus (ABV)
—	*Bicaudaviridae*	dsDNA	Spindle-shaped with two tails	**Acidianus two-tailed virus (ATV)**
—	*Clavaviridae*	dsDNA	Bacilliform	*Aeropyrum pernix* bacilliform virus 1 (APBV1)
—	*Fuselloviridae*	dsDNA	Spindle-shaped	*Sulfolobus* spindle-shaped virus 1 (SSV1), Acidianus spindle-shaped virus 1(ASV1), s 1
—	*Globuloviridae*	dsDNA	Icosahedral, lipid envelope	*Pyrobaculum* spherical virus (PSV), *Thermoproteus tenax* spherical virus 1 (TTSV1)
—	*Guttaviridae*	dsDNA	Droplet-shaped	*Sulfolobus newzealandicus* droplet-shaped virus (SNDV)
—	*"Pleolipoviridae"*	ssDNA, dsDNA	No capsid, lipid envelope	*Halorubrum* pleomorphic virus 1 (HRPV-1)
—	*"Sphaerolipoviridae"*	dsDNA	Icosahedral, internal lipid membrane	*Haloarcula hispanica* virus SH1, *Natrinema* virus SNJ1, P23-77
—	*"Spiraviridae"*	ssDNA	Hollow cylinder	*Aeropyrum* coil-shaped virus (ACV)
—	*"Turriviridae"*	dsDNA	Icosahedral, turreted, internal lipid membrane	**Sulfolobus turreted icosahedral virus (STIV)**

have corresponding electron micrographs. Even Genome Announcements, an online journal from the American Society for Microbiology devoted to the publication of new genomes, has streamlined the publication process such that images are not allowed. Thus, when electron micrographs are obtained for a particular phage, as was the case with GRCS, a staphylococcal phage belonging to the *Podoviridae* that was recently sequenced by my group, the images are relegated to "unpublished data" status (Swift, Nelson 2014).

Perhaps most problematic of all, the hierarchical classification system employed by the ICTV is based on vertical transmission of genetic characteristics, whereas it is well documented that horizontal exchange within large shared genetic pools has bestowed a level of genomic mosaicism in phages not seen in any other organism (Hendrix et al. 1999; Pedulla et al. 2003; Casjens 2005; Reyes et al. 2012), thereby further muddling the concepts of family, genus, and species. This mosaicism plagues taxonomists by sometimes preventing creation of new taxa at the family and order level, while at other times hindering or blocking assign-

ment of phage to a particular family. These issues are evident in that only three of the ten phage families are assigned to an order in the current ICTV release, and even within the well-defined *Podoviridae* (order *Caudovirales*), one third of the species (14 out of 44) are not assigned a genus.

A watershed of alternative taxonomic ideas

All cellular organisms possess ribosomes that translate nucleic acid sequences into the amino acid sequences of proteins. The RNA components of these ribosomes, and likewise the genes that produce them, are extremely well conserved and, as such, provide a basis for phylogenetic study of all cellular life forms. Indeed, analysis of prokaryotic 16S ribosomal RNA (rRNA) allowed Carl Woese and George Fox to distinguish Archaea as the third domain of life (Woese, Fox 1977; Woese, Kandler, Wheelis 1990). Thus rRNA is the basis for modern taxonomic classification within all three domains. However, this classification tactic cannot be applied to phages, nor indeed to any other viruses, because none encode their own ribosomes. Even worse news: examination of 105 fully sequenced phage genomes available in 2002 indicated that there is no single protein marker that is conserved in a majority of phage genomes (Rohwer, Edwards 2002).

By the early 2000s, the lack of a universally shared genetic locus (i.e., a 16S rRNA equivalent), the shortcomings of the ICTV classification scheme, and the extensive acquisition of new genomic data prompted many phage researchers to not only question the ICTV classification system, but to suggest new approaches. Four groups independently published alternative classification schemes, all based on analysis of the increasingly available genomic data.

Forest Rohwer and Rob Edwards put forth the Phage Proteomic Tree (PPT), a proteome-based classification system that groups phages relative to their near neighbors as well as in the context of all other phages (Rohwer, Edwards 2002). Their method used a distance matrix generated with the BLASTP and PROTDIST programs to analyze the relationships between the predicted proteomes of 105 phages. The resultant trees that showed relationships based on individual phage proteins, entire genomes, and phage groups were generally congruent with the ICTV families but also resolved several anomalies of the ICTV system. For example, the PPT reclassified phage P22 (*Podoviridae*) in the λ-like *Siphoviridae* sub-family. Likewise, the PRD1 phage, which is classified as *Tectiviridae* by the ICTV due to its internal lipid membrane, was moved to the PZA-like (now φ29-like) group within the *Podoviridae* based on its proteome which, in turn, reflects the protein-primed DNA replication machinery shared by members of that group (*see page 8-14*). (In this book, an updated PPT including 1220 genomes is used to classify the featured phages.)

About the same time, a second classification scheme was proposed that employed a genomic analysis focused exclusively on the structural gene module of the phages (Proux et al. 2002). The rationale for this approach stems from the observation that the structural gene module is the most conserved module in dairy phage in the family

Siphoviridae. These conserved proteins were thus assumed to represent the ancestral taxonomic fingerprint, and thus the inclusion of non-structural genes in the analysis could mask these relationships. Dot plots comparing numerous temperate lactococcal phages at both the DNA and protein sequence level revealed graded relatedness. Even in the absence of detectable sequence relatedness, synteny between their structural gene maps provided a basis for their classification. Such comparative genomic analyses were used to define four species of lactococcal phage belonging to two genera based specifically on the structural genes involved in head morphogenesis. In contrast, similar comparisons of all non-structural genes yielded no defining characteristics that could be used to discern different species.

A subsequent manuscript from the same group further supports the use of structural genes for comparative genomics, in this case within the *Myoviridae*. In this report they showed extensive sequence identity between the structural genes of *Lactobacillus* phage LP65 and those of *Bacillus* phage SPO1, as well as the related *Listeria* phage A511 and *Staphylococcus* phage K (Chibani-Chennoufi et al. 2004). However, further analysis of the structural genes from these related phages indicated that the SPO1-like genus (now called the *Spounavirinae* sub-family in the latest ICTV classification) shares more similarity with the λ-like *Siphoviridae* than with other genera of the *Myoviridae*, and hence may represent a bridge between the *Myoviridae* with their contractile tails and the *Siphoviridae* with their non-contractile tails.

Yet a third viewpoint was offered in the early 2000s by a group at the Pittsburgh Bacteriophage Institute when they suggested that it may be impossible to have a strictly hierarchical taxonomic system for phages given the extent of their genetic mosaicism, a result of their active horizontal gene transfer (Lawrence, Hatfull, Hendrix 2002). In their model, the top taxonomic levels would still follow the hierarchical ICTV approach with viruses first being divided into "domains" according to their nucleic acid content, then further partitioned into "divisions" based on defining morphological characteristics (e.g., filamentous phages distinct from tailed phages). Below the level of division, three basic tenets would guide further classification. First, one or more loosely defined "cohesion mechanisms" should be similar among all members of a group. Second, to provide an evolutionary basis, all members of a taxonomic cluster should show significant sequence similarity, preferably based on whole genome comparisons. Third, the phage may simultaneously belong to multiple groups based on the first two criteria. This web-like taxonomy affords a flexibility that is not found in any of the other hierarchical approaches. Nonetheless, while the reticulate nature at the core is the strength of the scheme, it is also a weakness as layers of complexity are added with each phage that is cross-referenced into two or more groups.

A provocative fourth method for phage classification, based on neither morphology or gene/protein sequence, was proposed by Dennis Bamford in 2003 (Bamford 2003). His "primordial soup" hypothesis postulates that significant folding of biological macromolecules preceded the appearance of the first life forms, and therefore we should look to conserved protein folds in structural proteins, such as the major capsid proteins, to define viral lineages. While at the surface this approach sounds very similar to the proteomic tree of Rohwer and Edwards or the structural gene approach taken by Proux, it is distinctly different. While those methods rely on homology evident in pairwise sequence alignment of genes or proteins, Bamford compares the tertiary structure of proteins in search of common folds. Protein fold does not necessarily correlate to primary sequence. For example, proteins that are 88% identical in sequence homology have been shown to display completely different folds and functions (He et al. 2008). Most usefully, protein tertiary structure can be evolutionarily conserved long after sequence similarity has been lost. That complex tertiary folds have been found to be shared by viruses of all three domains of life suggests that these folds—and likewise these viral lineages—predate the di-

vergence of the domains. As an example, Bamford points to nearly identical topologies of the major capsid protein from the bacteriophage PRD1 and that of an adenovirus, despite any detectable sequence homology between these viruses (*see page 6-51*). Since widespread evaluation of Bamford's idea would require vast structural protein datasets (crystallography or NMR coordinates), this is currently more of a hypothesis than an actual working method. Nevertheless, as a protein biochemist, I find it very intriguing. Perhaps future generations of phage taxonomists will not be performing BLAST and PFAM bioinformatics searches on genes and proteins as they do today, but rather will be calculating the root-mean-square deviations (RMSD) between aligned alpha-carbon positions of capsid proteins in the Protein Data Bank (PDB).

It may be advantageous at times to use multiple approaches, as exemplified by the description of the new genus *Viunalikevirus* (Adriaenssens et al. 2012). Seven members of the *Myoviridae* family (*Salmonella* phages ViI, SFP10, and USH19, *Escherichia* phages CBA120 and PhaxI, *Shigella* phage phiSboM-AG3, and *Dickeya* phage LIMEstone1) lack all of the genes associated with outer baseplate proteins and the long tail fibers characteristic of *Myoviridae*. Comparative genomics revealed several distinguishing features common to these seven, e.g., genome size and organization, gene synteny, replacement of thymine by a modified uracil, and the presence of four tailspike proteins instead of the long tail fibers characteristic of phage T4. Electron microscopy confirmed the presence of multiple star-like tailspike projections and an absence of long tail fibers for several of these phages. Thus, a combination of morphology, genomics, unique nucleic acid and structural features, and genome organization led to the description of a new genus, the *Viunalikevirus*, named after the phage ViI archetype.

Since the early 2000s when the above taxonomic approaches were first postulated, refinements on these approaches and new computational methods have populated the literature. One new tool is the use of protein clusters (PCs) to organize the ORFans (viral genes of unknown function that are unrelated to other known genes) that dominate most viral genomes (Yooseph et al. 2007). PCs also allowed a recent revision in the estimates of the still largely unexplored viral sequence space that yielded a value three orders of magnitude lower than the bold extrapolations of a decade ago (Ignacio-Espinoza, Solonenko, Sullivan 2013).

Another alternative organization scheme is the use of phage orthologous groups, or POGs, based on the concept of evolutionary conservation of ortholog function (Kristensen et al. 2010). Significantly, it is claimed that many viral taxa contain POGs that can serve as diagnostic signatures for a given taxon despite the fluidity of the viral pangenome (Kristensen et al. 2013). Such signature genes must meet the following criteria: (1) they are present in most or all members of the taxon; (2) they are never or only very rarely observed outside of the taxon; (3) they are not present in prokaryotic genomes except within an identifiable prophage; (4) preferably they are only present as a single copy in the viral genome.

Researchers have also determined in recent years that gene order and position within a genome can be just as taxonomically valuable as gene content itself (Li, Halgamuge, Tang 2008). This is particularly useful for genomes that display high levels of horizontal gene transfer, such as phage. Lastly, some investigators are assessing tetranucleotide usage deviation (TUD) patterns as a metric for determining phage ancestral relationships since TUD patterns in Bacteria had been previously shown to yield phylogenetic relationships similar to those derived from 16S rRNA analysis (Pride et al. 2006). This research found that, likely due to host influences, phages with a similar host range carried similar genomic signatures in the form of their TUD patterns.

A word about nomenclature

One cannot discuss phage taxonomy without a parallel discussion of nomenclature. Ever since the discovery of phage, their nomenclature has been seemingly as lawless as the Wild West. Phages were often named using various Greek letters (λ, γ, etc.) or simple numeric codes (N4, C_1, etc.), while for others, names were borrowed from people (Twort, etc.) or reflected the whimsical musings of their discoverers (IronMan, SweetiePie, KittenMittens, MisterCuddles, etc.). When some names are derived from simple utilitarian means and others reflect the creativity of phage researchers, the resulting systemic breakdown is a hindrance to phage taxonomy efforts, particularly when different names are used in different labs.

To illustrate the point from my own perspective, let me discuss my trials and tribulations determining the pedigree of the streptococcal C_1 phage I have worked with for 15 years (Fig. 3). This phage was first isolated from a Milwaukee, WI, sewage treatment plant in 1925 by Paul Clark and Alice Clark at the University of Wisconsin (Clark, Clark 1926). It was therefore referred to in the literature as the "Clark" phage, but also as the "sludge phage," it being from sewage sludge (Shwartzman 1927). In 1934, as a result of this particular bacteriophage being able to infect streptococcal strain 563, the sludge phage was renamed B563, for Bacteriophage of strain 563 (Evans 1934). Almost 25 years later, Dick Krause, then at The Rockefeller Institute, renamed it C_1, to imply an exquisite specificity for group C streptococci (Krause 1957). An aliquot of the phage stock also made its way to Japan, where the Chugai Pharmaceutical Co., Ltd., investigated potential anti-tumor properties of the phage (Takagaki et al. 1974). When the phage was found to have none, it was deposited in the American Type Culture Collection (ATCC) as ATCC 21597B. I have subsequently sequenced the Rockefeller C_1 phage (Nelson et al. 2003) and the ATCC 21597B phage (unpublished) and they are indeed identical. To further complicate the issue, there are no less than five distinct phages named "C1" in the literature (Kropinski, Prangishvili, Lavigne 2009).

To address the issue of nomenclature, a group of scientists intimately associated with the ICTV published a position paper calling for a rational new system for phage nomenclature (Kropinski, Prangishvili, Lavigne 2009). Under this proposal, a phage name would begin with a prefix, either vB for a bacterial virus or vA for an Archaeal virus. This would be followed by a three-letter host abbreviation similar to those used for restriction enzymes (e.g., Eco for *Escherichia coli*, Sau for *Staphylococcus aureus*), then a single letter family designation (e.g., P for *Podoviridae*, S for *Siphoviridae*, M for *Myoviridae*), and lastly a phage-specific designation. For the latter, a lab-specific or common name can be used, although it is recommended that Greek letters, Roman numerals, and superscripts be avoided. Thus, under this premise, the phage C_1 that I

work with that infects *S. dysgalactiae* and belongs to the *Podoviridae* would be named vB_SdyP_C1 and the aforementioned MisterCuddles would be vB_MsmS_MisterCuddles since it is a member of the *Siphoviridae* that infects *Mycobacterium smegmatis*. Granted, this nomenclature is a bit cumbersome, and traditions run deep in the phage community. Both of those factors may lead to slow acceptance of this proposal or the outright refusal to change some of the well-known traditional names. Nevertheless, this nomenclature system is the first in the 100 year history of phage to codify the nomenclature and therefore should be given its due attention.

Figure 3. A regional PPT created for this book to show *Streptococcus* phage C_1 and it neighbors. For information on the PPT, see Appendix A3 (*page 8-8*).

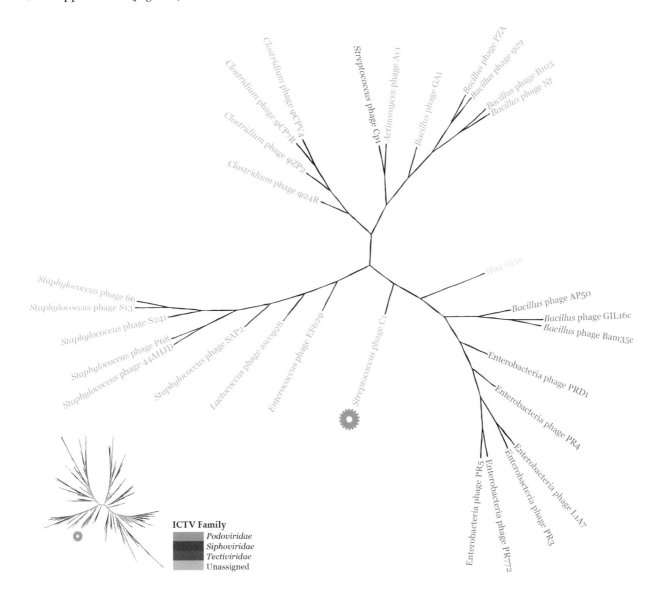

The way forward

A 2013 update to the ICTV definition of "species" states that the properties that define a species will now be determined by various study groups of the ICTV and may include, but are not limited to, natural and experimental host range and the degree of relatedness of their genomes or genes (Adams et al. 2013). While it is exciting that the ICTV is finally willing to embrace genomic data, it is not clear how these changes will be implemented, which analysis methods will be chosen, whether the issue of mosaicism can be resolved, and if the use of genomic data will be restricted to the species level or will ultimately be used for genus, family, or order level classifications.

In the end, despite all the controversy surrounding phage taxonomic schemes, it may not make much difference which approach is taken or whether multiple approaches are used together. Notwithstanding a few outliers, several of which are detailed above, all of the alternative genome-based methods ultimately cluster the vast majority of phages within the same order, family, and genus as currently assigned by the ICTV. Nonetheless, changes need to be made as we move forward. The ICTV database (ICTVdb) is no longer being maintained or updated, and with the speed at which new viral genomes are being sequenced, it is unlikely to get back on track. In a very frank and open commentary, Adrian Gibbs calls for the end of the exploration era of viral taxonomy (Gibbs 2013). He maintains that the higher taxa designations are uninformative and calls for an emphasis on the description of monophyletic groups or even individual viruses, instead. Second, he calls for movement of the ICTVdb and all associated metadata files to a public, open access forum, such as the Encyclopedia of Life (http://eol.org) or Wikipedia (http://www.wikipedia.org), where crowd sourcing by groups of interested scientists directly engaged with the virus(es) described can best curate the entry to satisfy all stakeholders. In my opinion, this is a much better long-term solution than relying on annual subcommittee meetings of the ICTV to deal with the increasing volume of new entries. To a limited degree, the phage community is already embracing the open access concept for some of the more famous phage. For example, see the entry on phage λ (http://en.wikipedia.org/wiki/Lambda_phage). As the rate of data acquisition continues to escalate, it will require a wide-based community effort, including some innovative strategies, to keep pace.

References

Ackermann, HW. 2011. Bacteriophage taxonomy. Microbiol Australia 32:90-94.

Adams, MJ, EJ Lefkowitz, AM King, EB Carstens. 2013. Recently agreed changes to the International Code of Virus Classification and Nomenclature. Arch Virol 158:2633-2639.

Adriaenssens, EM, HW Ackermann, H Anany, et al. 2012. A suggested new bacteriophage genus: "Viunalikevirus". Arch Virol 157:2035-2046.

Baltimore, D. 1971. Expression of animal virus genomes. Bacteriol Rev 35:235-241.

Bamford, DH. 2003. Do viruses form lineages across different domains of life? Res Microbiol 154:231-236.

Botstein, D, I Herskowitz. 1974. Properties of hybrids between *Salmonella* phage P22 and coliphage lambda. Nature 251:585-589.

Bradley, DE. 1967. Ultrastructure of bacteriophage and bacteriocins. Bacteriol Rev 31:230-314.

Casas, V, F Rohwer. 2007. Phage metagenomics. Methods Enzymol 421:259-268.

Casjens, SR. 2005. Comparative genomics and evolution of the tailed-bacteriophages. Curr Opin Microbiol 8:451-458.

Chibani-Chennoufi, S, ML Dillmann, L Marvin-Guy, S Rami-Shojaei, H Brussow. 2004. *Lactobacillus plantarum* bacteriophage LP65: A new member of the SPO1-like genus of the family Myoviridae. J Bacteriol 186:7069-7083.

Clark, PF, AS Clark. 1926. A "bacteriophage" active against a hemolytic streptococcus. J Bacteriol 11:89.

d'Herelle, FH. 1917. Sur un microbe invisible antagoniste des bacilles dysenteriques. Comptes Rendu. Acad. Sci. (Paris) 165:373-375.

Evans, AC. 1934. *Streptococcus* bacteriophage: A study of four serological types. Public Health Rep 49:1386-1401.

Fan, X, L Xie, W Li, J Xie. 2014. Prophage-like elements present in Mycobacterium genomes. BMC Genomics 15:243.

Ferretti, JJ, WM McShan, D Ajdic, et al. 2001. Complete genome sequence of an M1 strain of *Streptococcus pyogenes*. Proc Natl Acad Sci USA 98:4658-4663.

Gibbs, AJ. 2013. Viral taxonomy needs a spring clean; its exploration era is over. Virol J 10:254.

He, Y, Y Chen, P Alexander, PN Bryan, J Orban. 2008. NMR structures of two designed proteins with high sequence identity but different fold and function. Proc Natl Acad Sci USA 105:14412-14417.

Hendrix, RW, MCM Smith, RN Burns, ME Ford, GF Hatfull. 1999. Evolutionary relationships among bacteriophages and prophages: All the world's a phage. Proc Natl Acad Sci USA 96:2192-2197.

Ignacio-Espinoza, JC, SA Solonenko, MB Sullivan. 2013. The global virome: Not as big as we thought? Curr Opin Virol 3:566-571.

King, AMQ, E Lefkowitz, MJ Adams, EB Carstens, editors. 2011. *Virus Taxonomy: Ninth Report of the International Committee on Taxonomy of Viruses:* Elsevier.

Krause, RM. 1957. Studies on bacteriophages of hemolytic streptococci. I. Factors influencing the interaction of phage and susceptible host cell. J Exp Med 106:365-384.

Kristensen, DM, L Kannan, MK Coleman, YI Wolf, A Sorokin, EV Koonin, A Mushegian. 2010. A low-polynomial algorithm for assembling clusters of orthologous groups from intergenomic symmetric best matches. Bioinformatics 26:1481-1487.

Kristensen, DM, AS Waller, T Yamada, P Bork, AR Mushegian, EV Koonin. 2013. Orthologous gene clusters and taxon signature genes for viruses of prokaryotes. J Bacteriol 195:941-950.

Kropinski, AM, D Prangishvili, R Lavigne. 2009. Position paper: The creation of a rational scheme for the nomenclature of viruses of Bacteria and Archaea. Environ Microbiol 11:2775-2777.

Lawrence, JG, GF Hatfull, RW Hendrix. 2002. Imbroglios of viral taxonomy: Genetic exchange and failings of phenetic approaches. J Bacteriol 184:4891-4905.

Li, J, SK Halgamuge, SL Tang. 2008. Genome classification by gene distribution: An overlapping subspace clustering approach. BMC Evol Biol 8:116.

Linnaeus, C. 1758. Systema naturae per regna tria naturae, secundum classes, ordines, genera, species, cum characteribus, differentiis, synonymis, locis. Holmiae: Laur. Salvii.

Luria, SE, M Delbruck, TF Anderson. 1943. Electron microscope studies of bacterial viruses. J Bacteriol 46:57-67.

Mann, NH. 2005. The third age of phage. PLoS Biol 3:e182.

Mokili, JL, F Rohwer, BE Dutilh. 2012. Metagenomics and future perspectives in virus discovery. Curr Opin Virol 2:63-77.

Nelson, D. 2004. Phage taxonomy: We agree to disagree. J Bacteriol 186:7029-7031.

Nelson, D, R Schuch, S Zhu, DM Tscherne, VA Fischetti. 2003. Genomic sequence of C1, the first streptococcal phage. J Bacteriol 185:3325-3332.

Ortmann, AC, B Wiedenheft, T Douglas, M Young. 2006. Hot crenarchaeal viruses reveal deep evolutionary connections. Nat Rev Microbiol 4:520-528.

Pawlowski, A, I Rissanen, JK Bamford, M Krupovic, M Jalasvuori. 2014. *Gammasphaerolipovirus*, a newly proposed bacteriophage genus, unifies viruses of halophilic archaea and thermophilic bacteria within the novel family *Sphaerolipoviridae*. Arch Virol 159:1541-1554.

Pedulla, ML, ME Ford, JM Houtz, et al. 2003. Origins of highly mosaic mycobacteriophage genomes. Cell 113:171-182.

Pietilä, MK, P Laurinmäki, DA Russell, C-C Ko, D Jacobs-Sera, SJ Butcher, DH Bamford, RW Hendrix. 2013a. Insights into head-tailed viruses infecting extremely halophilic archaea. J Virol 87:3248-3260.

Pietilä, MK, P Laurinmäki, DA Russell, C-C Ko, D Jacobs-Sera, RW Hendrix, DH Bamford, SJ Butcher. 2013b. Structure of the archaeal head-tailed virus HSTV-1 completes the HK97 fold story. Proc Natl Acad Sci USA 110:10604-10609.

Pietilä, MK, E Roine, L Paulin, N Kalkkinen, DH Bamford. 2009. An ssDNA virus infecting archaea: a new lineage of viruses with a membrane envelope. Mol Microbiol 72:307-319.

Prangishvili, D. 2013. The wonderful world of archaeal viruses: A personal experience. Annu Rev Microbiol 67:565-585.

Pride, DT, TM Wassenaar, C Ghose, MJ Blaser. 2006. Evidence of host-virus co-evolution in tetranucleotide usage patterns of bacteriophages and eukaryotic viruses. BMC Genomics 7:8.

Proux, C, D van Sinderen, J Suarez, P Garcia, V Ladero, GF Fitzgerald, F Desiere, H Brussow. 2002. The dilemma of phage taxonomy illustrated by comparative genomics of Sfi21-like Siphoviridae in lactic acid bacteria. J Bacteriol 184:6026-6036.

Reyes, A, NP Semenkovich, K Whiteson, F Rohwer, JI Gordon. 2012. Going viral: Next-generation sequencing applied to phage populations in the human gut. Nat Rev Microbiol 10:607-617.

Rohwer, F, R Edwards. 2002. The phage proteomic tree: A genome-based taxonomy for phage. J Bacteriol 184:4529-4535.

Shwartzman, G. 1927. Studies on *Streptococcus* bacteriophage: I. A powerful lytic principle against hemolytic streptococci of Erysipelas origin. J Exp Med 46:497-509.

Swift, SM, DC Nelson. 2014. Complete genome sequence of *Staphylococcus aureus* phage GRCS. Genome Announc 2: e00209-14.

Takagaki, Y, Y Sugawara, S Suzuki, H Ogawa, A Yamamoto. 1974. Method of production of anti-tumor substances. USPTO, editor (USA: Chugai Pharmaceutical Co Ltd).

Thomas Jr., CA, J Abelson. 1966. The isolation and characterization of DNA from bacteriophages. In: GL Cantoni, DR Davies, editors. *Procedures in Nucleic Acid Research*: Harper & Row. p. 553-561.

Twort, FW. 1915. An investigation on the nature of ultra-microscopic viruses. Lancet ii:1241-1246.

Wildy, P. 1971. *Classification and nomenclature of viruses: First report of the International Committee on Nomenclature of Viruses.* Basel: S. Karger.

Woese, CR, GE Fox. 1977. Phylogenetic structure of the prokaryotic domain: The primary kingdoms. Proc Natl Acad Sci USA 74:5088-5090.

Woese, CR, O Kandler, ML Wheelis. 1990. Towards a natural system of organisms: Proposal for the domains Archaea, Bacteria, and Eucarya. Proc Natl Acad Sci USA 87:4576-4579.

Yooseph, S, G Sutton, DB Rusch, et al. 2007. The Sorcerer II global ocean sampling expedition: Expanding the universe of protein families. PLoS Biol 5:e16.

10^{31}

phage virions on Earth

Calculated assuming 10 virions for
each of the 10^{30} microbes on Earth.

Whitman, WB, DC Coleman, WJ Wiebe.
1998. Prokaryotes: The unseen majority.
Proc Natl Acad Sci USA 95:6578-6583.

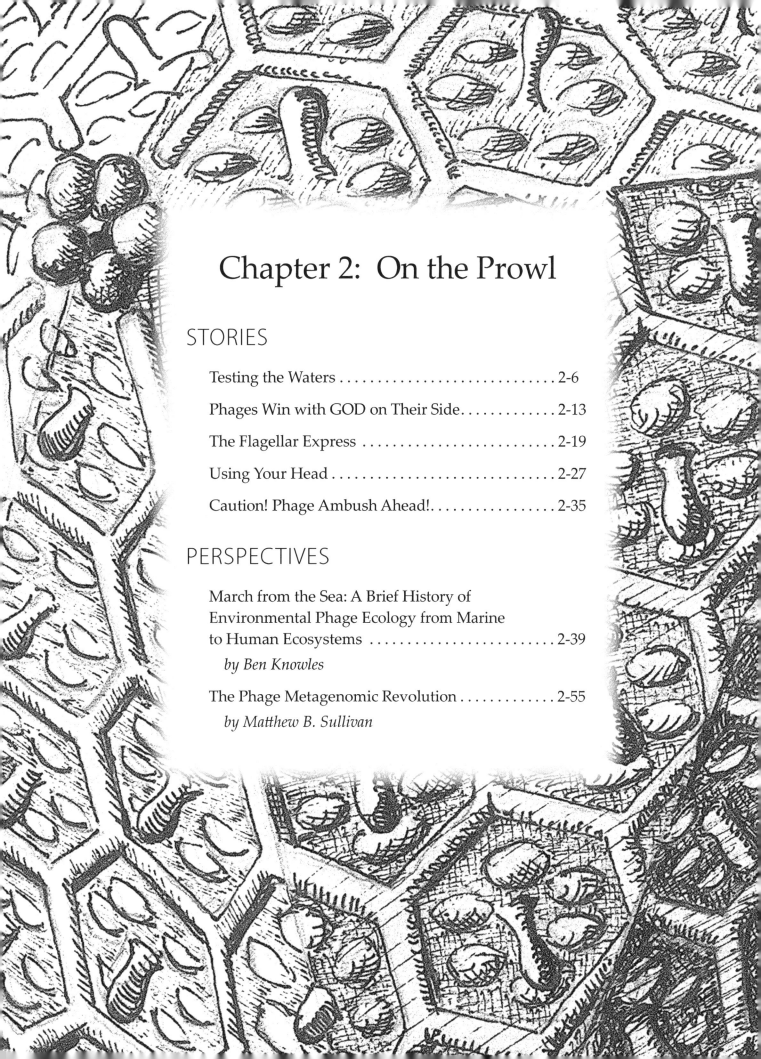

Chapter 2: On the Prowl

STORIES

PERSPECTIVES

Enterobacteria Phage RB49

a Myophage that senses environmental cues to decide when to extend its long tail fibers

Genome

dsDNA; linear

164,018 bp

279 predicted ORFs; 0 RNAs

Encapsidation method

Packaging; T = 13 capsid

Common host

Escherichia coli

Habitat

Mammalian intestines

Lifestyle

Lytic

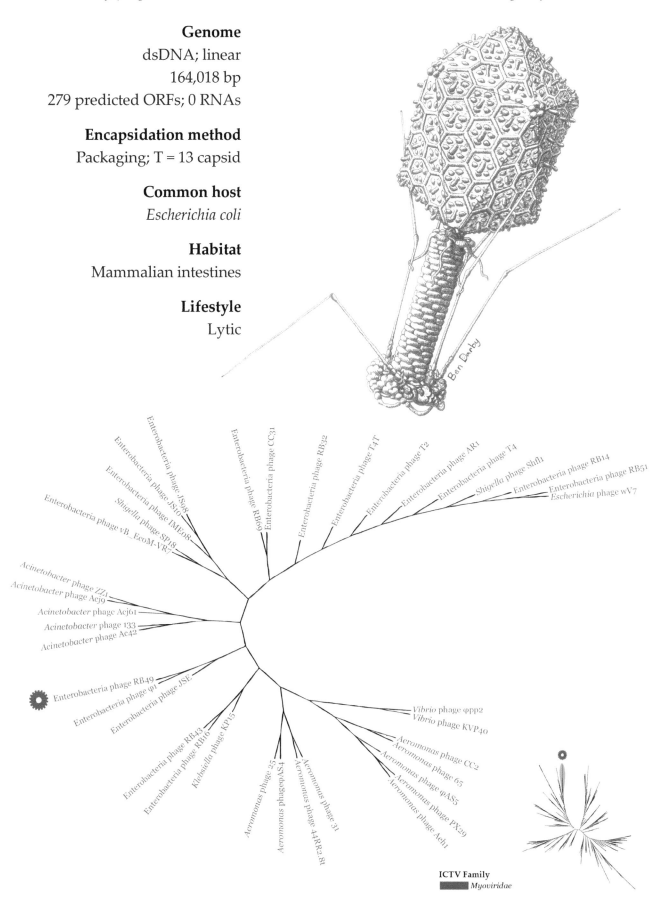

ICTV Family

Myoviridae

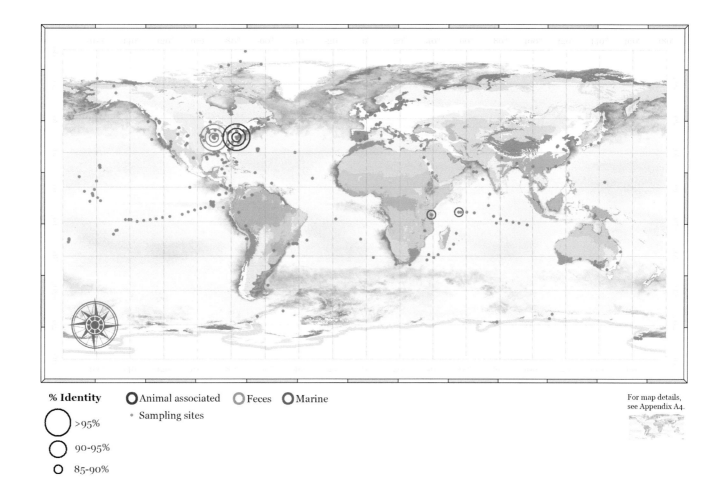

% Identity   ○ Animal associated   ○ Feces   ○ Marine
                 • Sampling sites

○ >95%

○ 90-95%

○ 85-90%

For map details,
see Appendix A4.

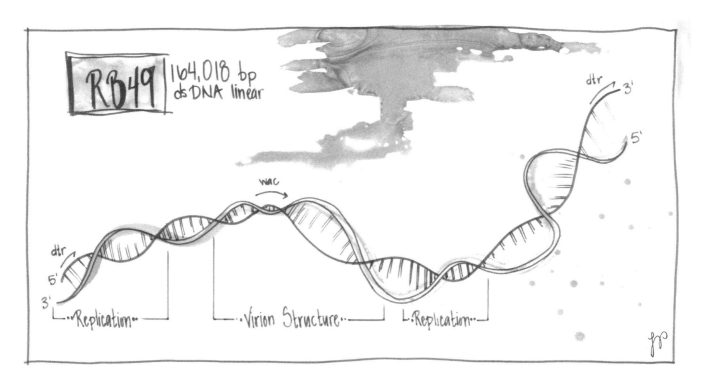

Enterobacteria Phage RB49

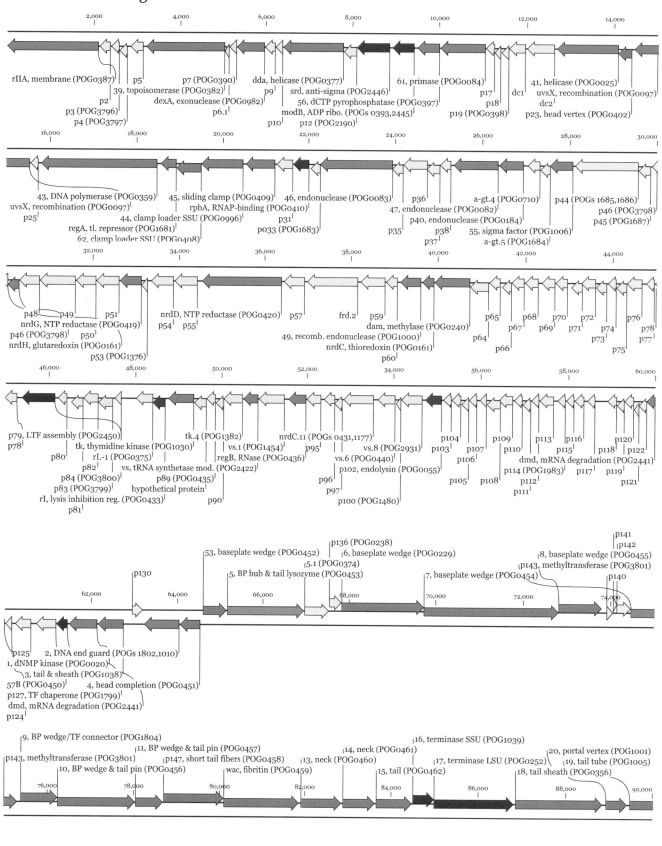

On the Prowl | Entry | Takeover | Replication Lytic | Replication Lysogenic | Assembly | Escape | Transfer RNA | Non-coding RNA | Multiple | Unknown

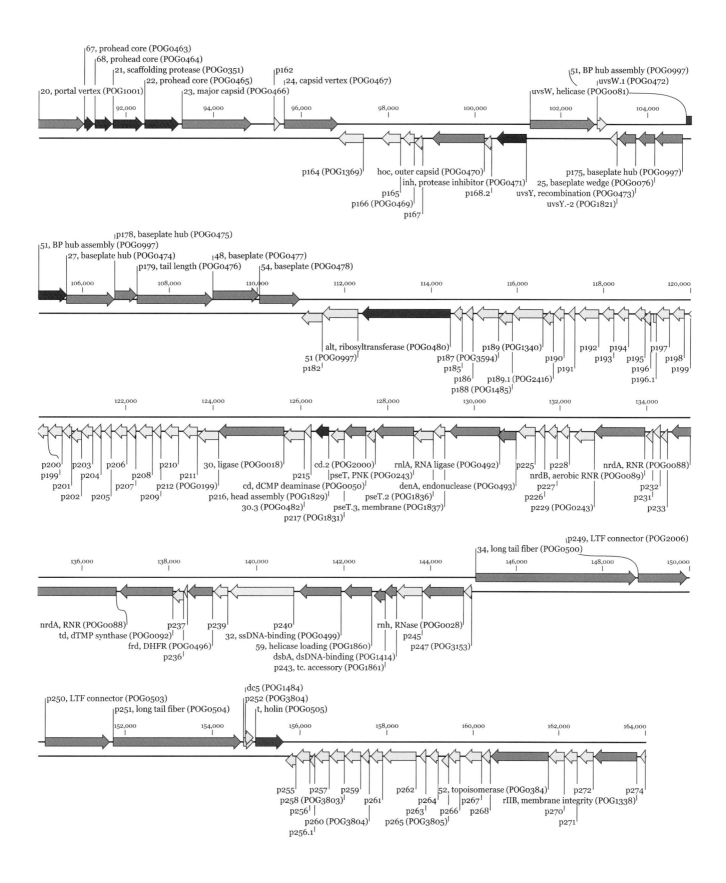

Testing the Waters

Merry Youle

Mise-en-Scène: *All life forms sense key environmental cues and then respond appropriately. Phages are no exception. They keep tabs on the external environment, some then choosing to promote or delay adsorption depending on conditions. For these phages, to extend their tail fibers or not to extend: that is the question. Temperate phages are even more sophisticated, weighing factors such as host physiology and abundance as they make their lysis/lysogeny decision.*

The iconic image of a phage seen on T-shirts and coffee mugs is that of a T4 **virion** tumbling through the milieu, its six 'claws' outstretched, poised for a deadly encounter with a hapless *E. coli*. However, such images can be misleading. Consider a more restrained possibility: a phage holding most of its tail fibers close to its tail or head, gingerly extending just one at a time to test the waters. This demure strategy offers some advantages. When extended, the tail fibers are more susceptible to damage (Kellenberger et al., 1965) and they slow virion diffusion. More importantly, there is no need for all six to be deployed to search for prey, as one extended fiber surveys almost as large a volume as does six. So which strategy do the phages choose: travel with all tail fibers extended, a few, or none?

While on the prowl, Podophage T7 extends individual tail fibers sequentially, just one at a time, to scout for prey (Hu et al., 2013). When it contacts a potential host, it walks along the cell surface like a six-legged dancer lightly balancing on only one leg at a time (http://www.youtube.com/watch?v=Gy42CoyqKjE). Each fiber in turn binds reversibly, and only weakly, but even weak interactions can provide enough 'gravity' to keep the phage exploring the surface rather than drifting away. This approach decreases the search space from three dimensions to two. When by chance a tail fiber encounters T7's specific **receptor**, walking comes to a halt. Now all six tail fibers bind and soon an infection is underway. Is T7 the exception or the norm?

Whiskers

Consider the T4-like phages, phages such as RB49. Being Myophages, their situation is a bit different. Their tail is a complex macromolecular machine,

typically about 144 nm long and composed of at least 430 polypeptide chains. Each tail bears three sets of fibrous structures: six long tail fibers (LTFs) essential for host recognition and the initial reversible **adsorption**; six short tail fibers (STFs) required for irreversible adsorption; and six whiskers. As their name suggests, the whiskers are located at the phage 'neck' and are short, only 53 nm long. They don't interact directly with the host surface, but nevertheless they play a key role when on the prowl for a host.

These whiskers are stiff bristles, each one built from three parallel molecules of the Wac (<u>w</u>hisker <u>a</u>ntigen <u>c</u>ontrol) protein (Efimov et al., 1994). Although simpler than the LTFs, they nevertheless comprise three distinct regions. The middle 80% of the protein chain is a coiled coil α-helical structure that constitutes most of the length of the bristle (Letarov et al., 2005). The C-terminal domain at the distal end of each chain serves as a **foldon** that ensures correct folding and trimerization. The N-terminal domains of all the whiskers form a wheel-like collar around the neck, with the domains of adjacent whiskers linked by one copy of an unidentified protein (Kostyuchenko et al., 2005). This arrangement spaces the whiskers evenly and anchors them to the capsid.

The whiskers are put to work right away to assist with the last assembly step: the attachment of the LTFs. This maneuver is a bit of a trick. Try to picture docking one end of a ~144 nm long LTF to the baseplate of a preassembled virion within the crowded cytoplasm of the host cell. These phages align each LTF for attachment by using both ends of a whisker to 'grasp' it at specific locations (Kostyuchenko et al., 2005).

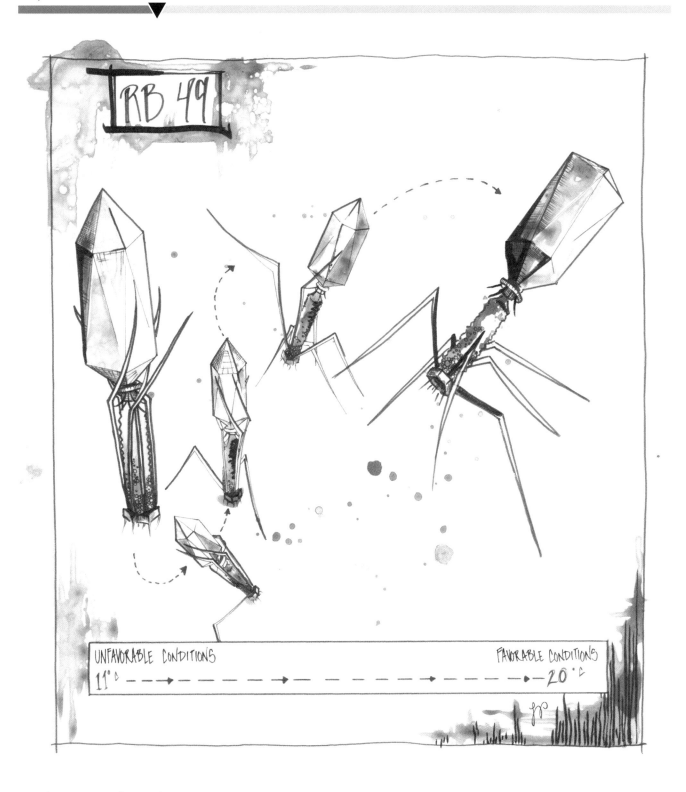

Environmental sensing

After host lysis, freshly minted RB49 virions set off into the world to repeat the cycle of carnage. Should we imagine them adrift in search of new hosts with all six tail fibers displayed? Or, T7-like, with most LTFs held close? Because it has whiskers, RB49 can choose. It adaptively retracts or extends its LTFs depending upon the environmental conditions it encounters. If the phage judges the environment to be adverse, its whiskers hold the LTFs in the retracted position where they form a 'jacket' around the tail sheath, slightly overlapping the head. Such introverted virions are not infective (Kellenberger et al., 1965). This also shelters the LTFs from damage.

What environmental conditions do these phages monitor? For one, they perform a litmus test, retracting their LTFs when the pH drops to 5 or below (Kellenberger et al., 1965). Likewise they consider 0.10 M salt hospitable, but retract their LTFs if the salt concentration decreases to 0.01 M (Conley and Wood, 1975). If the temperature drops from 20° C to 11° C, they respond by retracting. These are reversible responses, not permanent inactivation. When favorable conditions return, they unfurl their LTFs and infectivity is restored.

Some T4 strains have a more refined mechanism that tests the environment for a specific compound required for infectivity. These phages keep their LTFs retracted by binding them to their tail sheath until they 'sense' the presence of the cofactor (Brenner et al., 1962). For one such phage (T4B), a single molecule of tryptophan per LTF is sufficient to disrupt this binding and allow LTF extension (Kellenberger et al., 1965). It is likely that other phages use different cofactors when hunting in the intestinal milieu.

Shades of gray

A dynamic picture emerges for RB49 and the many other T4-like phages. When on the prowl, if conditions are unfavorable, their whiskers hold the LTFs close, thereby preventing adsorption. However, this need not be an all-or-none response. Perhaps when in the gut, influenced by multiple environmental signals, RB49 might take a cue from the tryptophan-requiring T4 strains and modulate its response. Depending on the tryptophan concentration, those strains extend one, two, three, or more LTFs. Even when denied tryptophan, only 80-85% retract all their LTFs, which still leaves 15-20% one-legged virions able to contact a host (Kellenberger et al., 1965).

Whiskers are typically described in the literature as "rudimentary" sensory devices, implying they are primitive or undeveloped. In actuality, they are a sophisticated and economical mechanism enabling 'inert' virions to respond adaptively to diverse external clues. They raise the question: Are T4 and its relatives sentient beings?

Cited references

Brenner, S., Champe, S., Streisinger, G., and Barnett, L. (1962). On the interaction of adsorption cofactors with bacteriophages T2 and T4. Virology 17, 30-39.

Conley, M.P., and Wood, W.B. (1975). Bacteriophage T4 whiskers: A rudimentary environment-sensing device. Proc Natl Acad Sci USA 72, 3701-3705.

Efimov, V.P., Nepluev, I.V., Sobolev, B.N., Zurabishvili, T.G., Schulthess, T., Lustig, A., Engel, J., Haener, M., Aebi, U., and Venyaminov, S.Y. (1994). Fibritin encoded by bacteriophage T4 Gene *wac* has a parallel triple-stranded α-helical coiled-coil structure. J Mol Biol 242, 470-486.

Hu, B., Margolin, W., Molineux, I.J., and Liu, J. (2013). The bacteriophage T7 virion undergoes extensive structural remodeling during infection. Science 339, 576-579.

Kellenberger, E., Bolle, A., Boy De La Tour, E., Epstein, R., Franklin, N., Jerne, N., Reale-Scafati, A., Sechaud, J., Bendet, I., and Goldstein, D. (1965). Functions and properties related to the tail fibers of bacteriophage T4. Virology 26, 419-440.

Kostyuchenko, V.A., Chipman, P.R., Leiman, P.G., Arisaka, F., Mesyanzhinov, V.V., and Rossmann, M.G. (2005). The tail structure of bacteriophage T4 and its mechanism of contraction. Nat Struct Mol Biol 12, 810-813.

Letarov, A., Manival, X., Desplats, C., and Krisch, H. (2005). gpwac of the T4-type bacteriophages: Structure, function, and evolution of a segmented coiled-coil protein that controls viral infectivity. J Bacteriol 187, 1055-1066.

Recommended review

Leiman, P., F Arisaka, M van Raaij, V Kostyuchenko, A Aksyuk, S Kanamaru, M Rossmann. 2010. Morphogenesis of the T4 tail and tail fibers. Virol J 7:355.

10^{24}

productive viral infections
per second on Earth

Hendrix, RW. 2010. Recoding in bacteriophages.
in JF Atkins, RF Gesteland, editors. *Recoding:
Expansion of Decoding Rules Enriches Gene Expression.*
Springer. p. 249-258.

Bordetella Phage BPP-1

a Podophage that generates diverse receptor-binding proteins by targeted mutagenesis

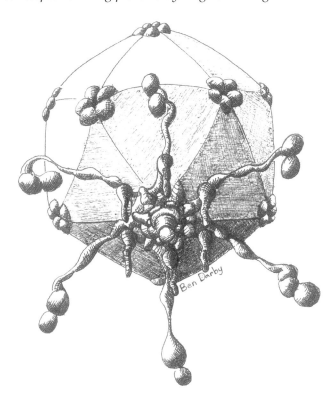

Genome
dsDNA; linear
42,493 bp
49 predicted ORFs; 0 RNAs

Encapsidation method
Packaging; T = 7 capsid

Common host
Bordetella bronchiseptica

Habitat
Host-associated; mammals

Lifestyle
Temperate

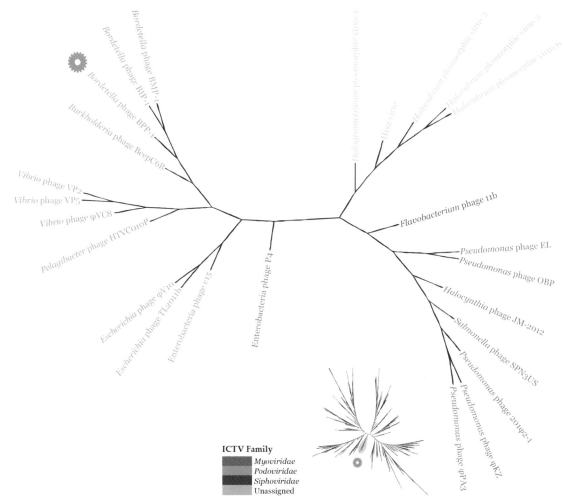

ICTV Family
Myoviridae
Podoviridae
Siphoviridae
Unassigned

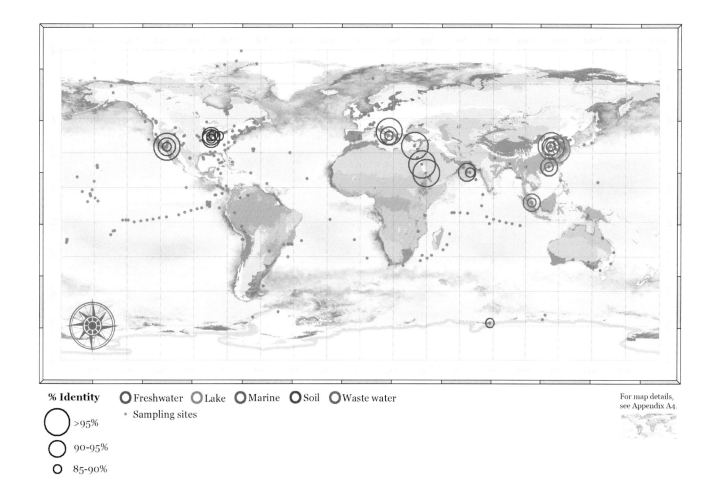

% Identity
>95%
90-95%
85-90%
Freshwater
Lake
Marine
Soil
Waste water
Sampling sites
For map details, see Appendix A4.

BPP-1
42,493 bp
dsDNA linear
brt
template
mtd
5'
3'
5'
3'
Tropism Switching
Virion Structure
Replication

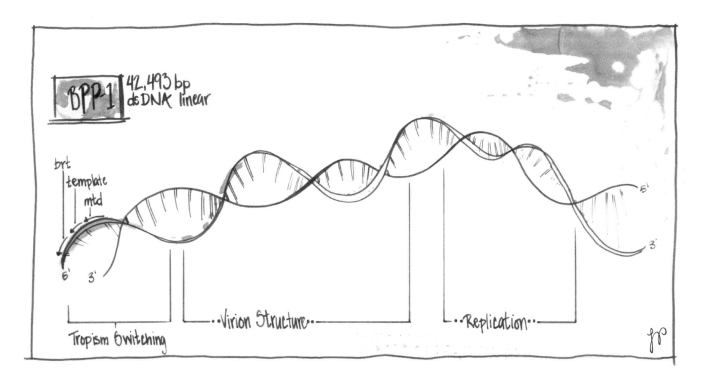

Bordetella Phage BPP-1

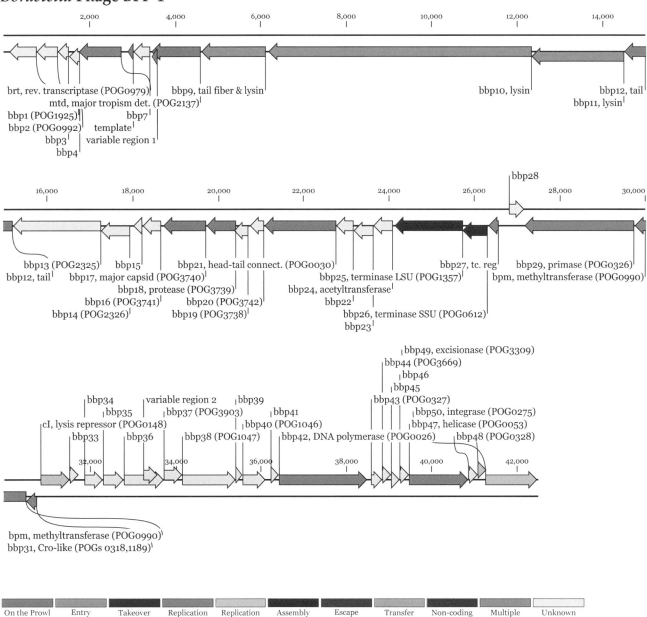

| On the Prowl | Entry | Takeover | Replication Lytic | Replication Lysogenic | Assembly | Escape | Transfer RNA | Non-coding RNA | Multiple | Unknown |

Phages Win with GOD on Their Side

Merry Youle

Mise-en-Scène: *We humans point proudly to the production of >10^{12} (perhaps 10^{14}) different immunoglobulins by our adaptive immune system. To accomplish this feat we rely on several mechanisms, including recombination and hypermutation. The* Bordetella *phage BPP-1 has a very economical and precisely targeted mechanism that creates 10^{12} variants of its receptor binding protein. Not bad for a 'simple' phage.*

Note: **G**eneration **O**f **D**iversity (GOD), an immunological term designating the molecular mechanisms that underlie the specificity of our innate and adaptive immune systems.

Every **virion** requires its specific **receptor** on the cell surface in order to recognize a potential host and launch an infection. What's a phage to do if its receptor disappears? No receptor, no infection, and soon no phage. 'Knowing' this, Bacteria often evolve a slightly modified variant of a phage's receptor that is unrecognizable to the phage but still functions somewhat for the host. Impaired function puts the host at a competitive disadvantage unless phages are decimating the competition. Moreover, a host relying on this tactic gains only a brief respite as phage mutants able to use the modified receptor quickly arise. Even worse for the bacterium, some phages have ways to rapidly generate diversity in their receptor recognition structures. Why not jettison the receptor entirely? Likely some Bacteria did, but we'd never know for certain. Any surface component currently serving as a receptor likely provides sufficient benefit, at least under some circumstances, to counter the cost of phage predation.

Receptors come, receptors go

Some receptors do come and go from the cell surface in the normal course of bacterial life, giving those Bacteria a temporary cloak of invisibility. To keep pace, some phages engage in **tropism** switching. One stratagem for alternating tropisms is to encode the genes that influence receptor binding within a gene cassette that can be inverted in the phage genome every so often. Genes for binding one receptor are transcribed only when the cassette is in one orientation; other genes yielding different receptor-binding capabilities are expressed when the cassette is flipped. As a result, phages with their cassette in

the different orientations can infect different hosts. Such invertible cassettes are found in phages Mu and P1, among others (Howe 1980; Iida 1984).

Useful variants

Instead of playing either/or, some phages routinely generate exceedingly diverse receptor-binding possibilities using targeted mutagenesis to modify their receptor-binding protein. This tactic was first discovered in BPP-1, a phage that infects Bacteria in the genus *Bordetella* (e.g., *B. pertussis* that causes whooping cough). These Bacteria regulate the activity of their virulence genes in response to a variety of environmental signals. During their virulent phase (termed Bvg$^+$ phase for *Bordetella virulence* genes), they express a suite of genes that enable them to infect a host. One of those genes encodes pertactin, an outer membrane protein used to adhere to tracheal epithelial cells. Phage BPP-1 uses pertactin as its receptor. Thus it infects Bvg$^+$ phase *Bordetella*, but not Bvg$^-$ cells.

Nevertheless, in every million BPP-1 progeny virions, there is one that infects Bvg$^-$ cells and another that infects cells in both phases. (Let's call those phages BPP-1/– and BPP-1/±, respectively.) Thus these BPP-1 virions fall into three types, each able to infect a different group of *Bordetella* hosts. Moreover, infection by a virion from any group yields some progeny from the other two groups as well, at rates that range from one in a thousand to one in a million. (Given that one routinely obtains 10^9 or more virions per ml when growing a phage with its host in liquid culture, such 'rare' variants are easily isolated.) The phages in all three groups are almost identical. They differ in just one protein,

Mtd (<u>M</u>ajor <u>t</u>ropism <u>d</u>eterminant), the protein at the tips of their tail fibers that binds to the receptor on the host. Moreover, all the differences among those phages are restricted to only twelve of Mtd's 381 amino acids. That restriction reflects both how the variants arise and how they influence receptor binding specificity.

One domain of the Mtd protein has the three-dimensional structure of a C-type lectin fold (Medhekar, Miller 2007). This protein fold contributes to cell adhesion or recognition activities in the immune system of many metazoans. In Mtd, this domain sits at the very end of each tail fiber with its receptor-binding pocket exposed and facing outward, ready to bind a host cell. All twelve of those key amino acids (see above) lie on the binding surface, thus directly influence receptor recognition. By changing one or more of those twelve, the phage can generate 10^{12} different proteins each with a potentially different binding capability. Let the bacterium try to hide by modifying its pertactin, and chances are some phage variant will arise that can bind the new version or that can use a different surface component as its receptor. Witness the BPP-1/– phages that switched from pertactin to some unknown receptor. Given the immense number of phages, even such seemingly improbable events help the phages win.

Behind the curtain

How does the phage generate this diversity? BPP-1 is equipped with a Diversity Generating Retroelement (DGR), the first one to be discovered (Liu et al. 2002). With this tool in hand, the phage focuses its efforts on the portion of the *mtd* gene that encodes the twelve key amino acids—the 3'-terminal 134 bp of *mtd*. Very close to this 3'-terminus resides an unaltered template copy of the 134 bp sequence, and adjacent to that template is the gene for the *Bordetella* reverse transcriptase (Brt). The template is transcribed into RNA, but unlike the case for the *mtd* gene, this RNA is not translated into protein. Instead, the Brt reverse transcriptase transcribes the RNA into complementary DNA, mutating some or all of the adenines in the process. These altered DNA copies sometimes mutate the *mtd* gene by replacing part or all of the corresponding 134 bp region.

Thus, this precisely targeted mutagenesis is restricted to the 134 bp region of the *mtd* gene and affects only adenine-containing codons. Furthermore, as a result of selection over evolutionary time, the only adenine-containing codons remaining in that region are those that encode the twelve critical amino acids at Mtd's receptor binding site. Mutation of other amino acids that might disrupt Mtd's structure or function is prevented. Pretty smart! When BPP-1 synthesizes its tail fiber proteins using a mutated *mtd* gene, the product is an altered Mtd protein with potentially different receptor binding abilities. Usually these mutations are neutral or deleterious, but occasionally they enable the phage to counter the evasion strategies of their host.

Beyond the phage lectin fold

DGRs are versatile mechanisms that are not limited to altering only proteins with the C-type lectin fold. DGRs found in phages within the human gut (*see page 2-35*) are predicted to also introduce diversity into proteins with an immunoglobulin-like fold (Minot et al. 2012). Nor are DGRs proprietary phage property. Bacteria also employ them, albeit with their own distinctive twists. For example, some Bacteria encode one template that services two genes, while others have multiple non-**homologous** DGRs (Doulatov et al. 2004) each with its own template, diversified gene, and reverse transcriptase. Thus, these diversity-generating weapons are used by both sides in the ongoing arms race between phages and their hosts.

Our own activities sometimes can influence this arms race in unexpected ways. The vaccines we have been using against *Bordetella* have inadvertently favored the rise of *B. pertussis* strains that have dispensed with pertactin entirely, thus eliminating BPP-1's customary receptor (Bodilis, Guiso 2013). Clearly, this loss won't baffle BPP-1 for long. This phage, equipped with capabilities for GOD, won't miss a beat.

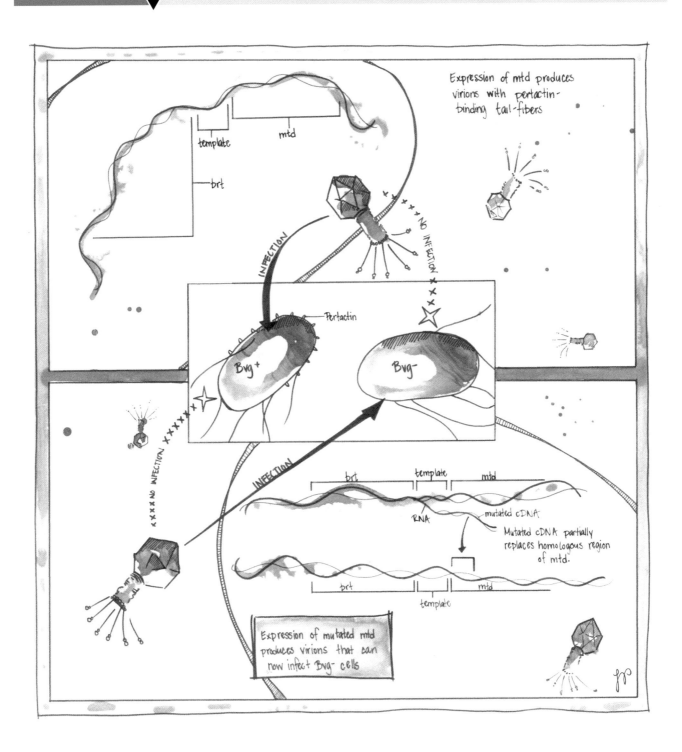

Cited references

Bodilis, H, N Guiso. 2013. Virulence of pertactin-negative *Bordetella pertussis* isolates from infants, France. Emerg Infect Dis 19:471.

Doulatov, S, A Hodes, L Dai, N Mandhana, M Liu, R Deora, RW Simons, S Zimmerly, JF Miller. 2004. Tropism switching in *Bordetella* bacteriophage defines a family of diversity-generating retroelements. Nature 431:476-481.

Howe, MM. 1980. The invertible G segment of phage mu. Cell 21:605-606.

Iida, S. 1984. Bacteriophage P1 carries two related sets of genes determining its host range in the invertible C segment of its genome. Virology 134:421-434.

Liu, M, R Deora, SR Doulatov, et al. 2002. Reverse transcriptase-mediated tropism switching in *Bordetella* bacteriophage. Science 295:2091-2094.

Medhekar, B, JF Miller. 2007. Diversity-generating retroelements. Curr Opin Microbiol 10:388-395.

Minot, S, S Grunberg, GD Wu, JD Lewis, FD Bushman. 2012. Hypervariable loci in the human gut virome. Proc Natl Acad Sci USA 109: 3962–3966.

Enterobacteria Phage χ

a Siphophage that lassos a rotating flagellum and rides it to the host cell surface

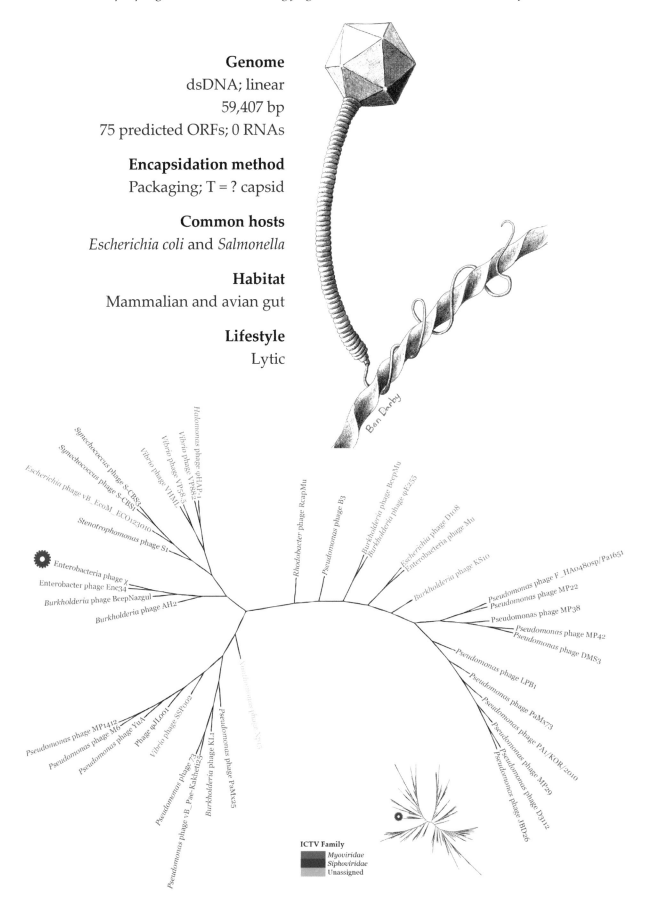

Genome
dsDNA; linear
59,407 bp
75 predicted ORFs; 0 RNAs

Encapsidation method
Packaging; T = ? capsid

Common hosts
Escherichia coli and *Salmonella*

Habitat
Mammalian and avian gut

Lifestyle
Lytic

Ben Darby

Synechococcus phage S-CBS3
Synechococcus phage S-CBS1
Escherichia phage vB_EcoM_ECO123010
Stenotrophomonas phage S1
Enterobacteria phage χ
Enterobacter phage Enc34
Burkholderia phage BcepNazgul
Burkholderia phage AH2

Halomonas phage φHAP-1
Vibrio phage YP8582
Vibrio phage VP58.5
Vibrio phage VHML

Rhodobacter phage RcapMu
Pseudomonas phage B3
Burkholderia phage BcepMu
Burkholderia phage φE255
Escherichia phage D108
Enterobacteria phage Mu
Burkholderia phage KS10
Pseudomonas phage F_HA0480sp/Pa1651
Pseudomonas phage MP22
Pseudomonas phage MP38
Pseudomonas phage MP42
Pseudomonas phage DMS3
Pseudomonas phage LPB1
Pseudomonas phage PaMx73
Pseudomonas phage PA1/KOR/2010
Pseudomonas phage MP29
Pseudomonas phage D3112
Pseudomonas phage JBD26

Pseudomonas phage MP1412
Pseudomonas phage M6
Pseudomonas phage YuA
Phage φJL001
Vibrio phage SSP002
Pseudomonas phage vB_Pae-Kakheti25
Burkholderia phage KL1
Pseudomonas phage PaMx25
Xanthomonas phage Xp15

ICTV Family
Myoviridae
Siphoviridae
Unassigned

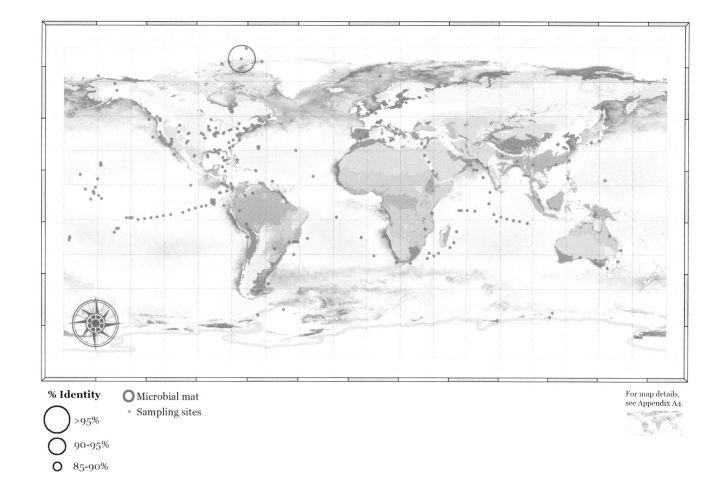

% Identity

⭕ >95%

⭕ 90-95%

⭕ 85-90%

⭕ Microbial mat

· Sampling sites

For map details,
see Appendix A4.

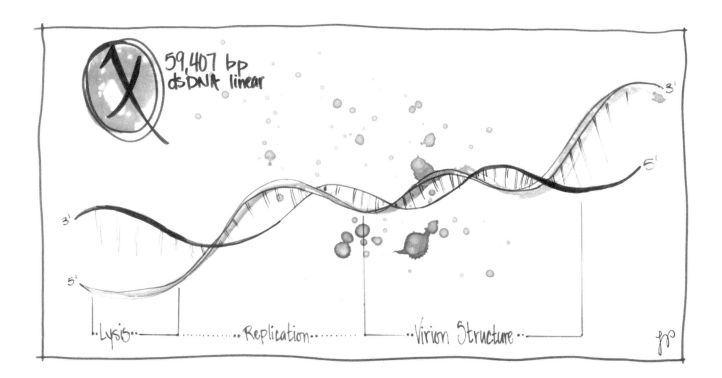

59,407 bp
dsDNA linear

3'

5'

3'

5'

··Lysis·· ··········Replication·········· ··Virion Structure··

Enterobacteria Phage χ

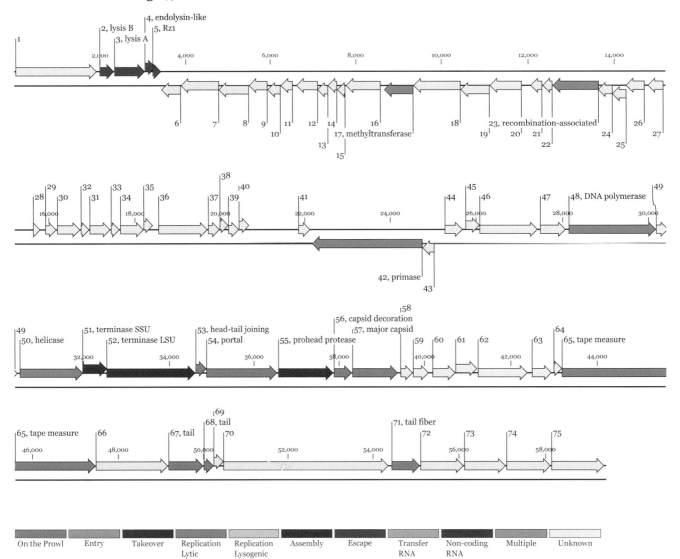

The Flagellar Express

Merry Youle

Mise-en-Scène: *Phages that use an essential cell surface structure as their receptor put their host in a no-win situation. Modifying that structure to block phage adsorption is apt to impair its function, thus apt to decrease host fitness. Flagella are one such structure that is skillfully exploited by flagellotropic phages. As a further benefit, flagella present a large target, thereby increasing the probability of a successful phage-host collision.*

An *E. coli* swims along at a brisk 20 µm sec^{-1} through a sea of phages, propelled by its bundle of rotating **flagella** trailing behind. No sitting duck, this bacterium, but rather a challenging quarry worthy of a skilled hunter. While this speedy travel has its advantages, the numerous long flagella employed offer vulnerable targets for any phage able to throw a lasso. Here comes such a phage now! Phage χ curls a tail fiber around a rotating flagellum and reaps a fast ride to the cell surface. There it locates its specific **receptor** nearby, binds irreversibly, and introduces its DNA. Infection!

Flagella are widespread in the Bacteria, and flagellotropic phages, like χ, abound. These phages include both Siphophages and Myophages, all characterized by having curly or coiled tail fibers. (The head filaments of *Caulobacter* phage [*see page 2-22*] are a kinky variation on the theme.) Bacteria could escape infection by these phages if only they would get rid of their flagella. When grown at 42° C, *E. coli* does just that and eludes χ attachment (Schade, Adler, Ris 1967). Whirring *E. coli* in a blender for two minutes has a similar, but only temporary, effect. Flagellar filaments are sheared off, but flagella quickly regrow. In a few minutes there are short stubs, and χ once again attaches, albeit more slowly (Schade, Adler, Ris 1967).

Runs & tumbles

Flagellotropic phages such as χ use only rotating flagella (those of *E. coli* rotate at >100 rpm) and the direction of rotation matters (Samuel et al. 1999). All the flagellar propellers on a bacterium rotate in the same direction, but that direction alternates every 1-2 sec as part of the cell's chemotactic response. During counter-clockwise (CCW) rotation, the spinning helical flagella bundle together and propel the cell forward—a 'run.' Clockwise (CW) turning disperses the bundle and the cell 'tumbles' without making any headway. The bacterium moves preferentially toward an attractant, such as food, by increasing the length of the runs; tumbles provide an opportunity to reorient as needed. χ requires periods of CCW rotation (i.e., runs) for infection.

Nuts & bolts

A phage tail fiber wrapped around a flagellum is analogous to a right-handed nut screwed onto a threaded bolt. To thread a nut onto a bolt, you turn the nut CW (as viewed from the end of the bolt) or you can turn the bolt CCW. Turn either in the wrong direction and the nut falls off the end. When a flagellum rotates CCW, a χ phage 'nut' threads rapidly toward the cell surface; CW rotation sends the phage in the opposite direction. Mutant hosts that rotate mostly CW reduce χ infection efficiency. Threading a phage nut onto a flagellum also requires that the phage's tail fiber fit into the helical grooves that run along the surface of the flagellum from its base to tip. Mutations that alter the dimensions of the groove reduce phage infectivity.

Whether the phage makes it to the cell or is spun off the end of the flagellum depends on where along the flagellum it attaches and how soon afterwards the motor reverses. Given a moderate 100 rpm rotation and a ~50 nm pitch of the surface helical groove of the flagellum, χ could travel a few micrometers in one second. A typical CCW run lasts one to two seconds, which gives χ a good shot at making it all the way home (Samuel et al.

1999). If the first attempt fails, the expelled phage can likely grab onto a nearby filament when the bacterium starts the next run and try again.

A rhizosphere adaptation

Warning! You can't judge a bacterium by its flagellum. Most bacterial flagella are built from very similar flagellin proteins and adopt very similar filament structures. Thus the same phage nut fits all, or at least many. A phage might ride a rotat-

ing flagellum to the cell surface only to discover that it has 'bolted' to the wrong doorstep: this cell is not a host. However, there are some bacterial flagella that are recognizably different. One type, known as complex flagella, enables Bacteria such as *Agrobacterium* sp. H13-3 to motor through the viscous rhizosphere. These relatively rigid flagella are built from three different flagellins instead of just one, making their surface a complex pattern of ridges and grooves (Yen, Broadway, Scharf 2012).

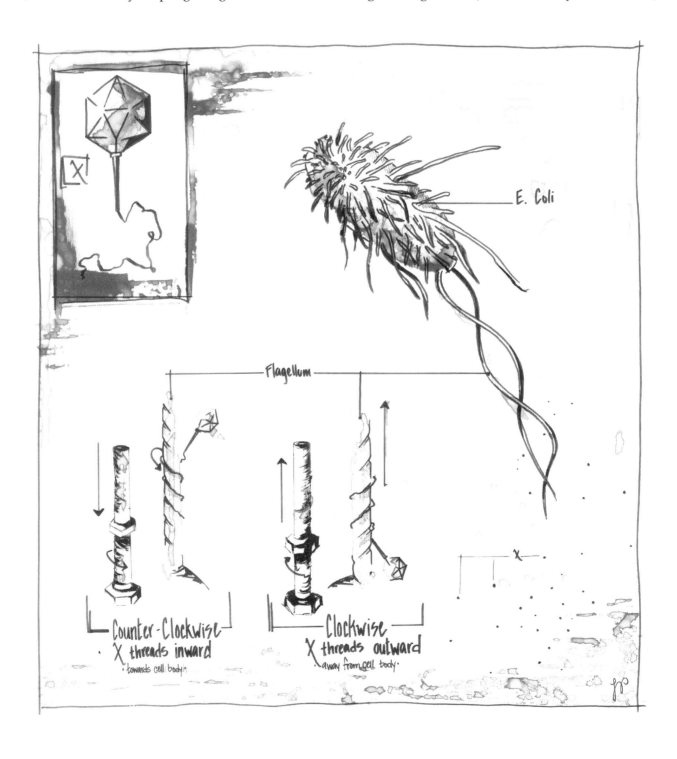

E. Coli

Flagellum

Counter-Clockwise
X threads inward
·towards cell body·

Clockwise
X threads outward
·away from cell body·

Moreover, they rotate in only one direction: CW. Despite these differences, flagellotropic phages such as Myophage 7-7-1 exploit these rotating flagella to efficiently infect this *Agrobacterium*. Instead of a long, kinky χ-style tail fiber, 7-7-1's tail ends with ten short (16 nm) fibers—a different style 'nut' for this different flagellar 'bolt.'

No escape

Reflect for a moment on the plight of the *E. coli* that, at the start of this story, was swimming through a phage-infested milieu. The churning of the flagellar propellers may create a vortex drawing the phage close. Every forward run may be whisking a phage or two directly to the cell surface, poised to launch an infection. Phage-resistant mutants can be readily isolated in the lab, but they usually lack full-length, functional flagella—a defense with too high a fitness cost to be maintained in the real world. Thus these phages seem to have a free ride to their host cells. However, arriving at the door is only the first step. The Bacteria can still block adsorption or interfere with phage replication at any of a number of later steps—until some shrewd phage counters their move and regains the advantage.

Cited references

Samuel, AD, TP Pitta, WS Ryu, PN Danese, EC Leung, HC Berg. 1999. Flagellar determinants of bacterial sensitivity to χ phage. Proc Natl Acad Sci USA 96:9863-9866.

Schade, SZ, J Adler, H Ris. 1967. How bacteriophage χ attacks motile bacteria. J Virol 1:599-609.

Yen, JY, KM Broadway, BE Scharf. 2012. Minimum requirements of flagellation and motility for infection of *Agrobacterium* sp. strain H13-3 by flagellotropic bacteriophage 7-7-1. Appl Environ Microbiol 78:7216-7222.

Caulobacter Phage φCbK

a Myophage that lassos a rotating flagellum with its long head filament

Genome
dsDNA; linear
215,710 bp
338 predicted ORFs; 26 RNAs

Encapsidation method
Packaging; T = 7 capsid

Common host
Caulobacter crescentus

Habitat
Oligotrophic aquatic environments

Lifestyle
Lytic

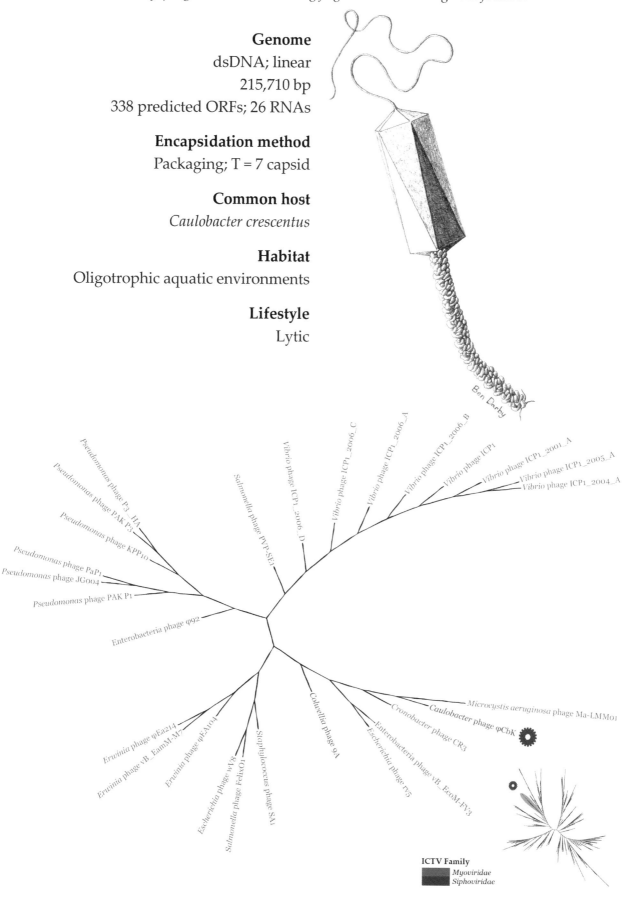

Ben Darby

Pseudomonas phage P3_HA
Pseudomonas phage PAK P3
Pseudomonas phage KPP10
Pseudomonas phage PaP1
Pseudomonas phage JG004
Pseudomonas phage PAK P1
Enterobacteria phage φ92
Salmonella phage PVP-SE1
Vibrio phage ICP1_2006_D
Vibrio phage ICP1_2006_C
Vibrio phage ICP1_2006_A
Vibrio phage ICP1_2006_B
Vibrio phage ICP1
Vibrio phage ICP1_2001_A
Vibrio phage ICP1_2005_A
Vibrio phage ICP1_2004_A
Microcystis aeruginosa phage Ma-LMM01
Caulobacter phage φCbK
Cronobacter phage CR3
Enterobacteria phage vB_EcoM-FV3
Escherichia phage rv5
Colwellia phage 9A
Staphylococcus phage SA1
Salmonella phage FelixO1
Escherichia phage wV8
Erwinia phage φEA104
Erwinia phage vB_EamM-M7
Erwinia phage φEa214

ICTV Family
Myoviridae
Siphoviridae

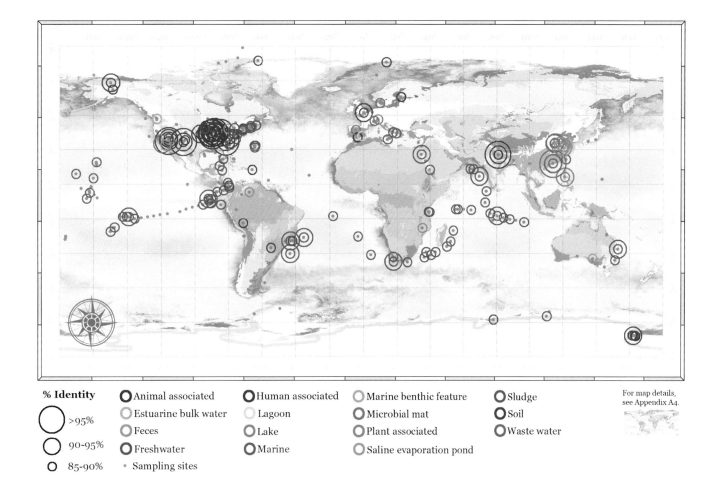

% Identity

○ >95%

○ 90-95%

○ 85-90%

● Sampling sites

○ Animal associated
○ Estuarine bulk water
○ Feces
○ Freshwater

○ Human associated
○ Lagoon
○ Lake
○ Marine

○ Marine benthic feature
○ Microbial mat
○ Plant associated
○ Saline evaporation pond

○ Sludge
○ Soil
○ Waste water

For map details,
see Appendix A4.

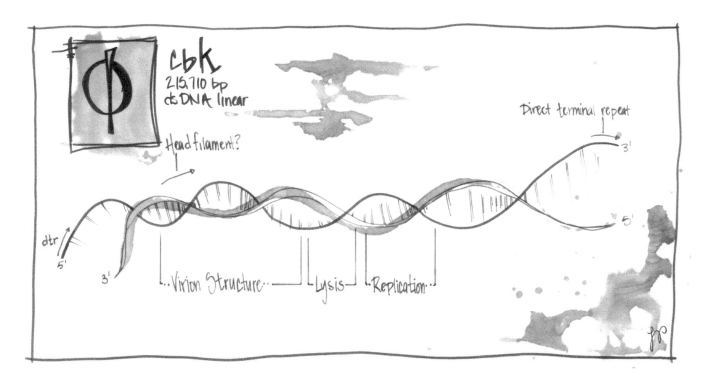

CbK
215,710 bp
dsDNA linear

Head filament?

Direct terminal repeat

3'

5'

dtr

5'

3'

···Virion Structure··· ···Lysis··· ···Replication···

Caulobacter Phage φCbK

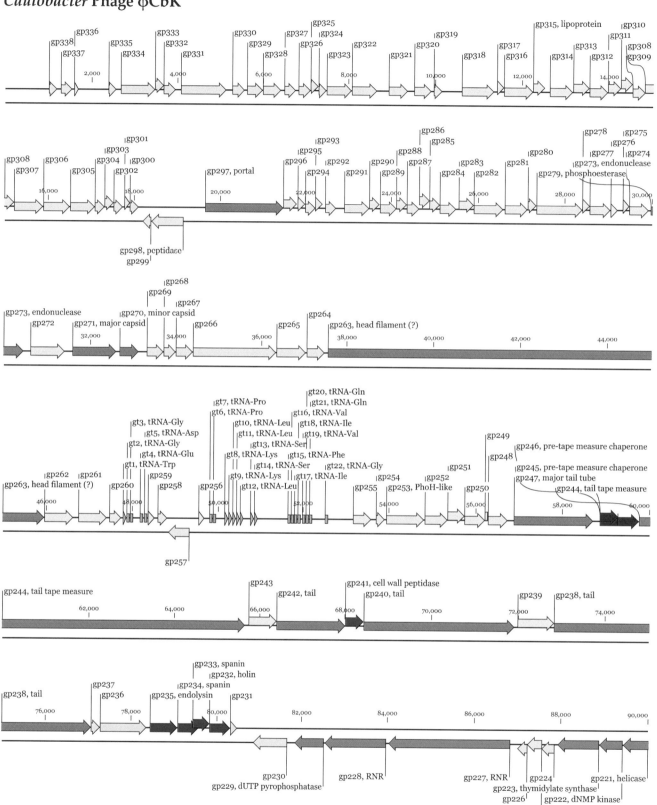

On the Prowl   Entry   Takeover   Replication Lytic   Replication Lysogenic   Assembly   Escape   Transfer RNA   Non-coding RNA   Multiple   Unknown

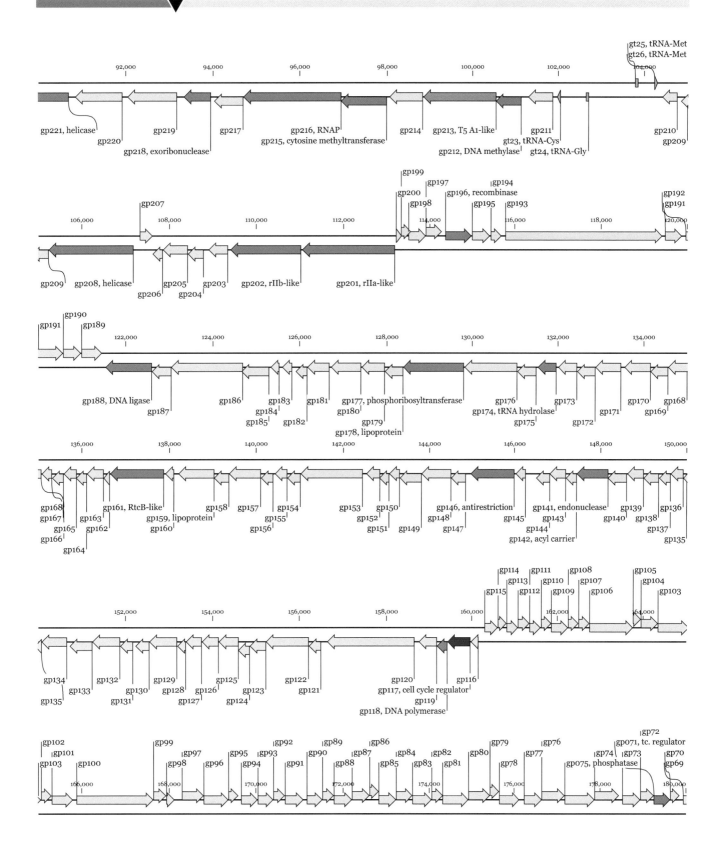

Continued next page

Caulobacter **Phage** φCbK *continued*

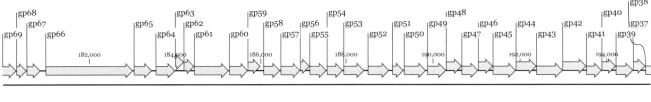

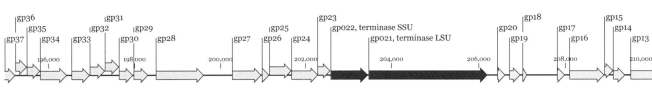

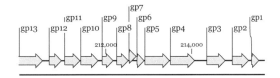

| On the Prowl | Entry | Takeover | Replication Lytic | Replication Lysogenic | Assembly | Escape | Transfer RNA | Non-coding RNA | Multiple | Unknown |

Using Your Head

Merry Youle

Mise-en-Scène: *Microbes can acquire temporary immunity to a phage by not expressing the phage receptor under some conditions or during some stages in their life cycle. Undaunted, some phages make do with efficiently preying on the susceptible members of the population. Such is the case with flagellotropic phages that infect the motile swarmer cells of* Caulobacter crescentus. *An alternative coping strategy, related in another story, is for the phage to rapidly modify its receptor specificity (see page 2-13).*

Note: Some of the research related in this story was performed using phage φCb13, a *Caulobacter* phage that, like φCbK, possesses a head filament. For simplicity, the story refers only to φCbK.

The life cycle of *Caulobacter crescentus* offers a fleeting opportunity for an adept phage hunter. This protean bacterium alternates between two forms: stalked cells that attach to a surface and motile **swarmer cells**. The swarmer cells have a single **flagellum** and several **pili** located at one pole—structures that the stalked cells eschew. Both of these structures on other Bacteria are exploited by phages as their entryway. For example, pili serve as receptors for the filamentous Ff phages of *E. coli* (*see page 7-17*); flagella are the site of initial adsorption for numerous phages, including phage χ (*see page 2-19*). These flagellotropic phages are readily spotted under the EM by the telltale kinks or curls in their tail fibers—evidence that they are adapted to wrap around a rotating flagellum as the first step in a successful attack.

A headfirst approach

There is one group of flagellotropic phages that does not abide by the normal protocol. These are the *C. crescentus* Myophages φCb13, φCbK, and their kin. They don't seem to know their head from their tail. Unique among all known phages, they have a flexible filament extending ~200 nm from their heads, from the 'top' vertex opposite the tail to be precise (Leonard et al. 1972). It is this filament that these phages wrap around the bacterium's rotating flagellum (Guerrero-Ferreira et al. 2011). The wrapped filament forms a 'nut' that threads along the surface grooves of the rotating flagellar 'bolt.' Depending on the direction of rotation, the phage nut is spun headfirst toward the cell surface or in the opposite direction.

Tail-first adsorption

The phage still needs to bring its tail fiber into contact with its specific receptors—the pilus portals located at the cell pole near the base of the flagellum. Phage φCbK's method is to make strategic use of the host's swimming mode. These swarmer cells switch the direction of rotation of their flagellum approximately every five seconds. CW rotation pushes the cell forward with the flagellum at the stern; CCW propels the cell backwards with the flagellum in the lead (Ely et al. 1986). Picture φCbK swinging from a flagellum by its head filament. As the flagellum rotates CCW, the phage spins along the flagellum toward the cell pole. At the same time, the cell is moving, flagellum first, through the water. Resistance from the water sweeps the dangling phage tail towards the cell, putting the tail fiber in close proximity to the pilus portals. When contact is made, irreversible adsorption follows. An infection has begun.

The flagellar benefit

φCbK can infect only the motile swarmer stage of *C. crescentus* because the pilus portals that are essential for infection are not present in the stalked cells. The flagellum, on the other hand, is not indispensable for infection. Rather it serves to increase the probability of the phage tail contacting the actual receptors on this moving target. If, through mutation, *C. crescentus* loses the ability to construct or rotate its flagellum, phage adsorption efficiency drops two- to three-fold (Guerrero-Ferreira et al. 2011). In the competitive environment, that makes a huge difference to the phage.

An odd filament

All φCbK-like phages have head filaments—a unique hallmark of the group. Likely those filaments are built from one protein, gp263, formerly known as gp76 (Gill et al. 2012), that has also been found only in these phages and is unrelated to any known proteins. Could gp263 do the job? These filaments must be flexible; the high glycine content (21%) of this large protein (2,799 amino acids) indicates a very flexible structure. These filaments must recognize and adhere to cells; gp263 contains a carbohydrate-binding domain that could serve this purpose. Genes involved in virion morphogenesis are often clustered in the genome; gene 263

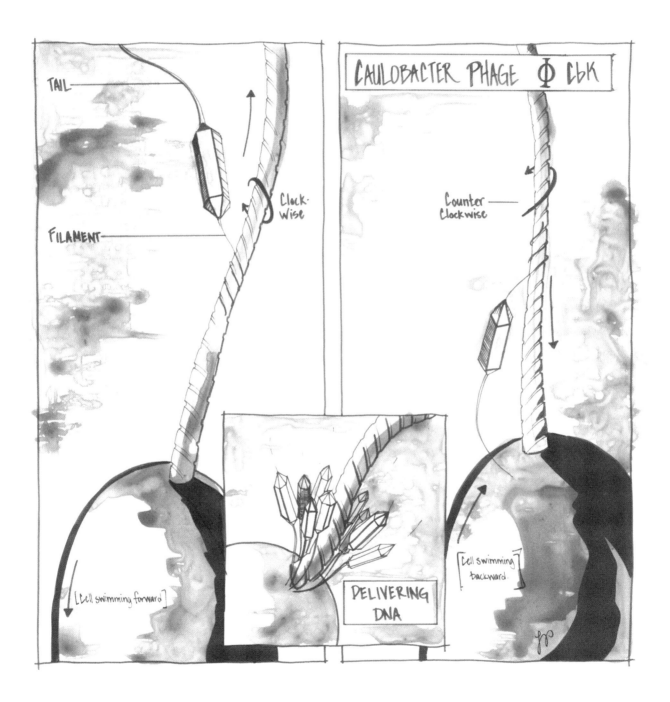

resides in that genomic neighborhood. So gp263 is the favored candidate, but direct confirmation of its role is yet to come.

At this point, not enough is known about these odd filaments to make heads or tails of them. How does a capsid assemble two unique vertices that are poles apart? So far there are no clues as to how the filament assembles or attaches to the capsid. One wonders if it might be an over-grown decoration protein or perhaps a modified tail fiber, but no sequence **homology** to either is evident. Surely there is a novel story here to be uncovered by a researcher who uses their head.

Cited references

Ely, B, CJ Gerardot, DL Fleming, SL Gomes, P Frederikse, L Shapiro. 1986. General nonchemotactic mutants of *Caulobacter crescentus*. Genetics 114:717-730.

Gill, JJ, JD Berry, WK Russell, L Lessor, DA Escobar-Garcia, D Hernandez, A Kane, J Keene, M Maddox, R Martin. 2012. The *Caulobacter crescentus* phage φCbK: Genomics of a canonical phage. BMC Genomics 13:542.

Guerrero-Ferreira, RC, PH Viollier, B Ely, JS Poindexter, M Georgieva, GJ Jensen, ER Wright. 2011. Alternative mechanism for bacteriophage adsorption to the motile bacterium *Caulobacter crescentus*. Proc Natl Acad Sci USA 108:9963-9968.

Leonard, K, A Kleinschmidt, N Agabian-Keshishian, L Shapiro, J Maizel Jr. 1972. Structural studies on the capsid of *Caulobacter crescentus* bacteriophage φCbK. J Mol Biol 71:201-216.

Enterobacteria Phage RB51

a Myophage that ambushes Bacteria in the mucus layers of metazoans

Genome

dsDNA; linear

168,394 bp

274 predicted ORFs; 9 RNAs

Encapsidation method

Packaging; T = 13 capsid

Common host

Escherichia coli

Habitat

Sewage, vertebrate intestines

Lifestyle

Lytic

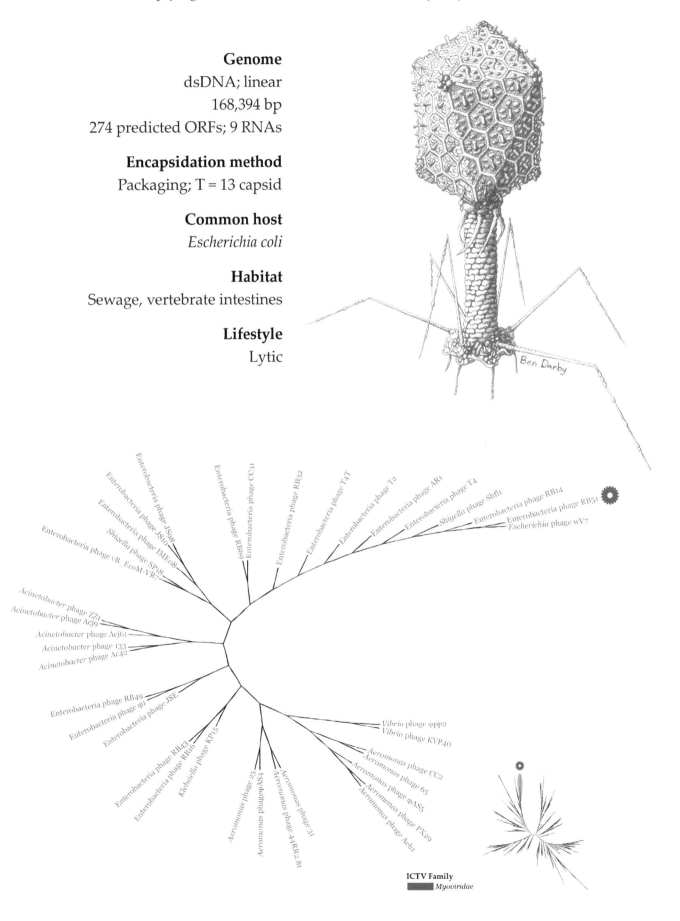

Ben Darby

ICTV Family

Myoviridae

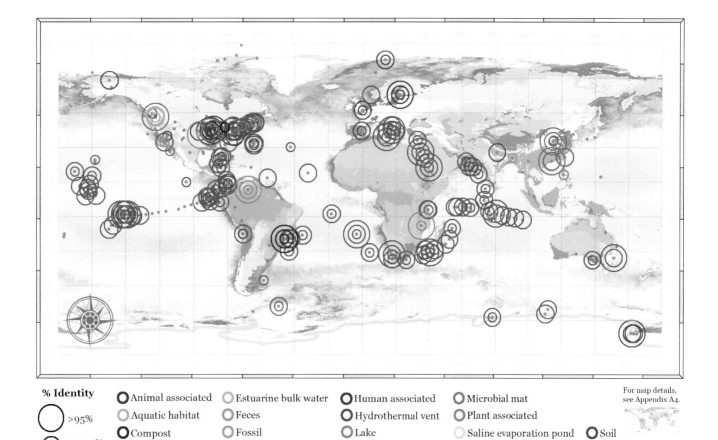

% Identity

- ○ >95%
- ○ 90-95%
- ○ 85-90%
- · Sampling sites

○ Animal associated	○ Estuarine bulk water
○ Aquatic habitat	○ Feces
○ Compost	○ Fossil
○ Cultured	○ Freshwater

○ Human associated	○ Microbial mat
○ Hydrothermal vent	○ Plant associated
○ Lake	○ Saline evaporation pond
○ Marine	○ Sediment

○ Soil	
○ Waste water	

For map details,
see Appendix A4.

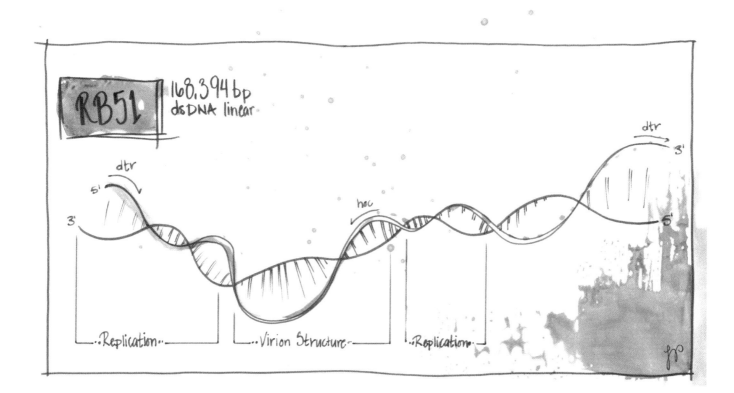

RB51   168,394 bp
dsDNA linear

dtr
5'
3'
dtr
3'
5'
hoc

···Replication···   ···Virion Structure···   ···Replication···

Enterobacteria Phage RB51

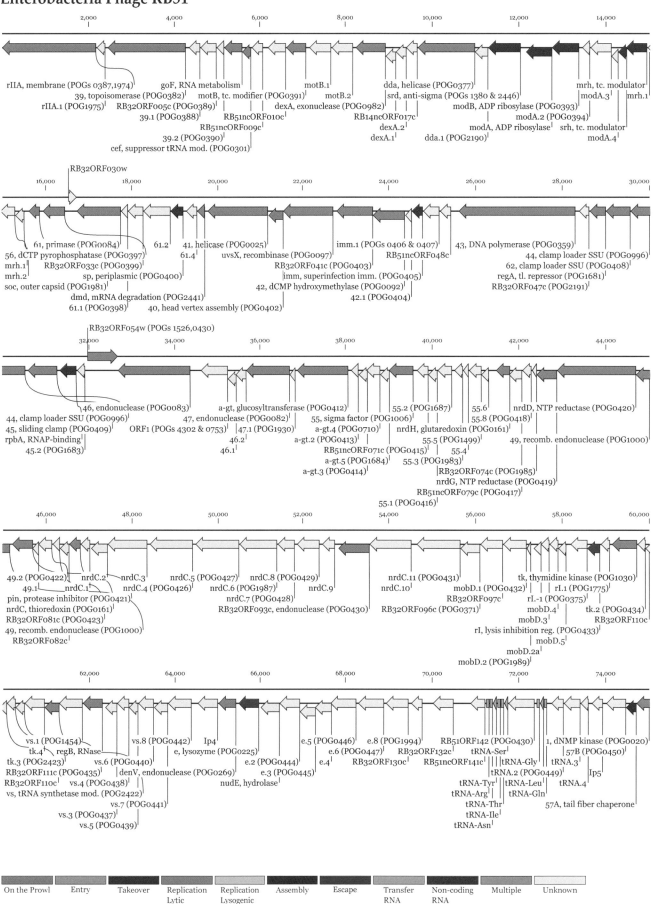

On the Prowl | Entry | Takeover | Replication Lytic | Replication Lysogenic | Assembly | Escape | Transfer RNA | Non-coding RNA | Multiple | Unknown

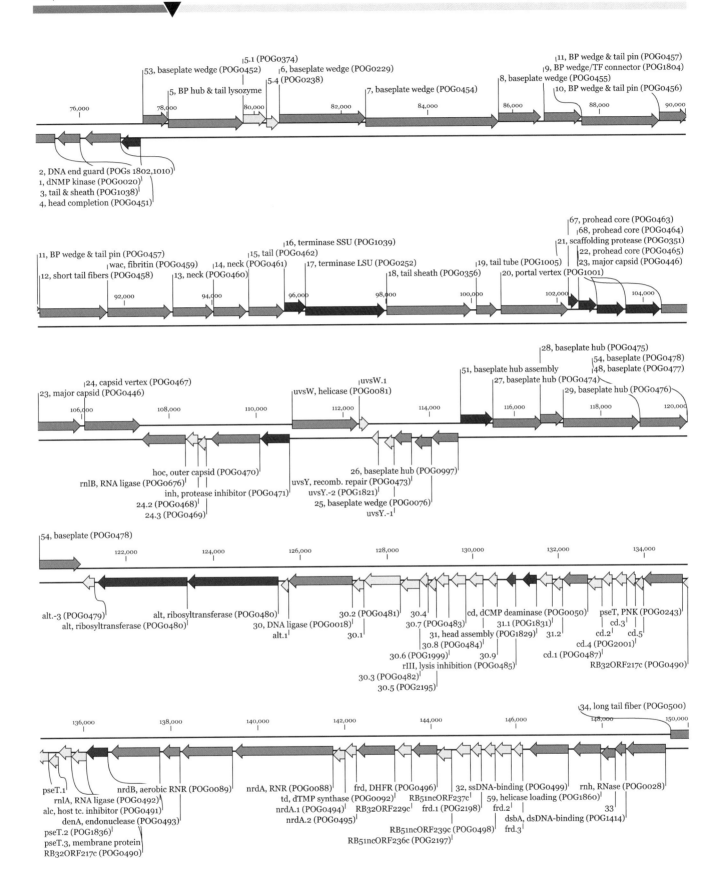

Continued next page

Enterobacteria Phage RB51 *continued*

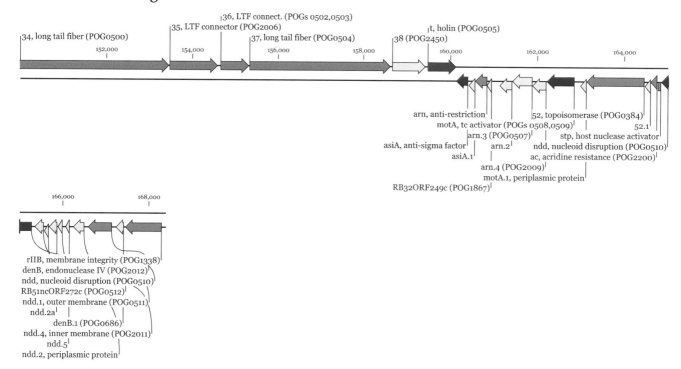

Caution! Phage Ambush Ahead!

Merry Youle

Mise-en-Scène: *Although phages are non-motile, they have strategies for positioning themselves to ambush prey in favorable locations such as the Bacteria-rich metazoan mucus layers. The continual sloughing of mucus that protects the epithelium from invading pathogens also quickly casts off many phages. Despite the challenges, some phages have adapted to exploit this opportunity by both adhering to and diffusing into the mucus.*

Note: Much of the research related in this story was performed using phage T4, but T4-like RB51 possesses a similar Hoc protein with 90% amino acid identity to T4's Hoc.

Ceteris paribus, the more prey, the better the hunting. But can phages hunt? Granted, their virions are not equipped for locomotion, and they don't actively stalk their prey. But they do hang out and wait where Bacteria congregate—the proven tactic of the ambush predator. And what better place for an ambush than the Bacteria-rich **mucosal surfaces** of animals? Despite the protective mucus, such surfaces are inherently vulnerable to bacterial infection because they must be thin and permeable to carry out their physiological functions (e.g., food adsorption in the gut, gas exchange in the lung). The T4-like phages, among many others, exploit this opportunity and turn the viscous mucus into a killing field.

The opportunity

Bacteria find mucus hospitable and congregate there in greater abundance than in the surrounding milieu (Barr et al. 2013). Mucus offers them glycans to eat, a structured environment, and for the motile pathogens among them, proximity to the underlying epithelial cells. Where abundant prey graze, predators abound. There is also an unusually large number of phages here. Whereas in most environments phage **virions** outnumber Bacteria by roughly ten-to-one, in these mucus layers that ratio increases to more than forty-to-one (Barr et al. 2013). More virions mean a bacterium is more likely to encounter one—typically about 14 times more likely than if it traversed the same distance but avoided the mucus (J. Barr, personal communication). Fortunately for the phages, day after day, the congenial mucus lures Bacteria into their ambush.

Maintaining your ambush position

Successful establishment of an ambush position in the mucus is not simple for a phage. At first glance, it might seem sufficient for a virion to adhere to the outermost layer of mucus and wait for an incoming bacterium to pass by. That tactic would not work because the surface layer is transitory. The underlying epithelium continually secretes mucus. The new mucus pushes the existing overlying layers upward and prompts the outermost zone to slough off. While this mucus conveyor belt evicts invading pathogens, it also indiscriminately sloughs any phage virions stuck to the outer mucus.

One way to counter this flux is for the phage to diffuse deeper into the mucus faster than the mucus is flowing outward. The sticky molecular network of the mucus dissuades most invaders. However, phage virions are small enough to pass through the glycoprotein mesh that makes mucus viscous. The sticky mucus also uses hydrophobic bonds to protect the epithelium from bits of passing debris, but these virions evade that trap by having a dense array of charged amino acid groups on their surface (Cone 2009).

What's needed is just the right amount of stick-to-it-iveness. If a phage virion adheres too weakly to the mucus, it won't stick around long enough to set up an ambush. If it adheres too strongly, it won't diffuse inward from the surface fast enough to avoid eviction by the mucus conveyor belt. The phage whose virion adheres just right will hunt slightly longer in the prey-rich mucus and still

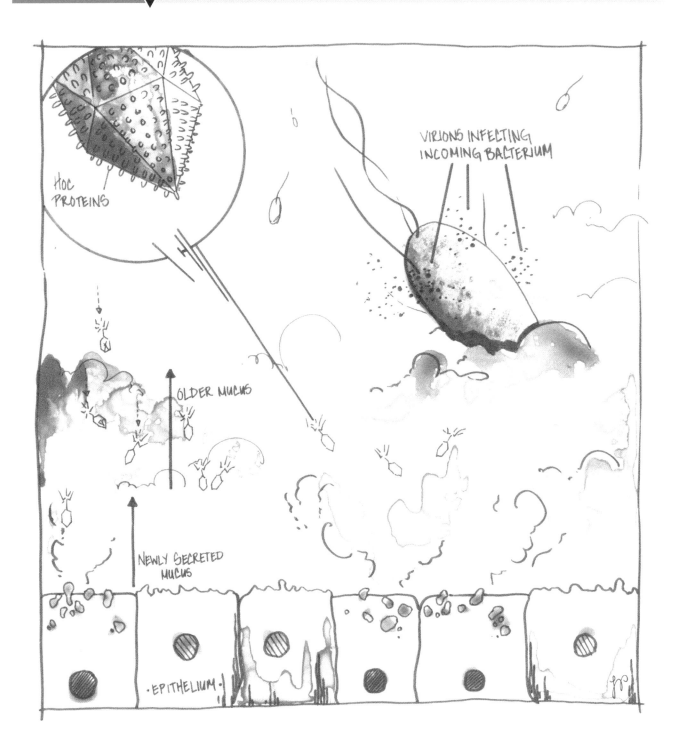

sometimes diffuse slowly inward to a more en- during position.

How to cling to mucins

To adhere to mucus requires adherence to its main macromolecular components, the mucins. These are giant glycoproteins up to 10^3–10^6 kDa that re- semble a 'bottle brush' in structure. The central rod of the brush is formed by a long, wormlike polypeptide chain composed of hydrophobic re- gions that alternate with domains rich in serine

and threonine residues. The numerous serines and threonines bear *O*-linked glycan chains that extend outward up to 5 nm to form the 'brush bristles' that are the likely sites for phage binding.

How might an inert virion hold onto a glycan? In the cellular world, cells often use proteins to recognize and adhere to glycans or other specific carbohydrates on the surface of other cells. Some of these binding proteins employ immunoglob- ulin-like (Ig-like) domains for these specific gly-

Label within image: MUCIN PROTEINS, GLYCANS, LYSED BACTERIUM

can-binding duties. Phages, too, encode proteins with Ig-like domains. Of the tailed phages whose genomes have been sequenced, ~25% encode Ig-like domains, all of which are located in structural proteins and thus are likely to be exposed on the surface of the virion (Fraser et al. 2006). Those exhibited on the tail might bind to host surface polysaccharides to assist with phage adsorption, but those displayed on the capsid, such as the Hoc (Highly antigenic Outer Capsid) protein, likely serve some other purpose.

In phage T4 and T4-like phages (e.g., phage RB51) 155 copies of Hoc are bound to specific locations on each capsid as a decoration protein (Fokine et al. 2011). Each copy contains one domain that adheres to the capsid and three immunoglobulin-like (Ig-like) domains that together form a short rod projecting out from the capsid surface (Sathaliyawala et al. 2010). Hoc's function was difficult to determine since none of these phages need it in order to replicate under lab conditions. Outside the lab, Hoc makes itself useful, particularly for phage communities hunting near mucosal surfaces (Barr et al. 2013). By weakly binding glycans in the mucus, Hoc slows diffusion of these phage virions just enough to keep them in the mucus layer while they prepare their ambush (Barr et al. 2013).

These clingy phages face another difficulty. Mucus contains several different mucins, each bearing many different glycans. Moreover, mucin glycosylation patterns differ from one metazoan species to another and from one mucosal surface to another within each animal. For even more variety, when challenged by a microbial infection, the epithelium changes its current pattern to confound the pathogens (Linden et al. 2008). Key to phage success here is to strike a balance between specificity and promiscuity in glycan binding. Better to bind weakly to a variety of mammalian glycans, rather than strongly to any particular one. This is indeed Hoc's modus operandi. The phages also have a way to specifically adapt glycan-binding proteins like Hoc to accommodate different glycans. Like cellular life, phages have found both C-lectin folds and Ig-like domains to be useful structures for embodying this variability. To introduce the variation, phages in the mucus-associated environment of the human gut are equipped with a mechanism for targeted mutagenesis (*see page 2-13*) that might specifically alter the binding sites (Minot et al. 2012).

Penetrating deeper

If a phage could navigate to a position deeper within the mucus layer, it could stick around longer to exploit the plentiful Bacteria foraging there. Indeed, a phage can do that by hitching a ride inside a moving bacterium. While replicating inside a motile host, a lytic phage can traverse considerable distance since the host will continue to swim normally until the moment of lysis (*see page 7-5*). If the bacterium travels farther into the mucus layer, then when the progeny virions are released they will start their hunt in an advantageous location. Temperate phages that travel as prophages can fare even better. If their bacterial host is a commensal that resides closer to the epithelium, the prophage has a perpetual home in a good locale.

Collateral benefits to the metazoan

Lytic phages are in the business of killing Bacteria. Acting in their own self-interest, some have perfected the art of the ambush to prey upon the bacterial multitude congregating on mucosal surfaces. So doing, they also pay back the metazoan that supplied the mucus by protecting its vulnerable mucosal surfaces from bacterial invasion. Like all metazoans, we are holobionts that rely on our bacterial, archaeal, and phage partners. Long live the holobiont!

Cited references

Barr, JJ, R Auro, M Furlan, KL Whiteson, ML Erb, J Pogliano, A Stotland, R Wolkowicz, AS Cutting, KS Doran, P Salamon, M Youle, F Rohwer. 2013. Bacteriophage adhering to mucus provide a non-host-derived immunity. Proc Natl Acad Sci USA 110:10771-10776.

Cone, RA. 2009. Barrier properties of mucus. Adv Drug Deliv Rev 61:75-85.

Fokine, A, MZ Islam, Z Zhang, VD Bowman, VB Rao, MG Rossmann. 2011. Structure of the three N-terminal immunoglobulin domains of the highly immunogenic outer capsid protein from a T4-like bacteriophage. J Virol 85:8141-8148.

Fraser, JS, Z Yu, KL Maxwell, AR Davidson. 2006. Ig-like domains on bacteriophages: A tale of promiscuity and deceit. J Mol Biol 359:496-507.

Linden, S, P Sutton, N Karlsson, V Korolik, M McGuckin. 2008. Mucins in the mucosal barrier to infection. Mucosal Immunol 1:183-197.

Minot, S, S Grunberg, GD Wu, JD Lewis, FD Bushman. 2012. Hypervariable loci in the human gut virome. Proc Natl Acad Sci USA 109:3962-3966.

Sathaliyawala, T, MZ Islam, Q Li, A Fokine, MG Rossmann, VB Rao. 2010. Functional analysis of the highly antigenic outer capsid protein, Hoc, a virus decoration protein from T4-like bacteriophages. Mol Microbiol 77:444-455.

Recommended reviews

Barr JJ, M Youle, F Rohwer. 2013. Innate and acquired bacteriophage-mediated immunity. Bacteriophage 3(3):e25857.

Cone, RA. 2009. Barrier properties of mucus. Adv Drug Deliv Rev 61:75-85.

March from the Sea: A Brief History of Environmental Phage Ecology from Marine to Human Ecosystems

Ben Knowles[†]

Abstract: *Underlying the functioning of every ecosystem is a riotous carnival of phages, long overlooked because we could not detect them. Powered by the methodological advances in recent decades, environmental phage ecology marched from obscure neglect to mainstream science. With each new set of observations came yet another set of questions that then drove the next round of exploration. Initially this played out mainly in the more tractable marine ecosystem, the site of significant advances in various aspects of phage ecology: phage abundances in the 1980s and 1990s, lytic and lysogenic dynamics in the 1990s, phage-host networks and metabolic functionality in the 2000s. Here was laid the foundational knowledge demonstrating phage importance to diverse aspects of life ranging from biogeochemistry to community metabolism and host evolution. The heavy marine bias of our knowledge is giving way now as phage researchers explore other, less salty environments. One such environment is the human body, with its individual phage populations in numerous unique micro-environments. Viewing human health from the perspective of phage ecology promises to have profound impacts on our understanding of disease and treatment. Overall, the history of phage ecology convinced many that invisibility does not mean insignificance, and that our understanding of our world is still wonderfully incomplete.*

[†]Department of Biology, San Diego State University, San Diego, CA
Email: benjaminwilliamknowles@gmail.com

You won't find what you've never seen

It seems strange to us now, with our current conception of the ocean as a bustling metropolis swarming with microbes and viruses and protists (and the occasional whale), but until recently the ocean was viewed as a sparsely populated desert. This perception came from the work of pioneering environmental microbiologists who concluded there were very low numbers of microbes in the sea based on what they could grow in the lab. That they saw anything at all is amazing. From its foundation, environmental microbiology was hindered by the necessity for researchers to invent basic techniques from scratch. For example, in the 1940s researchers had to resort to developing their own culture media in order to study or count microbes. However, while culture-based approaches have allowed researchers to domesticate and study environmental microbes in the lab for almost a century, using counts of cultured microbes to assess microbial abundance in the ocean suffers profoundly from the fact that only a tiny fraction (approximately 1%) could be cultured with contemporary methods. As a result of this unrecognized bias, now known as the Great Plate

Count Anomaly, researchers radically underestimated marine microbial abundance for most of the last century.

In science, direct measurements should trump indirect estimates. However the perception of vacant oceans was so thoroughly entrenched between the 1940s and 1970s that direct microscopy evidence contradicting indirect culture-based enumerations was discounted (Jannasch, Jones 1959). Unfortunately for phage biology, this seeming dearth of microbes in the sea led researchers to conclude that marine phage could not possibly be ecologically significant as there were not enough prey to support them—a view formally codified by the father of marine microbiology, Claude ZoBell, in his defining 1946 textbook (ZoBell 1946). Locked within this paradigm, environmental phage ecology was relegated for several more decades to lab-based studies of only some phages that could be cultured (see the work of Moebus below).

Advances with isolates

Although today we would not call it *environmental* phage biology due to its dependence on lab-based

Figure 1. The microbes (large objects) and virus-like particles (dots) found in 1/8th of a microliter of seawater from the coral reef at Guam. Genomic content was stained using Sybr Gold (Noble, Fuhrman 1998) and imaged at ~800X magnification. Credit: Ben Knowles.

microbial and phage domestication, important work was conducted between the 1940s and 1980s on phage-host pairs from diverse environments using culture-based approaches. For example, the *Pseudoalteromonas* phage PM2 was isolated and cultured from seawater off the Chilean coast in the late 1960s and was the subject of hundreds of publications (Mannisto et al. 1999). Studies using phage and bacterial isolates from aquatic systems provided insights into ecological questions such as the effects of phage infection on host virulence (Barksdale, Pappenheimer 1954), impact of temperature on phage adsorption (Seeley, Primrose 1980), specificity of phage host selection (Markel, Fowler, Eklund 1975), and community composition of polluted waters (Tartera, Jofre 1987). Similar questions were investigated in soil (Crosse,

Hingorani 1958; Kowalski et al. 1974) and even in the leaves of apple trees (Ritchie, Klos 1977). Answers gained during this period would allow researchers to better understand what they were observing when phage ecology became truly environmental in the 1980s.

Discoveries by epifluorescence

Despite these culture-based advances, the field yearned for a means to scrutinize phage and microbes in their natural environment. That revolutionary advance arrived in the late 1970s with the advent of an epifluorescence microscopy technique that allowed researchers to directly count microbes in seawater (Hobbie, Daley, Jasper 1977). Everything changed: there were microbes everywhere and in massive abundances, approximately

10^6 microbes per ml throughout the oceans (reviewed in [Wommack, Colwell 2000; Suttle 2007]). The prevailing worldview was proven to be radically inaccurate.

Although this brought marine microbes to the attention of the scientific community, phage were unable to claim their place in the emerging microbial paradigm in the 1980s because phage virions and phage genomes were too small to be detected using contemporary visualization techniques. However, given the immense numbers of microbes just discovered, the indications were strong that there must also be abundant phage preying on them. Given this reasonable expectation, the hunt was on over the next decade for a method to accurately quantify phage in natural waters.

At first things did not seem promising when initial direct counts yielded only 10^4 viruses per ml in the ocean (Torrella, Morita 1979), 100 times less than the abundance of their hosts. Perhaps ZoBell was right after all. Environmental phage ecology limped out of the 1970s in obscurity. It took another decade before reliable counts by Øivind Bergh and colleagues, published in a 1989 paper with the understated title of *High Abundance of Viruses found in Aquatic Environments*, reported 10^8 viruses per ml of seawater (Bergh et al. 1989). Using epifluorescence microscopy techniques refined by Jed Fuhrman and Rachel Noble in 1998, it is now possible to rapidly count the viruses accurately and robustly in environmental samples without the need for laborious electron microscopy work (Fig. 1, [Noble, Fuhrman 1998]). Counts conducted with this approach, used in myriad studies all over the world, have revealed that phages are the most abundant organisms on the planet identified to date. Ocean surveys have shown average viral abundances of approximately 10^7 per ml in the ocean; thus marine phages outnumber their hosts by a factor of ten. Although this is fairly well accepted, there is no consensus as to why this ratio occurs, or why microbial and phage abundances in the ocean tend to be magically consistent, varying by only an order of magnitude or so across very different conditions.

Rediscovering How to Count Environmental Phage

A year in the lab will save you a day in the library, as the saying goes. This is wonderfully exemplified in the decade-long search between 1977 (Hobbie, Daley, Jasper 1977) and 1989 (Bergh et al. 1989) for a robust way to count viral abundances in the aquatic environment. In 1989 Bergh and colleagues published what they thought was a new technique—concentrating virions by ultracentrifugation and then counting them by transmission electron microscopy. Actually, Norman Anderson (Anderson et al. 1967) had developed that methodology approximately two decades earlier, but he had published outside the purview of environmental phage biologists in a book about viral impacts on drinking water. Marine researchers only became aware of his work when he was invited by Curtis Suttle to the American Society for Limnology and Oceanography conference in the early 2000s where they spoke with him in person.

By the end of the 1980s it was clear that the ocean is nothing like a desert.[1] An average ml of seawater contains the population of urban Rio de Janeiro in phage alone, not to mention thousands of protists and millions of bacteria and other microbes. And in that average ml of seawater these organisms are moving around, scavenging particles, photosynthesizing, reproducing, eating each other while trying to avoid being eaten, with many dying every day. Every ml of seawater is a riotous Carnival. No one would consider Rio de Janeiro a desert, and yet the perception of the open ocean

---

[1] This is a metaphorical desert. Even real deserts are not deserted, and bear little actual resemblance to metaphorical deserts. This reflects the main point of this section: life is more pervasive than we imagine it to be, figuratively or scientifically.

as such has been difficult to dispel from popular and scientific mindsets and is still taught in some universities today, perhaps due to the difficulty of comprehending such riotous activity in a minuscule world we cannot directly perceive.

Massive bacteriocide by rapacious phages

Oceanographic work throughout the early and mid-1980s showed that while protists kill a large proportion of bacteria on a daily basis, a comparable amount of bacterial mortality remained unaccounted for. This was one of the deeper mysteries in marine science for over a decade. Phage were not considered credible suspects in this killing until it was shown how supremely abundant they are in the sea. At that point, researchers turned to oceanographic radiolabeled uptake techniques developed in microbial ecology to ask whether phage could be the unknown agents in this massive daily bacteriocide. They then rapidly showed that not only do phage kill as many Bacteria as do grazing protists in some locations (Fuhrman, Noble 1995), thereby causing a decline in bacterial productivity (Suttle, Chan, Cottrell 1990), but they can also infect up to 70% of the bacterial community (Proctor, Fuhrman 1990), and kill up to ~50% of Bacteria daily at some sites (Jiang, Paul 1994). That amounts to over 500,000 phage-mediated bacteriocidal events per ml per day and equates to approximately 10^{23} viral infection events per second in the global ocean, lytic and lysogenic infections combined (Suttle 2007). Viral infection is surely the most frequently occurring interaction between organisms on the planet. Given an estimated 10^{30} phage in the ocean, one phage in every hundred thousand is actively infecting a host every minute. Busy.

Further, radiotracer research in the early 1990s turned up something unexpected: the absence of phages was sometimes associated with large reductions in microbial productivity (reviewed in [Wilhelm, Suttle 1999]). It makes intuitive sense that phage-induced *reductions* of bacterial productivity arise via lysis of significant portions of the host community, but phage bacteriocide leading to *increased* bacterial production did not. And yet

the observation was widespread in empirical and modeled data from a number of research groups around the world. It was quickly suggested by researchers such as Frede Thingstad and Farooq Azam (who had described the microbial loop a decade earlier [Azam et al. 1983]), that this enhanced productivity was fueled by the cellular debris and metabolites liberated by host lysis—a process labeled the viral shunt by Steven Wilhelm and Curtis Suttle in 1999 after almost a decade of conjecture and the gathering of evidence.

The viral shunt model has far ranging implications for ecosystem function (reviewed in [Suttle 2007; Weitz, Wilhelm 2012]). According to this model, bacterial lysis liberates organic carbon and other nutrients (e.g., nucleotides, amino acids) that are immediately recycled through the microbial community instead of being devoured as intact cells by protists that are, in turn, eaten by larger organisms. Released nutrients that are not readily assimilated into microbial metabolism such as refractory dissolved organic carbon are lost from the productive surface waters, falling as marine snow to deep ocean communities.

The price of success

In 2000, Frede Thingstad published the "kill-the-winner" hypothesis that modeled the effects of specialist phage predation and generalist protistan bactivory on bacterial communities (Thingstad 2000). He described predation by phages as scaling with the abundance of their specific prey according to an idealized Lotka-Volterra model. In contrast, generalist protistan predators were considered to consume all prey as they encounter them. In the model, microbial species that become common (the 'winners') face ever-increasing phage predation pressure as they become more numerous. Likewise protistan predation also grows as the increasing abundance of a prey species brings higher encounter rates. Thus, abundant species face compounded protistan and phage predation, while rare species are in low-density numerical refugia. However, the specificity of phage predation combined with the rapid increase of the phage population on the heels of the increase in their specific prey leads ultimately to a collapse of the prey population, quickly followed by the marked decline in the phage predators dependent on them. This rise and fall does not perturb the protist population that simply transitions to consuming the new dominant microbe.

This constant turnover of dominant species implied in the kill-the-winner model leads to increased diversity of microbial communities by promoting increased evenness among prey populations and by intermittent periodic predatory pressure on individual species. The ability of the kill-the-winner hypothesis to elegantly explain phenomena such as the commonly observed 10:1 virus:microbe ratio and bacteria-phage coevolutionary diversification led to its adoption by phage ecologists. However, confirmation of this hypothesis awaited the advent of metagenomics. It is now known that while the overarching oscillations in dominant and rare microbial organisms suggested by kill-the-winner occur, this cycling occurs at the strain level, rather than at the level of species as originally envisioned in the model (Rodriguez-Brito et al. 2010). Thus there can be constant diversity in an environment at the species level while the strain composition of those species cycles as evolving phage predation tactics are countered by ongoing innovations in host resistance mechanisms. The kill-the-winner hypothesis ultimately helps clarify the key tradeoff of microbial life: how to maximize reproduction by balancing competitive acquisition of resources and defense against infection (discussed further in [Thingstad et al. 2014]). Being a successful bacterium is not a secure occupation.

Phage lysogeny and transduction

When the massive magnitude and impact of phage predation were established at the birth of phage ecology in the early 1990s, predation became fixed—probably erroneously—in our minds as the dominant role of phage. Phage became the killers of the sea akin to their pathogenic viral cousins in the hospitals. However, research on the less overt effects of marine phage lysogeny, almost solely by John Paul and colleagues, formed a counter plot

to that lytic-centric commentary. In a series of papers throughout the early and mid-1990s, they reported the results of adding the induction agent mitomycin C to seawater sampled from a variety of regions ranging from coral reefs to the open ocean (reviewed in [Paul et al. 2002]). They found that ~30% to 50% of the microbes in the ocean are lysogens (as defined by carrying a prophage that responds to mitomycin C induction). In some samples, it looked like all the microbes harbored at least one inducible prophage. Lysogeny was particularly high in stable and oligotrophic areas such as the open ocean and lower in disturbed, seasonal, and eutrophic areas. This has been borne out by studies in temperate (Maurice et al. 2010), but not Antarctic, lakes (Laybourn-Parry, Marshall, Madan 2007). However, high rates of lysogeny observed in soil show that eutrophic conditions and lysogeny are not mutually exclusive (Ghosh et al. 2008), and in fact lysogeny may be favored in soil compared to aquatic environments (Marsh, Wellington 1994). What determines the preponderance of lytic versus lysogenic activity in an environmental community? What drives the relative abundances of lytic versus temperate phages and what determines the lysogenic/lytic decisions made by the latter are still very controversial, with varied results from studies in different systems, possibly due to the widespread reliance on artifact-prone DNA-damaging induction. Also of import, when Paul and colleagues investigated rates of transduction (virus-mediated gene exchange) in marine environments, they estimated that 10^{14} transduction events occur every year in Tampa Bay alone (Jiang, Paul 1998a). That extrapolates to 10^{28} base pairs of DNA being transferred every year in the global ocean (Paul et al. 2002).

While the prevalence of lysogeny and transduction are both high in the marine environment, demonstrating their effect is very challenging and lends itself to underestimation. For example, when researchers study lytic dynamics, there are abundant phage to count as an outcome to successful infection. Lysogenic infection may have no similarly overt signature, appearing at one level of observation as if there was no infection at all. This

Life as a Phage

Life in the cellular landscape: Some phages are virulent (lytic), some temperate. Although virulent phage may reside in the host cell for a time, manipulating host metabolism, their infection is not sustained. They invade, replicate, and lyse the host, spewing progeny into the extracellular milieu in a spray of host metabolites and cellular debris. Temperate phages are more nuanced. Upon infection, they may either act lytically or postpone lysis. In the latter case, they maintain as a prophage, usually integrated in the host chromosome, thus forming a union with the host—a lysogen. This process of lysogenic conversion can lead to a gain of function in the host cell such is observed in the increase in virulence of *Vibrio cholerae* when infected by CTXφ phage (Waldor, Mekalanos 1996). Ultimately, all temperate infections will end either in the prophage deciding (canonically due to DNA damage to the host) to excise from the host genome, a process called induction, or in inactivation of the prophage by mutation such that it cannot excise or replicate. If the host is stressed, when should a prophage opt to abandon ship and go lytic (canonical induction)? Will there be time to replicate before the ship goes down? And if so, will the progeny broadcasted to the extracellular milieu find a fit host? The other option for the prophage is to cling to the 'wreckage' of the stressed host and weather the storm because conditions are not favorable for phage reproduction or finding new hosts. Either option seems intuitively sound as a strategy, but to date our methods have relied on stress-induced induction to determine how many Bacteria in an environment are lysogens, despite the prevalence of the latter being unknown.

Both lytic and lysogenic phage have the capacity to transfer genes horizontally between microbial cells. When a lytic phage genome enters a cell but is, for example, cleaved by the cell's patrolling restriction endonucleases (*see page 4-10*) its genome fragments can recombine with the host genome, thus passing genetic material to the host by generalized transduction. For temperate phages, imprecise excision can lead to the prophage genome picking up some extra DNA from the adjacent region of the host chromosome and subsequently delivering that along with the phage DNA into the next host cell where those genes can possibly be incorporated into the host genome. This form of phage-mediated horizontal gene transfer is called specialized transduction.

While prophage replication is limited to the reproductive rate of their host, i.e., merely doubling with each host generation, lytic infection by a single phage yields ~20-400 progeny per infectious cycle (Wommack, Colwell 2000). However, lytic phage must then brave the extracellular milieu while prophage may retain more secure lodgings within their hosts. This suggests that phage may simplistically be placed within the ecological paradigm of r- and K-selected lineages, with lytic phage demonstrating more r-selected tradeoffs and temperate phage being more K-selected (Suttle 2007; Zhang et al. 2011; Keen 2014).

Life in the extracellular milieu: Phage usually do not remain viable for long outside the cell, with infectivity decreasing ~1% to 80% per hour in various environments due to factors such as the presence of particulate organic material, temperature, and most significantly, sunlight (Wommack, Colwell 2000). Interestingly, phages retain infectivity longer in their native environments compared to exotic ones (Wommack, Colwell 2000). The ultimate challenge of life in the milieu is finding a host. Phage are thought to encounter hosts by chance as a result of random Brownian motion, although other factors may be involved in environments such as the mucus layer protecting metazoan epithelial surfaces (Barr et al. 2013). Thus, phage predation is sensitive to the concentration of suitable hosts present. Robin Wilcox and Jed Fuhrman found that when seawater was diluted by a factor of approximately twenty, lytic activity was halted (Wilcox, Fuhrman 1994), presumably because phage degraded before encountering a host. Residence in the milieu is a race to find a host before exposure sets in. It is not easy being a phage.

makes phage success hard to assess and define. This decoupling between infection and observable effect (i.e., host lysis) is apparent in that although lysogeny is rampant in the ocean, spontaneous lysogen induction accounts for only 0.02% of detectible free phage (Jiang, Paul 1998b). Although the work of Paul and colleagues allowed us a glimpse into the world of phage mediated gene flow and lysogeny, it would take the advent of metagenomic sequencing a decade later to really see these processes directly.

Enter the sequencer

In 1999, the Pittsburgh Phage Group led by Roger Hendrix and Graham Hatfull published a paper with the oft-quoted subtitle "all the world's a phage" (Hendrix et al. 1999). Sequence homologies observed across a large number of phage genomes, both lysogenic and lytic, indicated a high degree of relatedness between disparate phage groups. Further, they suggested not only that the tailed phages comprise one lineage, but also that phage genomes may be mosaic constructs that over evolutionary time recruit elements from shared gene pools resulting from host overlap (*see page 5-55*). Genes were on the move between phages that shared hosts. Phage genomes, transmutable through mosaicism and recombination, reflected a more complex shared evolutionary past. The power of comparative genomics has made the field of phage biology more exciting and, from the perspective of a Darwinian vertical heritability, stranger than ever.

Inspired to extend the work of the Pittsburgh Phage Group to environmental phages, the San Diego Phage Group proudly announced the sequencing of the 'first' marine phage genome in 2000, that of Roseophage SIO1, a predator of the *Roseobacter* genus (Rohwer et al. 2000). In fact, this was not such a major leap forward for the field, as the *actual* first marine phage genome and proteome had been published the year before by Dennis Bamford and colleagues, that of *Pseudoalteromonas* phage PM2 (Kivelä et al. 1999; Mannisto et al. 1999). Once the news reached San Diego, Roseophage reluctantly exchanged its gold medal for bronze. Analysis of

its genome had revealed evolutionary linkages between seemingly unrelated phages from different environments. Clearly environmental boundaries evident to us do not constrain phages or their genes. Now that the ability to sequence environmental phages was established, the push was on to expand sequencing approaches from cultured strains like SIO1 and PM2 to environmental communities. This stimulated the development of one of the most important tools in phage ecology today: shotgun metagenomics.

Sequencers with shotguns

Research done in the 1990s that amplified and sequenced conserved phage genes demonstrated that environmental phage were diverse, but such studies were limited to groups of closely related phages (reviewed in [Short, Suttle 1999]). Intriguing as these findings were, this approach could not be extended to include less related phages as phage are of radically disparate lineages. They have no universal gene in common, nothing comparable to the handy ribosomal DNA with which the three domains of cellular life were resolved (Woese, Kandler, Wheelis 1990). The very first shotgun metagenome ever, a virome published in 2002 (years ahead of the first microbial metagenome) by the San Diego Phage Group, provided our first direct look at the diversity of viruses in any environment (Breitbart et al. 2002). The diversity observed was incredible: up to 7,000 phage types in 200 L of surface seawater. For comparison, there are only 5,000 to 6,000 species of mammals on the entire planet. Phage diversity was also radically patchy, varying from ~300 to 7,000 viral types species per 200 L in different samples from nearby locations.

While the phages were locally diverse, some—or at least some phage genes—are widely dispersed (Angly et al. 2006), found all over the world in almost every biome imaginable, as demonstrated for two highly conserved DNA polymerases named Hector and Paris (Breitbart, Miyake, Rohwer 2004). This echoed earlier reports of related algal virus polymerases in antipodal samples from Antarctica and British Columbia (Short,

Suttle 2002). Some phage are so cosmopolitan that members of the phage communities in lakes, sediment, and soil can also infect marine microbes (Sano et al. 2004). However, there is also evidence of localized adaptation that precludes phage from infecting potential hosts from neighboring trees while allowing infection of hosts from within the same tree (Koskella et al. 2011). Tree to tree infective exclusion in a world that allows marine-soil permissiveness of phage infection boundaries to phage infection is not intuitive. This patchiness makes assessing the size of the global virome non-trivial and suggests that we cannot estimate the total phage diversity on Earth by assessing the diversity in various biomes and then calculating their sum. Ultimately, it appears that global phage diversity is not as large as the findings of Breitbart and colleagues (Breitbart et al. 2002) would suggest (*see page 2-60* and [Ignacio-Espinoza, Solonenko, Sullivan 2013]). Armed with shotgun metagenomics as we are now, we can look back on sixty years of being shackled to culture-based observations and appreciate the formidable challenges faced by culture-bound researchers like Moebus. Working within those constraints, he found that phage could not infect microbes isolated from even less than 200 miles away, whereas now, examining viral communities directly, we know that phage are infective across biomes and across the globe.

Archaeal viruses in the shadows

Science has acknowledged the high abundance of bacteriophage hosts in the environment since the 1970s, but archaeal abundances remain underestimated. Even though we now realize that Archaea are not constrained to extreme environments, such as acidic hot springs, solar salterns, and hydrothermal vents in the perennially dark marine depths, their roles in environmental processes (Chaban, Ng, Jarrell 2006; Lipp et al. 2008), including the human environment (Probst, Auerbach, Moissl-Eichinger 2013) are generally disregarded. As a result, while our appreciation for phage and their ecological significance has bloomed, archaeal viruses remain a mostly unexplored realm promising surprising rewards for those who dare to enter. For example, the most

---

Culturing at Sea with Karl-Heinz Moebus

When limited to culture-based techniques, researchers are inclined to investigate a limited number of pet strains of microbes and phage in the lab. However, despite the constraints of culture-based techniques throughout the 1980s, Karl-Heinz Moebus was asking experimental questions using isolates from mixed communities. Further, while most of his peers were working on domesticated strains in the lab, he was doing these very modern experiments at sea. For example, in the early 1980s, working with Nattkemper he conducted experiments wherein they combined phage and microbes from different sides of the Atlantic (Moebus, Nattkemper 1981). They found that bacteria were more prone to infection from autochthonous than exotic phage. This suggested that in terms of phage-host infection networks, there were functionally different populations in regions separated by as little as 200 miles. The Atlantic was apparently a complex patchwork of distinct communities. If, as envisioned by Baas Becking, "everything is everywhere," the environment was strongly selective on an unexpected scale. It is a reflection of how phage ecology was viewed in the 1980s that this fascinating and crucial insight, extending biogeographic questions to phage communities at a cross-ocean scale, was not published in mainstream journals. Although their findings suffered from culture bias, their work had profound potential to spark interest in phage ecology a decade before Bergh's 1989 paper. Moebus, who lamented the lack of interest in marine phage and championed their importance (Moebus 1980), was ahead of his time.

extreme instance of genomic mosaicism known to date is from an archaeal virus found almost by chance in a hot, acidic lake by a research team led by Ken Stedman (Diemer, Stedman 2012). This virus, a chimera of sequences from DNA and RNA viruses, has radically broadened the known scope of mosaicism in viral genomes. Further analysis suggested that this phenomenon may be widespread, but overlooked, in other environments.

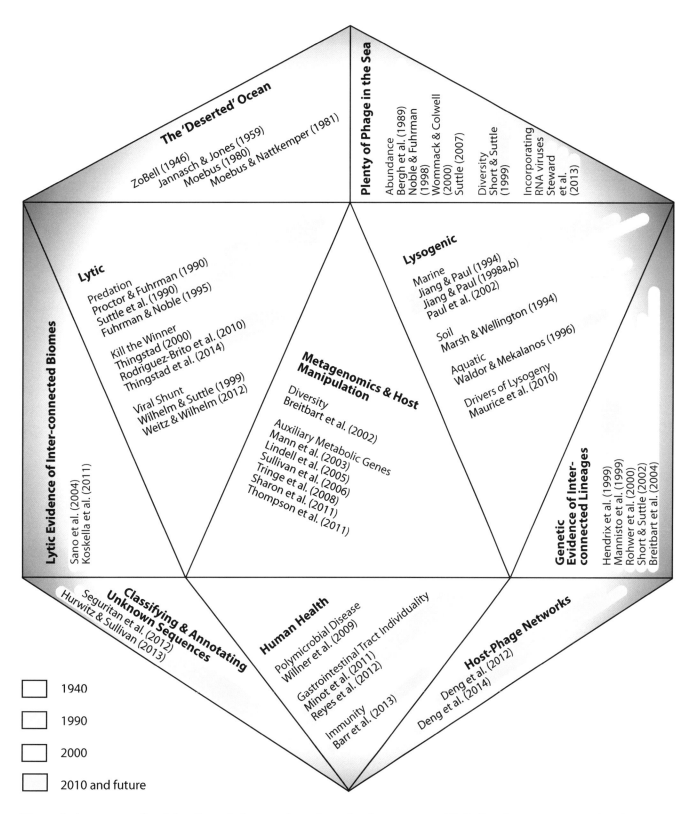

Figure 2. A conceptualized timeline of phage ecology from the desert ocean of ZoBell to the environmental and metagenomic analyses of today. Adjacent triangles are conceptually related.

The Dark Matter

Phage metagenomics also disclosed the genetic composition of phage communities, and things got even more interesting. The vast majority of the sequences in the first viral metagenome were unknown, i.e., were reads that do not have recognizable similarity to any genes in the predominantly microbial genome databases microbiologists had laboriously created (Breitbart et al. 2002). Despite the dramatic increase in the number of sequenced genomes now included in the databases, the percentage of phage unknowns has not decreased significantly. Phage genomes definitely remain the deepest reservoir of unknown functional potential on earth—the beckoning genetic Dark Matter.

Phage metabolism

The first viral metagenome also showed that phage genomes do not encode only 'phage' genes required for genome replication and virion assembly. Phages also carry some unexpected, but clearly recognizable, genes of known metabolic function. First discovered were the photosynthesis genes (*psbA* and *psbD*) found in a cyanophage

that encode the key photosystem II reaction center proteins D1 and D2, respectively (Mann et al. 2003). Soon thereafter, others confirmed that not only were these genes present in many lytic cyanophages (Sullivan et al. 2006), in at least some cases their expression during infection helped to maintain photosynthesis for the duration and thus enhanced phage fecundity (Lindell et al. 2005; Clokie et al. 2006). Similarly, phage genomes also contain photosystem I genes functionally organized in cassettes that, when combined, probably encode a whole photosystem I reaction complex (Sharon et al. 2009). These complexes may be mosaic constructs from different hosts (Mazor et al. 2012). Not only are these phage genes encoding metabolic proteins, but some at least travel back-and-forth between phage and host over evolutionary time. While sojourning within a phage, they evolve more rapidly and under different selection pressures, sometimes yielding improved functionality (Lindell et al. 2004; Frank et al. 2013). Subsequent metagenomic work led by Sullivan found an almost ubiquitous prevalence of *psbA* in cyanophage isolates, with half of the strains inves-

Table 1. Environments probed using metagenomics by members of the San Diego Phage Group throughout the 2000s, ranked by publication date of the first paper for that environment.

Environment Sampled	Biome	Reference
Coastal & open ocean	Marine	Breitbart et al. 2002
		Angly et al. 2006
Feces	Human	Breitbart et al. 2003
		Breitbart et al. 2008
		Reyes et al. 2010
Sediment	Marine	Breitbart et al. 2004
Blood	Human	Breitbart, Rohwer 2005
Soil	Terrestrial	Fierer et al. 2007
Stromatolite & thrombolite	Marine	Desnues et al. 2008
Coral reef	Marine	Dinsdale et al. 2008b
Coral-associated	Marine	Marhaver, Edwards, Rohwer 2008
		Thurber et al. 2008
Lung	Human	Willner et al. 2009
Mosquito	Human-Associated	Willner, Thurber, Rohwer 2009
Fish farm & solar saltern	Aquatic	Rodriguez-Brito et al. 2010

tigated having both *psbA* and *psbD* (Sullivan et al. 2006). Moreover, their data provided evidence of repeated host-phage and phage-phage exchange of these genes, further demonstrating the genetic connections between phages with overlapping host ranges.

At the time of their publication, many metagenomes are made publically available and archived in standard formats. As a result, there now exists an abundance of metagenomic data accessible to any ecologist, data that readily lends itself to meta-analysis and synthesis (*see page 2-61*). Although still hampered by the large proportion of phage genes without functional annotations in reference databases, these datasets nevertheless allow us to probe molecular facets of phage ecology and behavior in previously unimaginable depth. Since the mid-2000s, for example, there has been a string of papers characterizing the diversity and function of some of the metabolic genes that phage carry around in their genomes, largely based on publically available sequences (reviewed in [Breitbart et al. 2007; Breitbart 2012]). Phage manipulate not only photosynthesis but numerous other aspects of host metabolism to their own benefit by encoding functional auxiliary metabolic genes (Breitbart 2012), e.g., genes involved in energetic and nucleotide metabolism through the pentose phosphate pathway (Thompson et al. 2011), post-translational protein modification by peptide deformylase (Sharon et al. 2011; Frank et al. 2013), and in the deep sea, where sulfur metabolism is of crucial importance, elemental sulfur oxidation (Ananth-araman et al. 2014). The conclusion is clear: phage play a significant and varied role (Dinsdale et al. 2008a) in microbial metabolic processes, likely complimenting and extending microbial function in every environment examined, and we've only just begun digging.

Connecting the ecological dots

While environmental phage ecology has made great strides since its nascence in the late 1980s, exciting challenges remain. As a science of connections between unseen organisms, progress in this field is clearly tied closely to methodological inno-vation. There are currently a slew of new methods available that will enable us to address interesting but previously unanswerable questions in the near future.

Metagenomic characterization of environmental phage communities is still hampered by the high percentage of unknown reads. This must be remedied in some manner before we can see the full scope of phage genetic potential. Given the inherent challenges here, there has been a strong incentive to develop alternative approaches to assign putative functions to some of these unknowns. Methods used so far include protein clustering (Hurwitz, Sullivan 2013) and artificial neural networks trained on known datasets (Seguritan et al. 2012). Researchers have also refined high throughput means such as viral tagging (*see page 2-66*) to identify host-phage pairs (Deng et al. 2012), helping us understand host specificity and overlap, and thus map the common genetic pools connecting phages. Determination of phage host ranges is essential for charting routes for gene flow via host overlap in the environment as well as determining the degree of their host specificity in natural environments. Much of this work uses innovative tagging techniques coupled to flow cytometry to sort infected cell-phage pairs that can then be sequenced using new DNA amplification techniques (Duhaime et al. 2012).

Metagenomics goes rogue

After publishing the first virome using marine samples, the San Diego Phage Group went on an interdisciplinary, cross-country environmental phage metagenome excursion. They sequenced viromes from a wild diversity of environments (Table 1). By the conclusion of this scientific March from the Sea, the value of culture-independent metagenomics for phage exploration in all conceivable environments was established.

The human body is also an environment, or rather multiple environments, all of which are amenable to investigation by ecological metagenomic methods. For example, Yijun Ruan, one of the primary biomedical researchers who identified and characterized SARS in Singapore in 2003, has used metagenomics techniques to explore the composition, ecology, and clinical impacts of microbes and viruses in air and wastewater (Tringe et al. 2008; Rosario et al. 2009). Metagenomics characterized the time course for acquisition and development of our gut virome, with the diversity of the phage community being low in newborns and increasing with time (Breitbart et al. 2008). Further, it has been shown that while our gastrointestinal microbiota are remarkably functionally conserved between individuals, the resident phage communities we each house are functionally far more unique (Reyes et al. 2010), and appear to respond to diet (Minot et al. 2011).

Metagenomics also revealed that phages exist in human environments that researchers and clinicians had thought to be sterile, environments such as the healthy human lung. The lung seemed to be sterile because no bacterial isolates from lungs were culturable. Culture-independent methods such as metagenomics have overturned this view and demonstrated that lungs, too, are ecological landscapes with niches, succession, and other ecological properties (Willner et al. 2009). The parallels between the 'sterile' lung and the 'desert' ocean of Claude ZoBell are striking and, to a potential patient, alarming. The appreciation of the human body as an ecosystem housing microbial and viral communities creates a new paradigm for understanding disease and also new possibilities for disease treatment. Viewing polymicrobial diseases such as cystic fibrosis in an ecological context leads directly to innovative clinical protocols (Conrad et al. 2013) with exciting potential to improve human health.

In our quest for health, phage may already be our partners in diverse ways. For example, some phages use Ig-like domains of capsid decoration proteins to adhere to the mucus membranes that protect our tissues from the environment (Barr et al. 2013). There they take advantage of the greater abundance of Bacteria that makes for higher rates of host encounter and lysis—thus reproductive

success for the phage and potentially reduced pathogen invasion for us (*see page 2-35*). Even though there is no more 'self' a system than the immune system, part of our immune system that guards mucosal surfaces such as those found in the gut and lungs may be harnessing a non-human-derived immunity through phage. More overtly, in this era of antibiotic-resistant bacterial infections, we will almost certainly witness a renewed commitment to the development of phage therapies (*see page 4-46*), now guided by the knowledge and methodologies resulting from decades of marine phage research. From the ocean of marine phage dynamics to phage-host interactions within the human ecosystem, the march goes on.

The future

Environmental phage ecology has made enormous strides in the past 25 years and is positioned now to tackle many remaining exciting challenges. While earlier observational studies brought awareness of the vast scope of phage activity in nature, we currently lack a mechanistic understanding of many important processes. For example, while observations suggest that factors such as microbial abundance, perturbation, season, and biogeography are determinants of phage abundance, rigorous experimental evidence detailing their interacting roles is scarce. While marine virus:microbe ratios appear to follow depth and latitudinal trends, the ultimate drivers of these, too, remain unknown. Similarly, despite research confirming abundant lysogeny in the marine environment, the ecological factors that select for temperate phage, that determine their choice of lysis or lysogeny during each infection, and that cause their ultimate induction are all unknown. Experimentally sorting out this, and other aspects of phage ecology, is especially challenging as phage affects host that then affects phage in an endless rotation of complex circular interactions.

Phage ecology has had a strong marine bias with less frequent excursions into environments such as sediments, soils, lakes, hot springs, animals, and plants. Facilitated by the rapid decline of sequencing costs, deeper probing of more biomes lies ahead. The results obtained will complement, expand, and rewrite much of what we have learned from the marine environment. Shifting the focus from individual biomes to the global scale we see yet more unexplored territory. That some phages (or phage genes) are already known to be present in what we perceive as different environments raises numerous questions about phage mobility both globally and between biomes, as well as about phage gene mobility between different hosts. These investigations have the power to reconfigure our understanding of the ecological divisions and connections in nature. Boundaries we perceive or intellectually construct are not always real.

Even within the marine environment there is a major pro-DNA bias to our understanding of phage. Double-stranded DNA phages, especially the tailed phages (order *Caudovirales*) dominate the sequence space we have investigated so far. The RNA viruses that comprise a large proportion—perhaps half, perhaps even more—of marine viruses remain often overlooked (Steward et al. 2013). We cannot say our vision of marine phage ecology is complete until all phage genotypes are incorporated into our science. This raises methodological challenges as RNA viruses are much more difficult to study and both RNA and single-stranded DNA genomes are not compatible or optimal with many of our most common methods.

Although they are arguably outside the scope of bacteriophage ecology, archaeal viruses currently represent a large gap in our understanding of en-

vironmental viral dynamics. Archaeal viruses are some of the most genomically diverse viruses on the planet and their virion architectures are unimaginably creative, but their ecological impacts remain almost completely unknown. Deeper probing of archaeal viruses could eliminate a large hole in our knowledge of viral ecology and increase our appreciation of the impact and diverse forms of viral genetic exchange in the environment.

Finally, research on the human holobiont as an ecosystem that has only just begun has exciting potential. Work to date has altered our understanding of human immune function and suggested other potential roles of phage in human development and health. The abundance and diversity of human-associated phages is radically altering our perception of humanness and individuality. This is sure to be a controversial and insightful field going forward, yet one with promise to improve human life by harnessing ecological knowledge gained in environmental systems (Reyes et al. 2012).

Acknowledgements

This manuscript benefitted greatly from readthroughs and feedback from Linda Kelly, Cynthia Silveira, Kevin Green, Yan Wei Lim, and Emma George. It would not have been possible (or readable) at all without profound input from Aaron Hartmann and Savannah Sanchez. Most of all, I am deeply grateful to Forest Rohwer for his mentorship and the opportunity to write this chapter, and to Merry Youle for transforming my somewhat swampy thoughts into sleek, shiny prose.

References

Anantharaman, K, MB Duhaime, JA Breier, KA Wendt, BM Toner, GJ Dick. 2014. Sulfur oxidation genes in diverse deep-sea viruses. Science 344:757-760.

Anderson, NG, GB Cline, WW Harris, JG Green. 1967. Isolation of viral particles from large fluid volumes. In: G Berg, editor. Transmission of Viruses by the Water Route: Interscience Publishers. p. 75-88.

Angly, FE, B Felts, M Breitbart, et al. 2006. The marine viromes of four oceanic regions. PLoS Biol 4:e368.

Azam, F, T Fenchel, JG Field, JS Gray, LA Meyerreil, F Thingstad. 1983. The ecological role of water-column microbes in the sea. Mar Ecol Prog Ser 10:257-263.

Barksdale, WL, AM Pappenheimer. 1954. Phage host relationships in nontoxigenic and toxigenic diphtheria-bacilli. J Bacteriol 67:220-232.

Barr, JJ, R Auro, M Furlan, et al. 2013. Bacteriophage adhering to mucus provide a non-host-derived immunity. Proc Natl Acad Sci USA 110:10771-10776.

Bergh, O, KY Borsheim, G Bratbak, M Heldal. 1989. High abundance of viruses found in aquatic environments. Nature 340:467-468.

Breitbart, M. 2012. Marine viruses: Truth or dare. Ann Rev Mar Sci 4:425-448.

Breitbart, M, B Felts, S Kelley, JM Mahaffy, J Nulton, P Salamon, F Rohwer. 2004. Diversity and population structure of a near-shore marine-sediment viral community. Proc Biol Sci 271:565-574.

Breitbart, M, M Haynes, S Kelley, et al. 2008. Viral diversity and dynamics in an infant gut. Res Microbiol 159:367-373.

Breitbart, M, JH Miyake, F Rohwer. 2004. Global distribution of nearly identical phage-encoded DNA sequences. FEMS Microbiol Lett 236:249-256.

Breitbart, M, F Rohwer. 2005. Method for discovering novel DNA viruses in blood using viral particle selection and shotgun sequencing. Biotechniques 39:729-736.

Breitbart, M, P Salamon, B Andresen, JM Mahaffy, AM Segall, D Mead, F Azam, F Rohwer. 2002. Genomic analysis of uncultured marine viral communities. Proc Natl Acad Sci USA 99:14250-14255.

Breitbart, M, LR Thompson, CA Suttle, MB Sullivan. 2007. Exploring the vast diversity of marine viruses. Oceanography (Wash D C) 20:135-139.

Chaban, B, SY Ng, KF Jarrell. 2006. Archaeal habitats--from the extreme to the ordinary. Can J Microbiol 52:73-116.

Clokie, MR, J Shan, S Bailey, Y Jia, HM Krisch, S West, NH Mann. 2006. Transcription of a 'photosynthetic' T4-type phage during infection of a marine cyanobacterium. Environ Microbiol 8:827-835.

Conrad, D, M Haynes, P Salamon, PB Rainey, M Youle, F Rohwer. 2013. Cystic fibrosis therapy: a community ecology perspective. Am J Respir Cell Mol Biol 48:150-156.

Crosse, JE, MK Hingorani. 1958. Method for isolating *Pseudomonas morsprunorun* phages from the soil. Nature 181:60-61.

Deng, L, A Gregory, S Yilmaz, BT Poulos, P Hugenholtz, MB Sullivan. 2012. Contrasting life strategies of viruses that infect photo- and heterotrophic Bacteria, as Revealed by viral tagging. Mbio 3:e00373-00312.

Deng, L, JC Ignacio-Espinoza, AC Gregory, BT Poulos, JS Weitz, P Hugenholtz, MB Sullivan. 2014. Viral tagging reveals discrete populations in *Synechococcus* viral genome sequence space. Nature advance online publication.

Desnues, C, B Rodriguez-Brito, S Rayhawk, et al. 2008. Biodiversity and biogeography of phages in modern stromatolites and thrombolites. Nature 452:340-343.

Diemer, GS, KM Stedman. 2012. A novel virus genome discovered in an extreme environment suggests recombination between unrelated groups of RNA and DNA viruses. Biol Direct 7:13.

Dinsdale, EA, RA Edwards, D Hall, F Angly, M Breitbart, JM Brulc, M Furlan, C Desnues, M Haynes, L Li. 2008a. Functional metagenomic profiling of nine biomes. Nature 452:629-632.

Dinsdale, EA, O Pantos, S Smriga, et al. 2008b. Microbial ecology of four coral atolls in the Northern Line Islands. Plos One 3:1-17.

Duhaime, MB, L Deng, BT Poulos, MB Sullivan. 2012. Towards quantitative metagenomics of wild viruses and other ultra-low concentration DNA samples: A rigorous assessment and optimization of the linker amplification method. Environ Microbiol 14:2526-2537.

Frank, JA, D Lorimer, M Youle, P Witte, T Craig, J Abendroth, F Rohwer, RA Edwards, AM Segall, AB Burgin. 2013. Structure and function of a cyanophage-encoded peptide deformylase. Isme Journal 7:1150-1160.

Fierer, N, M Breitbart, J Nulton, et al. 2007. Metagenomic and small-subunit rRNA analyses reveal the genetic diversity of bacteria, archaea, fungi, and viruses in soil. Appl Environ Microbiol 73:7059-7066.

Fuhrman, JA, RT Noble. 1995. Viruses and protists cause similar bacterial mortality in coastal seawater. Limnol Oceanogr 40:1236-1242.

Ghosh, D, K Roy, KE Williamson, DC White, KE Wommack, KL Sublette, M Radosevich. 2008. Prevalence of lysogeny among soil bacteria and presence of 16S rRNA and trzN genes in viral-community DNA. Appl Environ Microbiol 74:495-502.

Hendrix, RW, MCM Smith, RN Burns, ME Ford, GF Hatfull. 1999. Evolutionary relationships among diverse bacteriophages and prophages: All the world's a phage. Proc Natl Acad Sci USA 96:2192-2197.

Hobbie, JE, RJ Daley, S Jasper. 1977. Use of nuclepore filters for counting bacteria by fluorescence microscopy. Appl Environ Microbiol 33:1225-1228.

Hurwitz, BL, MB Sullivan. 2013. The Pacific Ocean Virome (POV): A marine viral metagenomic dataset and associated protein clusters for quantitative viral ecology. Plos One 8:12.

Ignacio-Espinoza, J, SA Solonenko, MB Sullivan. 2013. The global virome: Not as big as we thought? Curr Opin Virol 3:566-571.

Jannasch, HW, GE Jones. 1959. Bacterial populations in sea water as determined by different methods of enumeration. Limnol Oceanogr 4:128-139.

Jiang, SC, JH Paul. 1994. Seasonal and diel abundance of viruses and occurrence of lysogeny/bacteriocinogeny in the marine environment. Mar Ecol Prog Ser 104:163-172.

Jiang, SC, JH Paul. 1998a. Gene transfer by transduction in the marine environment. Appl Environ Microbiol 64:2780-2787.

Jiang, SC, JH Paul. 1998b. Significance of lysogeny in the marine environment: Studies with isolates and a model of lysogenic phage production. Microb Ecol 35:235-243.

Keen, EC. 2014. Tradeoffs in bacteriophage life histories. Bacteriophage 4:e28365.

Kivela, HM, RH Mannisto, N Kalkkinen, DH Bamford. 1999. Purification and protein composition of PM2, the first lipid-containing bacterial virus to be isolated. Virology 262:364-374.

Koskella, B, JN Thompson, GM Preston, A Buckling. 2011. Local biotic environment shapes the spatial scale of bacteriophage adaptation to bacteria. Am Nat 177:440-451.

Kowalski, M, GE Ham, Frederic.Lr, IC Anderson. 1974. Relationship between strains of *Rhizobium japonicum* and their bacteriophages from soil and nodules of field-grown soybeans. Soil Sci 118:221-228.

Laybourn-Parry, J, WA Marshall, NJ Madan. 2007. Viral dynamics and patterns of lysogeny in saline Antarctic lakes. Polar Biol 30:351-358.

Lindell, D, JD Jaffe, ZI Johnson, GM Church, SW Chisholm. 2005. Photosynthesis genes in marine viruses yield proteins during host infection. Nature 438:86-89.

Lindell, D, MB Sullivan, ZI Johnson, AC Tolonen, F Rohwer, SW Chisholm. 2004. Transfer of photosynthesis genes to and from *Prochlorococcus* viruses. Proc Natl Acad Sci USA 101:11013-11018.

Lipp, JS, Y Morono, F Inagaki, K-U Hinrichs. 2008. Significant contribution of Archaea to extant biomass in marine subsurface sediments. Nature 454:991-994.

Mann, NH, A Cook, A Millard, S Bailey, M Clokie. 2003. Marine ecosystems: Bacterial photosynthesis genes in a virus. Nature 424:741-741.

Mannisto, RH, HM Kivela, L Paulin, DH Bamford, JKH Bamford. 1999. The complete genome sequence of PM2, the first lipid-containing bacterial virus to be isolated. Virology 262:355-363.

Marhaver, KL, RA Edwards, F Rohwer. 2008. Viral communities associated with healthy and bleaching corals. Environ Microbiol 10:2277-2286.

Markel, DE, MJ Fowler, C Eklund. 1975. Phage-host specificity texts using *Levinea* phages and isolates of *Levinea* spp and *Citrobacer Freundii*. Int J Syst Bacteriol 25:215-218.

Marsh, P, EMH Wellington. 1994. Phage-host interactions in soil. FEMS Microbiol Ecol 15:99-107.

Maurice, CF, T Bouvier, J Comte, F Guillemette, PA del Giorgio. 2010. Seasonal variations of phage life strategies and bacterial physiological states in three northern temperate lakes. Environ Microbiol 12:628-641.

Mazor, Y, I Greenberg, H Toporik, O Beja, N Nelson. 2012. The evolution of photosystem I in light of phage-encoded reaction centres. Philos Trans R Soc Lond B Biol Sci 367:3400-3405.

Minot, S, R Sinha, J Chen, H Li, SA Keilbaugh, GD Wu, JD Lewis, FD Bushman. 2011. The human gut virome: Inter-individual variation and dynamic response to diet. Genome Res 21:1616-1625.

Moebus, K. 1980. A method for the detection of bacteriophages from ocean water. Helgol Mar Res 34:1-14.

Moebus, K, H Nattkemper. 1981. Bacteriophage sensitivity patterns among bacteria isolated from marine waters. Helgol Mar Res 34:375-385.

Noble, RT, JA Fuhrman. 1998. Use of SYBR Green I for rapid epifluorescence counts of marine viruses and bacteria. Aquat Microb Ecol 14:113-118.

Paul, JH, MB Sullivan, AM Segall, F Rohwer. 2002. Marine phage genomics. Comp Biochem Physiol B Biochem Mol Biol 133:463-476.

Probst, AJ, AK Auerbach, C Moissl-Eichinger. 2013. Archaea on human skin. Plos One 8:e65388.

Proctor, LM, JA Fuhrman. 1990. Viral mortality of marine bacteria and cyanobacteria. Nature 343:60-62.

Reyes, A, M Haynes, N Hanson, FE Angly, AC Heath, F Rohwer, JI Gordon. 2010. Viruses in the faecal microbiota of monozygotic twins and their mothers. Nature 466:334-338.

Reyes, A, NP Semenkovich, K Whiteson, F Rohwer, JI Gordon. 2012. Going viral: Next-generation sequencing applied to phage populations in the human gut. Nat Rev Microbiol 10:607-617.

Ritchie, DF, EJ Klos. 1977. Isolation of *Erwinia amylovora* bacteriophage from aerial parts of apple trees. Phytopathology 67:101-104.

Rodriguez-Brito, B, LL Li, L Wegley, et al. 2010. Viral and microbial community dynamics in four aquatic environments. ISME J 4:739-751.

Rohwer, F, A Segall, G Steward, V Seguritan, M Breitbart, F Wolven, F Azam. 2000. The complete genomic sequence of the marine phage Roseophage SIO1 shares homology with nonmarine phages. Limnol Oceanogr 45:408-418.

Rosario, K, C Nilsson, YW Lim, YJ Ruan, M Breitbart. 2009. Metagenomic analysis of viruses in reclaimed water. Environ Microbiol 11:2806-2820.

Sano, E, S Carlson, L Wegley, F Rohwer. 2004. Movement of viruses between biomes. Appl Environ Microbiol 70:5842-5846.

Seeley, ND, SB Primrose. 1980. Effect of temperature on the ecology of aquatic bacteriophages. J Gen Virol 46:87-95.

Seguritan, V, N Alves, M Arnoult, A Raymond, D Lorimer, AB Burgin, P Salamon, AM Segall. 2012. Artificial neural networks trained to detect viral and phage structural proteins. Plos Comput Biol 8:22.

Sharon, I, A Alperovitch, F Rohwer, M Haynes, F Glaser, N Atamna-Ismaeel, RY Pinter, F Partensky, EV Koonin, YI Wolf. 2009. Photosystem I gene cassettes are present in marine virus genomes. Nature 461:258-262.

Sharon, I, N Battchikova, EM Aro, C Giglione, T Meinnel, F Glaser, RY Pinter, M Breitbart, F Rohwer, O Beja. 2011. Comparative metagenomics of microbial traits within oceanic viral communities. ISME J 5:1178-1190.

Short, SM, CA Suttle. 1999. Use of the polymerase chain reaction and denaturing gradient gel electrophoresis to study diversity in natural virus communities. Hydrobiologia 401:19-32.

Short, SM, CA Suttle. 2002. Sequence analysis of marine virus communities reveals that groups of related algal viruses are widely distributed in nature. Appl Environ Microbiol 68:1290-1296.

Sullivan, MB, D Lindell, JA Lee, LR Thompson, JP Bielawski, SW Chisholm. 2006. Prevalence and evolution of core photosystem II genes in marine cyanobacterial viruses and their hosts. Plos Biology 4:1344-1357.

Suttle, CA. 2007. Marine viruses - major players in the global ecosystem. Nat Rev Microbiol 5:801-812.

Suttle, CA, AM Chan, MT Cottrell. 1990. Infection of phytoplankton by viruses and reduction of primary productivity. Nature 347:467-469.

Tartera, C, J Jofre. 1987. Bacteriophages active against *Bacteroides fragilis* in sewage polluted waters. Appl Environ Microbiol 53:1632-1637.

Thingstad, TF. 2000. Elements of a theory for the mechanisms controlling abundance, diversity, and biogeochemical role of lytic bacterial viruses in aquatic systems. Limnol Oceanogr 45:1320-1328.

Thingstad, TF, S Vage, JE Storesund, RA Sandaa, J Giske. 2014. A theoretical analysis of how strain-specific viruses can control microbial species diversity. Proc Natl Acad Sci USA 111:7813-7818.

Thompson, LR, Q Zeng, L Kelly, KH Huang, AU Singer, J Stubbe, SW Chisholm. 2011. Phage auxiliary metabolic genes and the redirection of cyanobacterial host carbon metabolism. Proc Natl Acad Sci USA 108:E757-E764.

Thurber, RLV, KL Barott, D Hall, et al. 2008. Metagenomic analysis indicates that stressors induce production of herpes-like viruses in the coral *Porites compressa*. Proc Natl Acad Sci USA 105:18413-18418.

Torrella, F, RY Morita. 1979. Evidence by electron-micrographs for a high incidence of bacteriophage particles in the waters of Yaquina Bay, Oregon - Ecological and taxonomical implications. Appl Environ Microbiol 37:774-778.

Tringe, SG, T Zhang, XG Liu, et al. 2008. The airborne metagenome in an indoor urban environment. Plos One 3:10.

Waldor, MK, JJ Mekalanos. 1996. Lysogenic conversion by a filamentous phage encoding cholera toxin. Science 272:1910-1914.

Weitz, JS, SW Wilhelm. 2012. Ocean viruses and their dynamical effects on microbial communities and biogeochemical cycles. F1000 Biol Rep 4:17.

Wilcox, RM, JA Fuhrman. 1994. Bacterial viruses in coastal seawater: lytic rather than lysogenic production. Mar Ecol Prog Ser 114:35-35.

Wilhelm, SW, CA Suttle. 1999. Viruses and Nutrient Cycles in the Sea - Viruses play critical roles in the structure and function of aquatic food webs. Bioscience 49:781-788.

Willner, D, M Furlan, M Haynes, R Schmieder, FE Angly, J Silva, S Tammadoni, B Nosrat, D Conrad, F Rohwer. 2009. Metagenomic analysis of respiratory tract DNA viral communities in cystic fibrosis and non-cystic fibrosis individuals. Plos One 4:e7370.

Willner, D, RV Thurber, F Rohwer. 2009. Metagenomic signatures of 86 microbial and viral metagenomes. Environ Microbiol 11:1752-1766.

Woese, CR, O Kandler, ML Wheelis. 1990. Towards a natural system of organisms - Proposal for the domains Archaea, Bacteria, and Eucarya. Proc Natl Acad Sci USA 87:4576-4579.

Wommack, KE, RR Colwell. 2000. Virioplankton: Viruses in aquatic ecosystems. Microbiol Mol Biol Rev 64:69-114.

Zhang, Y, C Huang, J Yang, N Jiao. 2011. Interactions between marine microorganisms and their phages. Chin Sci Bull 56:1770-1777.

ZoBell, CE. 1946. Marine microbiology, a monograph on hydrobacteriology. Waltham, Mass.: Chronica Botanica Company.

The Phage Metagenomic Revolution

Matthew B. Sullivan[†]

Abstract: *Phages in nature are abundant and important as modulators of microbial population structure and metabolic outputs, yet quantifying their impacts in complex and interacting communities remains a major challenge. Fortunately, phage ecological methods have now advanced from counting 'dots' to, for instance, linking phages to their hosts in a population- and genome-based framework. Metagenome-enabled methodologies have led to the realization that phages directly manipulate microbial metabolisms through encoding their own 'auxiliary metabolic genes.' Other applications have organized the vast unknown phage sequence space into countable protein clusters and demonstrated that cyanophage genome sequence space is sufficiently structured to allow populations to be counted. These latter advances in particular allow the field to leverage and test decades of ecological and evolutionary theory to accelerate progress not only for phage research but for the fields of ecology and evolution, as well. It is time for studies of Earth's micro- and nanoscale ecosystem inhabitants to begin leading the life sciences!*

[†] Department of Ecology and Evolutionary Biology, The University of Arizona, Tucson AZ
Email: mbsulli@email.arizona.edu
Website: http://eebweb.arizona.edu/Faculty/mbsulli/

"This changes everything!" I remember saying this to phylogeneticist Ken Halanych while sitting in a Woods Hole classroom back in 2000. Most people were still elated from having survived the big 'Y2K' scare—the fear pandemic that computers everywhere might implode and all human existence would melt down to chaos! But I was talking about a groundbreaking paper that had just come out in *Limnology and Oceanography* (Rohwer et al. 2000). It was by a relatively little-known postdoc, Forest Rohwer, in which he presented the first marine phage genome and its ecological context. They reported that marine phages share features (genes) with non-marine phages, that phosphate scavenging genes appear critical for their survival in P-limited marine waters, and that virion structural proteins might be unrecognizable in the genomes of environmental phages. Looking back, their findings foretold much that we would slowly tease apart in the decade following.

Genome envy

To me, this paper meant so much because I was in my second year of PhD training at MIT and the Woods Hole Oceanographic Institution, and I had just isolated my first marine cyanobacterial phages (cyanophages) of marine *Prochlorococcus*. While I was productively chugging through basic phage characterization of a select few isolates, I yearned for more. I realized, having just read Rohwer's paper, that what I wanted was a genomic map for some of my cyanophages like the one he had for Roseophage SIO1. The power of genomics was alluring; it would be so informative, particularly for my environmental phages that lacked the foundation of decades of knowledge accumulation and genetic tool development. With a genome sequence in hand, you immediately could start thinking about what that phage was doing and what it might look like (getting electron micrographs of environmental phages can be challenging). You could even develop hypotheses about why that particular phage was successful in the environment. Not all the answers are there in the genome alone, of course, as most phage genes are 'unknown,' but still pieces and parts of the story are typically apparent in the average genome. Moreover, the genomic novelty added by each new environmental phage isolate that was sequenced allowed the first predictions of the size of the global pool of *phage* genomes (the global virome). In 2003 a bold estimate—this also from Rohwer—reckoned that it might comprise two billion proteins (Rohwer 2003). This number was an extrapolation from a scant 14 mycobacteriophage genomes, but the possibility of phage sequence space being that large was intriguing. It would make phages the largest source of genetic diversity on Earth.

But I was serious. I did want genomes for my four new *Prochlorococcus* phages. I approached my PhD adviser, Penny Chisholm, about this and she immediately recognized the challenges. At that stage, I was hardly capable of amassing the required quantities of these phages; non-optimal (in hindsight) culturing conditions meant that we were lucky if we could get 1 nanogram of DNA—a thousand-fold less than was needed for genome sequencing at that time! So Penny set up an opportunity for me to actually visit and collaborate with Forest—I was so nervous !!!—to learn how to sequence, annotate, and make sense of phage genome sequences. Even more daunting, for that mission I was allotted two weeks.

Upon arrival, Forest and I had coffee and chatted, and I was having a great time. He thought about all things phage, and in a totally offbeat way compared to the more microbe-centric world I normally lived in. It was great to finally get to put phage first. Anyone who knows Forest will not be surprised that, rather than work directly on my cyanophage genomes (he hates photosynthesis anyway), he convinced me that we should do two other things instead. First, we should make a web page that walked folks through the steps of getting a phage genome sequenced. (Now, over a decade later, the *Guide to Phage Genomics* is still a top Google hit (http://www.sci.sdsu.edu/PHAGE/guide.html). I loved that he would set me to doing this, as in the process I actually did really think through the isolate-to-genome process. Moreover, that we were providing a community resource exemplified a founding principle of the Luria and Delbruck school of phage biology: the creation of resources for the common scientific good. They had reasoned that rather than scientists competing with each other, the best way to study phage biology—since it is so hard—would be for the field to share anecdotal information and the subtleties of lab protocols so that we could all make more progress towards our particular research goals. I loved it. The second thing Forest suggested I do was help write a marine phage genomics review, my first paper as a PhD student. This was such an early stage in the field and there was so little to review that I knew every publication on the topic inside and out.

Eventually, Forest did send me down the path of learning how to do environmental phage genomics. The first step was on my way back East when I made a detour through Madison, Wisconsin, to spend a week learning how to make clone libraries from nothing—eventually branded as nanocloning. This training was with David Mead, president and founder of what was then a much smaller Lucigen Corporation, and I learned so much. David was incredibly patient with me, and I was soon on my way towards getting genomes of four marine cyanophages. Next step on the path was to figure out how to sequence, an operation that back then was neither easy or cheap. Fortunately, Penny smoothed the way for me as she had gotten a Community Sequencing Program grant funded. This meant that we sent our nanoclone libraries for each phage to the Department of Energy's Joint Genome Institute for Sanger sequencing and assembly. What came back was a finished genome of high quality. It was wonderful to step back in at that point as manually closing the genome could have been a *lot* of work, particularly with so little DNA. I was lucky.

Is this annotation for real?

So, by 2002 I had genomes for all four of my first cyanophages. Now it was time to visit Forest again. On this visit, though, I had my own genomes to look at, and one of Forest's grad students (Mya Breitbart) was there to hold my hand through the process. Back then annotating a phage genome was a manual, brute force process. Mya was amazing, helping me through all the tough spots. I then flew back to MIT with annotated genomes in hand and with new ideas swirling around in my head about who these phages are and what they might be doing. I was so struck by the fact that these cyanophages, isolated as they were from the middle of the low-nutrient, open ocean using marine cyanobacteria as a host, strongly resembled T4-like and T7-like phages of heterotrophic *E coli* isolated from sewage. This was cool—a universality, in a sense—from sewage to open ocean—and some commonalities among the phages infecting bacteria.

Ahh, but there was this problem, or at least I thought it was a problem. The issue was that these cyanophage genome assemblies contained a core photosynthesis gene (*psbA*), as well as other cyanobacterial genes. This was perplexing. I assumed these were 'assembly artifacts,' and I struggled with trying to correct them since I knew so little about the assembly process at that stage. In thinking this through, I shared these findings one day with Debbie Lindell, a post-doc with Penny at MIT at the time, and Debbie went through the roof. "Oh, this is cool," she said. "This might be real! What if the phages need these photosynthesis genes because their hosts are photosynthetic?" Wow, I thought, now that *would* be cool. And of course, a few years of work with Debbie and work by Nick Mann and Martha Clokie in marine *Synechococcus* cyanophages—the first to report photosynthesis genes in cyanophages (Mann et al. 2003)—showed that this was exactly what was going on. Thus was launched the 'photosynthetic phage' paradigm. Labs around the world found that cyanophages had indeed picked up these core photosynthesis genes and that evolution had kept them around since they offered the phages a fitness advantage by helping them to make more phages (Lindell et al. 2004; Lindell et al. 2005; Zeidner et al. 2005; Sullivan et al. 2006; Bragg, Chisholm 2008). The rationale is that, for optimal replication, cyanophages need to be able to repair or prevent degradation of the photosynthetic machinery so that photosynthesis keeps going throughout their infection. Debbie had been on top of this from the outset, but there I was, thinking these genes were assembly artifacts in the cyanophage genomes.

Of course, while we were delighting in building an understanding of photosynthetic phages one genome at a time, Mya and Forest had been busy with bigger things. They were using nanocloning to sequence genomic fragments from entire viral communities (Breitbart et al. 2002). This was the birth of viral metagenomics or viromics. Now things would really get interesting.

Phage metabolism is not an oxymoron

By this time, technological innovation had increased our understanding of the importance of microbes in nature. Microbes, we now realize,

control the flow of energy and nutrients on Earth by encoding nanomachines (enzymes) that move molecules through energy consuming and yielding redox reactions (Falkowski, Fenchel, Delong 2008). Closer to home for us, microbes matter to humans in diverse ways. Our gut-associated microbial communities are now credited for helping us to stay skinny (or not) amidst the obesity epidemic (Gross 2013), and even to enjoy (and digest) sushi (Hehemann et al. 2010; Hehemann et al. 2012). On the maintenance of health side, microbes help humans to resist illness (Clemente et al. 2012; Pflughoeft, Versalovic 2012) through prevention of invasion by pathogens (Xie et al. 2012; Maltby et al. 2013) and maintenance of our immune system (Hooper, Littman, Macpherson 2012). If microbes are so important, then surely their phages must be too. Already by 1989 we knew that phages were abundant in the oceans, being ten times as numerous as microbial cells almost everywhere you looked. The challenge was to advance from knowing they are abundant to understanding what their roles in nature are. We also realized that likely <1% of the microbes in the sea were culturable (Rappe, Giovannoni 2003), so we could only be doing worse with culturing marine phages. So, enter metagenomics—finally a tool to access the unculturable majority.

Plenty of questions awaited just such a tool. For example, do phages directly manipulate other microbial metabolisms besides photosynthesis? At that time, the idea was that phage genomes contained genes for host takeover, genes that countered host defenses or encoded virion structural proteins, and genes to manage lysis or lysogeny. But the phage photosynthesis story suggested that phages might encode and use many genes that we typically associate with microbial metabolisms, and that these genes might improve phage fitness (*see page 5-12*). Viromes offered the opportunity to explore this possibility through identifying novel auxiliary metabolic genes (AMGs) carried by the phages (Breitbart et al. 2007).

Oded Beja led some of the early virome research that expanded our understanding of AMGs with

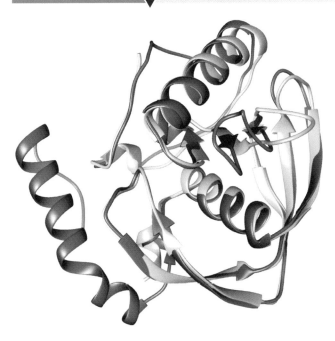

Figure 1. (*See page 5-12 for the underlying story*.) Overlay of 3D ribbon diagrams of the crystal structures of PDF proteins from the *Arabidopsis thaliana* chloroplast (maroon, PDB ID: 3CPM) and *Synechococcus* phage S-SSM7 (mustard, PDB ID: 3UWA). Note the α-helix (left) in the *A. thaliana* protein that is missing from the phage protein. Structural data were downloaded from the Protein Data Bank (www.rcsb.orb/pdb). Ribbon diagrams were re-colored and aligned in Chimera (Pettersen et al. 2004).

his renewed look at the cyanophage photosynthesis story. His first advance concerned the already-discovered core PSII reaction center genes that were identifiably either 'phage' or 'microbial.' Approximately 60% of those he found in *microbial* metagenomes were phage, and these phage versions were expressed in natural samples (Sharon et al. 2007). This meant that not only were these photosynthesis genes common in cyanophage genomes, but that cyanophage infection of wild microbes could be boosting photosynthesis using phage-encoded PSII core reaction center genes. Open questions remain as to whether these transcripts lead to proteins that are localized in the photosynthetic membranes (hard to imagine they would not), and what fraction of the photosynthetic proteins in intact *microbial* photosystems are phage-encoded. Beja made a second major contribution, again using a metagenomic approach, when he shifted the focus from the core PSII photosynthesis genes to those of the PSI reaction center (Sharon et al. 2009) and found that these genes were also prevalent among phages. Then, in

2011, he identified many additional virus-encoded AMGs in marine metagenomes, the most abundant of which were peptide deformylases (Fig. 1; *see page 5-12*).

Indeed, phage metabolic capabilities extend far beyond photosynthesis. We now know from viromic studies that phages directly modulate host cycling and uptake of nutrients (phosphate, nitrogen, sulfur), as well as nearly all host central carbon metabolism (Dinsdale et al. 2008; Sharon et al. 2011; Hurwitz, Hallam, Sullivan 2013; Anantharaman et al. 2014). Clearly phages are much more than killing machines. Given that at any point in time some one-third to one-half of the microbial cells in the oceans are infected by phages and that phage infection drastically alters cellular metabolic outputs, clearly phages "manipulate the marine environment" (Rohwer, Thurber 2009) well beyond anything we had dreamed about before metagenomic sequencing became available.

Towards a phage census

By studying AMGs, early metagenomic studies were beginning to answer "what are phages doing?" in various environments. Of course, we also sought to answer "who is there?" Both of these questions are simpler to deal with in better characterized systems or longer-established fields; they represent major challenges when applied to environmental phages. Viewing taxonomy from my ecology-minded perspective, it was no fun to know that more than one million phages existed in a milliliter of seawater, but that for a PhD one needed to consider spending five years studying one or a few culturable isolates. Which one do you choose, and on what basis? In contrast, metagenomes that could include the DNA from a whole community of phages would provide the kind of data needed to better answer "who is there." But there was a problem in that only a small fraction of a metagenome could be attributed to a known phage (most frequently one of the tailed phages—the *Myoviridae*, *Podoviridae*, and *Siphoviridae*), which left 63-93% of the reads unassigned (Hurwitz, Sullivan 2013). Rather than settle for the very limited taxonomy attainable using database mapping, Florent Angly set out to directly estimate the number of different phages in a seawater sample. To do this, he assembled metagenomic reads *in silico* to yield contigs. The idea was that metagenomic sequencing of diverse phage communities would yield relatively low coverage for any particular genotype, and thus assembly would produce short contigs. Conversely, less diverse communities would result in longer contigs. By combining mathematical models with the observed spectrum of contig lengths, Florent came up with predictions ranging from hundreds to hundreds of thousands of 'genotypes' in pooled seawater samples (Angly et al. 2005; Angly et al. 2006). Unfortunately, these predictions were far from precise, often spanning as they did orders of magnitude, presumably because only the most abundant phages were sampled. As Bart Haegeman and Joshua Weitz showed for microbial diversity estimates (Haegeman et al. 2013), these es-

timators are only accurate when rare populations are well sampled.

These were good starting places for thinking about taxonomy in phage communities, but to advance beyond descriptive studies, viromics had to get past some issues and move towards quantitative ecology. First there was the assembly issue: sequence a community of phages and you get nothing but tiny fragments of phage genomes, not sizable contigs. This is because phage community diversity is so high that sequencing leads to just a few reads from even the most abundant phages in a complex environmental sample. Second, viromics methods suffered from inefficiencies and biases at many steps along the path from sample to sequence. Methods for obtaining phage concentrates were inefficient, different purification steps were used in different virome preps, and, having so little initial phage DNA, researchers often re-

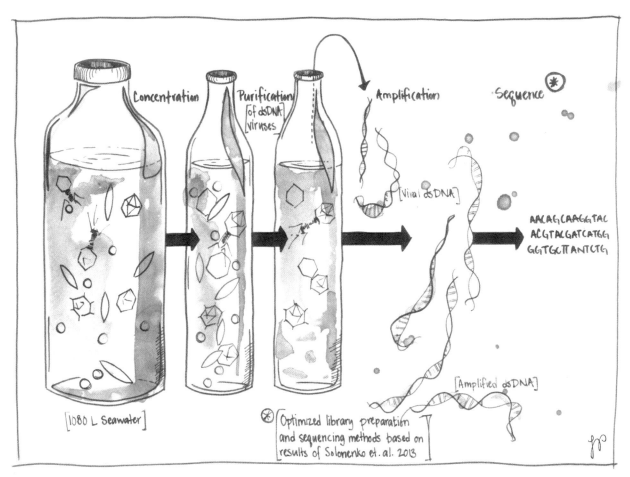

Figure 2. The sample-to-sequence viral metagenomic pipeline, well tuned for dsDNA viruses. Steps include VLP concentration by $FeCl_3$ precipitation (John et al. 2011), purification by CsCl plus DNase (Hurwitz et al. 2013), linker amplification (Duhaime et al. 2012), and sequencing (Solonenko et al. 2013).

sorted to DNA amplification. On this latter point, the amplification enzymes were later revealed to have biases that were both systematic (preferring circular ssDNA) and stochastic (over-representing the first piece of DNA encountered at exceptionally low DNA concentrations). Thus, as the field attempted to advance towards a quantitative science, some dirty laundry had to be cleaned up (and who *wants* to do methods optimizations?!@#!@). Third, a bonus issue we have not yet tackled as a field, is that we're not capturing ssDNA or RNA phages in most viromes produced to date, and have no equivalent quantitative viromics methodologies for these phages.

Fortunately, after four years of methods optimizations, we now have a sample-to-sequence pipeline (Duhaime, Sullivan 2012; Solonenko, Sullivan 2013) (see Fig. 2) that is finely tuned for dsDNA phages. Briefly, it uses a chemistry-based method to reproducibly and efficiently concentrate phages (John et al. 2011), empirical data to guide the choice of purification options (Hurwitz et al. 2013), and linker-amplification methods that are near quantitative for dsDNA phages (Duhaime et al. 2012). This dsDNA sample-to-sequence viromic pipeline can create reproducible and quantitative viromes from miniscule amounts of DNA (<100 femtograms), and has been tested across myriad library preparation methods and sequencing platforms (Solonenko et al. 2013). Once we had such a well-characterized viromic pipeline in-hand, what the field really needed was a dataset for phages comparable to that provided for microbes by the Global Ocean Survey (GOS) (Rusch et al. 2007). Through some forward-thinking funding opportunities from the DOE Joint Genome Institute and the Gordon and Betty Moore Foundation, we were able to use this quantitative sample-to-sequence pipeline to generate such a dataset. This new viromic dataset, the Pacific Ocean Virome (POV), consists of 32 systematically and quantitatively prepared viromes originating from diverse locations and depths in the Pacific Ocean (Hurwitz, Sullivan 2013).

Just as GOS did for microbial ecology a half-decade earlier, the POV dataset has led to major discoveries in marine virology (Fig. 3). In 2013 alone, the marine cyanophages, long considered the most abundant phages in the sea, relinquished that title to the pelagiphages infecting *Pelagibacter ubiquitans* or SAR11 (Zhao et al. 2013) and the SAR116 phages (Kang et al. 2013) that now rank as number one and number two in abundance among the marine phages. Sadly, of course, this relegated my cyanophages from most abundant (Williamson et al. 2008) to third most abundant, and dropping. At the other end of the abundance spectrum, the POV and other virome datasets were enabling discovery of the rare virosphere, i.e., phages that seem to be ubiquitous, but not abundant, such as Karin Holmfeldt's *Bacteriodetes* phages (Holmfeldt et al. 2013). The simple availability of such a large-scale, quantitatively-prepared virome dataset opens up all kinds of opportunities for placing new phage isolates into ecological context, as well as for making discoveries about how phages and their hosts interact (e.g., the central carbon metabolism AMG story mentioned above).

Estimating the unknown

Developing a large-scale, quantitative and systematic viromic dataset was a big step forward, but now all this data brought us face-to-face with the next problem. Because only a tiny fragment of phage sequence space has been examined with traditional methods (e.g., cultured isolates, whole genome sequencing), sequence similarity searches showed that many of the sequenced reads in a new virome—often 70–90%!—were new to science (Hurwitz, Sullivan 2013). Given that there are only about a thousand reference genomes in public databases and given the likely level of phage diversity in a single seawater sample (Angly et al. 2006), it is not surprising that much of what we observe in environmental viromes is novel. But how much unknown sequence space is there left to explore? Back in 2003, a younger and still black-haired (not dyed) Forest Rohwer suggested that the size of the global virome was something like two billion proteins (extrapolating from 14 mycobacteriophage genomes to the world—only in phage ecology!) and thus represented the largest reservoir of unexplored sequence space (Rohwer 2003). Here, a decade later, we had generated enough viromic data to take another crack at estimating the global virome. Indeed, with much more data available for guesstimating the scope of this sequence space, some revision seemed to be in order. We suggested downsizing the global virome estimate some three orders of magnitude to a few million proteins (Ignacio-Espinoza, Solonenko, Sullivan 2013). This reduces the back-of-the-envelope cost for functionally annotating the global virome from 17 trillion dollars (Rohwer, personal communication) to about 17 billion dollars. Now this is becoming approachable. For comparison, an estimated $0.98 billion is being spent to collect a few grams of orbiting asteroid dust (the OSIRIS-REx project according to Wikipedia), when for only about 20 times that we could explore the 'phage dark matter' here on Earth and determine the function of all phage proteins—my druthers (sorry NASA!).

Organizing the unknown into clusters

With all the phage unknowns out there, it is comforting to know that even today, without functional annotations for every phage protein, there are good ways to organize the unknown. Currently, the most promising method follows in the footsteps of Shibu Yooseph who handled the same problem for the original microbial GOS data (Yooseph et al. 2007) by implementing arbitrary, sequence-based clustering methods to create 'bins' into which new sequences can be deposited. These

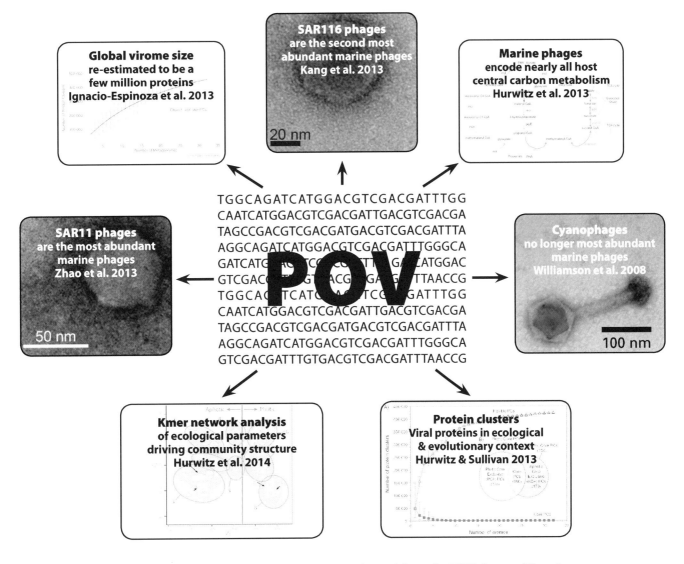

Figure 3. Some of the key findings already derived from the POV dataset (Hurwitz et al. 2013) that highlight the value of this data resource.

methods have been applied to viromes by Bonnie Hurwitz and Simon Roux (Roux et al. 2012; Hurwitz, Sullivan 2013), and with great results. In this way, even without knowing anything about the function of the proteins themselves, researchers can quickly delineate what and where particular kinds of proteins (i.e., protein clusters) have been previously observed, while also being able to estimate both rank abundance curves of protein clusters and overall community diversity.

This is a big step forward for anyone wanting to know which proteins, annotated or not, are under- or overrepresented along ecological gradients or under evolutionary selection. Both kinds of information can be discerned from this organizing of viromes into protein clusters, and together they can serve as a foundation for exploring fundamen-

tal ecological and evolutionary questions about phages. Granted, the diversity of phage protein clusters is not phage diversity; nevertheless, it does get us one step closer to having a comparative phage community diversity metric since rarefaction or collector curves as well as rank abundance curves can be developed for protein clusters. I see this protein clustering framework as the birth of a 'gene ecology' wherein the fundamental ecological unit is the gene or protein rather than the organism (which is still too complex to be tractable). That we can now discern abundance patterns and sequence variation within and between protein clusters allows us to leverage decades of ecological and evolutionary theory to rapidly bring phage ecology up to speed with more traditional ecological disciplines, and this, in turn, will undoubtedly lead to new theory. Imagine having a phage ecol-

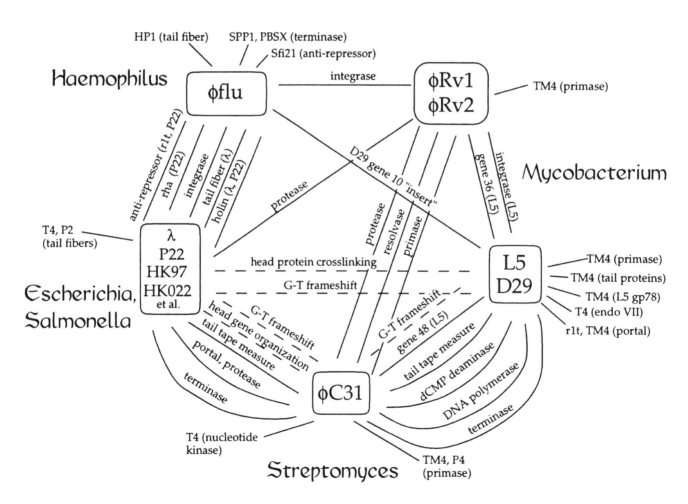

Figure 4. Sequence connections among phages and prophages, which served as the initial highlight of rampant mosaicism across phage genomes. Solid and dotted lines represent sequence similarities and gene organization/function commonalities, respectively. Closely related phages are boxed and bacterial hosts are shown at the perimeter of the web. Reprinted from Hendrix et al. 1999 with permission of the publisher. Copyright (1999) National Academy of Sciences, U.S.A.

ogy discipline that even hard-core ecologists will appreciate. We're not too far off, with hints emerging that we may already be there for cyanophages.

Protein clustering should also prove to be of great interest to mechanistically-minded, reductionist researchers. For example, protein cluster counts from around the world will identify proteins that are abundant in nature, many of which will undoubtedly have completely unknown function. These might be attractive experimental targets. Having their gene sequence, one can make the protein and then assay for function—the classical combination of genetic and biochemical approaches for gene characterization. In addition, although not fully utilized yet, you could imagine that the extensive environmental metadata associated with these phage protein clusters would help form hypotheses about protein function and possible interacting protein partners. I see this mattering to a 'reductionist' because, if you're going to devote five years to a protein, you might as well be characterizing one that matters in the real world.

Beyond clustering

Nevertheless, to move forward we will have to get beyond protein clustering. For example, returning to the topic of taxonomy, we had found the idea of comparing phage communities, when so much of the available data remained taxonomically unassigned, hard to stomach. Protein clustering helped, but on average only 50-70% of a new virome will self-cluster or map to existing protein cluster databases. So, there is work to be done before we can compare two viromes in toto. On this front, Bonnie Hurwitz and Anton Westwald recently blended viromics with network analytics by leveraging sequence characteristics (the frequency of 'kmers') of every read in each virome. These data became input into a modeling framework derived from social network analyses that can rapidly scale to place ever-growing large-scale datasets into a unified statistical visualization framework (think about all that next time you are on Facebook). What they found, using every read from every virome, were patterns across the POV dataset that provided evidence as to which ecological parameters were driving phage community structure (Hurwitz et al. 2014). Likely this is just the first instance of a brand new, data-driven, discovery science, but it offers a glimpse of what the "math guys" (Forest's words) could do if they engage with biologists who now have big data with which to play. This new era is exciting, but environmental virology, like most disciplines suffering from data deluge, will need creative and innovative solutions to complicated analytical problems.

Further, these sequence datasets are also foundational for emerging methods designed to link phages with their hosts—another major challenge in phage ecology. Already these sequences are being employed to design probes for microscopy methods (e.g., phageFISH [Allers et al. 2013]) and primers for PCR-based methods (e.g., microfluidic digital PCR [Tadmor et al. 2011]). Both enable researchers to detect particular phage genes inside individual host cells in a way that will enable them to determine lineage-specific phage mortality rates in complex communities. Computational methods are also emerging whereby phage sequence data are better linked to their hosts. First, microbial community DNA in clone libraries that target large genome regions (e.g., ~40kb fosmids) can be screened to identify the phage signal. Having such large genome fragments then allows researchers to utilize sequence composition to identify possible hosts for the phages (Mizuno et al. 2013). Second, single cell genomic sequencing projects offer a means to explore microbial dark matter by pre-screening for a "barcode gene" of interest to enrich for under-explored groups of bacteria (only about half of bacterial phyla are known by more than a single gene!). However, such data can also be mined for phages since ~1/3 of microbes in nature are infected at any given time (Roux, et al. 2014). Thus, these *in silico* methods can lead to much-needed new phage reference genomes to better map phages across the tree of life which also serve as hooks for more extensive ecological and evolutionary study of specific phage-host interactions from community genome datasets. This new toolkit will enable strain-specific resolution of phage-host infection dynamics, even in complex natural communities.

Phage species: Myth or reality?

Another challenge in metagenomics today is to determine at what percent identity cut-off is a metagenomic read really mapping to a reference genome. This seems so simple, but I believe that the fact that this is so troublesome and ambiguous reflects our naive picture of the underlying population structure of phage genome sequence space in nature. If phage genomes are exchanging genes at high rates (i.e., are rampantly mosaic), as has long been thought to be the case, then we would expect that such gene exchange would be so destructive to 'species' boundaries that it would stymie any genome-based taxonomic efforts (Lawrence, Hatfull, Hendrix 2002). If this were indeed true, then our ability to study phages in the environment would be greatly curtailed; after all, what should we count? This paradigm dates back to Roger Hendrix's pioneering work with Siphovirus genomes where he observed genome fragments with high identity across different genomes (Hendrix et al. 1999). These observations led to the hypothesis, famously termed the moron accretion hypothesis, that phages must be exchanging genes at a relatively high rate to yield their mosaic genome structure (Fig. 4; *see page 5-48*). Again, if this is universally true, then it is bad news for our field.

To place this in context, at around the same time *microbial* evolutionary biologists were also detecting high rates of horizontal gene transfer (Ochman, Lawrence, Groisman 2000), which led some to paint a picture of a web of life, rather than a tree (Doolittle 1999)—ideas for which Doolittle recently won Canada's famed Herzberg Medal of Science. I hope this helps you see that this is powerful stuff, with big implications for both microbes and phages.

To me (and others) it was clear at this point that although both microbes and their phages exchange genes among themselves at rates that are higher than those observed in eukaryotes, there might still be hope that microbial and phage genome sequence space in the environment was structured enough to provide taxonomic boundaries. In microbes, for example, Kostas Konstantinidis and Jim Tiedje proposed that the average nucleotide

identity (ANI) of shared genes in microbial genomes could be used to delineate species (Konstantinidis, Tiedje 2005). When Kostas mapped metagenomic reads against reference genomes, he clearly showed apparent break points in sequence identity (Caro-Quintero, Konstantinidis 2012). This suggested boundaries to variation within and between microbial populations—i.e., population structure. Recently, Eric Alm and Martin Polz generated large-scale environmental and isolate-based *Vibrio* datasets that they then subjected to formal population genetics tests to evaluate speciation and selection (reviewed in [Polz, Alm, Hanage 2013]). These studies make a compelling case that gene flow in microbes is akin to sex in eukaryotes: it is much more frequent within than between species and it serves to delineate essentially reproductively isolated populations.

But, enough about these pesky microbes. Does similar population structure exist among phages in the environment? For many years, it did not look likely. Mya Breitbart showed that ocean phages seemed to share access to a global gene pool as inferred from identical marker gene sequences amplified from ocean samples from around the world (Breitbart, Miyake, Rohwer 2004; Breitbart, Rohwer 2005). Similarly, mycobacteriophages isolated from many different soil samples (though never more than two phages per site) exhibited rampant mosaicism with many detectable gene flow events. Nevertheless, those mycobacteriophages could be assigned to clusters based on genome-wide comparisons. Furthermore, Forest along with Rob Edwards showed that BLAST-based mapping of *in silico*-generated phage genome fragments to the Phage Proteomic Tree (Rohwer, Edwards 2002) assigned the reads to the correct genome >95% of the time (Edwards, Rohwer 2005). This suggested some genome structure or boundaries between genomes, at least among the phages with sequenced genomes at that time. Similarly, in studies of T4-like cyanophages, there was evidence that clusters of marker gene sequences had annually repeatable ecological patterns (Marston, Amrich 2009) and that single-gene phylogenies of most (83%) of the T4-like cyanophage core genes were statistically indistin-

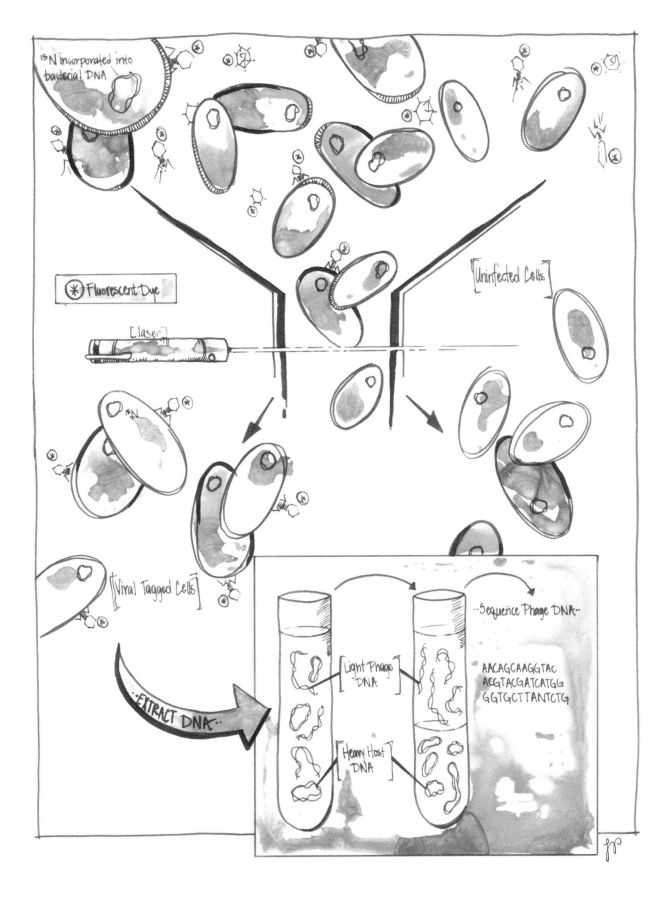

Figure 5. The viral tagging (VT) protocol. Fluorescently-labeled viruses (FLVs) are mixed with cultured 'bait' host cells whose DNA was prelabeled with stable isotopes. Cells with adsorbed FLVs are detected and sorted by a shift in fluorescence, the DNA extracted, and the cellular DNA separated from phage DNA by density gradient centrifugation.

guishable from those of the concatenated supertree or species tree (Ignacio-Espinoza, Sullivan 2012). These findings suggested population structure in the naturally occurring T4-like cyanophages.

Thus, the effort to infer population structure from available sequences, which had been identified as a key goal for the field over a decade ago, remained challenging and unresolved. This was due, in part, to the difficulty of deconvoluting the true (and still unknown) population structure given the relatively uncharacterized sampling strategy (never more than two phages per site, most often only one). Reflecting this, almost every phage genome paper published in the last decade at some point in their story reverted to the default assumption that phage genomes are rampantly mosaic. In the words of those studying some of the most intensively sequenced phage genomes (see below), any structure was likely artifact due to improper sampling of a genetic continuum.

> *Given the bias and extreme sparseness of sampling, it is still possible in my view that the global population is a smooth genetic continuum, with the apparent structure an artefact arising from the sampling.* — R. Hendrix (Hendrix 2003)

> – and –

> *The global population of mycobacteriophages would seem more likely to form a continuum of relationships, and the observed clusters may emerge from biases imposed by the isolation procedures.* — G. Hatfull (Hatfull 2010)

> – and –

> *Organization into clusters and subclusters should not be interpreted as representing any well-defined boundaries between different types of viruses, but rather a reflection of incomplete sampling of a large and diverse population of viruses occupying positions on a broad spectrum of multidimensional relationships … Clusters thus do not represent lineages per se, but do provide a convenient means of representing the heterogeneity of the currently sequenced phages.* — G. Hatfull (Hatfull 2012)

Again, I saw this as bad news for our field. This was because to advance marine phage ecology from a descriptive to a quantitative and predictive science would require discrete biological units (i.e., species) to count. Isolate-based genomics was suggesting such discrete units might not exist. Hints to the contrary had been accumulating (see above). What we really needed, we reasoned, was to figure out a way to deeply sample the diversity of phages in complex communities. Only with such deep sampling might we finally be able to evaluate whether phage genome sequence space was structured or existed as some continuum whereby 'counting' could never happen at the genomic level.

Getting a handle on phage population structure

To this end, Phil Hugenholtz had the idea to develop a relatively anonymous and high-throughput way of screening millions of naturally-occurring phages for those phages that are associated with a particular host cell of interest (Fig. 5). This method, named viral tagging or VT, was so high-throughput that it could allow us to evaluate the population structure of phage genome sequence space in complex communities (Deng et al. 2013; Deng et al. 2014). Specifically, for VT we take wild phages and stain them, wash the stain away to create fluorescently-labeled viruses (FLVs) sensu (Noble, Fuhrman 2000), and then mix these FLVs with a cultured 'bait' host. Cells that adsorb the FLVs can be detected and sorted by a shift in fluorescence. Cellular DNA can be separated from phage DNA by density gradient centrifugation following pre-labeling of the cellular DNA with stable isotopes.

Li Deng and Cesar Ignacio-Espinoza applied this method to a coastal Pacific Ocean phage community to select ~10^7 cyanophages specific for a marine *Synechococcus* strain (WH7803) and sequenced this phage DNA to yield VT metagenomic data (Deng et al. 2014). This scale of data proved invaluable for elucidating natural phage population structure, at least for the abundant T4-like cyanophages in the sample. Specifically, the resulting population genome landscape plot (derived from comparison of the average nucleotide identity, or ANI, of the VT metagenome contigs against reference genomes) revealed that T4-like cyanophages had a discrete population structure (Fig. 6). This means that these populations could be counted, and their abundances estimated from the metagenomic coverage (number of reads per population).

Further, the VT metagenomic data provide large swaths of genomic sequence for each population, data that is valuable for two reasons. First, comparative genomic studies could use such data to assess gene-based ecological drivers that might enable one population to outcompete another. Second, the scale of this VT data could help answer the simple question of when a metagenomic read should be assigned to a reference genome or not. Here, for

example, it looks like ANIs derived from T4-like cyanophage populations are nearly all >99%— for both 'core' (shared by all) and 'flexible' (sporadically distributed) genes. Thus those metagenomic reads mapping at 40, 50, even 80 or 90% identity are unlikely to be derived from the same population as the reference genome, likely being instead just the 'best hit' available in databases.

Taking this a step further, Ann Gregory sequenced 142 cyanophage genomes from cultures randomly isolated using the same host from the VT source waters described above (Gregory et al. submitted). Ann reasoned that having whole genomes would allow her to assess whether these naturally-occurring cyanophage populations fit the traditionally accepted definition of 'species' that would require

Figure 6. Viral genome sequence space is clustered in nature, at least for this host at this time and site. This figure was first published in Deng et al. 2014 and is reproduced here with the permission of the journal.

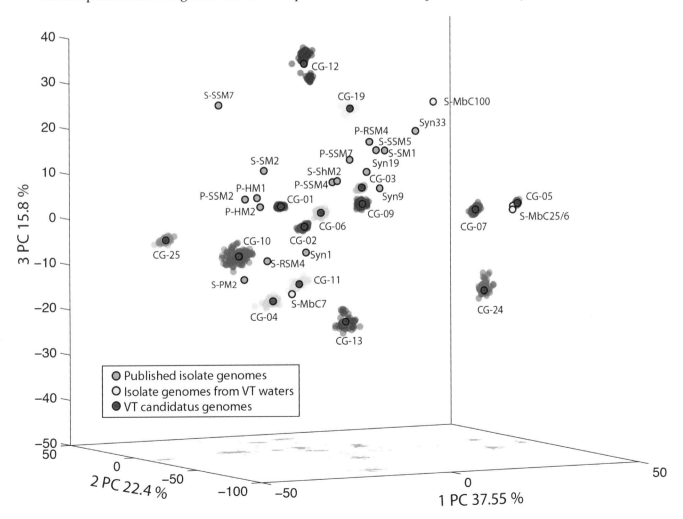

gene flow to be greater within than between populations. Indeed this was the case; these cyanophage populations appear to represent phage species. This expands the VT foundation of discrete population structure to enable more advanced ecology. Here she chose to address a fundamental ecological question of how phage populations survive in stochastic environments, and found that these cyanophage species appear to survive the dynamic marine environment through bet hedging. Now, this is starting to sound like a genome-based foundation for doing real phage ecology!

Although clearly just a first look at phage population structure in the environment, these findings open an important window through which to observe phages and their interactions in nature. Specifically, this enables population- and genome-based, lineage-specific, host-linked phage ecology. An obvious open question remains as to how generalizable such patterns or population structures are across phage types, hosts, and sample sites. However, if these data are at all representative of naturally-occurring phage population structure beyond the cyanophages, then they represent a watershed moment for phage ecology that will allow the field to rapidly advance by leveraging decades of ecological and evolutionary theory. These are exciting times!

Not long ago, Carl Woese and Nigel Goldenfield described how genomics and metagenomics were leading to biology's next revolution (Goldenfeld, Woese 2007). They argued this fundamental new window into life on earth would alter the very foundations of 'biology,' and that phages would play a prominent role both as agents of horizontal gene transfer between microbes and as mediators of microbial mortality and function. Woese saw the true form of the tree of life decades before others (Woese, Fox 1977), and it appears here that his crystal ball was spot on in recognizing the central role of phages in driving the microbial engines that power planet Earth. Indeed, the phage metagenomic revolution has landed and its time to unlock the hidden secrets of our nanoscale Earthlings.

Acknowledgements

Christine Schirmer for help with figures; the Tucson Marine Phage Lab and the Gordon and Betty Moore Foundation for taking a crazy leap of faith on a young faculty member hoping to do some new science; and for years of stimulating conversations: Forest Rohwer, Mya Breitbart, Jennifer Brum, Jed Fuhrman, Steven Hallam, Phil Hugenholtz, Virginia Rich, Joshua Weitz, Mark Young, the Chisholm Lab, the Tucson Marine Phage Lab, and more.

References

Allers, E, C Moraru, M Duhaime, E Beneze, N Solonenko, J Barerro-Canosa, R Amann, MB Sullivan. 2013. Single-cell and population level viral infection dynamics revealed by phageFISH, a method to visualize intracellular and free viruses. Environ Microbiol:2306-2318.

Anantharaman, K, MB Duhaime, JA Breier, K Wendt, BM Toner, GJ Dick. 2014. Sulfur oxidation genes in diverse deep-sea viruses. Science 344:757-760.

Angly, F, B Rodriguez-Brito, D Bangor, P McNairnie, M Breitbart, P Salamon, B Felts, J Nulton, J Mahaffy, F Rohwer. 2005. PHACCS, an online tool for estimating the structure and diversity of uncultured viral communities using metagenomic information. BMC Bioinformatics 6:41.

Angly, FE, B Felts, M Breitbart, et al. 2006. The marine viromes of four oceanic regions. PLoS Biol 4:e368.

Bragg, JG, SW Chisholm. 2008. Modelling the fitness consequences of a cyanophage-encoded photosynthesis gene. PLoS One 3:e3550.

Breitbart, M, JH Miyake, F Rohwer. 2004. Global distribution of nearly identical phage-encoded DNA sequences. FEMS Microbiol Lett 236:249-256.

Breitbart, M, F Rohwer. 2005. Here a virus, there a virus, everywhere the same virus? Trends Microbiol 13:278-284.

Breitbart, M, P Salamon, B Andresen, JM Mahaffy, AM Segall, D Mead, F Azam, F Rohwer. 2002. Genomic analysis of uncultured marine viral communities. Proc Natl Acad Sci USA 99:14250-14255.

Breitbart, M, LR Thompson, CS Suttle, MB Sullivan. 2007. Exploring the vast diversity of marine viruses. Oceanography 20:353-362.

Caro-Quintero, A, KT Konstantinidis. 2012. Bacterial species may exist, metagenomics reveal. Environ Microbiol 14:347-355.

Clemente, JC, LK Ursell, LW Parfrey, R Knight. 2012. The impact of the gut microbiota on human health: An integrative view. Cell 148:1258-1270.

Deng, L, A Gregory, S Yilmaz, BT Poulos, P Hugenholtz, MB Sullivan. 2013. Contrasting life strategies of viruses that infect photo- and heterotrophic bacteria, as revealed by viral tagging. MBio 4:e00516-00512.

Deng, L, JC Ignacio-Espinoza, A Gregory, B Poulos, JS Weitz, P Hugenholtz, MB Sullivan. 2014. Viral tagging reveals discrete populations in *Synechococcus* viral genome sequence space. Nature 513:242-245.

Dinsdale, EA, RA Edwards, D Hall, et al. 2008. Functional metagenomic profiling of nine biomes. Nature 452:629-632.

Doolittle, WF. 1999. Phylogenetic classification and the universal tree. Science 284:2124-2199.

Duhaime, M, MB Sullivan. 2012. Ocean viruses: Rigorously evaluating the metagenomic sample-to-sequence pipeline. Virology 434:181-186.

Duhaime, MB, L Deng, BT Poulos, MB Sullivan. 2012. Towards quantitative metagenomics of wild viruses and other ultra-low concentration DNA samples: A rigorous assessment and optimization of the linker amplification method. Environ Microbiol 14:2526-2537.

Edwards, RA, F Rohwer. 2005. Viral metagenomics. Nat Rev Microbiol 3:504-510.

Falkowski, PG, T Fenchel, EF Delong. 2008. The microbial engines that drive Earth's biogeochemical cycles. Science 320:1034-1039.

Goldenfeld, N, C Woese. 2007. Biology's next revolution. Nature 445:369.

Gross, M. 2013. Does the gut microbiome hold clues to obesity and diabetes? Curr Biol 23:R359-362.

Haegeman, B, J Hamelin, J Moriarty, P Neal, J Dushoff, JS Weitz. 2013. Robust estimation of microbial diversity in theory and in practice. ISME J 7:1092-1101.

Hatfull, GF. 2010. Mycobacteriophages: Genes and genomes. Annu Rev Microbiol 64:331-356.

Hatfull, GF. 2012. The secret lives of mycobacteriophages. Adv Virus Res 82:179-288.

Hehemann, JH, G Correc, T Barbeyron, W Helbert, M Czjzek, G Michel. 2010. Transfer of carbohydrate-active enzymes from marine bacteria to Japanese gut microbiota. Nature 464:908-912.

Hehemann, JH, AG Kelly, NA Pudlo, EC Martens, AB Boraston. 2012. Bacteria of the human gut microbiome catabolize red seaweed glycans with carbohydrate-active enzyme updates from extrinsic microbes. Proc Natl Acad Sci USA 109:19786-19791.

Hendrix, RW. 2003. Bacteriophage genomics. Curr Opin Microbiol 6:506-511.

Hendrix, RW, MC Smith, RN Burns, ME Ford, GF Hatfull. 1999. Evolutionary relationships among diverse bacteriophages and prophages: All the world's a phage. Proc Natl Acad Sci USA 96:2192-2197.

Holmfeldt, K, N Solonenko, M Shah, K Corrier, L Riemann, NC Verberkmoes, MB Sullivan. 2013. Twelve previously unknown phage genera are ubiquitous in global oceans. Proc Natl Acad Sci USA 110:12798-12803.

Hooper, LV, DR Littman, AJ Macpherson. 2012. Interactions between the microbiota and the immune system. Science 336:1268-1273.

Hurwitz, BH, MB Sullivan. 2013. The Pacific Ocean Virome (POV): A marine viral metagenomic dataset and associated protein clusters for quantitative viral ecology. PLoS One 8:e57355.

Hurwitz, BH, A Westwald, JR Brum, MB Sullivan. 2014. Modeling ecological drivers in marine viral communities using comparative metagenomics and network analyses. Proc Natl Acad Sci USA in press.

Hurwitz, BL, L Deng, BT Poulos, MB Sullivan. 2013. Evaluation of methods to concentrate and purify ocean virus communities through comparative, replicated metagenomics. Environ Microbiol 15:1428-1440.

Hurwitz, BL, SJ Hallam, MB Sullivan. 2013. Metabolic reprogramming by viruses in the sunlit and dark ocean. Genome Biol 14:R123.

Ignacio-Espinoza, JC, SA Solonenko, MB Sullivan. 2013. The global virome: Not as big as we thought? Curr Opin Virol 3:566-571.

Ignacio-Espinoza, JC, MB Sullivan. 2012. Phylogenomics of T4 cyanophages: Lateral gene transfer in the "core" and origins of host genes. Environ Microbiol 14:2113-2126.

John, SG, CB Mendez, L Deng, B Poulos, AK Kauffman, S Kern, J Brum, MF Polz, EA Boyle, MB Sullivan. 2011. A simple and efficient method for concentration of ocean viruses by chemical flocculation. Environ Microbiol Rep 3:195-202.

Kang, I, HM Oh, D Kang, JC Cho. 2013. Genome of a SAR116 bacteriophage shows the prevalence of this phage type in the oceans. Proc Natl Acad Sci USA 110:12343-12348.

Konstantinidis, KT, JM Tiedje. 2005. Towards a genome-based taxonomy for prokaryotes. J Bacteriol 187:6258-6264.

Lawrence, JG, GF Hatfull, RW Hendrix. 2002. Imbroglios of viral taxonomy: Genetic exchange and failings of phenetic approaches. J Bacteriol 184:4891-4905.

Lindell, D, JD Jaffe, ZI Johnson, GM Church, SW Chisholm. 2005. Photosynthesis genes in marine viruses yield proteins during host infection. Nature 438:86-89.

Lindell, D, MB Sullivan, ZI Johnson, AC Tolonen, F Rohwer, SW Chisholm. 2004. Transfer of photosynthesis genes to and from *Prochlorococcus* viruses. Proc Natl Acad Sci USA 101:11013-11018.

Maltby, R, MP Leatham-Jensen, T Gibson, PS Cohen, T Conway. 2013. Nutritional basis for colonization resistance by human commensal *Escherichia coli* strains HS and Nissle 1917 against *E. coli* O157:H7 in the mouse intestine. PLoS One 8:e53957.

Mann, NH, A Cook, A Millard, S Bailey, M Clokie. 2003. Bacterial photosynthesis genes in a virus. Nature 424:741.

Marston, MF, CG Amrich. 2009. Recombination and microdiversity in coastal marine cyanophages. Environ Microbiol 11:2893-2903.

Mizuno, CM, F Rodriguez-Valera, NE Kimes, R Ghai. 2013. Expanding the marine virosphere using metagenomics. PLoS Genet 9:e1003987.

Noble, RT, JA Fuhrman. 2000. Rapid virus production and removal as measured with fluorescently labeled viruses as tracers. Appl Environ Microbiol 66:3790-3797.

Ochman, H, JG Lawrence, EA Groisman. 2000. Lateral gene transfer and the nature of bacterial innovation. Nature 405:299-304.

Pettersen, EF, TD Goddard, CC Huang, GS Couch, DM Greenblatt, EC Meng, TE Ferrin. 2004. UCSF Chimera—a visualization system for exploratory research and analysis. J Comput Chem 25:1605-1612.

Pflughoeft, KJ, J Versalovic. 2012. Human microbiome in health and disease. Annu Rev Pathol 7:99-122.

Polz, MF, EJ Alm, WP Hanage. 2013. Horizontal gene transfer and the evolution of bacterial and archaeal population structure. Trends Genet 29:170-175.

Rappe, MS, SJ Giovannoni. 2003. The uncultured microbial majority. Annu Rev Microbiol 57:369-394.

Rohwer, F. 2003. Global phage diversity. Cell 113:141.

Rohwer, F, R Edwards. 2002. The Phage Proteomic Tree: A genome-based taxonomy for phage. J Bacteriol 184:4529-4535.

Rohwer, F, A Segall, G Steward, V Seguritan, M Breitbart, F Wolven, F Azam. 2000. The complete genomic sequence of the marine phage Roseophage SIO1 shares homology with nonmarine phages. Limnol Oceanogr 45:408-418.

Rohwer, F, RV Thurber. 2009. Viruses manipulate the marine environment. Nature 459:207-212.

Roux, S, A Hawley, MT Beltran, M Scofield, P Schwientek, R Stepanauskas, T Woyke, S Hallam, M Sullivan. Ecology and evolution of viruses infecting uncultivated SUP05 bacteria as revealed by single-cell and environmental genomics. eLife:10.7554/eLife.03125.

Roux, S, F Enault, A Robin, V Ravet, S Personnic, S Theil, J Colombet, T Sime-Ngando, D Debroas. 2012. Assessing the diversity and specificity of two freshwater viral communities through metagenomics. PLoS One 7:e33641.

Rusch, DB, AL Halpern, G Sutton, et al. 2007. The Sorcerer II global ocean sampling expedition: Northwest Atlantic through eastern tropical Pacific. PLoS Biol 5:e77.

Sharon, I, A Alperovitch, F Rohwer, M Haynes, F Glaser, N Atamna-Ismaeel, RY Pinter, F Partensky, EV Koonin, YI Wolf, N Nelson, O Beja. 2009. Photosystem I gene cassettes are present in marine virus genomes. Nature 461:258-262.

Sharon, I, N Battchikova, EM Aro, C Giglione, T Meinnel, F Glaser, RY Pinter, M Breitbart, F Rohwer, O Béjà. 2011. Comparative metagenomics of microbial traits within oceanic viral communities. ISME J 5:1178-1190.

Sharon, I, S Tzahor, S Williamson, M Shmoish, D Man-Aharonovich, DB Rusch, S Yooseph, G Zeidner, SS Golden, SR Mackey, N Adir, U Weingart, D Horn, JC Venter, Y Mandel-Gutfreund, O Beja. 2007. Viral photosynthetic reaction center genes and transcripts in the marine environment. ISME J 1:492-501.

Solonenko, SA, JC Ignacio-Espinoza, A Alberti, C Cruaud, S Hallam, K Konstantinidis, G Tyson, P Wincker, MB Sullivan. 2013. Sequencing platform and library preparation choices impact viral metagenomes. BMC Genomics 14:320.

Solonenko, SA, MB Sullivan. 2013. Preparation of metagenomic libraries from naturally occurring marine viruses. Methods Enzymol 531:143-165.

Sullivan, MB, D Lindell, JA Lee, LR Thompson, JP Bielawski, SW Chisholm. 2006. Prevalence and evolution of core photosystem II genes in marine cyanobacterial viruses and their hosts. PLoS Biology 4:e234.

Tadmor, AD, EA Ottesen, JR Leadbetter, R Phillips. 2011. Probing individual environmental bacteria for viruses by using microfluidic digital PCR. Science 333:58-62.

Williamson, SJ, DB Rusch, S Yooseph, et al. 2008. The Sorcerer II global ocean sampling expedition: Metagenomic characterization of viruses within aquatic microbial samples. PLoS One 3:e1456.

Woese, CR, GE Fox. 1977. Phylogenetic structure of the prokaryotic domain: The primary kingdoms. Proc Natl Acad Sci U S A 74:5088-5090.

Xie, H, J Hong, A Sharma, BY Wang. 2012. *Streptococcus cristatus* ArcA interferes with *Porphyromonas gingivalis* pathogenicity in mice. J Periodontal Res 47:578-583.

Yooseph, S, G Sutton, DB Rusch, et al. 2007. The Sorcerer II global ocean sampling expedition: Expanding the universe of protein families. PLoS Biol 5:e16.

Zeidner, G, JP Bielawski, M Shmoish, DJ Scanlan, G Sabehi, O Beja. 2005. Potential photosynthesis gene recombination between *Prochlorococcus* and *Synechococcus* via viral intermediates. Environ Microbiol 7:1505-1513.

Zhao, Y, B Temperton, JC Thrash, MS Schwalbach, KL Vergin, ZC Landry, M Ellisman, T Deerinck, MB Sullivan, SJ Giovannoni. 2013. Abundant SAR11 viruses in the ocean. Nature 494:357-360.

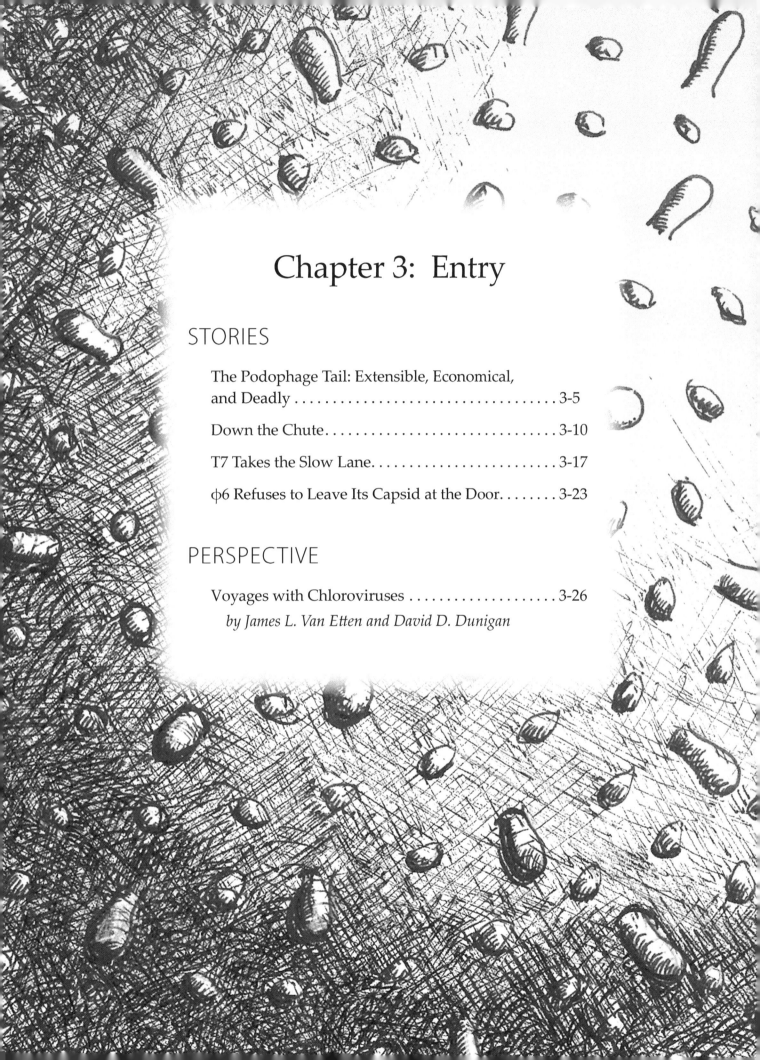

Chapter 3: Entry

STORIES

PERSPECTIVE

    by James L. Van Etten and David D. Dunigan

Enterobacteria Phage T3

a Podophage that, upon adsorption, ejects proteins from within its capsid to extend its stubby tail

Genome
dsDNA; linear
38,208 bp
55 predicted ORFs; 0 RNAs

Encapsidation method
Packaging; T = 7 capsid

Common host
Escherichia coli

Habitat
Mammalian intestines &
sewage

Lifestyle
Lytic

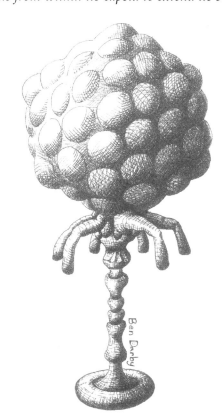

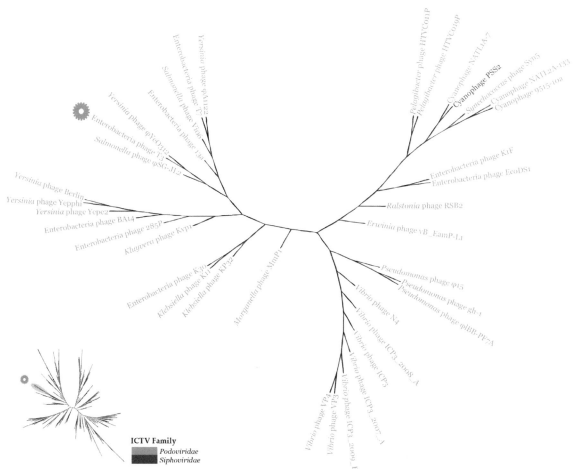

ICTV Family
Podoviridae
Siphoviridae

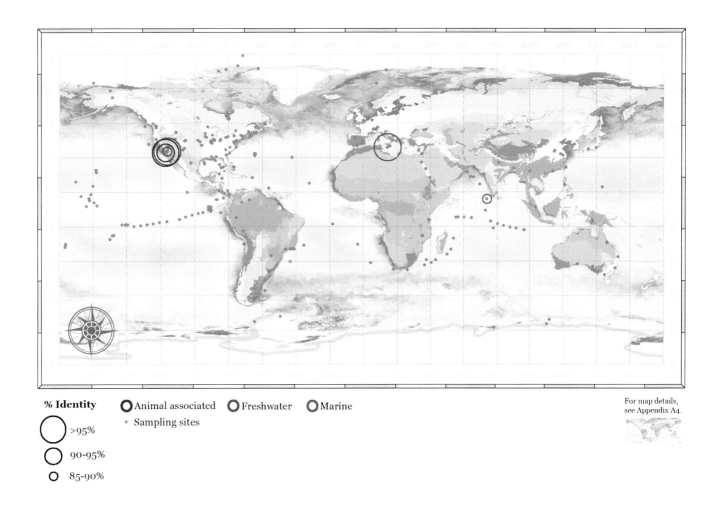

% Identity

◯ Animal associated   ◯ Freshwater   ◯ Marine

    • Sampling sites

◯ >95%

◯ 90-95%

○ 85-90%

For map details,
see Appendix A4.

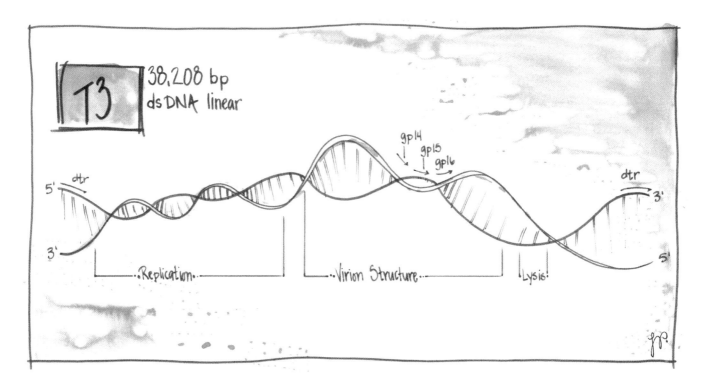

Enterobacteria Phage T3

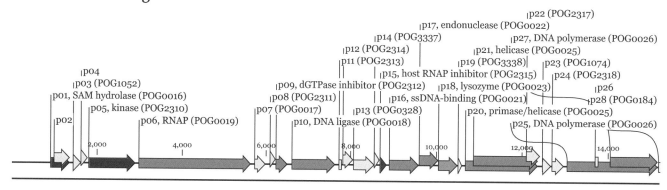

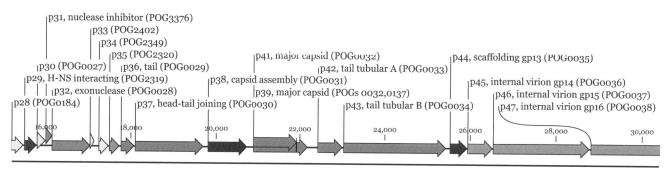

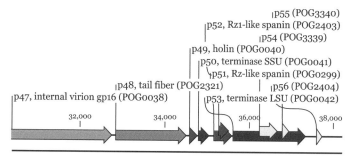

On the Prowl	Entry	Takeover	Replication Lytic	Replication Lysogenic	Assembly	Escape	Transfer RNA	Non-coding RNA	Multiple	Unknown

▼

The Podophage Tail: Extensible, Economical, and Deadly

Heather Maughan & Merry Youle

Mise-en-Scène: *According to the classical view of phage infection, virion proteins are discarded and only the genome enters the host cell. However, most (if not all) phages package essential proteins inside their virions and deliver them along with the genome. Some are enzymes to jump-start replication, others are genome bodyguards to deflect or inactivate host defenses. Yet others assist during genome delivery by digesting peptidoglycan barricades, forming a protective tube for genome passage, or escorting the genome into the cell. Phages with Gram-negative bacterial hosts use these proteins in innovative ways to bridge the periplasmic moat.*

Note: The research related in this story was performed using phage T7, but phage T3 also encodes **homologs** (69-97% identity) of these three tail proteins of T7.

Gram-negative Bacteria pose a particular challenge to a phage seeking entry. For such a phage to deliver its genome and launch an infection it must cross not one, but two, membranes, and traverse the nuclease-infested, **peptidoglycan** maze of the **periplasm** in between. The long-tailed Myophages and Siphophages are well equipped to bridge this ~24 nm stretch. When docked to the outer membrane, their tail tip easily reaches to the cell membrane and beyond. But this length comes at a cost, both in genes and cellular resources. Myophage T4, for instance, sports a tail that is 120 nm long and 25 nm wide (De Rosier, Klug 1968), comprised of at least 435 protein monomers representing 20 or more different proteins (Rossmann et al. 2004). The stubby-tailed Podophages that infect *E. coli* have developed a simpler—but equally deadly—mechanism. Concealed within their capsids is the minimal equipment needed to channel their DNA into the cytoplasm.

The Podophage solution

Measuring only 23 nm long, T7's tail cannot quite span the ~24 nm separating *E. coli*'s inner and outer membranes. In actuality, T7 does not lack an adequate tail; it simply postpones assembling its full length until it is needed. Until then, the structural proteins for the tail extension prudently wait, precisely positioned just inside the capsid between the genome and the exit door. As soon as T7 adsorbs irreversibly to its **lipopolysaccharide** (LPS) receptor on the outer membrane, these proteins set to work constructing a tail extension. The final extended tail is 40-55 nm long with a 3-4 nm diameter

central channel (Serwer et al. 2008)—a conduit for easy delivery of T7's 2 nm diameter double-stranded genomic DNA all the way to the cytoplasm.

For this tail extension upon arrival, T7 uses multiple copies of three **internal proteins** that it carries in a visible core near the capsid exit portal: gp14, gp15, and gp16. Closest to the portal and likely the first to exit are the ten molecules of gp14. They localize to the host's outer membrane where they form a channel through that first barrier. Following them out of the capsid are eight copies of gp15 and four of gp16 that are released into the periplasm. There gp16's **muralytic** N-terminal domain digests a path through the host peptidoglycan to the cell membrane (Moak, Molineux 2004). Together gp15 and gp16 form the tube extension that spans the periplasm and penetrates the cell membrane (Kemp, Garcia, Molineux 2005; Chang, Kemp, Molineux 2010). Assembly completed, the phage genome now has a protected passageway from capsid to cytoplasm, sheltered from the nucleases in the periplasm.

Efficiency

Compared to the extravagant Myophage tail, this T7 solution scores high for efficiency. The basic tail on a T7 virion is built from two main tail proteins, 12 copies of one and six of the other. Adding to that the 22 internal protein molecules that participate in the tail extension gives a total of 40 protein molecules—about one tenth the number in a T4 tail. Moreover, not only do gp15 and gp16 form the channel through which T7's DNA passes, but together they form a molecular motor that ratchets the initial portion of the genome into the cell—the first phase of a three-step genome entry operation (*see page 3-17*; [Kemp, Garcia, Molineux 2005; Chang, Kemp, Molineux 2010]). Once genome delivery is complete, their work done, these channel-forming proteins disappear from the scene. The membrane reseals without any leakage of ions or collapse of membrane potential, both of which betray the act of infection by some other phages. The T7 genome is now face-to-face with the next weapon in the host's defense arsenal (*see page 4-17*).

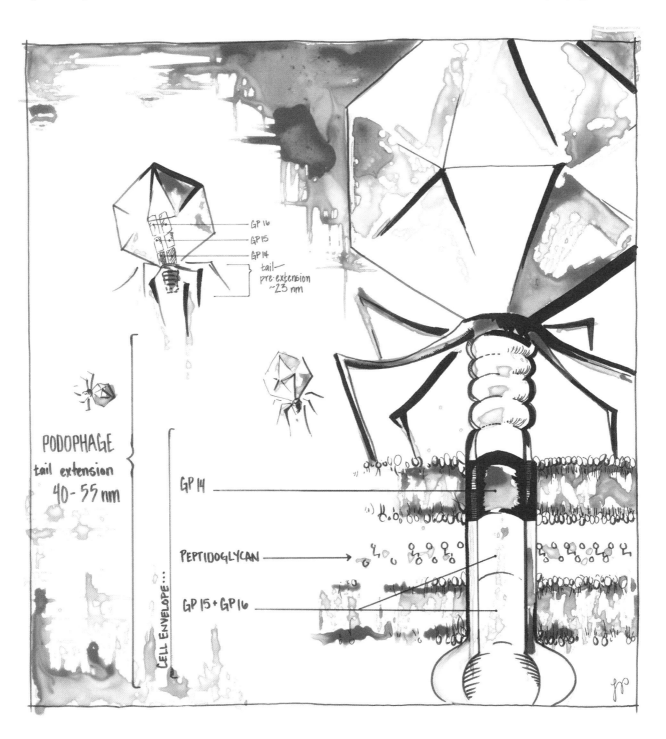

GP 16
GP 15
GP 14
tail pre-extension ~23 nm

PODOPHAGE
tail extension
40-55 nm

CELL ENVELOPE...

GP 14

PEPTIDOGLYCAN

GP 15 + GP 16

Cited references

Chang, CY, P Kemp, IJ Molineux. 2010. Gp15 and gp16 cooperate in translocating bacteriophage T7 DNA into the infected cell. Virology 398:176-186.

De Rosier, D, A Klug. 1968. Reconstruction of three dimensional structures from electron micrographs. Nature 217:130-134.

Kemp, P, LR Garcia, IJ Molineux. 2005. Changes in bacteriophage T7 virion structure at the initiation of infection. Virology 340:307-317.

Moak, M, IJ Molineux. 2004. Peptidoglycan hydrolytic activities associated with bacteriophage virions. Mol Microbiol 51:1169-1183.

Rossmann, MG, VV Mesyanzhinov, F Arisaka, PG Leiman. 2004. The bacteriophage T4 DNA injection machine. Curr Opin Struct Biol 14:171-180.

Serwer, P, ET Wright, KW Hakala, ST Weintraub. 2008. Evidence for bacteriophage T7 tail extension during DNA injection. BMC Res Notes 1:36.

Recommended review

Casjens, SR, IJ Molineux. 2012. Short noncontractile tail machines: Adsorption and DNA delivery by Podoviruses. In: MG Rossmann, VB Rao, editors. *Viral Molecular Machines*: Springer. p. 143-179.

Enterobacteria Phage PRD1

a Podophage that delivers its genome through a lipid-protein tail tube formed after adsorption

Note: PRD1 is classified as a member of the family Tectiviridae by the ICTV based on its virion morphology, as a close relative of STIV, PM2, PBCV-1, and Adenovirus based on the structure of its major capsid protein, and as a φ29-like Podophage by the proteomic analysis used to generate the PPT.

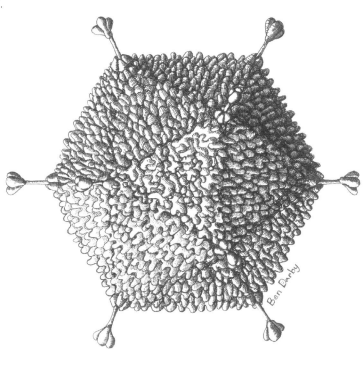

Genome

dsDNA; linear

14,927 bp

31 predicted ORFs; 0 RNAs

Encapsidation method

Packaging; T = 25 capsid

Common host

Escherichia coli harboring a conjugative plasmid

of the N, W, or P incompatibility group

Habitat

Vertebrate intestines and other locations

Lifestyle

Lytic

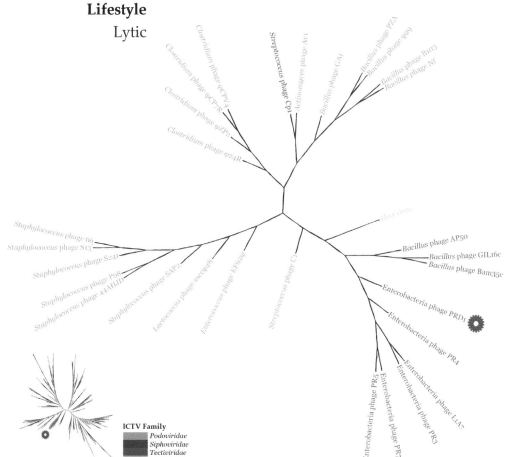

ICTV Family	
	Podoviridae
	Siphoviridae
	Tectiviridae
	Unassigned

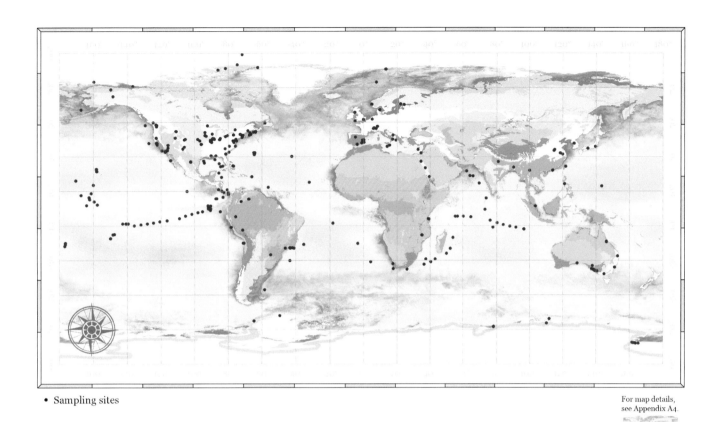

• Sampling sites

For map details,
see Appendix A4.

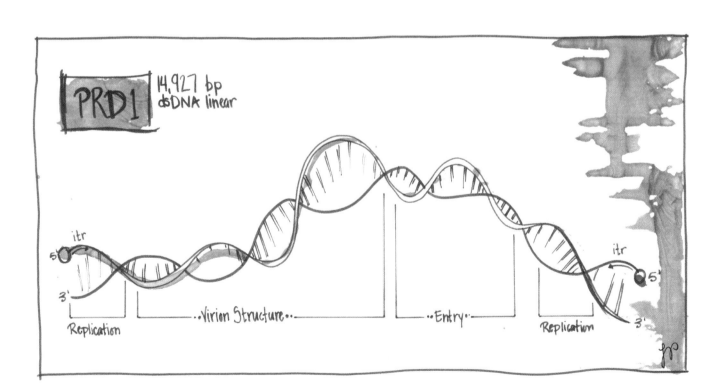

Enterobacteria Phage PRD1

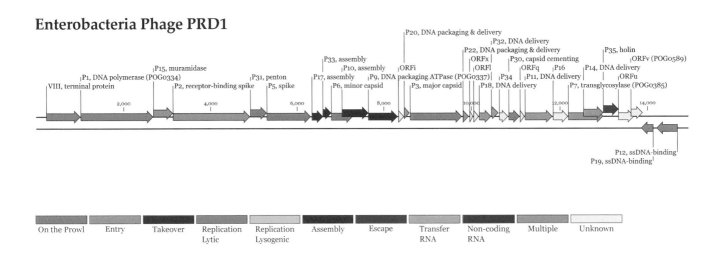

Down the Chute

Merry Youle

Mise-en-Scène: *By now,* **virions** *of more than 6000 phages have been observed using the electron microscope. The overwhelming majority are members of the order Caudovirales, the tailed phages. Tails are seemingly useful. Their distal fibers recognize and adsorb to potential hosts; the tail tube conveys the DNA safely between capsid and cytoplasm. Nevertheless, there are several families of tailless icosahedral or rod-shaped phages. Investigation of how they manage without a tail has revealed that some form a 'tail tube' for DNA delivery after adsorption. The Tectiviruses, for example, transform the protein-rich membrane lining the inside of their capsid into a delivery tube that spans the envelope of their Gram-negative hosts.*

Note: Although the *Tectiviridae* are recognized as a family in morphology-based taxonomies; the proteomic approach embodied in the Phage Proteomic Tree places them as a sub-family within the φ29-like Podophages. Nevertheless, the term Tectivirus remains useful to identify this group of phages that use a protein-primed mechanism of DNA replication and whose virions contain an internal membrane.

Tectivirus PRD1 is a tailless wizard. Lacking even a stubby Podophage tail, it nevertheless delivers its DNA chromosome across the double-membrane **envelope** of a broad range of Gram-negative Bacteria. Its magic formula? When it has irreversibly adsorbed to a host, its 'tail' assembles and pops out from the icosahedral capsid. Moreover, unlike other phages that construct a tail on the fly from proteins on board (e.g., Podophages T3 and T7 [*see page 3-5*] and Microphage φX174 [Sun et al. 2013]), PRD1 fashions its tail tube from its protein-rich internal lipid membrane.

Before adsorption

A PRD1 virion arriving on the scene presents a pedestrian icosahedral capsid, 66-70 nm in diameter, composed of a major capsid protein (P3)

and reinforced by a cementing protein (P30). A touch of flair is provided by the two spike proteins (P2 and P5) that project outward from its vertices, the base of each firmly anchored to the underlying capsid. Moreover, all PRD1 vertices are not created equal. Eleven of the twelve are adapted for host recognition and sport two spikes each (Huiskonen, Manole, Butcher 2007). The unique twelfth vertex is equipped with different proteins that perform different tasks. This vertex is the portal for DNA transit—transported inward during packaging and outward for delivery during infection. Thus, the packaging ATPase is restricted to this vertex (Gowen et al. 2003), as are two other proteins (P20 and P22) involved in genome delivery (Strömsten, Bamford, Bamford 2003).

A PRD1 virion also contains lipids. You might imagine that this Tectiphage membrane would be an external envelope destined to fuse with a host membrane during infection, as is the case with the Cystophages (*see page 3-23*) and Siphophages infecting Mollicutes (*see page 5-37*), but 'tis not so. Instead this membrane lines the inside of the protein capsid and forms an icosahedral vesicle surrounding the DNA. The additional proteins that stud the lipid membrane bring the total number of different phage proteins in the virion to 18 (Butcher, Manole, Karhu 2012).

Adsorption

For its receptor, PRD1 uses a structure on the host cell surface that is encoded not by the host chromosome but by a resident conjugative **plasmid** (of the N, W, or P incompatibility group; [Rydman, Bamford 2002b]). These plasmids direct the construction of numerous mating pair formation sites on the surface of the cell which they use when 'secreting' their single-stranded DNA during mating. PRD1 uses these structures merely as a receptor, opting to handle its own genome delivery.

When on the prowl, tailed phages can use their tail and tail fibers to walk across the cell surface (*see page 2-6*). Without a tail, PRD1 can't walk, so it rolls instead. A spike protein at each of the vertices acts like a tail fiber in that it recognizes its receptor in the outer membrane (OM) of a potential host, then binds to it weakly and reversibly. Picture an icosahedron rolling along on the OM, one vertex spike after another adhering just long enough and firmly enough to keep the virion from drifting away. When a favorable roll aligns the portal vertex with a receptor, PRD1 binds irreversibly (Strömsten, Bamford, Bamford 2003). All typical phage behavior so far.

The birth of a tail tube

Although lacking the usual tail, PRD1 still needs a way to deliver its genome across a bacterial cell envelope. Its solution is to assemble an unusual internal membrane and then, during infection, restructure it into a tube. While this membrane has the typical bilayered structure, it is also studded with approximately ten different phage-encoded proteins (Rydman, Bamford 2002a) that combined comprise ~50% of the membrane volume (Peralta et al. 2013). The lipid components are a selected subset of host lipids (Laurinavičius et al. 2004) that differ in their proportions between the inner and outer leaflets, and moreover those in the outer leaflet are grouped rather than randomly distributed (Cockburn et al. 2004). Two membrane proteins (P20 and P22) link the membrane specifically to the unique vertex (Strömsten, Bamford, Bamford 2003). After the DNA is packaged it exerts an estimated pressure of 45 atm on the membrane. Due to virion geometry, the tension is greatest at the vertices, making them metastable. All the players are in place, poised for the dramatic events to come.

Irreversible adsorption by the unique vertex triggers the loss of the spike complex from that vertex and from several others nearby. In an instant, the interior of the virion is open to the environment. This sudden change in the osmolarity within the virion initiates a dramatic remodeling of the membrane vesicle. While still enclosing the DNA, the lipid pouch changes shape, likely through self-assembly of its protein components. The icosahedral sac transforms into an icosahedral sac with a tube protruding through the portal of the unique vertex. Even empty **procapsids** form equivalent tubes *in vitro*—evidence that the instructions directing this metamorphosis are contained in the assembled structure of the protein capsid and membrane.

This newly sprouted 'tail tube' is impressive. It is long enough (average 50 nm) to deliver the genome through a Gram-negative cell envelope (Peralta et al. 2013). Likewise its interior channel (4.5 nm in diameter) is ample for passage of a double-stranded DNA chain (~2.6 nm diameter). Its walls, assembled from lipids and a protein lattice, are almost 5 nm thick. The protruding conical tip of the tube is securely sealed. Thus formed, the tail tube is well equipped to safely deliver the linear DNA across the **periplasm** and into the host cytoplasm.

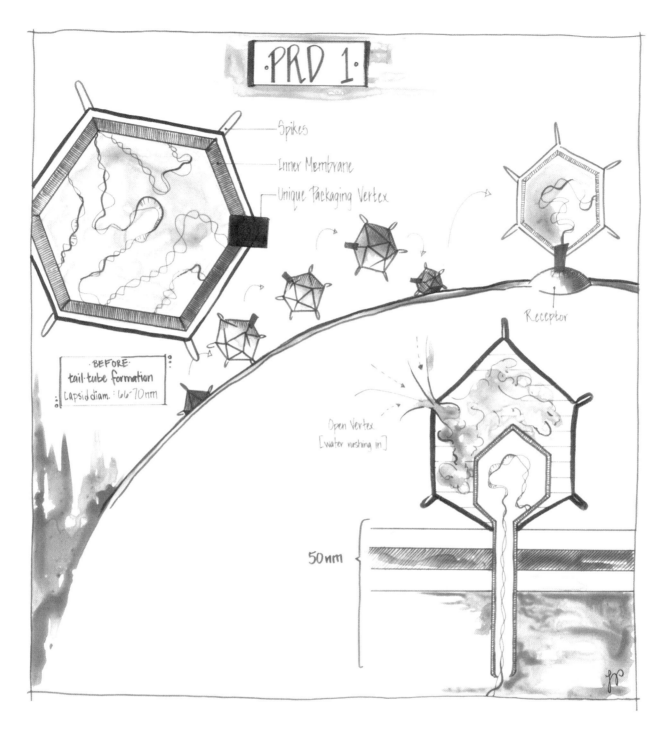

Spikes

Inner Membrane

Unique Packaging Vertex

BEFORE:
tail-tube formation
Capsid diam.: 66-70nm

Receptor

Open Vertex
[water rushing in]

50nm

Genome delivery

Forming the tube and translocating the DNA is a complex affair requiring at least five PRD1 membrane proteins (Grahn, Daugelavičius, Bamford 2002). During the first step in delivery, a strongly adhesive membrane protein in the tail tube interacts with the host's OM to allow the tube to pass through (Bamford, Mindich 1982). Next step: penetrate the **peptidoglycan** layer. The openings in a typical peptidoglycan mesh are 4-6 nm (Rydman, Bamford 2002b), but the outside diameter of the fat tail tube is 14 nm. A membrane protein (P7) with lytic transglycosylase activity steps in here and digests a path through (Rydman, Bamford 2002a). The tail tube enters the cytoplasm, the distal tip opens on cue, and the DNA passes into the cytoplasm. Although the DNA may get its initial push out of the capsid from the internal capsid pressure, the rest of the oomph is likely from osmotic forces. The membrane vesicle in the capsid, now emptied of its DNA, deflates to less than half its previous size (Peralta et al. 2013).

Intellectual property

Can only Tectiviruses deliver their genomes via a lipoprotein tail tube? One would expect that other phage families with internal membranes and linear genomes might do likewise. So far only two ICTV bacteriophage families are known to possess internal membranes, the *Tectiviridae* and the *Corticoviridae*, but only the Tectiviruses have linear genomes. The one identified Corticovirus, PM2, does not use a tail tube and groups with the P2-like Myophages in the PPT. To deliver its circular genome, it disintegrates its capsid to expose the membrane, which then fuses with the host's outer membrane (Cvirkaitė-Krupovič et al. 2010). Internal membranes are found in viruses from all domains of life, and it seems worth looking for similar tubes in archaeal and eukaryotic viruses (Peralta et al. 2013). One candidate is already on the scene: Mimivirus (*see page 6-41*).

A broader question is whether the use of lipoprotein tubes to deliver cargo into a cell is exclusively a viral trick. The formation and function of these viral tubes bring to mind the tunneling nanotubes that form from eukaryotic cellular membranes and that facilitate cell-to-cell communication for many cell types (Gurke, Barroso, Gerdes 2008). The cellular versions are grander in scale, with diameters of 50 to 200 nm and lengths up to several cell diameters. Nevertheless, like the phage version, they, too, form de novo in a few minutes and carry cargo between cells—cellular cargo that includes vesicles and even organelles. This similarity may be a case of independent use of similar materials at hand to solve similar problems, or there may have been theft by one group or the other. More detailed investigation will be required before either party can be prosecuted for violation of intellectual property rights.

Cited references

Bamford, DH, L Mindich. 1982. Structure of the lipid-containing bacteriophage PRD1: Disruption of wild-type and nonsense mutant phage particles with guanidine hydrochloride. J Virol 44:1031-1038.

Butcher, SJ, V Manole, NJ Karhu. 2012. Lipid-containing viruses: Bacteriophage PRD1 assembly. In: MG Rossmann, VB Rao, editors. *Viral Molecular Machines*: Springer. p. 365-377.

Cockburn, JJ, NG Abrescia, JM Grimes, GC Sutton, JM Diprose, JM Benevides, GJ Thomas, JK Bamford, DH Bamford, DI Stuart. 2004. Membrane structure and interactions with protein and DNA in bacteriophage PRD1. Nature 432:122-125.

Cvirkaitė-Krupovič, V, M Krupovič, R Daugelavičius, DH Bamford. 2010. Calcium ion-dependent entry of the membrane-containing bacteriophage PM2 into its *Pseudoalteromonas* host. Virology 405:120-128.

Gowen, B, JK Bamford, DH Bamford, SD Fuller. 2003. The tailless icosahedral membrane virus PRD1 localizes the proteins involved in genome packaging and injection at a unique vertex. J Virol 77:7863-7871.

Grahn, AM, R Daugelavičius, DH Bamford. 2002. Sequential model of phage PRD1 DNA delivery: Active involvement of the viral membrane. Mol Microbiol 46:1199-1209.

Gurke, S, JF Barroso, H-H Gerdes. 2008. The art of cellular communication: Tunneling nanotubes bridge the divide. Histochem Cell Biol 129:539-550.

Huiskonen, JT, V Manole, SJ Butcher. 2007. Tale of two spikes in bacteriophage PRD1. Proc Natl Acad Sci USA 104:6666-6671.

Laurinavičius, S, R Käkelä, P Somerharju, DH Bamford. 2004. Phospholipid molecular species profiles of tectiviruses infecting Gram-negative and Gram-positive hosts. Virology 322:328-336.

Peralta, B, D Gil-Carton, D Castaño-Díez, A Bertin, C Boulogne, HM Oksanen, DH Bamford, NG Abrescia. 2013. Mechanism of membranous tunnelling nanotube formation in viral genome delivery. PLoS Biol 11:e1001667.

Rydman, PS, DH Bamford. 2002a. The lytic enzyme of bacteriophage PRD1 is associated with the viral membrane. J Bacteriol 184:104-110.

Rydman, PS, DH Bamford. 2002b. Phage enzymes digest peptidoglycan to deliver DNA. ASM News 68.

Strömsten, NJ, DH Bamford, JK Bamford. 2003. The unique vertex of bacterial virus PRD1 is connected to the viral internal membrane. J Virol 77:6314-6321.

Sun, L, LN Young, X Zhang, SP Boudko, A Fokine, E Zbornik, AP Roznowski, IJ Molineux, MG Rossmann, BA Fane. 2014. Icosahedral bacteriophage ϕX174 forms a tail for DNA transport during infection. Nature 505:432-435.

Enterobacteria Phage T7

a Podophage that delivers its genome in a slow, highly-controlled fashion, thereby evading a host defense

Genome
dsDNA; linear
39,937 bp
60 predicted ORFs; 0 RNAs

Encapsidation method
Packaging; T = 7 capsid

Common host
Escherichia coli

Habitat
Mammalian intestines & sewage

Lifestyle
Lytic

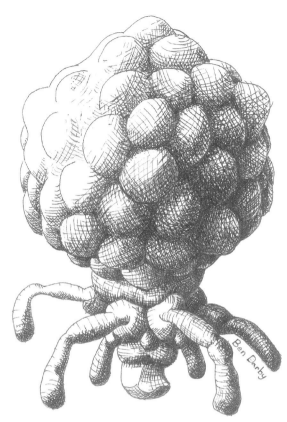

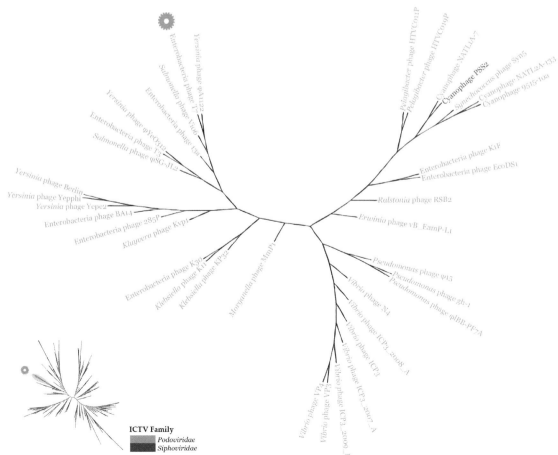

ICTV Family
Podoviridae
Siphoviridae

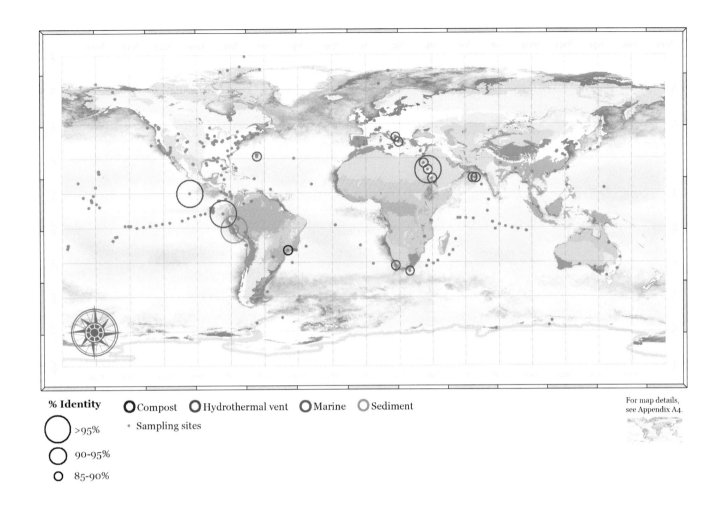

% Identity

○ Compost   ○ Hydrothermal vent   ○ Marine   ○ Sediment

○ >95%

• Sampling sites

○ 90-95%

○ 85-90%

For map details,
see Appendix A4.

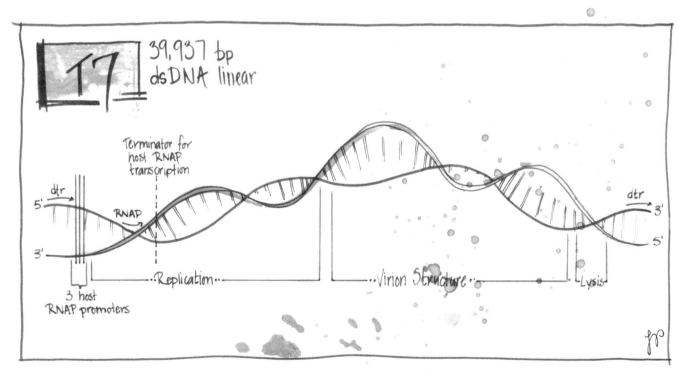

39,937 bp
dsDNA linear

Terminator for
host RNAP
transcription

dtr
5'
RNAP
3'

3 host
RNAP promoters

···Replication···

···Virion Structure···

Lysis

dtr
3'
5'

Enterobacteria Phage T7

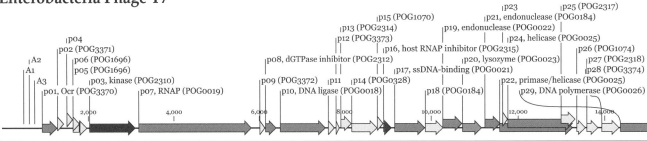

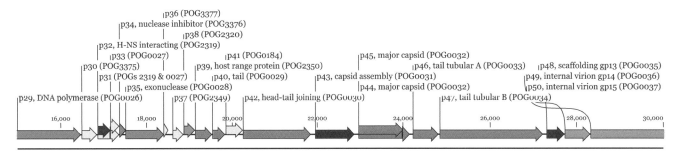

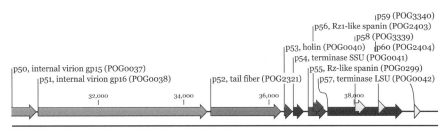

On the Prowl	Entry	Takeover	Replication Lytic	Replication Lysogenic	Assembly	Escape	Transfer RNA	Non-coding RNA	Multiple	Unknown

T7 Takes the Slow Lane

Heather Maughan & Merry Youle

Mise-en-Scène: *Triumphant genome delivery mechanisms are intimately interwoven with other aspects of the phage life cycle. For example, the success of Podophage T7's slow, three-step method depends on T7's genome organization, its tactics for regulating gene transcription, and the precise loading of capsids with proteins to be delivered along with the genome. Perhaps these requirements explain why most known members of the* Caudovirales *eschew T7's enzyme-dependent delivery in favor of the archetypal osmotically-driven process.*

Once a phage has secured a host, how quickly can it deliver its genome? Some phages travel in the fast lane. Phage λ, no laggard, transfers its 48.5 kbp genome at ~162 bp/s (Van Valen et al. 2012); T5 transfers its 121 kbp a tad faster at ~202 bp/s (Lanni 1968). Both are left in the dust by speedy T4; its 168.9 kbp enter at ~4,223 bp/s (Kalasauskaite, Grinius 1979). Some other phages take their sweet time. For example, Podophage T7 plods along in the slow lane clocking only 55-74 bp/s. At this rate, 9-12 minutes—roughly one third of its infection period—is spent delivering its 39.9 kbp genome. How does T7 set this slow pace? It orchestrates the transfer in three controlled steps, each carried out by a different molecular motor: first the gp15/gp16 transmembrane channel, then the host RNA polymerase (RNAP), and lastly T7's own RNAP.

The ratchet

Immediately upon adsorption, T7's **internal proteins** gp15 and gp16 exit the **capsid** and assemble the tail extension needed for this Podophage's stubby tail to reach the cell membrane (*see page 3-5*). Not only do these proteins construct the channel for T7's DNA delivery, but together they form the molecular motor, powered by the host's membrane potential, that starts at the 'left' end of the genome and ratchets in the first ~850 bp (Kemp, Garcia, Molineux 2005; Chang, Kemp, Molineux 2010). One motor protein, gp16, then clamps down to prevent further DNA translocation by this mechanism. Since the ratchet operates at 70 bp/s, this step requires about 12 s. In the process, it enables transcription of early phage genes and sets the stage for the next step.

Shifting gears

The remaining ~39 kbp (98%) of the T7 genome is pulled in by the über-powerful molecular motor of transcribing RNA polymerase (RNAP) (Wang et al. 1998). The first RNAP on the scene is the relatively sluggish host enzyme that T7 cons into assisting with its lethal invasion. T7 provides three promoters on that initial 850 bp that are recognized by the host's RNAP for transcription of the first phage genes. This transcription forcibly overcomes gp16's clamp and draws T7's DNA into the cell at 40-50 bp/s (Garcia, Molineux 1995; Molineux 2001). RNAP works its way along the T7 genome transcribing the early phage genes in the first ~7 kbp until halted by a transcription terminator. One of the genes transcribed encodes T7's faster RNAP, which is promptly synthesized. The newly-made phage RNAP then pulls in the remaining 32 kbp at a comparatively brisk 250 bp/s.

The slow lane to survival

Why amble in so slowly? Slow is one way to dodge a host defense, specifically *E. coli*'s **restriction endonucleases** (REs; *see page 4-10*) that recognize and cleave invading DNA. While T7's DNA enters the cytoplasm, it mysteriously eludes patrolling REs for approximately six minutes—roughly half the duration of genome delivery (Moffatt, Studier 1988). This buys T7 time to muster a permanent defense by synthesizing Ocr (<u>O</u>vercome <u>c</u>lassical <u>r</u>estriction) that inhibit REs for the duration of the infection (*see page 4-17*; [Roberts et al. 2012]).

Opting for life in the slow lane is not a bad idea when the fast lane leads to a dead end.

Cited references

Chang, CY, P Kemp, IJ Molineux. 2010. Gp15 and gp16 cooperate in translocating bacteriophage T7 DNA into the infected cell. Virology 398:176-186.

Garcia, LR, IJ Molineux. 1995. Rate of translocation of bacteriophage T7 DNA across the membranes of *Escherichia coli*. J Bacteriol 177:4066-4076.

Kalasauskaite, E, L Grinius. 1979. The role of energy-yielding ATPase and respiratory chain at early stages of bacteriophage T4 infection. FEBS Lett 99:287-291.

Kemp, P, LR Garcia, IJ Molineux. 2005. Changes in bacteriophage T7 virion structure at the initiation of infection. Virology 340:307-317.

Lanni, YT. 1968. First-step-transfer deoxyribonucleic acid of bacteriophage T5. Bacteriol Rev 32:227-242.

Moffatt, BA, FW Studier. 1988. Entry of bacteriophage T7 DNA into the cell and escape from host restriction. J Bacteriol 170:2095-2105.

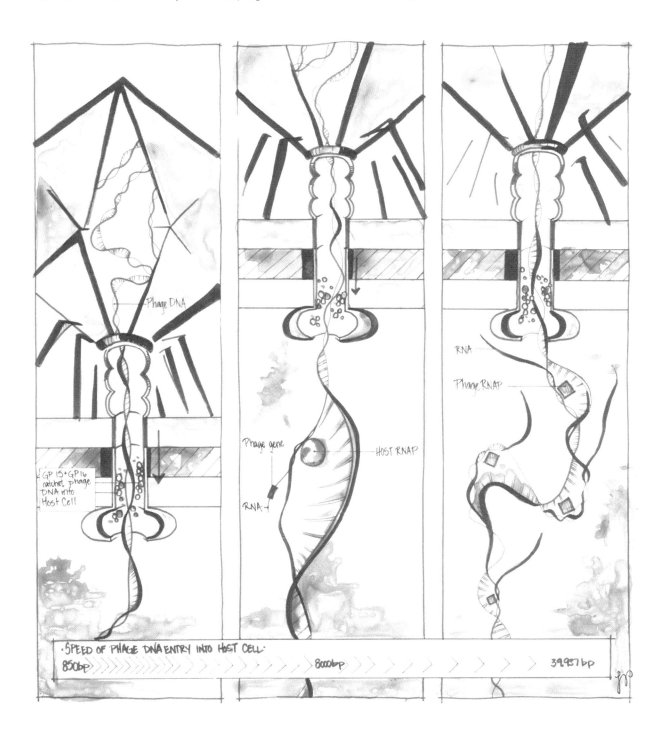

Molineux, IJ. 2001. No syringes please, ejection of phage T7 DNA from the virion is enzyme driven. Mol Microbiol 40:1-8.

Roberts, GA, AS Stephanou, N Kanwar, A Dawson, LP Cooper, K Chen, M Nutley, A Cooper, GW Blakely, DT Dryden. 2012. Exploring the DNA mimicry of the Ocr protein of phage T7. Nucleic Acids Res 40:8129-8143.

Van Valen, D, D Wu, YJ Chen, H Tuson, P Wiggins, R Phillips. 2012. A single-molecule Hershey-Chase experiment. Curr Biol 22:1339-1343.

Wang, MD, MJ Schnitzer, H Yin, R Landick, J Gelles, SM Block. 1998. Force and velocity measured for single molecules of RNA polymerase. Science 282:902-907.

Recommended review

Casjens, S.R. and I.J. Molineux. 2011. Short noncontractile tail machines: Adsorption and DNA delivery by Podoviruses. In: MG Rossmann, VB Rao, editors. *Viral Molecular Machines*: Springer. p. 143-179.

Pseudomonas Phage φ6

a Cystophage whose dsRNA genome remains inside a capsid throughout the phage's life cycle

Genome
dsRNA; linear; segmented
Segment L: 6,374 bp, 4 ORFs, 0 RNAs
Segment M: 4,063 bp, 4 ORFs, 0 RNAs
Segment S: 2,948 bp, 4 ORFs, 0 RNAs

Encapsidation method
Packaging; T = 13 & T = 2 capsids

Common host
Pseudomonas savastanoi pv. phaseolicola

Habitat
Host-associated; plant pathogens; plant leaf

Lifestyle
Lytic

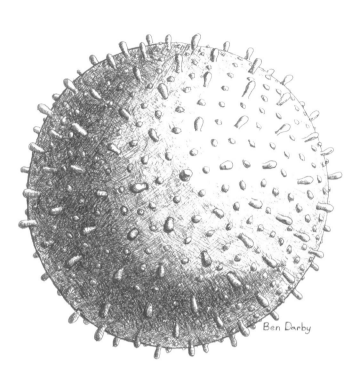

Ben Darby

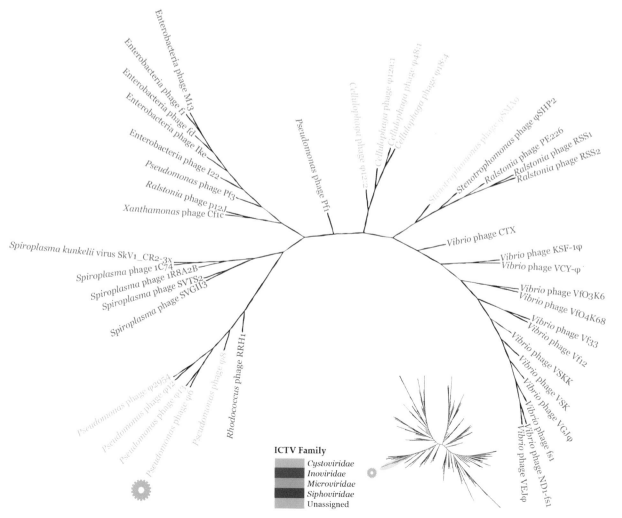

ICTV Family
Cystoviridae
Inoviridae
Microviridae
Siphoviridae
Unassigned

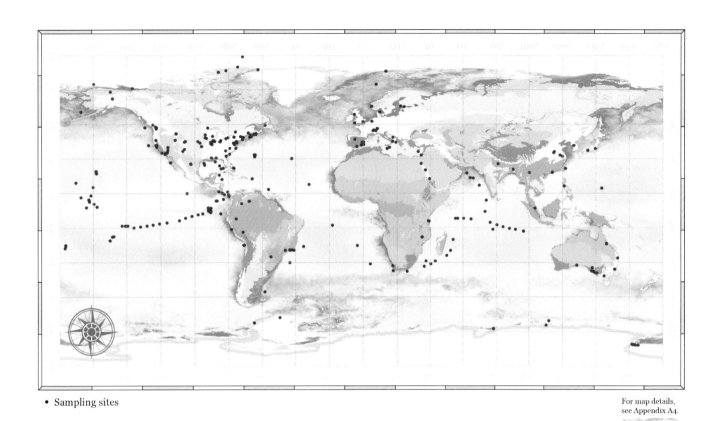

- Sampling sites

For map details,
see Appendix A4.

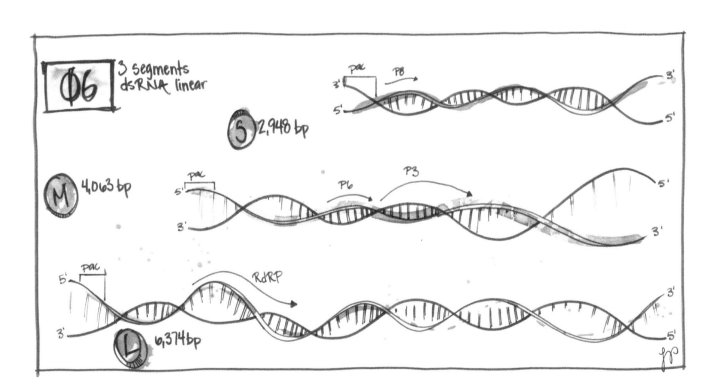

Pseudomonas Phage φ6

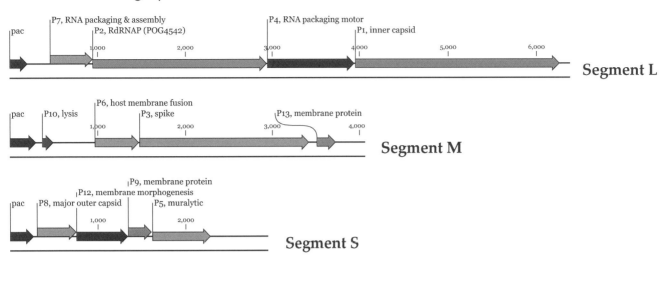

φ6 Refuses to Leave Its Capsid at the Door

Merry Youle

Mise-en-Scène: *Eukaryotic viruses often gain entry to a host cell through the cell's normal endocytic pathway. Even the giant Mimivirus successfully exploits this mechanism (see page 6-41). This route in is not generally an option for bacteriophages since Bacteria, with their cell wall, don't know how to endocytose. However, Cystophages, with their external lipid envelope, demonstrate that a bacterium can genetically adapt to take up a phage capsid by a process that looks very much like endocytosis.*

Hershey and Chase, in their classic 1952 publication, demonstrated that, during infection by phage T2 or T4, only the phage DNA genome enters the host cell; the protein capsid remains outside (Hershey, Chase 1952). Apparently phage φ6 missed that paper. Its oversight can be excused given that it speaks a different language, using double-stranded RNA (dsRNA), not DNA, for its genome. If a phage were foolish enough to deliver a naked dsRNA molecule, it would soon be dead, its genome cleaved by host ribonucleases. φ6 cleverly gift-wraps its dsRNA within a protein capsid and then dupes its host into receiving the package, capsid and all.

Arriving at the gate

The hoodwinked host in question is a Gram-negative plant pathogen, *Pseudomonas savastanoi* pv. phaseolicola. Extending 1-4 μm out from the *P. savastanoi* cell **envelope** are numerous **pili**—an inviting target for a passing phage. φ6 accepts the invitation. It uses the protein spikes (protein P3) that protrude from its capsid to adsorb to the side of a **pilus**. So far, so good, but now φ6 needs to migrate to the surface of the cell. This process is not straightforward because surrounding the cell's outer membrane (OM) is an obstructing 150 nm thick layer of exopolysaccharide (EPS). To get past this obstacle, φ6 exploits the host's behavior.

As part of its usual pathogenic activities, *P. savastanoi* alternately extends and retracts its pili. While the tip of the pilus is anchored to the surface of a leaf, retraction pulls the cell forward ('**twitching motility**'). Retraction also brings the attached phage through the EPS to the OM (Romantschuk, Bamford 1985). First obstacle cleared.

Commingling with the outer membrane

From this point on, φ6's entry is a striptease act. One by one, it sheds layers from its virion until ultimately its naked core particle reaches the cytoplasm. First to be jettisoned is the no-longer-needed spike protein P3. What remains is the complex icosahedral virion enveloped by a lipid layer. Although outer lipid layers are *de rigueur* for animal viruses, among the phages they are seen only in the Cystophages where they are acquired during virion assembly in the cytoplasm. The lipids for the membrane are pillaged from the host; all of the embedded membrane proteins that are essential for infection the phage encodes itself.

When the viral **envelope** contacts the OM, membrane protein P6 mediates their fusion; the virion envelope thus becomes part of the host's OM (Bamford, Romantschuk, Somerharju 1987). Stripped of its lipid cloak, the original 85 nm diameter virion is now a 56-58 nm nucleocapsid (NC) in the **periplasm**. Second barrier overcome.

Through the jungle

To reach the cell membrane, the sleek phage NC must traverse the periplasm, a jungle of **peptidoglycan** infested with nucleases. φ6 negotiates this obstacle course without help from its host. Exposed on the outer surface of the NC is an endopeptidase (P5) that digests a path through the peptidoglycan, enabling the NC to continue on its way (Mindich, Lehman 1979). Third obstruction cleared.

Internalization

Now the NC is face to face with the host's cell membrane. At this point, the φ6 NC has both an inner and an outer capsid surrounding its genome.

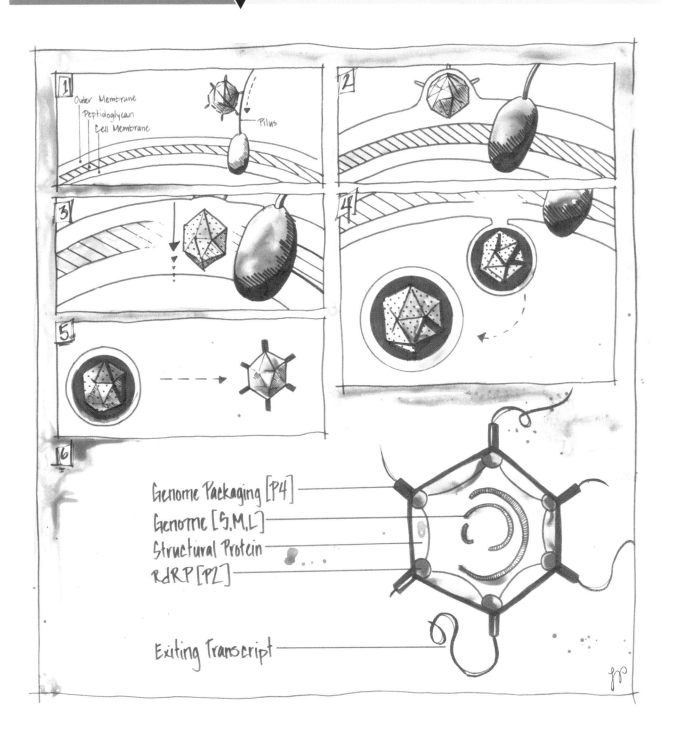

The outer layer is a simple lattice composed of a single protein, P8, the key player in the next step. P8 prompts a region of the cell membrane nearby to invaginate and form a deep pit surrounding the NC. The neck of the pit constricts to bud off a NC-containing vesicle from the cell membrane. This endocytosis-like process requires the normal membrane potential of an energetically-active host cell (Poranen et al. 1999). Fourth obstacle surmounted.

Let me out!

Our intrepid NC has made it past the cell membrane and into the host, but it is trapped inside a membrane vesicle. How it escapes is not known,

▼

but escape it does. In the process, it sheds its outer capsid and the intrinsic P8 proteins are degraded (Romantschuk, Olkkonen, Bamford 1988). All that remains now is a core particle—the inner capsid surrounding the genome. The final barrier has been conquered.

Replicating in private

The core particle (47 nm diameter) that arrives in the cytoplasm is a well equipped replication machine. Inside are the three molecules of dsRNA that comprise the φ6 genome, identified as the S, M, and L segments. Surrounding this is the dodecahedral inner capsid composed of structural protein P1. Embedded in the interior face of each of its twelve vertices is a single molecule of P2, the RNA-dependent RNA polymerase (RdRP). Using the genome segments as templates, these RdRPs transcribe **positive-sense RNAs** that exit from the capsid into the cell cytoplasm. There, some RNAs function as mRNA for translation of phage proteins by the host's machinery; others are packaged into progeny virions (*see page 6-25*) where multitalented RdRP adds the complementary **negative-sense** strands. Thus φ6 progresses through its entire life cycle without ever exposing its dsRNA genome to the host cytoplasm. φ6 can't run, but it can hide.

Cited references

Bamford, D, M Romantschuk, P Somerharju. 1987. Membrane fusion in prokaryotes: Bacteriophage φ6 membrane fuses with the *Pseudomonas syringae* outer membrane. EMBO J 6:1467-1473.

Hershey, AD, M Chase. 1952. Independent functions of viral protein and nucleic acid in growth of bacteriophage. J Gen Physiol 36:39-56.

Mindich, L, J Lehman. 1979. Cell wall lysin as a component of the bacteriophage φ6 virion. J Virol 30:489-496.

Poranen, MM, R Daugelavičius, PM Ojala, MW Hess, DH Bamford. 1999. A novel virus–host cell membrane interaction: Membrane voltage–dependent endocytic-like entry of bacteriophage φ6 nucleocapsid. J Cell Biol 147:671-682.

Romantschuk, M, DH Bamford. 1985. Function of pili in bacteriophage φ6 penetration. J Gen Virol 66:2461-2469.

Romantschuk, M, VM Olkkonen, DH Bamford. 1988. The nucleocapsid of bacteriophage φ6 penetrates the host cytoplasmic membrane. EMBO J 7:1821-1829.

Recommended review

Poranen, MM, R Daugelavicius, DH Bamford. 2002. Common principles in viral entry. Annu Rev Microbiol 56:521-538.

Voyages with Chloroviruses

James L. Van Etten[†] & David D. Dunigan[†]

Abstract: *The chloroviruses were discovered because of a casual conversation during a faculty party that led to a few simple experiments, and these led to a set of more complex and hypothesis-driven experiments. But the important breakthrough came when it was clear that the chloroviruses could be handled in much the same way as bacteriophages—they could be plaque-assayed with cultured zoochlorella—and that made all the difference. This chapter provides a personal account of some of the history and explorations with the chloroviruses including how they were discovered and some of their unusual properties, particularly those related to their large size. Looking back some 35 years, the discovery and characterization of the chloroviruses laid the groundwork for the finding of an ever-growing number of viruses generally referred to as 'giant viruses' or giruses, many of which replicate in protists. We suspect many more giants will be encountered as biologists continue to investigate lesser-known organisms in the Tree of Life. But, the complete story of the chloroviruses is certainly not yet written and many important chapters lie before us, including their role in aquatic systems, their diversity and distribution, and their evolutionary history.*

[†] Department of Plant Pathology and Nebraska Center for Virology, University of Nebraska – Lincoln, Lincoln, NE
Email: jvanetten1@unl.edu, ddunigan@unl.edu
Website: http://ncv.unl.edu/vanettenlab

This story of chlorovirus discovery began at a party that I[1] hosted in 1979. I was telling a colleague, Russ Meints, about an unusual dsRNA bacteriophage that I was working on (φ6, discovered by my colleague Anne Vidaver), when Russ mentioned a symbiotic alga he was working with might have a virus. Russ was studying symbiosis between the coelenterate *Hydra viridis* and a chlorella-like green alga (Fig. 1A). Specifically, he was interested in how the hydra recognized the symbiotic chlorellae (also called zoochlorellae) and accepted them as endosymbionts, distinct from free-living algae that it would take up and eat. In the course of his experiments, Russ and his colleague Kit Lee used electron microscopy to examine zoochlorellae that had recently been separated from the hydra, Russ at that time being unable to culture them independent of the hydra. At some point a visiting seminar speaker, Malcolm Brown[2], looked at the micrographs and mentioned to Russ that one alga looked as though it was infected with a virus. This

is the comment Russ was referring to when we were talking at the party. A couple of days later, I visited Russ's laboratory, also at the University of Nebraska-Lincoln (UNL) but on a separate campus, and we examined his 50 or more micrographs. Indeed, one alga contained a few icosahedral particles that looked like viruses (Fig. 1B). Russ had written on the back of the picture that this particular alga had been isolated from *Hydra* and then allowed to sit for a few hours before the sample was processed. No virus particles were seen in any of the other algal samples, including those zoochlorellae visualized while still within the hydra. Russ again commented that he was unable to culture the zoochlorellae. Therefore, we predicted that there might be a connection between the presence of the virus particles and the inability to grow the alga in culture, which prompted us to conduct a very simple experiment. We isolated zoochlorellae from the hydra, let them sit on the bench for up to 24 hours and then processed them for microscopy. By 24 hours, all of the algal cells showed signs of viral infection and the cells were dying (Meints et al. 1981). Clearly the viruses had something to do with the failed attempts to culture the algae.

Why was I interested in Russ's comment about an algal virus? In addition to my collaborative re-

---

[1] JLVE has studied the chloroviruses for over 35 years and DDD has worked with them for the past 12 years. In the text the first person comments refer to JLVE.

[2] Malcolm Brown was co-author of the first review that included the subject of algal viruses (Sherman and Brown, 1978). It focused primarily on cyanophages but contained several electron micrographs of virus-like particles that had been reported in a few algae.

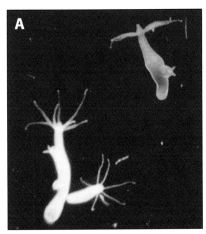 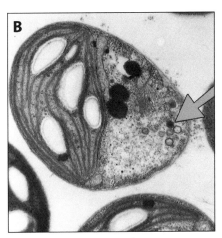 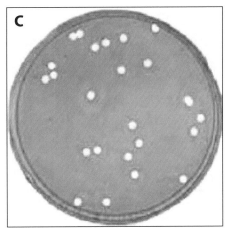

Figure 1. Chlorella cells and chloroviruses. (A) *Hydra viridis*: the green one contains symbiotic chlorella-like green algae that the white one lacks. (B) Electron micrograph of a chlorella cell taken 6 h after the alga was isolated from its symbiotic hydra host. Note the virus particles (arrow). (C) Plaques formed by virus PBCV-1 on a lawn of *Chlorella variabilis*. Sources: Fig. 1A from Russ Meints, with permission; Fig. 1B by Kit Lee (reprinted from Meints et al. 1981 with permission from Elsevier); Fig. 1C (reprinted from Van Etten, Dunigan 2012 with permission from Elsevier).

search project on phage φ6 with Anne Vidaver, my primary research effort at UNL for the previous 13 years (started in 1966) had focused on the physiology and molecular biology of fungal spore germination. During this same time period, considerable interest had developed in other labs concerning the discovery of dsRNA viruses in fungi (they were referred to as virus-like particles for many years because they were non-infectious). I was aware of this fungal virus research because many of these labs had requested φ6 dsRNA from us because its three segments (*see page 3-20*) were useful as molecular weight markers for the fungal virus studies. When Russ made his comment about having a possible algal virus, I was instantly interested for two reasons. First, I knew that very little was known about algal viruses; second, I assumed green algae were similar enough to fungi that, if they had viruses, the viruses would probably have dsRNA genomes. I was correct on the first point but completely wrong on the second, as subsequent experiments quickly established that the chloroviruses have very large dsDNA genomes.

Our discovery and characterization of the chloroviruses in the early 1980s led to a few research papers including one in *PNAS* (Van Etten et al. 1982). However, at the time the only way we could produce these viruses was to grow the host (either *Hydra viridis* or *Paramecium bursaria*) with their sym-

biotic algae, separate the zoochlorellae, and then wait about 24 h for the algae to lyse. We then isolated the released chloroviruses by conventional procedures and characterized them, which led to the discovery that they contained a large dsDNA genome and an internal lipid membrane. Since there was no infection system, I was close to terminating the project after the first few manuscripts. However, there were reports in the literature that some zoochlorellae could be cultured independent of their symbiotic host. The typical method to isolate these algae was to squash a hydra or a paramecium and then streak the released contents on culture plates. Occasionally, a green colony would appear two or three weeks later, along with many other microorganisms. Since none of these experiments were conducted with sterile hydra or paramecia, the green colonies could also be from contaminating surface algae. Anyhow, we obtained some of these algal cultures and discarded those that were badly contaminated with bacteria. By this time we had obtained four viruses from zoochlorellae, each with unique DNA restriction patterns, including three from different hydra isolates and one from a paramecium. We then tried to infect the most promising algal cultures with these four viruses. We were able to infect a *Chlorella* (originally named NC64A, now *Chlorella variabilis*) that had originally been isolated from a paramecium with our virus from the paramecium isolate.

This important discovery allowed us to grow the virus in culture independent of the paramecium[3].

Moving forward, we tried to plaque assay the virus on lawns of *C. varibilis*. I will always remember coming to work the next day and tilting the plates to take a premature look. I could barely see what looked like plaques!! (Fig. 1C)[4]. I was so excited that I made everyone in the building look at the plates—even people who did not know a plaque from a frog. This finding led to a nice paper in *Science* (Van Etten et al. 1983). Once we could plaque assay the virus and synchronously infect the cells, I decided to phase out my other two research projects, both of which were NIH supported, and spend the rest of my career studying the chloroviruses. This transition occurred over a two-year period as I had graduate students and postdocs working on the other projects. In fact, one of the graduate students, Geraldine Russo, was my last graduate student to work on fungi. Geraldine and I have been happily married for over 30 years.

Another discovery at about this time also contributed to my change in research focus. We asked: can we find plaque-forming chloroviruses in nature? To address this, we collected indigenous water from several sources and stored the samples in the cold room. A few weeks later, we filtered some of the water and plaque-assayed for viruses on *C. variabilis*. Well-defined plaques formed on most plates a couple of days later; however, my original thought was that we had somehow contaminated the samples. Fortunately, my technician, Dwight Burbank[5], looked at one of them and commented

that this water sample looked like it formed many tiny plaques, which I had missed. The plaque variants we had observed up to that point were relatively large; therefore, it was unlikely that these tiny plaques were due to contamination and that plaque-forming viruses on *C. variabilis* (referred to as NC64A viruses) probably existed in indigenous waters (Van Etten et al. 1985). In fact, 30 years later, we have four chlorovirus/host combinations that we work with and we know that chloroviruses are ubiquitous in nature with titers as high as 100,000 plaque-forming units (PFU) per ml of water. Typically the values are about 1 to 100 PFU per ml. A recent detailed review of the chloroviruses can be found at (Van Etten, Dunigan 2012)[6].

Restriction modification systems, but not where they should be

What happens when a graduate student doesn't tell his advisor about experiments that he plans to conduct? Well, sometimes he discovers something new and unexpected. Yuannan Xia, fresh from China and one of the first Chinese students to attend graduate school in the USA in the early 1980s, was my first graduate student to work on the chloroviruses. Shortly before his arrival at UNL, we had discovered that during infection by PBCV-1 (*Paramecium bursaria chlorella virus*), the genomic DNA of *C. variabilis* was degraded at about the same time that the viral DNA was beginning to replicate (Van Etten et al. 1984). How could these disparate events occur in the same cell? By that time we had evidence that the genomic DNAs of the numerous chloroviruses present in nature probably contained some methylated nucleotides,

---

[3] In our first papers on the hydra zoochlorellae viruses, we suggested that the viruses might exist in a lysogenic phase that was converted to a lytic phase when the algae were separated from the hydra. We have never gone back to investigate this possibility. However, we subsequently established that the viruses that appeared after separating zoochlorella from paramecia were clearly due to external infections by viruses that were present in the culture used to grow the paramecia.

[4] To get a nice green lawn of *C. variabilis* with plaques takes about 3 - 4 days (Fig. 1C).

[5] Many students, postdoctoral associates and senior scientists have participated in this research, but two research technicians, Dwight Burbank and James Gurnon, were in it for the long haul.

[6] JLVE has to acknowledge the many contributions of his former colleague Myron Brakke to both the φ6 and chlorovirus research. Myron was a well-respected USDA plant virologist stationed at UNL and was the person who developed sucrose density gradient centrifugation in the late 1940s and early 1950s. It was Myron who recognized how unusual φ6 was and when Russ and I showed him the first micrograph of the suspected chlorovirus, he said that the virus was huge, had to have a dsDNA genome and might have a lipid membrane. Myron would never let us include his name on any of the φ6 or chlorovirus papers because it was not in his USDA job description. He should have been a co-author on all of our φ6 papers and also all of the early chlorovirus publications.

▼

including both 5-methylcytosines and 6-N-meth- yladenines, because the viral genomes had different susceptibilities to restriction endonucleases (REs). This suggested that the chloroviruses might encode DNA methyltransferases (MTs).

Taking this one step farther, Yuannan suspected that the viruses might also encode a DNA RE that would selectively cleave the host genomic DNA, while the replicating viral DNA would be protected by being methylated in the RE recognition sequences. This is precisely what he discovered for the prototype chlorovirus PBCV-1, but he only showed the results to me after he had repeated the experiments several times (Xia et al. 1986)[7]. In the course of his PhD research, Yuannan discovered several more virus-encoded REs. This was at the time when a new realm of molecular biology was unfolding and REs were essential to manipulating DNA. Some of his REs have useful unique cleavage specificities and are still sold today by New England Biolabs. Meanwhile, Yuannan went on to have a successful career in biotechnology here in the USA and has recently retired.

Bacteria encode RE/MT systems to suppress virus infection (*see page 4-10*). Each RE is always associated with a cognate DNA MT that methylates the cellular DNA thereby rendering it resistant to the RE, while any foreign DNA that is not appropriately methylated is degraded by the RE. However, the algal virus system did not fit the canonical RE/MT model; the shoe was on the wrong foot, so to speak. Here it is the virus that encodes the MT/RE system and the process occurs in a eukaryotic cell.

To sequence or not to sequence, that was the question

As mentioned near the end of this chapter, the chloroviruses are believed to share an evolutionary ancestor with several other large dsDNA viruses including African Swine Fever Virus (ASFV; *see page 6-34*). ASFV, which is spread by a tick (*Or-*

nithodoros species), causes a lethal disease in domestic pigs and is quarantined in the USA. In the 1990s the Plum Island Animal Disease Center in New York State, a facility controlled by the U.S. Department of Agriculture, was the only place in the USA that was allowed to work on ASFV. The ASFV research effort was directed by Dan Rock, a former UNL faculty member and a friend. Dan was able to buy one of the early DNA sequencers and the instrument was installed at the University of Connecticut, where he had a courtesy appointment. He then used the instrument to sequence the ~185 kb genome of ASFV, an effort that took a couple of years at the time but now would only require a few hours. In about 1994, Dan called me and said that they had become quite proficient at DNA sequencing and wanted to know if we would be interested in having a portion of the PBCV-1 genome sequenced. Fortunately, Yu Li, a graduate student, had cloned most of the PBCV-1 genome into non-overlapping cosmids, so I immediately said "yes." The USDA/UConn group spent six months or so sequencing a 45 kb cosmid fragment from the left end of the PBCV-1 genome. The sequence was annotated at Plum Island by Gerald Kudish, who also had previously spent time at UNL, and a paper was published in *Virology* (Lu et al. 1995). Dan then agreed to sequence another 40 kb cosmid. After completing the sequencing and annotation of the second cosmid, he wanted to end the collaboration because there seemed to be no chlorovirus genes related to ASFV, a conclusion that proved to be wrong. However, I offered to pay the two technicians located at UConn from my NIH-funded grant if they would continue to sequence the PBCV-1 genome, which they did with Dan's permission. Because the cosmids were not overlapping, we did considerable manual sequencing of the genome concurrently at UNL, this being the work of graduate students Yanping Zhang, Quideng Que, Yu Li, and Susan Ropp. The result was that over a 30-month period we sequenced and annotated the 331 kb PBCV-1 genome and published the results in five *Virology* papers. Using 65 codons as the minimum required for a protein, we predicted that PBCV-1 encoded 377 proteins, 32% of which resembled something in

---

[7] We now know that host DNA degradation occurs almost immediately after virus infection (Agarkova et al., 2006) and that the REs, but not the MTs, are packaged in the PBCV-1 virion (Dunigan et al., 2012).

the databases. The virus was subsequently found to also encode 11 tRNA genes. At the time, PBCV-1 was the largest virus genome to be sequenced, superseding the 180 kb genome of vaccinia virus.

Our decision to take a non-hypothesis-driven approach and sequence the PBCV-1 genome was certainly questioned by some colleagues, including one Nobel Laureate who said, "better you than me." Even people in the lab thought it was a bit crazy. However, in retrospect, sequencing the PBCV-1 genome and *making the data publically available* opened up many new opportunities including the discovery of many virus-encoded proteins that had not been found previously in a virus genome. Some of these interesting genes are the subject of two more stories below. The PBCV-1 genome also encoded three types of introns making it the first virus known to encode more than one.

To put the PBCV-1 sequencing effort into perspective, we recently published the sequences of 40 more chloroviruses and included a comparative annotation of 41 chloroviruses in the same manuscript (Jeanniard et al. 2013). Collectively, the 41 viruses encode members of 632 protein families. Since any one virus encodes at most 400 proteins, the chloroviruses have tremendous genetic diversity. In recent years we have used that diversity to discover several aspects of chlorovirus biology.

What are the evolutionary roots of ion channels and biochemical minimalism?

Among the many new and unexpected genes found in the sequenced PBCV-1 genome was an ORF that contained a potassium ion (K^+) channel signature motif. Our now approximately 15-year effort to study the chlorovirus-encoded K^+ channel protein also began with a chance encounter. In about 1985 a German visitor, Werner Reisser, spent a few weeks in my lab to study the chloroviruses. Werner was interested in the symbiotic relationship between zoochlorella and the protist *Paramecium bursaria* and he had isolated a zoochlorella, called *Chlorella* Pbi, from a *P. bursaria* isolated in Germany. He was very disappointed to discover that the chloroviruses that infected *C. variabilis*

did not infect his *Chlorella* Pbi. Fortunately, when Werner returned to Göttingen University, he collected water from some of the local ponds and successfully plaque-assayed for virus on *Chlorella* Pbi (now named *Micractinium conductrix*) (Reisser, Becker, Klein 1986). This was interesting because, up to that time, our attempts to find viruses from European waters that infected *C. variabilis* had been unsuccessful. Werner's experiment led to the discovery of a second group of related chloroviruses, now referred to as Pbi viruses.

The story picks up again about 12 years later in Bergen, Norway, at the first meeting ever held on viruses that infect algae. About 25 scientists attended the meeting that was hosted by Gunnar Bratbak and his colleagues at the University of Bergen. The day before the meeting was to begin, the Norwegian air traffic controllers went on strike so all flights to Norway and within Norway were canceled and the start of the meeting was delayed a day. The participants at the meeting, including Russ Meints and myself, used all methods of transportation to get to Bergen; we took an $800 cab ride from Oslo to Bergen (two other people from Bergen shared the expenses). My plane from the States to Oslo was diverted to Sweden and a bus took us to Oslo. Russ's plane, coming from Oregon, was also diverted to Sweden and the airline bussed him to Oslo. Our meeting at the train station on a Saturday night in Oslo was totally by chance. We could not get a rental car and all of the trains were booked for the next couple of days—hence the cab ride.

At the meeting, a graduate student from Göttingen University, Barbara Plugge, presented results showing that the major capsid protein (MCP) of a Pbi virus was similar to the MCP in PBCV-1. Of course, this was very exciting to us. When asked what she planned to do next, she said that her advisor had moved to industry and that she was going to have to quit working on the chloroviruses because she had been assigned to a new lab that studied ion channels. My next comment was that our sequencing effort indicated that PBCV-1 might encode a K^+ channel protein. After subsequent dis-

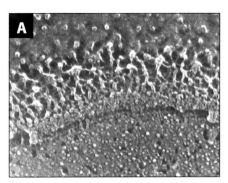 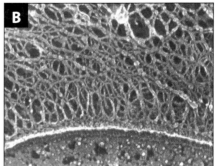 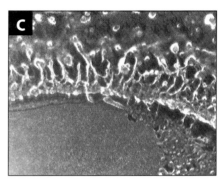

Figure 2. Cell wall surface of (A) uninfected *C. variabilis*, (B) 4 h after PBCV-1 infection, and (C) 4 h after PBCV-1 infection followed by incubation with hyaluronan lyase. This enzymatic removal of the hairy material indicates that it is composed of hyaluronan. Micrographs by John Heuser (Graves et al. 1999) and published with permission.

cussions, she received approval from her new advisor Gerhard Thiel (then at Göttingen, Germany, now at Darmstadt) to attempt to express the virus channel protein in *Xenopus* oocytes. In short order they demonstrated that the protein (now referred to as Kcv) formed a functional K⁺ channel in *Xenopus* oocytes and the findings were published in *Science* (Plugge et al. 2000). About the time the paper appeared, I was going to Europe for a meeting and contacted Gerhard about visiting his lab as well as the lab of Anna Moroni at the University of Milano, whom I noticed was conducting some of the ion channel experiments. Gerhard replied, "Just go to Italy. The food is better and Anna is my fiancée." I was strongly in favor of this suggestion because I had spent a wonderful year as an NSF postdoc working with Orio Ciferri at the University of Pavia, just south of Milano, in 1965-66. Accordingly, the three of us met in Italy. That meeting led to a 15 year working relationship and friendship with Gerhard and Anna, both of whom are card-carrying electrophysiologists.

The potassium ion channel protein encoded by PBCV-1 received considerable attention for a couple of reasons. With only 94 amino acids, it was the smallest protein known to form a K⁺ channel (the channel is a homotetramer) as well as being the first one found coded by a virus. At last count there were over 50 manuscripts published on Kcv, most of them trying to understand how Kcv in particular, and K⁺ channels in general, are regulated since they are very important in almost all aspects of cell physiology. Many of these studies lie outside the

expertise of our lab, so we, in collaboration with the two European labs, have focused our efforts on three other aspects of Kcv. (1) What is the biological function of the virus-encoded Kcv channel? We have evidence that the channel is located in the virus internal membrane and that K⁺ is released with the fusion of the viral membrane with the host plasma membrane during virus infection (see below). (2) What are the essential features of ion channels? We discovered over ten years ago that most of the NC64A chloroviruses encode 94 amino acid Kcv homologs. However, the amino acid sequences of some of these channels differ slightly and their physiological properties also differ, e.g., some K⁺ channels are blocked by Cs⁺ and some are not. Since only a few amino acids differ between channels with these two properties, site-directed mutagenesis can be used to determine which amino acids are critical for a particular channel property. This is a tremendous resource for studying the physiological properties of ion channels because nature has already selected for mutations that make functional channels (Gazzarrini et al. 2004; Kang et al. 2004). (3) What is the evolutionary origin of K⁺ channels? Evolutionary biologists have suggested that K⁺ channel proteins originated in bacteria, but, for several reasons, we have proposed that their evolutionary source was an ancestor of the chloroviruses (Thiel et al. 2013).

Chloroviruses must like sweets

Another surprise that resulted from sequencing the PBCV-1 genome was the discovery of several enzymes involved in sugar manipulations—most

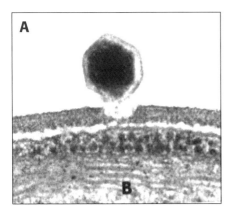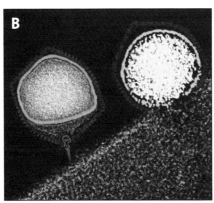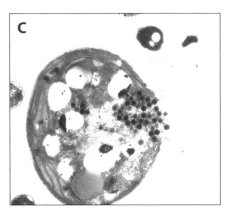

Figure 3. (A) Attachment of PBCV-1 to the algal wall and digestion of the wall at the point of attachment 1–3 min after addition of the virus. (B) Cryo-electron microscopy reconstructions of PBCV-1. (left) the initial phase of attachment via a spike structure (purple) at the unique vertex. (right) After binding and digestion of the wall, the particle begins to uncoat and release its DNA and associated proteins into the host cell.(C) Release of progeny PBCV-1 viruses 6–8 h post infection by host lysis resulting from localized rupture of the *C. variabilis* cell membrane and cell wall. Sources: Fig. 3A by Kit Lee (reprinted from Meints et al. 1984 with permission from Elsevier); Fig. 3B by Xinzheng Zhang and Michael Rossmann, © American Society for Microbiology; published on the cover *Journal of Virology* 108 (17) in 2011, and reprinted with permission; Fig. 3C by Kit Lee (reprinted from Meints et al. 1981 with permission from Elsevier).

unusual for viruses (Van Etten et al. 2010). When we sequenced the PBCV-1 genome, we had deposited the sequences in the databases as the papers were published[8]. In about 1996, I received a phone call from Paul DeAngelis who was at the University of Oklahoma Medical Center. Paul had just cloned hyaluronan synthetase from a vertebrate and when he compared it to sequences in the databases he obtained a strong hit with a gene from a chlorovirus. This was quite a surprise because hyaluronan had only been found in vertebrates and as an external capsule on a few pathogenic bacteria, for the latter presumably as a way to avoid their host's immune defenses. I was very interested in all that Paul had to say, but the best question I could think of was "what is hyaluronan?" Hyaluronan (also referred to as hyaluronic acid) is an extracellular matrix polysaccharide that consists of ~20,000 alternating residues of the sugars N-acetylglucoseamine and glucuronic acid. Three PBCV-1 encoded enzymes function in its synthesis: two are involved in synthesizing the two sugar moieties and the third, hyaluronan synthetase, makes the final product. When we sent the viral gene to Paul, he established that the protein had the predicted enzyme activity. This led to a

broader appreciation that the chloroviruses encode sugar-metabolizing enzymes and resulted in a nice paper in *Science* (DeAngelis et al. 1997).

Of course, our lab wondered if hyaluronan was somehow involved in chlorovirus replication. It was brought up in a lab meeting that some unpublished micrographs taken by John Heuser at Washington University in St. Louis about six years earlier might be relevant. John is well known for having developed quick-freeze deep-etch microscopy that enables one to view images in pseudo three dimensions. There is a long story on how we originally hooked up with John, but in about 1989 we had sent him PBCV-1 and its host *C. variabilis* in separate tubes. He mixed them together and examined the infected algae at various times using his technique. One unexpected result was that the cell walls of the infected alga became 'hairy' after infection (compare Fig. 2A with Fig. 2B). At the time, we had no explanation for these results and thought that maybe the walls were starting to unravel as the cells were getting ready to lyse and release infectious virions. However, after we discovered that PBCV-1 encoded a functional hyaluronan synthetase, we suspected that the hairy material might be hyaluronan. To test this, a postdoc, Mike Graves treated the infected cells with a hyaluronan-degrading enzyme (hyaluronidase). After this treatment, the fibers disappeared (Fig.

---

[8] In the mid 1990s we were not required to put sequences in the public databases. However, we made the decision to do this and as it turned out, this led to several research collaborations not mentioned in this chapter.

2C), so we knew that much, if not all, of the extracellular hairy material was due to hyaluronan (Graves et al. 1999). There are three other aspects of this story that also deserve to be mentioned.

First, after almost 20 years, we still have no idea why PBCV-1 encodes three enzymes involved in hyaluronan production—a very energy-demanding synthesis. Forming each disaccharide requires at least five ATPs; each of the thousands of strands exported through the cell wall consists, in turn, of thousands of disaccharides. Furthermore, the cell is going to lyse shortly after the appearance of the extracellular hyaluronan, so why 'waste' this much ATP?

Second, vertebrates synthesize hyaluronan at the plasma membrane and the product is pushed out of the cell into the surrounding extracellular matrix. Hyaluronan is also synthesized in the algal plasma membrane during PBCV-1 infection and pushed to the outside. However, in the case of the virus-infected algae the synthesized hyaluronan has to also pass through the algal cell wall. We have no idea how this is accomplished—a feat equivalent to pushing a thread through a furnace filter. One would expect the hyaluronan to bunch up underneath the wall, but it does not.

Third, not all chloroviruses encode a hyaluronan synthetase. In fact, some encode a chitin synthetase and produce chitin as an extracellular polysaccharide. Others encode both enzymes and produce both extracellular polysaccharides, while still others lack both genes and thus get by just fine without forming any extracellular polysaccharide on the surface of their algal host. Therefore, the biological function(s) of all of this activity remains unknown.

One more sweet story warrants brief mention. As is the case with many viruses, the PBCV-1 major capsid protein (MCP) is a glycoprotein. However, the chloroviruses have their own unique way of synthesizing the glycans and attaching them to their MCPs (Wang et al. 1993; De Castro et al. 2013). The typical virus that infects eukaryotic organisms adds and removes the MCP sugars as the protein transits through the endoplasmic reticu-

lum and the Golgi apparatus. The protein is then transferred to the plasma membrane of the cell and the newly forming virus particles bud though this region of the plasma membrane to acquire their glycoprotein coat. Consequently, these viruses only become infectious as they exit the cell. However, PBCV-1 differs in that it forms infectious virions inside the host cell and MCP glycosylation appears to occur in the cytoplasm by an unknown process. Furthermore, PBCV-1 encodes most, if not all, of the needed glycosylation machinery, including five putative glycosyltransferases. This process remains an active topic of research for us.

The split personality of the chloroviruses

All known hosts for the chloroviruses are zoochlorellae, i.e., green eukaryotic algae that live in a mutualistic association with their symbiotic hosts. Experiments many years ago demonstrated that their hosts benefited from the zoochlorellae because the algae carry out photosynthesis and export some of the newly synthesized sugars to the host. However, there remained the question as to what was the benefit to the algae. We suggest that, in at least some cases, the benefit is protection from virus infection.

Zoochlorellae have rigid cell walls like most Bacteria, but are otherwise typical eukaryotic cells. It is not surprising that viruses infecting zoochlorellae have adapted to harmonize with the cellular features of their hosts. If we think of virus replication in terms of three phases (attachment and entry, replication and maturation, exit and release), chlorovirus replication appears to employ themes originating in viruses that infect either Bacteria or eukaryotic organisms. Unlike other eukaryotic viruses, the chloroviruses attach to the outside of their host cell walls (Fig. 3A). The empty capsid remains attached to the wall after the contents of the virion (DNA and associated proteins) are released into the cell, much like what occurs in many bacteriophage infections. Additionally, the icosahedral chlorovirus PBCV-1 virion contains a unique feature at one vertex (Cherrier et al. 2009), a 'spike' that orients toward the algal cell wall during initial attachment (Fig. 3B) (Zhang et al. 2011). Both the spike structure and its orientation resemble the complex unique vertex of the tailed phages. In many of those phages, the spike, which is located in the tail, is required for wall penetration, while the tail channel serves as the conduit for viral DNA entry. However, PBCV-1 has no tail, only a spike, and the spike is too narrow to serve as a channel for DNA delivery. Presumably the spike is jettisoned once virus attachment is stabilized. Both bacteriophages and chloroviruses must penetrate the cell wall at the point of attachment. For this the chlorovirus virion is equipped with a cell wall-degrading activity that we call v-lysin and that is probably associated with the spike. However, cell wall degradation alone does not allow virus entry into the cell as it also must pass through the plasma membrane.

Some bacteriophages that use a bacterial surface protein as their receptor can be prompted to eject their DNA *in vitro* by contact with the isolated receptor protein. In contrast, attachment of PBCV-1 to isolated host cell walls does not result in DNA release, even after the wall has been digested at the point of attachment. Viral DNA and associated proteins are only released when the virion

Figure 4. The eukaryotic Tree of Life from a viral perspective. We adapted this eukaryotic tree by adding information about viruses known to infect these taxa. That the vast majority of these viruses have been isolated from animals and land plants reflects the research emphasis on organisms perceived to be important to humans. Ongoing exploration of other realms of this tree in recent years has demonstrated that these new taxa also support viruses. Remarkably, there are no known viruses associated with the taxa Excavates and Rhizaria at this time. Thus, we predict there are many novel viruses waiting to be discovered. Modified from Keeling et al. 2005 and published with permission.

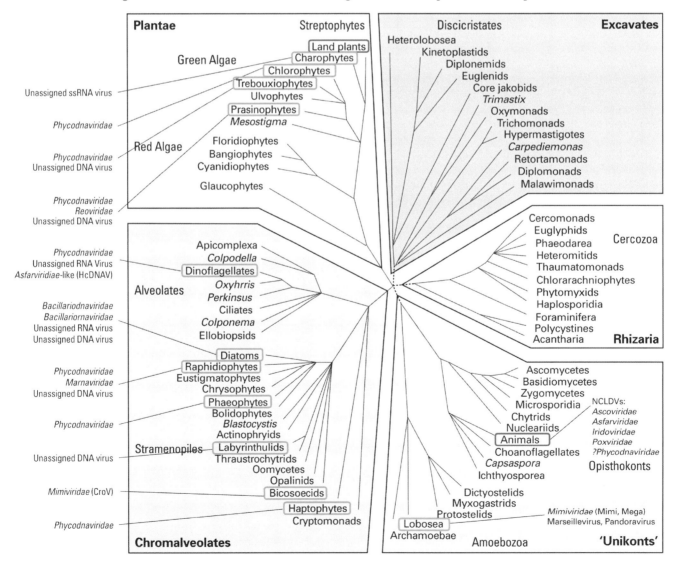

attaches to a susceptible living cell. The requisite attribute of the cell's physiological state is a normal membrane potential. Our working model for PBCV-1 infection is that after the spike has digested a hole through the host cell wall, fusion of the virion internal membrane with the host cell membrane releases the viral DNA and associated proteins into the cell. The fusion activates the viral K^+ channel (Kcv, described above) believed to be located in the virus membrane, which leads to the rapid depolarization of the host membrane (Frohns et al. 2006). The loss of membrane potential has many consequences for a cell, and, in the case of chlorovirus infection, these include a rapid efflux of K^+ (Neupärtl et al. 2008). This depolarization and efflux of K^+ reduces the high turgor pressure that is normally maintained by algal cells and that resists entry of virus DNA (Thiel et al. 2010). All of these obstacles to infection are very much like what bacteriophages experience, and likewise the tactics used by chloroviruses to overcome them. However, once inside the host cell, the chlorovirus lifestyle probably resembles that of eukaryotic viruses. Not much is known about PBCV-1 DNA replication and virion assembly, but the replication machinery, much of which is encoded by the virus, resembles that of other large dsDNA viruses that infect eukaryotes (*see page 6-41*). But then, once the virus has replicated and is ready to escape the host, these viruses start to resemble bacteriophages again.

The primary challenge for the virus at this point is to penetrate the plasma membrane and the cell wall and make its escape, and presumably this requires another type of cell wall-degrading activity (Fig. 3C). Very little is known about this event in the chlorovirus life cycle. Replication is completed within the cell and mature infectious virions accumulate starting about 45 min prior to cell lysis. Optimal timing of the lytic event would seem to depend on the action of a lysin-type activity at the optimal point in time and space to maximize the virus yield and the potential for gaining access to the next host (*see page 7-6*). From a population point of view, this timing depends not only on the rate of virus replication, but also on burst size.

Different chloroviruses have a range of replication rates (6-18 h) and burst sizes (10-500 PFU per cell). Exogenous factors (e.g., light, nutrients, antibiotics) can modulate these virus yields, but very little research has been done to understand the chlorovirus mechanisms for these events or what regulates their lysis clock function. In many bacteriophages, holin and holin-like proteins establish the lysis clock by accumulating within the plasma membrane until triggered, then quickly create a portal so that cell wall-degrading enzymes can access the cell wall (*see page 7-5*; Wang, Smith, Young 2000). However, no holin-like protein has been identified in the chloroviruses to date.

Another longstanding unknown about chlorovirus replication has been the source of the needed metabolic energy given that algal photosynthesis is significantly impaired shortly after infection (Thiel et al. 2010). Being photoheterotrophs, the zoochlorellae are not entirely dependent on photosynthesis for their replication, but virus replication in the dark results in a lower burst size, so there is a conundrum. The effect of virus infection on mitochondrial function has not been examined. We wonder if chloroviruses may provide some type of 'metabolic booster' like that found in certain cyanophages that augment photosynthesis. You never know with these viruses; there are always surprises.

The apparent mixed life style of the chloroviruses has made us wonder about the origin of these viruses. Evolutionarily, they are proposed to share a common ancestor with the other NCLDVs (see below). However, the origin of some of their genes is another story.

Most virologists believe that horizontal gene transfer between eukaryotic hosts and their viruses is the primary mechanism for gene flow between them. As obligate intracellular parasites, viruses experience intimate contact with their hosts while replicating, which leads to the prediction that the virus genome would resemble the host genome, and vice versa. However, chlorovirus genomes show very little evidence of gene acquisition from algal cells, with less than 2% of their genome ostensibly from that source (Jeanniard et al. 2013). Perhaps more surprising is that they contain a large number of genes with apparent bacterial origins, including genes from multiple taxa (Dunigan et al., unpublished). Such observations are not limited to the chloroviruses (Filée, Pouget, Chandler 2008), but they take on possible added significance for them in light of their seemingly interdenominational lifestyle. It may be that gene flow from bacterial-like origins has been important for the evolution of certain aspects of the chlorovirus infection cycle. For example, chloroviruses have the ability to overcome the fundamental barrier of the algal cell wall, in a manner much like the bacteriophages.

Where in the viral world?

Viruses in the family *Phycodnaviridae*, including the chloroviruses, together with those in the *Poxviridae*, *Iridoviridae*, *Ascoviridae*, *Asfarviridae*, *Mimiviridae*, and *Marseilleviridae* families, are referred to as Nuclear Cytoplasmic Large DNA Viruses (NCLDV) (Iyer, Aravind, Koonin 2001; Iyer et al. 2006). The NCLDVs are believed to have a common evolutionary ancestor, and thus a new virus order, *Megavirales*, has been proposed for them (Colson et al. 2013). Comparative analysis of 45 NCLDVs identified five genes present in all 45 and 177 additional genes that are present in at least two of the virus families (Yutin et al. 2009).

Lastly, although this book is primarily about bacteriophage, this chapter is to remind readers that the world of eukaryotic viruses is, likewise, mostly unexplored. Viruses infecting eukaryotic organisms from many kingdoms are known, but remarkably, not from all kingdoms. This uneven distribution of virus discovery is due to the fact that most virology is focused on only two groups—animals and crop plants—and thus illustrates the emphasis on "mission-directed research," rather than curiosity-driven discovery. Even in taxa with some virus representation (e.g., Plantae, Chromalveolates, Unikonts), the taxa are sparsely covered (Fig. 4). Therefore, there remain many opportunities to discover new viruses in a wide range of diverse hosts. For example, the taxa Excavates and Rhizaria have no reported viruses. Even within the Unikonts (that's us folks) there are only three taxa with known viruses: animals, fungi, and amoebae (Lobosea). Thus, we can confidently predict that there are not only many interesting bacteriophages that await discovery, but also many viruses that infect diverse eukaryotic cells. Perhaps what is required for these discoveries is a small group of curious people who will make simple observations, follow them with simple experiments, and then share food, drink, and ideas with friends.

Acknowledgements

This chapter is dedicated to Russ Meints, who was a co-discover of the chloroviruses with JLVE, and two technicians, the late Dwight Burbank and James Gurnon, who spent the last 35 years working on the viruses. Finally, we thank all the graduate students, undergraduate students, postdoctoral fellows, visiting scientists, and our many colleagues who have worked on the viruses. This research has been supported by grants from NIH, NSF, USDA, DOE, and the Stanley Medical Research Institute over the years.

References

Agarkova, IV, DD Dunigan, JL Van Etten. 2006. Virion-associated restriction endonucleases of chloroviruses. J Virol 80:8114-8123.

Cherrier, MV, VA Kostyuchenko, C Xiao, VD Bowman, AJ Battisti, X Yan, PR Chipman, TS Baker, JL Van Etten, MG Rossmann. 2009. An icosahedral algal virus has a complex unique vertex decorated by a spike. Proc Natl Acad Sci USA 106:11085-11089.

Colson, P, X De Lamballerie, N Yutin, S Asgari, Y Bigot, DK Bideshi, X-W Cheng, BA Federici, JL Van Etten, EV Koonin. 2013. "Megavirales", a proposed new order for eukaryotic nucleocytoplasmic large DNA viruses. Arch Virol 158:2517-2521.

De Castro, C, A Molinaro, F Piacente, JR Gurnon, L Sturiale, A Palmigiano, R Lanzetta, M Parrilli, D Garozzo, MG Tonetti. 2013. Structure of N-linked oligosaccharides attached to chlorovirus PBCV-1 major capsid protein reveals unusual class of complex N-glycans. Proc Natl Acad Sci USA 110:13956-13960.

DeAngelis, PL, W Jing, MV Graves, DE Burbank, JL Van Etten. 1997. Hyaluronan synthase of chlorella virus PBCV-1. Science 278:1800-1803.

Dunigan, DD, RL Cerny, AT Bauman, JC Roach, LC Lane, IV Agarkova, K Wulser, GM Yanai-Balser, JR Gurnon, JC Vitek. 2012. *Paramecium bursaria* chlorella virus 1 proteome reveals novel architectural and regulatory features of a giant virus. J Virol 86:8821-8834.

Filée, J, N Pouget, M Chandler. 2008. Phylogenetic evidence for extensive lateral acquisition of cellular genes by nucleocytoplasmic large DNA viruses. BMC Evol Biol 8:320.

Frohns, F, A Käsmann, D Kramer, B Schäfer, M Mehmel, M Kang, JL Van Etten, S Gazzarrini, A Moroni, G Thiel. 2006. Potassium ion channels of chlorella viruses cause rapid depolarization of host cells during infection. J Virol 80:2437-2444.

Gazzarrini, S, M Kang, JL Van Etten, S Tayefeh, SM Kast, D DiFrancesco, G Thiel, A Moroni. 2004. Long distance interactions within the potassium channel pore are revealed by molecular diversity of viral proteins. J Biol Chem 279:28443-28449.

Graves, MV, DE Burbank, R Roth, J Heuser, PL DeAngelis, JL Van Etten. 1999. Hyaluronan synthesis in virus PBCV-1-infected chlorella-like green algae. Virology 257:15-23.

Iyer, LM, L Aravind, EV Koonin. 2001. Common origin of four diverse families of large eukaryotic DNA viruses. J Virol 75:11720-11734.

Iyer, LM, S Balaji, EV Koonin, L Aravind. 2006. Evolutionary genomics of nucleo-cytoplasmic large DNA viruses. Virus Res 117:156-184.

Jeanniard, A, DD Dunigan, JR Gurnon, IV Agarkova, M Kang, J Vitek, G Duncan, OW McClung, M Larsen, J-M Claverie. 2013. Towards defining the chloroviruses: a genomic journey through a genus of large DNA viruses. BMC genomics 14:158.

Kang, M, A Moroni, S Gazzarrini, D DiFrancesco, G Thiel, M Severino, JL Van Etten. 2004. Small potassium ion channel proteins encoded by chlorella viruses. Proc Natl Acad Sci USA 101:5318-5324.

Keeling, PJ, G Burger, DG Durnford, BF Lang, RW Lee, RE Pearlman, AJ Roger, MW Gray. 2005. The tree of eukaryotes. Trends Ecol Evol 20:670-676.

Lu, Z, Y Li, Y Zhang, GF Kutish, DL Rock, JL Van Etten. 1995. Analysis of 45 kb of DNA located at the left end of the chlorella virus PBCV-1 genorne. Virology 206:339-352.

Meints, RH, K Lee, DE Burbank, JL Van Etten. 1984. Infection of a chlorella-like alga with the virus, PBCV-1: Ultrastructural studies. Virology 138:341-346.

Meints, RH, JL Van Etten, D Kuczmarski, K Lee, B Ang. 1981. Viral infection of the symbiotic chlorella-like alga present in *Hydra viridis*. Virology 113:698-703.

Neupärtl, M, C Meyer, I Woll, F Frohns, M Kang, JL Van Etten, D Kramer, B Hertel, A Moroni, G Thiel. 2008. Chlorella viruses evoke a rapid release of K+ from host cells during the early phase of infection. Virology 372:340-348.

Plugge, B, S Gazzarrini, M Nelson, R Cerana, J Van, C Derst, D DiFrancesco, A Moroni, G Thiel. 2000. A potassium channel protein encoded by chlorella virus PBCV-1. Science 287:1641-1644.

Reisser, W, B Becker, T Klein. 1986. Studies on ultrastructure and host range of a Chlorella attacking virus. Protoplasma 135:162-165.

Sherman, LA, RM Brown Jr. 1978. Cyanophages and viruses of eukaryotic algae. In: H Fraenkel-Conrat, R Wagner, editors. *Comprehensive Virology,* vol. 12: Plenum Press. p. 145-234.

Thiel, G, A Moroni, G Blanc, JL Van Etten. 2013. Potassium ion channels: Could they have evolved from viruses? Plant Physiol 162:1215-1224.

Thiel, G, A Moroni, D Dunigan, JL Van Etten. 2010. Initial events associated with virus PBCV-1 infection of Chlorella NC64A. In: U Lüttge, W Beyschlag, B Büdel, editors. *Progress in Botany.* Springer. p. 169-183.

Van Etten, JL, DE Burbank, J Joshi, RH Meints. 1984. DNA synthesis in a Chlorella-like alga following infection with the virus PBCV-1. Virology 134:443-449.

Van Etten, JL, DE Burbank, D Kuczmarski, RH Meints. 1983. Virus infection of culturable chlorella-like algae and dlevelopment of a plaque assay. Science 219:994-996.

Van Etten, JL, DE Burbank, AM Schuster, RH Meints. 1985. Lytic viruses infecting a chlorella-like alga. Virology 140:135-143.

Van Etten, JL, DD Dunigan. 2012. Chloroviruses: Not your everyday plant virus. Trends Plant Sci 17:1-8.

Van Etten, JL, JR Gurnon, GM Yanai-Balser, DD Dunigan, MV Graves. 2010. Chlorella viruses encode most, if not all, of the machinery to glycosylate their glycoproteins independent of the endoplasmic reticulum and Golgi. Biochim Biophys Acta 1800:152-159.

Van Etten, JL, RH Meints, D Kuczmarski, DE Burbank, K Lee. 1982. Viruses of symbiotic Chlorella-like algae isolated from *Paramecium bursaria* and *Hydra viridis*. Proc Natl Acad Sci USA 79:3867-3871.

Wang, I-N, Y Li, Q Que, M Bhattacharya, LC Lane, WG Chaney, JL Van Etten. 1993. Evidence for virus-encoded glycosylation specificity. Proc Natl Acad Sci USA 90:3840-3844.

Wang, I-N, DL Smith, R Young. 2000. Holins: The protein clocks of bacteriophage infections. Annual Rev Microbiol 54:799-825.

Xia, Y, D Burbank, L Uher, D Rabussay, JL Van Etten. 1986. Restriction endonuclease activity induced by PBCV-1 virus infection of a Chlorella-like green alga. Molecular and cellular biology 6:1430-1439.

Yutin, N, YI Wolf, D Raoult, EV Koonin. 2009. Eukaryotic large nucleo-cytoplasmic DNA viruses: Clusters of orthologous genes and reconstruction of viral genome evolution. Virol J 6:223.

Zhang, X, Y Xiang, DD Dunigan, T Klose, PR Chipman, JL Van Etten, MG Rossmann. 2011. Three-dimensional structure and function of the *Paramecium bursaria* chlorella virus capsid. Proc Natl Acad Sci USA 108:14837-14842.

10^{16}

phage virions in your gut

Calculated assuming 5×10^{12} virions/g of fecal matter and 2 kg fecal matter present

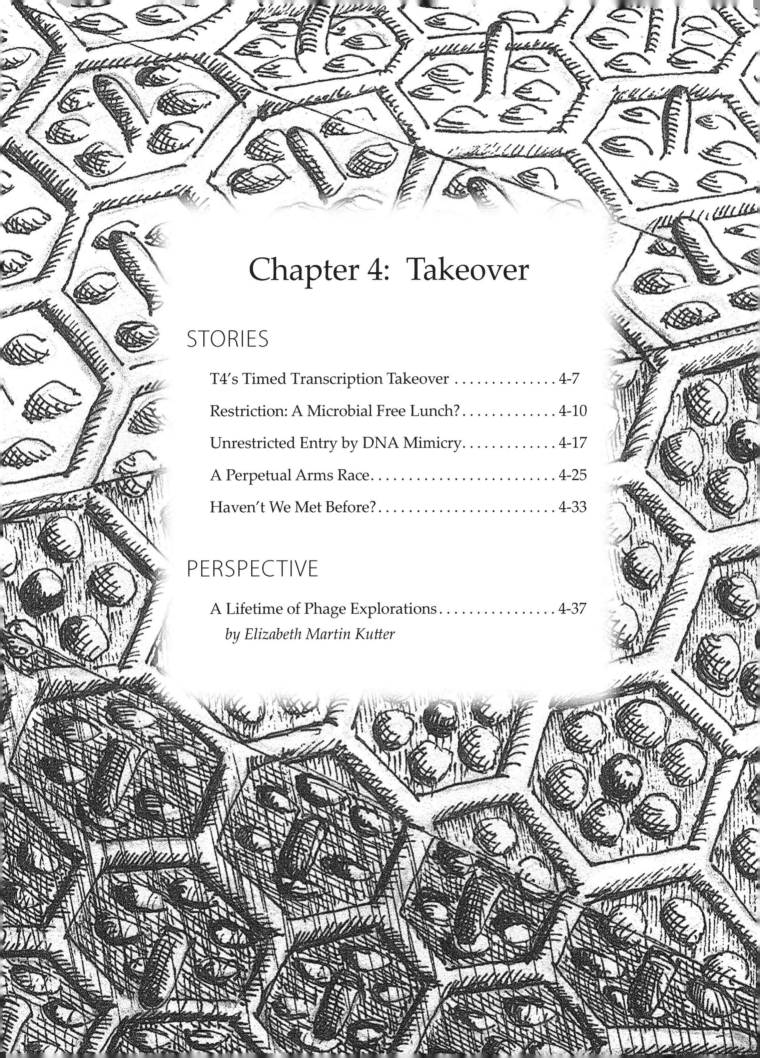

Chapter 4: Takeover

STORIES

PERSPECTIVE

Enterobacteria Phage T4

a Myophage that swiftly diverts the host's transcription machinery and degrades the host chromosome

Genome
dsDNA; linear
168,903 bp
280 predicted ORFs; 8 RNAs

Encapsidation method
Packaging; T = 13 capsid

Common host
Escherichia coli

Habitat
Mammalian intestines & sewage

Lifestyle
Lytic

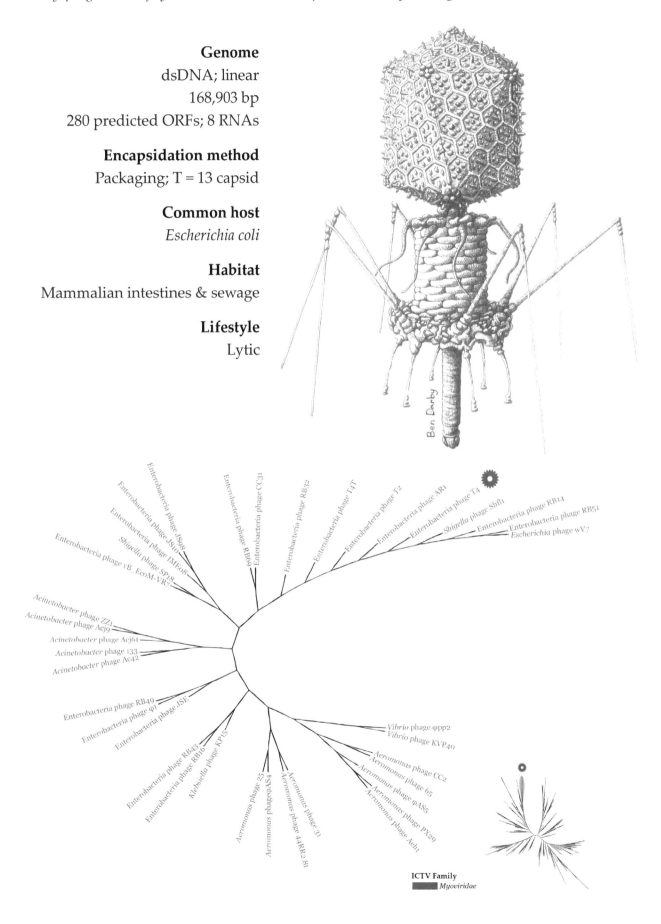

ICTV Family
Myoviridae

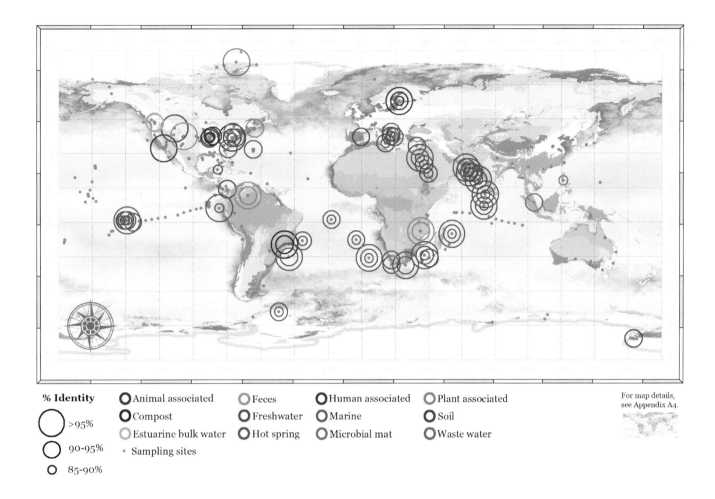

% Identity

○ >95%

○ 90-95%

○ 85-90%

○ Animal associated   ○ Feces   ○ Human associated   ○ Plant associated

○ Compost   ○ Freshwater   ○ Marine   ○ Soil

○ Estuarine bulk water   ○ Hot spring   ○ Microbial mat   ○ Waste water

· Sampling sites

For map details, see Appendix A4.

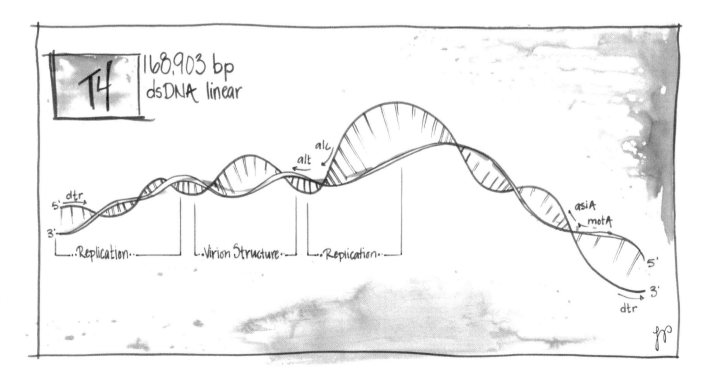

Enterobacteria Phage T4

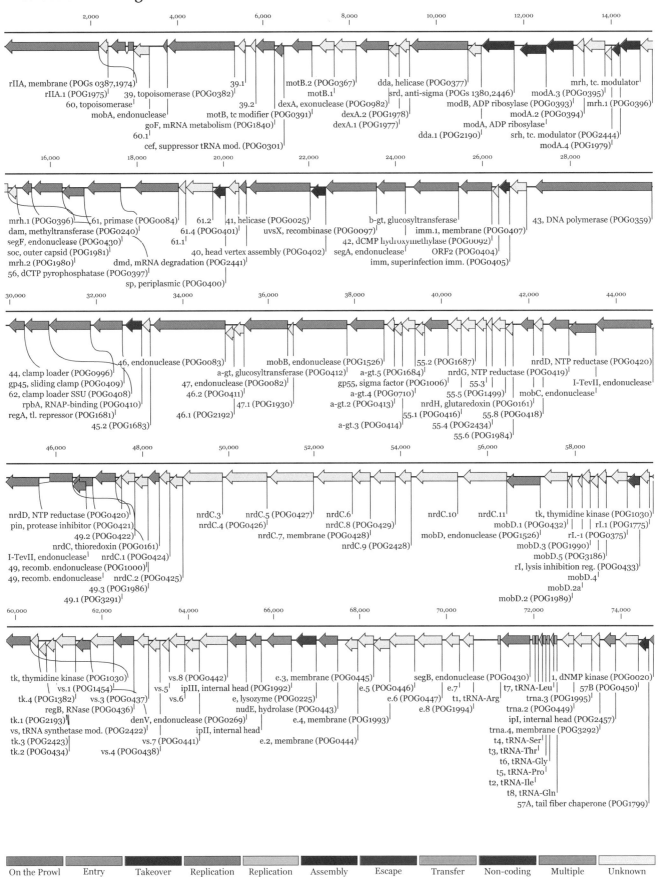

rIIA, membrane (POGs 0387,1974)
rIIA.1 (POG1975)
60, topoisomerase
mobA, endonuclease
goF, mRNA metabolism (POG1840)
60.1
cef, suppressor tRNA mod. (POG0301)

39, topoisomerase (POG0382)
39.1
39.2
motB, tc modifier (POG0391)

motB.2 (POG0367)
motB.1
dexA, exonuclease (POG0982)
dexA.2 (POG1978)
dexA.1 (POG1977)

dda, helicase (POG0377)
srd, anti-sigma (POGs 1380,2446)
modB, ADP ribosylase (POG0393)
modA, ADP ribosylase
dda.1 (POG2190)
srh, tc. modulator (POG2444)
modA.4 (POG1979)

mrh, tc. modulator
modA.3 (POG0395)
mrh.1 (POG0396)
modA.2 (POG0394)

mrh.1 (POG0396)
dam, methyltransferase (POG0240)
segF, endonuclease (POG0430)
soc, outer capsid (POG1981)
mrh.2 (POG1980)
56, dCTP pyrophosphatase (POG0397)

61, primase (POG0084)
61.2
61.4 (POG0401)
61.1
dmd, mRNA degradation (POG2441)
sp, periplasmic (POG0400)

41, helicase (POG0025)
uvsX, recombinase (POG0097)
40, head vertex assembly (POG0402)
segA, endonuclease

b-gt, glucosyltransferase
imm.1, membrane (POG0407)
42, dCMP hydroxymethylase (POG0092)
ORF2 (POG0404)
imm, superinfection imm. (POG0405)

43, DNA polymerase (POG0359)

44, clamp loader (POG0996)
gp45, sliding clamp (POG0409)
62, clamp loader SSU (POG0408)
rpbA, RNAP-binding (POG0410)
regA, tl. repressor (POG1681)
45.2 (POG1683)

46, endonuclease (POG0083)
47, endonuclease (POG0082)
46.2 (POG0411)
47.1 (POG1930)
46.1 (POG2192)

a-gt, glucosyltransferase (POG0412)
a-gt.4 (POG0710)
a-gt.2 (POG0413)
a-gt.3 (POG0414)

mobB, endonuclease (POG1526)
a-gt.5 (POG1684)
gp55, sigma factor (POG1006)
nrdH, glutaredoxin (POG0161)
55.1 (POG0416)
55.4 (POG2434)
55.6 (POG1984)

55.2 (POG1687)
nrdG, NTP reductase (POG0419)
55.3
55.5 (POG1499)
mobC, endonuclease
55.8 (POG0418)

nrdD, NTP reductase (POG0420)
I-TevII, endonuclease

nrdD, NTP reductase (POG0420)
pin, protease inhibitor (POG0421)
49.2 (POG0422)
nrdC, thioredoxin (POG0161)
I-TevII, endonuclease
49, recomb. endonuclease (POG1000)
49, recomb. endonuclease
49.3 (POG1986)
49.1 (POG3291)

nrdC.1 (POG0424)
nrdC.2 (POG0425)

nrdC.3
nrdC.4 (POG0426)
nrdC.5 (POG0427)
nrdC.7, membrane (POG0428)

nrdC.6
nrdC.8 (POG0429)
nrdC.9 (POG2428)

nrdC.10
nrdC.11
mobD.1 (POG0432)
mobD, endonuclease (POG1526)
mobD.3 (POG1990)
mobD.5 (POG3186)
mobD.4
mobD.2a
mobD.2 (POG1989)

tk, thymidine kinase (POG1030)
rI.1 (POG1775)
rI.-1 (POG0375)
rI, lysis inhibition reg. (POG0433)

tk, thymidine kinase (POG1030)
vs.1 (POG1454)
tk.4 (POG1382)
regB, RNase (POG0436)
tk.1 (POG2193)
vs, tRNA synthetase mod. (POG2422)
tk.3 (POG2423)
tk.2 (POG0434)

vs.8 (POG0442)
vs.5
vs.3 (POG0437)
vs.6
denV, endonuclease (POG0269)
ipII, internal head
vs.7 (POG0441)
vs.4 (POG0438)

e.3, membrane (POG0445)
ipIII, internal head (POG1992)
e, lysozyme (POG0225)
nudE, hydrolase (POG0443)
e.4, membrane (POG1993)
e.2, membrane (POG0444)

segB, endonuclease (POG0430)
e.5 (POG0446)
e.6 (POG0447)
e.8 (POG1994)

e.7
t1, tRNA-Arg
t4, tRNA-Ser
t3, tRNA-Thr
t6, tRNA-Gly
t5, tRNA-Pro
t2, tRNA-Ile
t8, tRNA-Gln
57A, tail fiber chaperone (POG1799)

1, dNMP kinase (POG0020)
t7, tRNA-Leu
trna.3 (POG1995)
trna.2 (POG0449)
ipI, internal head (POG2457)
trna.4, membrane (POG3292)

57B (POG0450)

On the Prowl | Entry | Takeover | Replication Lytic | Replication Lysogenic | Assembly | Escape | Transfer RNA | Non-coding RNA | Multiple | Unknown

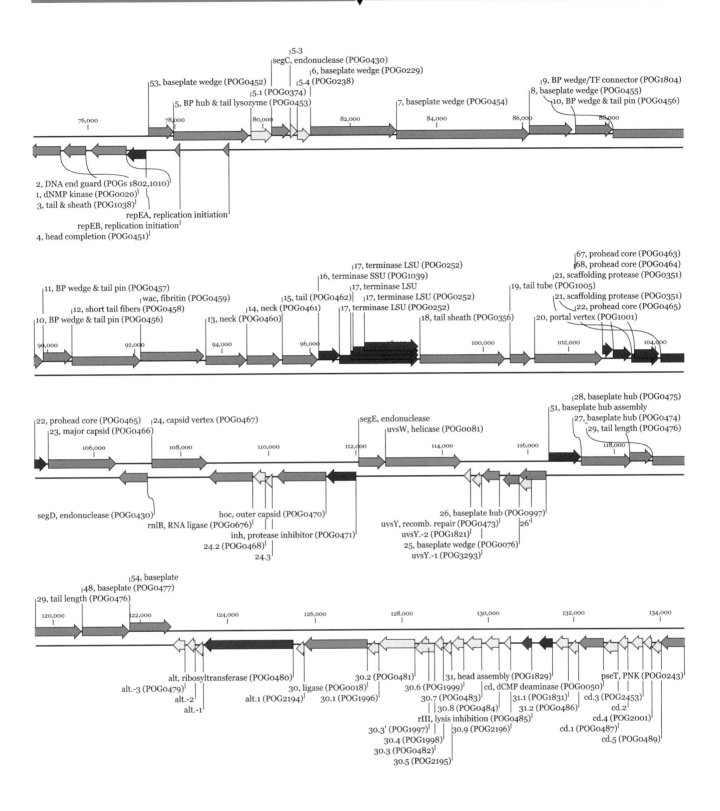

Continued next page

Enterobacteria Phage T4 *continued*

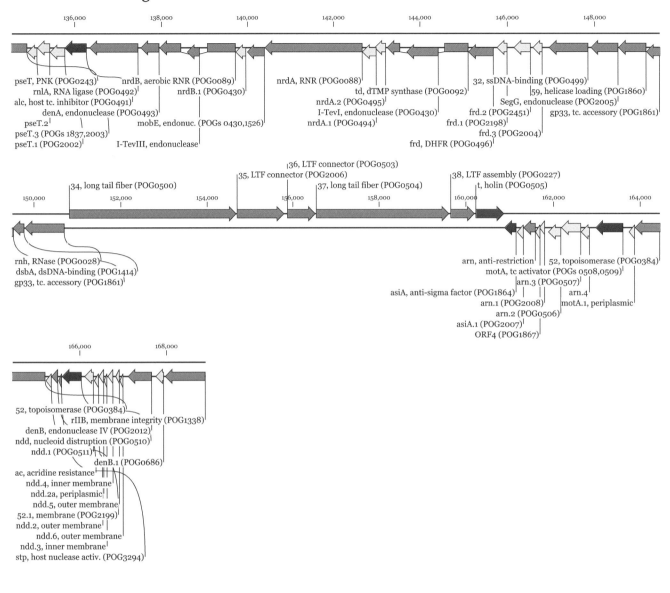

On the Prowl | Entry | Takeover | Replication Lytic | Replication Lysogenic | Assembly | Escape | Transfer RNA | Non-coding RNA | Multiple | Unknown

T4's Timed Transcription Takeover

Heather Maughan

Mise-en-Scène: *When a phage genome enters a host cell, it lands in a hive of metabolic activity that is organized and regulated for efficient reproduction and survival. To immediately kill that cell would be a simple matter requiring no particular skill, but that strategy would also be a dead end for the phage. The accomplished phage instead converts the cell into a **virocell**, a finely-tuned **virion** factory in which all the machinery is following orders from the phage. To that end, the old management has to be fired, the production line retooled for virion manufacture, existing resources recycled—all without disrupting the flow of needed energy and raw materials. The exquisitely precise, step-by-step takeover executed by phages such as T4 attests to a long-standing relationship with its host that has been evolving for countless generations.*

Oblivious of the phages drifting past, an *E. coli* bacterium contentedly grazes on carbohydrates in your intestine. Suppose that, as this particular *E. coli* is going about its business, a T4 virion delivers its DNA into the cell's cytoplasm. Being a relatively large phage that produces hundreds of progeny from a single host cell, T4 needs to rapidly seize control of all cellular resources. Within five minutes of arrival, T4 has diverted the host's transcription machinery from host genes to phage genes. This robs the cell of its defenses and facilitates the sequential expression of T4's early, middle, and late genes. So well orchestrated is T4's takeover that it is only about seven minutes later when the first infectious virions come off the assembly line; they'll wait ~15 minutes for more siblings to accumulate before collectively lysing the cell.

One minute post-infection: Hog the transcription machinery

Once inside its host, T4 immediately diverts many of the ~2,000 RNA polymerases (RNAPs) from host gene expression to transcription of early phage genes (Miller et al. 2003). Transcription involves promoters competing for attention from the promoter-binding σ⁷⁰ **subunit** of RNAP. T4's 39 early gene promoters successfully outcompete ~650 host promoters by having all the right nucleotides in all the right places, thereby usurping the RNAP supply. These nearly perfect promoters have so many RNAPs servicing them that a burst of rapidly repeated transcription of phage early

genes occurs within the first minute post infection (Miller et al. 2003; Hinton 2010).

Competitive superiority of its promoters is not T4's only trick; this phage is also equipped to modify the host's RNAP. Inside its capsid, and ejected into the host with its DNA, is the Alt protein that tweaks RNAP to increase transcription of early phage genes (Miller et al. 2003; Hinton 2010).

Two minutes post-infection: Render the host helpless

Despite T4's persuasive tactics, RNAPs have not completely abandoned host genes. Some of those genes have promoters that are as inviting as T4's, and yet others had already recruited their RNAP and their transcription continues. This is bad for T4. Not only does this reduce the number of copies of RNAP for the phage, but by now the host, aware of the invasion, could be transcribing anti-phage weaponry. Not to worry. Immediately upon arrival, T4 makes a protein (Alc) that quickly and selectively halts all ongoing transcription of host genes (*see page 4-41*; [Kashlev et al. 1993]). For this feat, T4 cleverly distinguishes unmodified host DNA from its own because it has replaced all of its own cytosines with glucosylated hydroxymethyl cytosines (*see page 4-25 and page 4-40*). This termination of host transcription frees up even more RNAPs for the phage, who by now has begun chopping up the host DNA with its newly made nucleases (EndoII and EndoIV). In these first two minutes T4 has not only foiled host plans for retaliation, it has also generated a pool of deoxyribonucleotides to use for genome replication.

Three minutes post-infection: Appropriate σ^{70}

Having already done away with host interference using its early gene products, T4 can turn its attention to gearing up for genome replication. This process requires production of its middle gene products (e.g., its **methyltransferase;** *see page 4-40*). T4 already got a head start on expression of these genes as some were transcribed by read-through of early genes. Now, to meet the heightened demand, it alters RNAP's promoter-binding preference to favor phage middle gene promoters over those for the early genes. Two early proteins (AsiA and MotA) are key here. By remodeling the 3D structure of σ^{70}, they coerce RNAP to service the middle promoters, thereby orchestrating the timely shift of transcription from early to middle proteins (Ouhammouch et al. 1995).

Five minutes post-infection: Mass produce phage genomes

To accommodate its burst size of 100-200 virions, T4 needs between 16 and 34 million deoxyribonucleotides. Never one to waste resources, T4 scavenges and reuses nucleotides liberated by degradation of the host's genome. But this source is far from enough. Recycling 100% of the 4,600 kbp genome of T4's favored host strain, *E. coli* B, can provide nucleotides for only 20-30 T4 genomes. Ergo, to make up for this shortfall, T4 makes some of its own nucleotides using a multi-enzyme complex of early gene products (*see page 4-45*). For raw materials, it steals the stash of nucleotide precursors the host prepared for its own use. The **ribonucleotides** it reduces to deoxyribonucleotides. The cytosine bases it hydroxymethylates and then glucosylates (Mathews 1993; Miller et al. 2003).

Seven minutes post-infection: Convert host into virions

At seven minutes T4 is poised to begin making proteins for virion production. Genome replication continues, joined now by transcription of late phage genes that kicks into high gear by 11-15 minutes post-infection. These genes encode the structural proteins of the capsid, tail, and tail fibers, as well as many proteins needed for virion assembly. Their transcription relies on at least three phage proteins that, once again, must redirect RNAP's efforts. Two proteins (gp33 and gp55) function together as a phage σ-subunit that steers RNAP away from the middle genes to the promoters of the late genes. Another protein (gp45) functions as a sliding clamp that tethers DNA polymerase to the DNA during replication. Moreover, as the clamp moves along with the DNA polymerase, it interacts with the phage σ-subunit to recruit RNAP to the late gene promoters (Geiduschek, Kassavetis 2010). Not surprisingly, T4 demonstrates excellent multi-tasking ability by coordinating production of phage genomes and virion structural components late in the infection cycle.

Twelve minutes post-infection: Prepare to flee!

With late gene transcription and DNA replication proceeding according to plan, T4's factory is now in full swing. Production is organized and calculated so as to keep the doomed host alive to provide energy and metabolites for virion production. Then, with enough virions assembled, T4 ends the game by lysing the host cell, thereby releasing its progeny into the environment to find a host in which to repeat the cycle. T4's transcription takeover has triumphed, completing a timed conversion of one bacterial cell into hundreds of virions now on the prowl.

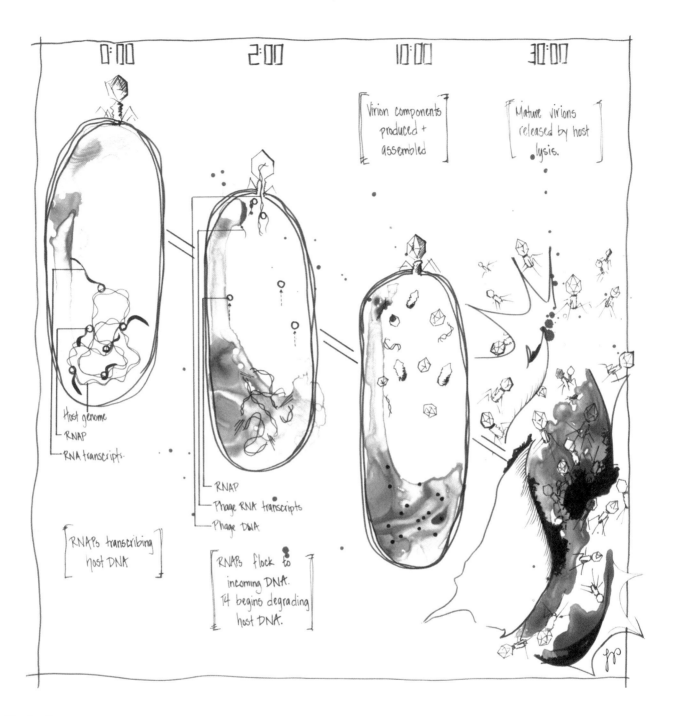

Cited references

Geiduschek, EP, GA Kassavetis. 2010. Transcription of the T4 late genes. Virol J 7:288.

Hinton, DM. 2010. Transcriptional control in the prereplicative phase of T4 development. Virol J 7:289.

Kashlev, M, E Nudler, A Goldfarb, T White, E Kutter. 1993. Bacteriophage T4 Alc protein: A transcription termination factor sensing local modification of DNA. Cell 75:147-154.

Mathews, CK. 1993. The cell-bag of enzymes or network of channels? J Bacteriol 175:6377-6381.

Miller, ES, E Kutter, G Mosig, F Arisaka, T Kunisawa, W Ruger. 2003. Bacteriophage T4 genome. Microbiol Mol Biol Rev 67:86-156.

Ouhammouch, M, K Adelman, SR Harvey, G Orsini, EN Brody. 1995. Bacteriophage T4 MotA and AsiA proteins suffice to direct *Escherichia coli* RNA polymerase to initiate transcription at T4 middle promoters. Proc Natl Acad Sci USA 92:1451-1455.

Recommended review

Mathews, C.K. 2010. Bacteriophage T4. In: *Encyclopedia of Life Sciences (ELS)*. John Wiley & Sons, Ltd. Chichester.

Restriction: A Microbial Free Lunch?

Merry Youle

Mise-en-Scène: *Billions of years ago microbes adapted a subset of nucleases to restrict phage infection. About forty years ago, molecular biologists adapted these restriction endonucleases (REs) to serve a multitude of purposes for research and beyond. Type II REs are especially helpful as they are single-function enzymes that consistently cleave DNA at a specific site. The library of characterized REs isolated from various microbes includes enzymes targeting hundreds of different DNA sequences. These have been absolutely essential for cutting and pasting together pieces of DNA in a test tube—the generation of recombinant DNA molecules. This methodology revolutionized research across the entire spectrum, from DNA sequencing to the development of pharmaceuticals and new approaches to agriculture and medicine. All of these benefits, and more to come, we owe to the ongoing arms race between the microbes and their phages.*

Microbes are unremittingly beset by mobile genetic elements of diverse forms, all of which use one trick or another to introduce their nucleic acid into the cell. Any microbial counter measure mobilized to destroy these invaders requires, at a minimum, the ability to distinguish self from non-self—the essence of innate immunity. One tactic used by almost all Bacteria and Archaea is to label their own DNA in some fashion, then destroy any untagged DNA that trespasses. Specifically, they carry a two-part **restriction-modification system** (R-M system) to *modify* their DNA and to *restrict* infection by degrading any unmodified DNA. This has spawned a microbial cottage industry producing diverse DNA modification enzymes, as well as an army of cognate restriction endonucleases (REs) that carry out the destruction. In each case, the modification enzyme recognizes and modifies a specific nucleotide in a string of 4-8 nucleotides. The RE checks the same sequence and cleaves the DNA if it has not been correctly modified on at least one of the two strands. Since these armaments are frequently exchanged between bacterial strains, an infecting phage may find itself face to face with a battery of endonucleases targeting diverse sites in its genome.

Modification by methylation

The most common method of DNA tagging is methylation. For this purpose, microbes express one or more dedicated **methyltransferases** (MTases), each of which adds a methyl group to one specific nucleotide—usually an adenine or cytosine—in each strand within its particular recognition sequence. Since these bulky methyl adjuncts extend into the major groove of the DNA helix, they do not interfere with normal base pairing. The preferred substrate for the methylation enzymes is often the hemimethylated DNA resulting from semi-conservative replication of bacterial and archaeal chromosomes. However, most can also modify fully unmethylated DNA such as newly replicated phage genomes. Thus when an infecting phage slips past the defenses, as some inevitably do, the progeny virions released at lysis will have been fully modified by that host's MTases. If those progeny infect a nearby cell that uses the same R-M system, the cell's REs will treat the phage DNA as 'self' and will not attack. To shrink this loophole, Bacteria maintain a diverse mixture of different R-M systems within a population. This increases the odds that there will be at least one RE in the next host able to nail these audacious phages.

Restriction strikes

Restriction is accomplished by the patrolling REs. Several thousand different REs are known that altogether recognize and attack hundreds of distinct target sites (termed **recognition sites** or restriction sites). Functionally they fall into a few groups. Type I REs are multifunctional proteins that also methylate one strand if the complementary strand is methylated, whereas the Type III REs link up

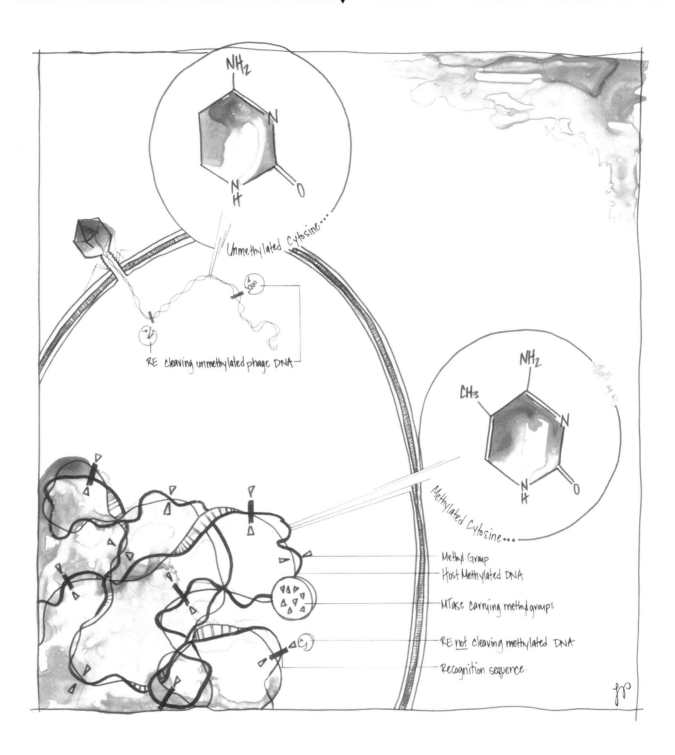

with an MTase to form the functional R-M complex. In Type II R-M systems, the RE and MTase are separate enzymes; these REs act independently and cleave DNA at a precise location within their recognition sequence each and every time. Lastly, a few Bacteria have a specialized Type IV RE that attacks only DNA with modified bases—a strategy to counter one of the phage anti-restriction strategies (*see page 4-25*).

REs recognize their target sequence and also evaluate its methylation state. To avoid attacking the host chromosome, they strike only when *both* strands are unmethylated. Since all REs cut both strands when restricting DNA, a single strike is death for the phage. But the abundant phages can succeed by playing a probability game. An incoming foreign genome containing recognition sequences that lack the sanctioned methylation pattern is a potential substrate for the host's MTases as well as host REs. There is always a small chance that the DNA will be methylated before it is restricted. When a phage genome does beat the odds and survive by getting methylated first, all of the progeny genomes released at **lysis** will be fully methylated and immune to all REs encoded by that host strain. Typical escape rates by this route range from one phage in a million to one in a hundred. The more restriction sites in the genome, the more likely it is that *one* will be found by an RE and cleaved before *all* have been methylated. To lower the rate of escape further, many Bacteria and Archaea carry multiple R-M systems, each with its own recognition sequence, each with its own low probability of phage escape. Multiple systems stack the odds even higher in favor of the host.

Artful dodging

If all this sounds grim for your favorite phage, fear not. The phages have a diverse array of counter measures to avoid restriction. These include eliminating or masking restriction sites, modifying the bases in their DNA (*see page 4-25*), inhibiting REs (*see page 4-17*), encoding their own methyltransferase, and stimulating the activity of the host methyltransferase, among others. Each phage defense is countered by some host innovation, and vice versa, maintaining a balance in which neither loses, and both win—long term.

Short term it is a different story. Restriction means end of game for the loser and a free nucleotide lunch for the winner.

Recommended reviews

Loenen, WA, EA Raleigh. 2014. The other face of restriction: Modification-dependent enzymes. Nucleic Acids Res 42:56-69.

Makarova, KS, YI Wolf, EV Koonin. 2013. Comparative genomics of defense systems in archaea and bacteria. Nucleic Acids Res 41:4360-4377.

Samson, JE, AH Magadán, M Sabri, S Moineau. 2013. Revenge of the phages: Defeating bacterial defences. Nature Rev Microbiol 11:675-687.

Wilson, GG, NE Murray. 1991. Restriction and modification systems. Annu Rev Genet 25:585-627.

4,111

restriction endonucleases targeting

397

different recognition sites in phage DNA

Data as of August 23, 2014.
Source: http://rebase.neb.com/rebase/statlist.html

Yersinia Phage φA1122

a T7-like Podophage that employs a DNA mimic to evade restriction endonucleases

Genome
dsDNA; linear
37,555 bp
50 predicted ORFs; 0 RNAs

Encapsidation method
Packaging; T = 7 capsid

Common host
Yersinia pestis

Habitat
Mammalian intestines

Lifestyle
Lytic

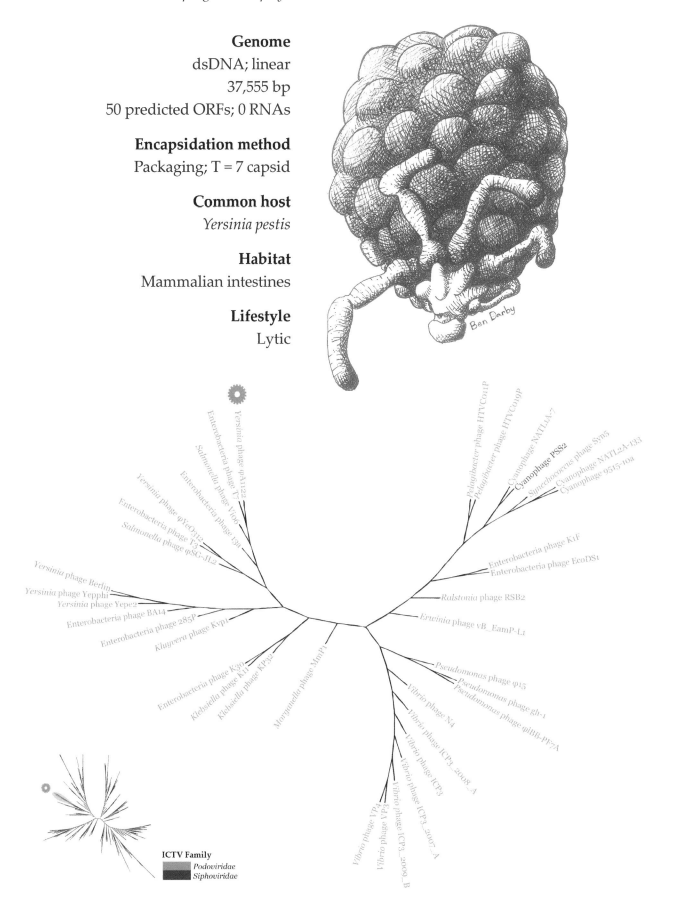

Ben Darby

ICTV Family
Podoviridae
Siphoviridae

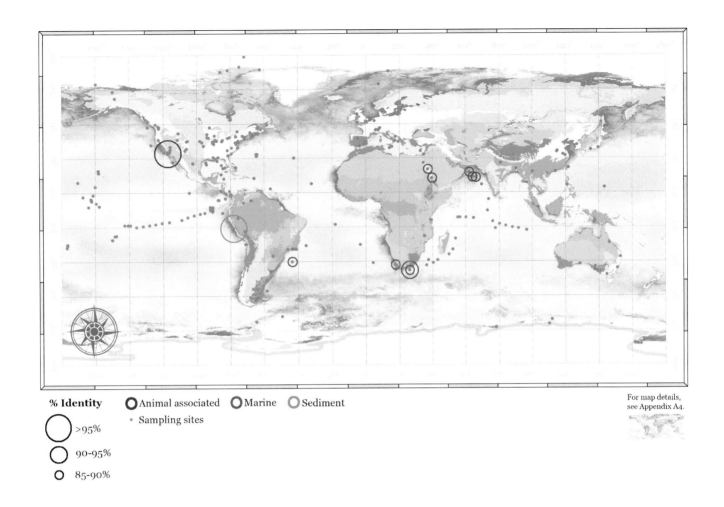

% Identity ◯ Animal associated ◯ Marine ◯ Sediment
• Sampling sites

◯ >95%

◯ 90-95%

○ 85-90%

For map details,
see Appendix A4.

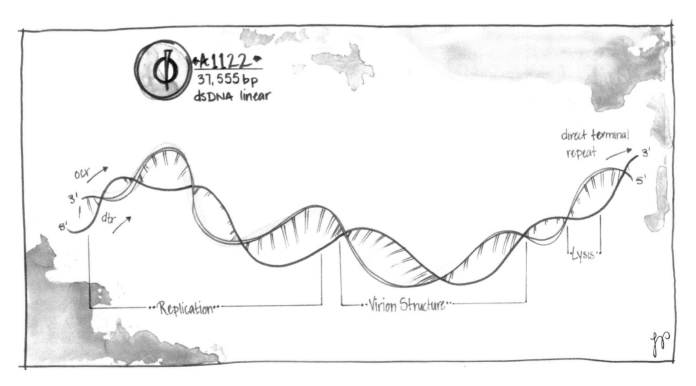

φA1122
37,555 bp
dsDNA linear

ocr

3'

5'

dtr

direct terminal
repeat   3'

5'

Lysis

Replication

Virion Structure

4-15

Yersinia Phage φA1122

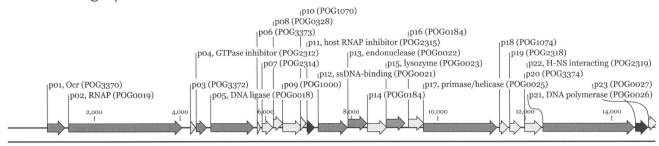

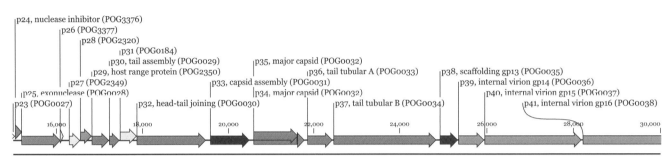

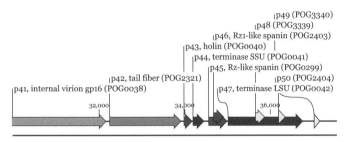

On the Prowl	Entry	Takeover	Replication Lytic	Replication Lysogenic	Assembly	Escape	Transfer RNA	Non-coding RNA	Multiple	Unknown

Unrestricted Entry by DNA Mimicry

Merry Youle

Mise-en-Scène: *Faced with lethal attack by host restriction endonucleases (REs), phages developed numerous tactics to disarm that host defense. The activity of REs, like that of other enzymes, can be curbed by molecules that compete with the normal substrate for access to the enzyme's active site. Such competitive inhibitors are often close structural analogs of the substrate. That each RE acts on a slightly different DNA substrate suggests that a different inhibitor would be required for each RE. The phages wisely evaded this requirement. Capitalizing on what is common to all RE substrates, they fashioned a generic one-size-fits-all DNA mimic that could inhibit whole RE families.*

Note: Most of the research related in this story was performed using coliphage T7, but *Yersinia* phage φA1122 also encodes the mimic, protein Ocr.

Unprotected, a double-stranded DNA phage genome entering an intended host will likely be quickly **restricted** and served up as lunch for the bacterium—nucleotides from heaven (*see page 4-10*). Suppose the host is one of the common lab strains of *E. coli* (e.g., K12, B). Each strain possesses some of *E. coli*'s roster of more than 600 **restriction endonucleases** (REs) that monitor many different recognition sequences for the sanctioned pattern of methylation. Even a sedulous phage can't shield or eliminate *every* potential **recognition site**. How to dodge them all with one neat trick? Some T7-like phages, including *Yersinia* phage φA1122, have a way: they inhibit two families of REs irrespective of their target sequence (Krüger, Schroeder 1981). The inhibitor is a protein that they synthesize straightaway upon arrival, a protein that impersonates a segment of DNA.

The first minutes

T7's genome does contain restriction sites that are recognized and that, if not methylated appropriately, will be cleaved by various REs on patrol in its host. Nevertheless, unmethylated T7 genomes infect common *E. coli* strains without being restricted. This dodging relies on a two-step maneuver. First, T7 delivers its genome slowly, spending 9-12 min in total (*see page 3-17*). For the first 6-7 min, some not yet identified mechanism—perhaps compartmentalization—protects its DNA from REs but does not block access or transcription by RNA polymerase (Moffatt, Studier 1988). This buys T7 time to synthesize proteins for its second line of genome defense. The first protein it **translates** upon arrival, Ocr (Overcome Classi-

cal Restriction; gp0.3), is on hand by 3-4 min post-infection (Studier 1975). It immediately intervenes and effectively inhibits all Type I and Type III REs for the duration of the infection. Without a functional *ocr* gene, T7 makes a quick lunch.

Mimicry

Ocr's success is rooted in its structure. Its active form is a 26 kDa banana-shaped dimer that resembles the shape and charge distribution of a segment of DNA. Moreover, this is a precisely bent banana, bent to match the bend in the DNA helix when bound to a Type I or Type III RE. Of Ocr's 116 amino acids, 34 are negatively-charged aspartic or glutamic acid residues strategically placed to mimic the charge distribution of DNA's phosphate backbone (Dunn et al. 1981; Walkinshaw et al. 2002). The net result is a protein that compellingly mimics DNA, specifically DNA that has been bent to fit readily in the active site of all Type I and Type III REs. So adroit is this protein that it binds even more readily to the RE enzymes than does DNA, thereby keeping the RE fully occupied (Atanasiu et al. 2001). It doesn't take many copies of Ocr to neutralize the host's defenses. The rapid synthesis of 10-40 Ocr dimers by the incoming phage is sufficient to counter the host's pool of a common Type I RE (EcoKI) (Zavilgelsky, Kotova, Rastorguev 2008).

Ocr can't fool all host REs. In particular, it is ineffective against the Type II restriction-modification systems that use separate enzymes to carry out methylation and restriction. Nonetheless, this small DNA mimic can fool some—actually many—of the REs all the time.

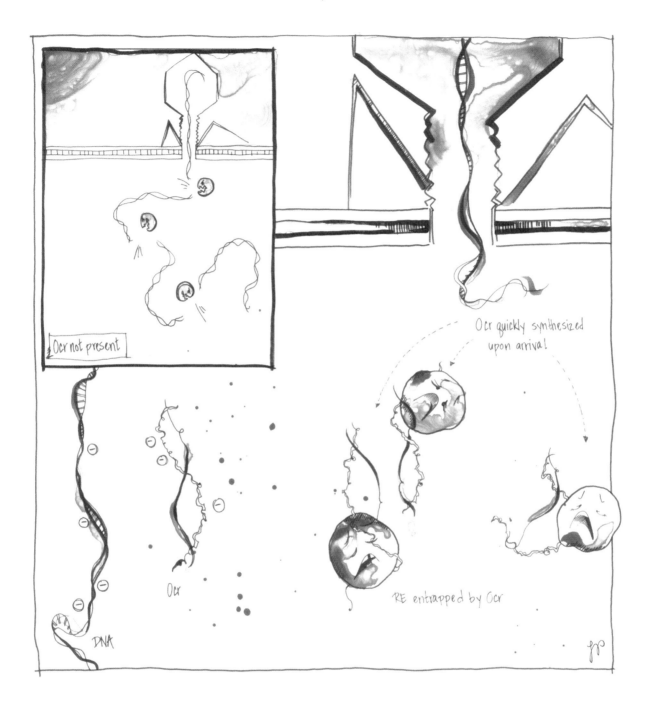

Cited references

Atanasiu, C, O Byron, H McMiken, S Sturrock, D Dryden. 2001. Characterisation of the structure of *ocr*, the gene 0.3 protein of bacteriophage T7. Nucleic Acids Res 29:3059-3068.

Dunn, JJ, M Elzinga, K-K Mark, FW Studier. 1981. Amino acid sequence of the gene 0.3 protein of bacteriophage T7 and nucleotide sequence of its mRNA. J Biol Chem 256:2579-2585.

Krüger, D, C Schroeder. 1981. Bacteriophage T3 and bacteriophage T7 virus-host cell interactions. Microbiol Rev 45:9-51.

Moffatt, BA, FW Studier. 1988. Entry of bacteriophage T7 DNA into the cell and escape from host restriction. J Bacteriol 170:2095-2105.

Studier, FW. 1975. Gene 0.3 of bacteriophage T7 acts to overcome the DNA restriction system of the host. J Mol Biol 94:283-295.

Walkinshaw, M, P Taylor, S Sturrock, C Atanasiu, T Berge, R Henderson, J Edwardson, D Dryden. 2002. Structure of Ocr from bacteriophage T7, a protein that mimics B-form DNA. Mol Cell 9:187-194.

Zavilgelsky, G, VY Kotova, S Rastorguev. 2008. Comparative analysis of anti-restriction activities of ArdA (Collb-P9) and Ocr (T7) proteins. Biochem (Mosc) 73:906-911.

Recommended review

Dryden, DTF. 2006. DNA mimicry by proteins and the control of enzymatic activity on DNA. Trends Biotechnol 24:378-382.

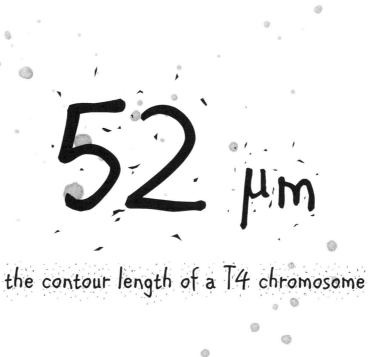

52 μm

the contour length of a T4 chromosome

Lang, D. 1970. Molecular weights of coliphages and coliphage DNA: III. Contour length and molecular weight of DNA from bacteriophages T4, T5 and T7, and from bovine papilloma virus. J Mol Biol 54:557-565.

Enterobacteria Phage RB69

a Myophage whose lineage has countered a long series of advances in host restriction endonucleases

Genome
dsDNA; linear
167,560 bp
275 predicted ORFs; 2 RNAs

Encapsidation method
Packaging; T = 13 capsid

Common host
Escherichia coli

Habitat
Mammalian intestines

Lifestyle
Lytic

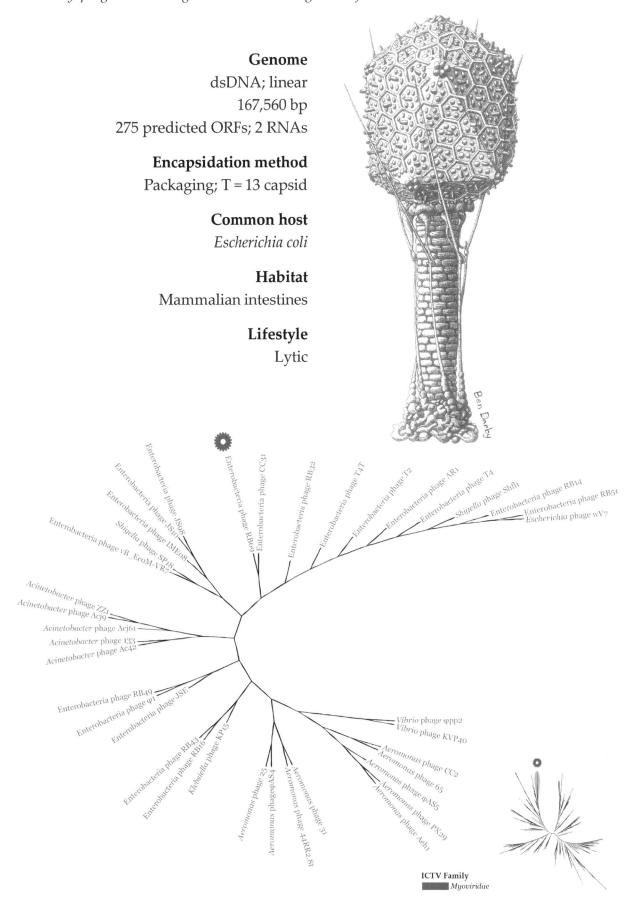

ICTV Family
Myoviridae

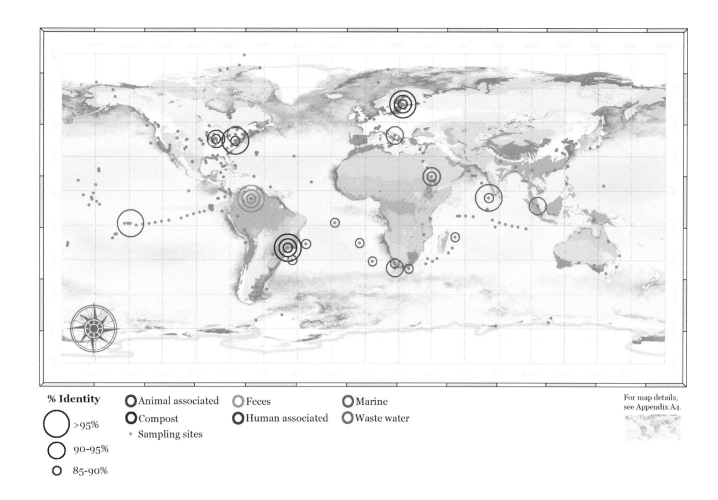

% Identity

○ >95%

○ 90-95%

○ 85-90%

○ Animal associated
○ Compost
• Sampling sites

○ Feces
○ Human associated

○ Marine
○ Waste water

For map details,
see Appendix A4.

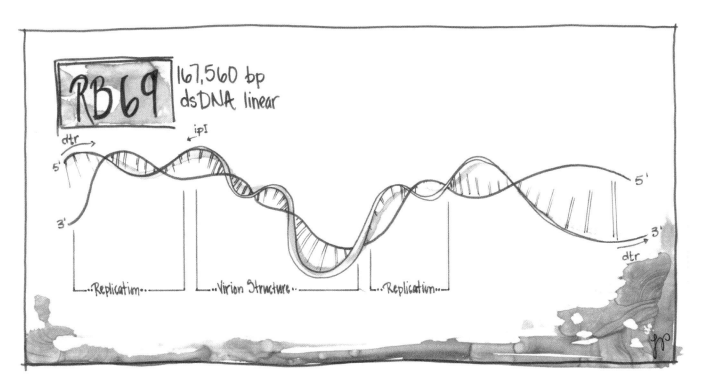

Enterobacteria Phage RB69

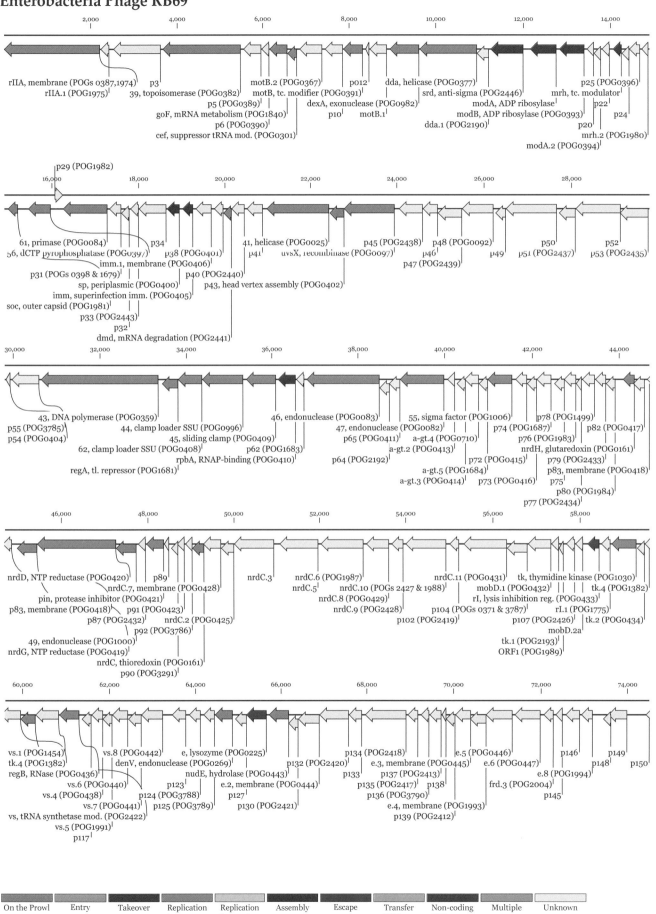

On the Prowl   Entry   Takeover   Replication   Replication   Assembly   Escape   Transfer   Non-coding   Multiple   Unknown
                                    Lytic   Lysogenic   RNA   RNA

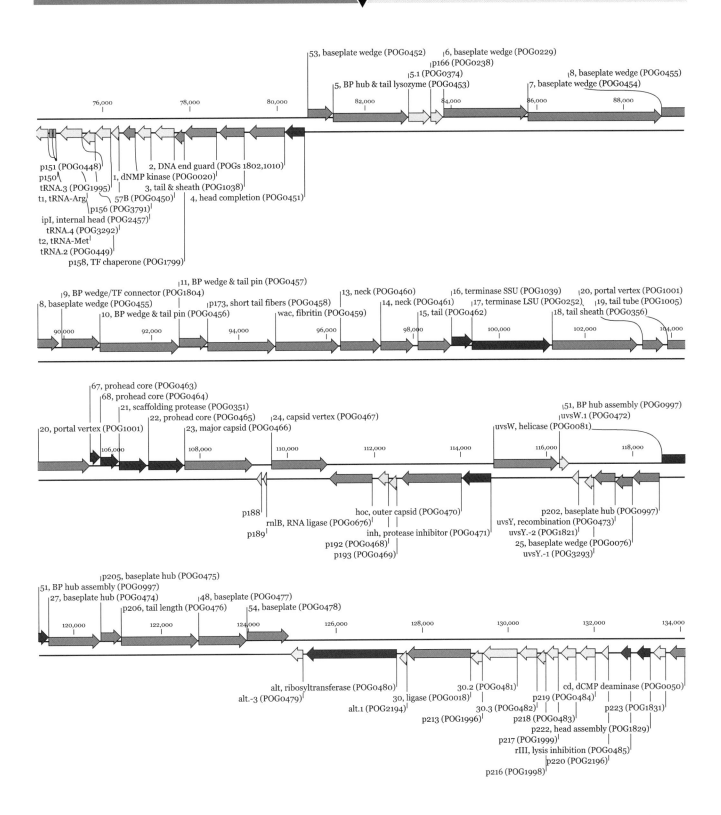

Continued next page

Enterobacteria Phage RB69 *continued*

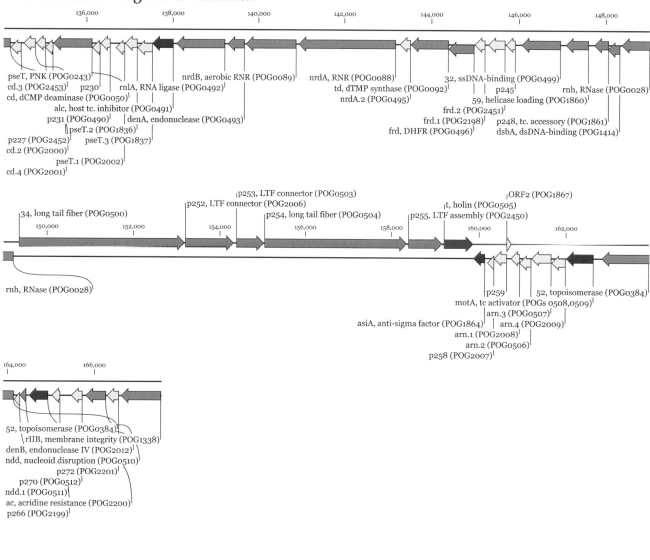

52, topoisomerase (POG0384)
rIIB, membrane integrity (POG1338)
denB, endonuclease IV (POG2012)
ndd, nucleoid disruption (POG0510)
p272 (POG2201)
p270 (POG0512)
ndd.1 (POG0511)
ac, acridine resistance (POG2200)
p266 (POG2199)

| On the Prowl | Entry | Takeover | Replication Lytic | Replication Lysogenic | Assembly | Escape | Transfer RNA | Non-coding RNA | Multiple | Unknown |

A Perpetual Arms Race

Merry Youle

Mise-en-Scène: *Evolutionary biologists followed their curiosity down a rabbit hole and soon arrived in Looking-Glass Land. A strange land indeed! As the Red Queen explained:* Now, here, you see, it takes all the running you can do, to keep in the same place. If you want to get somewhere else, you must run at least twice as fast as that! *When they returned from Looking-Glass Land to the real world, her words stuck with them. They seemed an apt description of the evolutionary arms race observed between predators and their prey—now termed Red Queen dynamics to acknowledge her insight. The term is eminently useful in Phageland where both phage and host continue, generation after generation, to run as fast as they can.*

Note: The following is a soundly-based hypothetical reconstruction of eons of history for one particular arms race between coliphage T4 and its *E. coli* hosts. Most of the research related in this story was performed using coliphage T4, but coliphage RB69 also glycosylates the HMCs in its DNA and encodes the inhibitor IP1.

Round 1: Once upon a time

E. coli swaggers into the ring armed with Type I, II, and III **restriction endonucleases** (REs), intent on converting any invading phage genome into a nucleotide snack (*see page 4-10*).

Despite such wicked defenses, an ancestral T-even phage could always eke out a meager living. **Restriction** is never absolute. Some arriving phage genomes are methylated by the host enzymes before the REs strike. Typically, depending on the type of phage, one in a hundred to one in a million infecting phages escapes restriction. Its progeny emerge fully methylated, thus immune to all the **restriction-modification (R-M) systems** encoded by that host.

Ceteris paribus, the more restriction sites in the genome, the more likely it is that *one* will be found by an RE and cleaved before *all* have been methylated. Thus, reducing the number of potential target sites increases the chance of phage survival. One way to do this without altering the information encoded is to methylate cytosines located within those sites. Such tactical methylation protected some coliphages from specific host REs until…

Round 2: Trading bases

… *E. coli* countered with R-M systems that specifically seek and destroy modified DNA, i.e., modification dependent systems. These are diverse families, all members of which are specialists in cleaving DNA that contains methylcytosine within its particular recognition site (Raleigh, Wilson 1986).

Although these new REs were formidable weapons collectively poised to attack many different genome sequences, the T-even phages evaded them all with one deft move. They substituted 5-hydroxymethylcytosine (HMC) for every cytosine in their genome (Wyatt, Cohen 1953). This sounds simple, but implementing this swap required several new phage capabilities (Warren 1980). First, the phages must exclude all cytosines from their own replicating DNA. For this they sabotage the supply of dCTP, the precursor for DNA synthesis, by expressing a dCTPase that converts dCTP and dCDP to dCMP. With typical phage frugality, they do not waste this dCMP. Instead they convert it into the HM dCTP needed for synthesis of their DNA. The usual DNA polymerases do not consider HM dCTP to be an acceptable nucleotide, but the DNA polymerase encoded by these phages does.

As a fringe benefit, this global replacement of genomic cytosine with HMC allowed the phage to distinguish between host DNA and phage DNA. The host DNA they now selectively dismantle using a pair of phage-encoded endonucleases that degrade only cytosine-containing DNA (Sadowski, Hurwitz 1969). They selectively transcribe their own DNA by modifying the host RNA polymerase to preferentially transcribe HMC-containing DNA (Kutter, Wiberg 1969; Kutter et al. 1981).

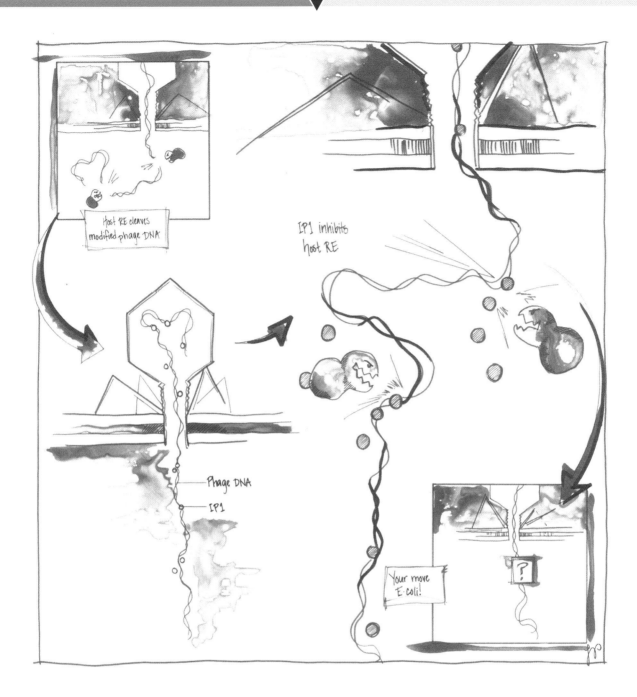

At this point, the T-even phages seem to be winning. They have neutralized the latest host defenses and, in so doing, have gained new options for efficient takeover of their host (*see page 4-7*).

Round 3: A sweeter phage

E. coli K-12's response? A new and enhanced restriction system that specifically cleaves HMC-containing DNA (Raleigh, Wilson 1986). These include the McrBC (<u>M</u>odified <u>c</u>ytosine <u>r</u>estriction) protein (Sutherland, Coe, Raleigh 1992). Under this new attack, the T-even phages sweeten their personalities. After the HMC bases are incorporated into phage DNA, phage-encoded enzymes (α-

and β-glucosyltransferases) attach a sugar to each one. Here the various T-evens display distinctive styles, differing in their choice of sugars, linkage, and percent of HMCs glucosylated. T4 dresses every HMC with glucose, 70% via an α linkage, 30% via a β linkage. This accomplished, T4 and its close relatives enjoy immunity to defenses such as McrBC, and are still resistant also to Type I, II, and III REs. Once again, T4 is one-up and can barge in boldly.

Round 4: Home delivery service

You can guess what happened next. Some pathogenic *E. coli* strains, e.g., CT596, developed a taste for sweetened DNA. This strain encodes a two-component restriction system, the functional unit being the heterodimer of two proteins: GmrS/GmrD (Bair, Black 2007). When a T-even phage arrives bearing glucosylated DNA, CT596 cleaves the invading DNA. This latest bacterial advance was built on existing restriction systems, now updated to cope with sweetened DNA. Many diverse Bacteria carry **homologs** of GmrS/GmrD, likely variants to counter the different sugars and linkages encountered in the DNAs of their phages. Much to the consternation of the phages, these useful genes hop between strains or species courtesy of mobile genetic elements (Black, Abremski 1974).

Some phages are stumped by this new defense, but not T4 or the related phage RB69. When they deliver their genome during infection, they send along protection in the form of ~360 copies of a GmrS/GmrD inhibitor known as **Internal Protein One**, or IP1 (Bair, Rifat, Black 2007). IP1 is one of three small basic internal proteins (IPs) that are encapsidated in hundreds of copies each along with T4's genome and that assist with DNA compaction. In addition, delivery of IP1 along with the genome also blocks these specific REs, a function not required when infecting most *E. coli* strains (Black, Abremski 1974). Of necessity, IP1 is a small protein, only 76 amino acids when packaged in the mature virion (Isobe, Black, Tsugita 1977). Moreover it is compact, so compact (30 × 30 × 15 Å) that it can exit with the DNA through the 30 Å portal and tail tube aperture without unfolding (Rifat et al. 2008). Also, whereas most protein virion components are made only late in infection, IP1 is synthesized at a very high rate by two min

after infection (Cowan et al. 1994), thereby providing increased nuclease protection for the phage.

Ongoing variation in *E. coli's* weaponry is countered by strategic diversification in the phage defense. The *ip1* locus in T4-related phages is in a highly variable region that often encodes several of these IPs that are made early as well as late in infection. Each IP might be suspected of targeting a different RE or modulating some other host function, in addition to assisting in phage DNA compaction. Despite their diversity, they all possess a highly conserved N-terminal capsid targeting sequence that gets them packaged inside the capsid for delivery with the DNA (Isobe, Black, Tsugita 1977; Repoila et al. 1994).

Round 5: On the defensive again

IP1 bought T4 temporary protection from the GmrS/GmrD attack, not permanent shelter. Both uropathogenic *E. coli* UT189 and avian pathogenic *E. coli* already have a counter strategy. Instead of separate GmrS and GmrD proteins, they make a single fused GmrSD protein that has retained the ability to restrict glucosylated phage DNA. These fusion proteins are, respectively, 89% and 93% identical to the heterodimeric GmrS/GmrD of CT596. They are both immune to IP1 (Rifat et al. 2008). It is only a matter of time before this innovation spreads to more Bacteria.

What will be T4's response? Some mutants able to escape the fusion proteins are now showing up, but their tactics are not yet known. One thing we do know for certain: T4 will counter *E. coli's* latest move and the arms race will go on and on and on…

Cited references

Bair, CL, LW Black. 2007. A type IV modification dependent restriction nuclease that targets glucosylated hydroxymethyl cytosine modified DNAs. J Mol Biol 366:768-778.

Bair, CL, D Rifat, LW Black. 2007. Exclusion of glucosyl-hydroxymethylcytosine DNA containing bacteriophages is overcome by the injected protein inhibitor IP1*. J Mol Biol 366:779-789.

Black, LW, K Abremski. 1974. Restriction of phage T4 internal protein I mutants by a strain of *Escherichia coli*. Virology 60:180-191.

Cowan, J, K d'Acci, B Guttman, E Kutter. 1994. Gel analysis of T4 prereplicative proteins. In: JD Karam, KN Kreuzer, DH Hall, editors. Molecular Biology of Bacteriophage: ASM Press. p. 520-527.

Isobe, T, LW Black, A Tsugita. 1977. Complete amino acid sequence of bacteriophage T4 internal protein I and its cleavage site on virus maturation. J Mol Biol 110:165-177.

Kutter, E, D Bradley, R Schenck, B Guttman, R Laiken. 1981. Bacteriophage T4 alc gene product: General inhibitor of transcription from cytosine-containing DNA. J Virol 40:822-829.

Kutter, EM, JS Wiberg. 1969. Biological effects of substituting cytosine for 5-hydroxymethylcytosine in the deoxyribonucleic acid of bacteriophage T4. J Virol 4:439-453.

Raleigh, EA, G Wilson. 1986. *Escherichia coli* K-12 restricts DNA containing 5-methylcytosine. Proc Natl Acad Sci USA 83:9070-9074.

Repoila, F, F Tétart, J Bouet, H Krisch. 1994. Genomic polymorphism in the T-even bacteriophages. EMBO J 13:4181-4192.

Rifat, D, NT Wright, KM Varney, DJ Weber, LW Black. 2008. Restriction endonuclease inhibitor IP1* of bacteriophage T4: A novel structure for a dedicated target. J Mol Biol 375:720-734.

Sadowski, PD, J Hurwitz. 1969. Enzymatic breakage of deoxyribonucleic acid I. Purification and properties of endonuclease II from T4 phage-infected *Escherichia coli*. J Biol Chem 244:6182-6191.

Sutherland, E, L Coe, EA Raleigh. 1992. McrBC: A multisubunit GTP-dependent restriction endonuclease. J Mol Biol 225:327-348.

Warren, R. 1980. Modified bases in bacteriophage DNAs. Annu Rev Microbiol 34:137-158.

Wyatt, G, SS Cohen. 1953. The bases of the nucleic acids of some bacterial and animal viruses: The occurrence of 5-hydroxymethylcytosine. Biochem J 55:774.

Recommended reviews

Labrie, SJ, JE Samson, S Moineau. 2010. Bacteriophage resistance mechanisms. Nature Rev Microbiol 8:317-327.

Samson, JE, AH Magadán, M Sabri, S Moineau. 2013. Revenge of the phages: Defeating bacterial defences. Nature Rev Microbiol 11:675-687.

10^{28}

bp of DNA transduced per year by marine phages

Paul, JH, MB Sullivan, AM Segall, F Rohwer.
2002. Marine phage genomics. Comp Biochem
Physiol B Biochem Mol Biol 133:463-476.

Streptococcus Phage 2972

a Siphophage that demonstrates one way to quickly dodge a CRISPR defense

Genome
dsDNA; linear
34,704 bp
44 predicted ORFs; 0 RNAs

Encapsidation method
Packaging; T = ? capsid

Common host
Streptococcus thermophilus

Habitat
Bovine mammary mucosa & raw milk

Lifestyle
Lytic

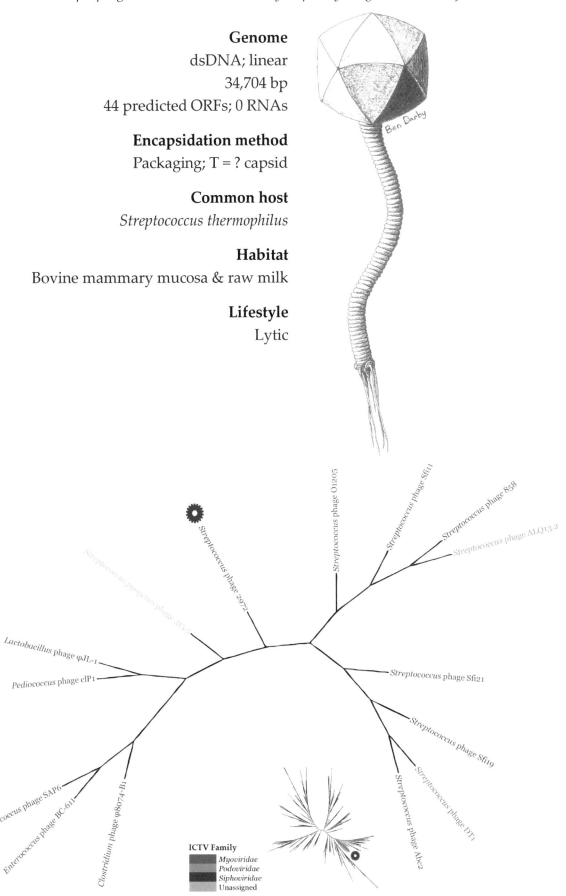

Ben Darby

Streptococcus phage O1205

Streptococcus phage Sfi11

Streptococcus phage 858

Streptococcus phage ALQ13.2

Streptococcus prophage phage φ1.3?

Streptococcus phage 2972

Lactobacillus phage φJL-1

Pediococcus phage clP1

Streptococcus phage Sfi21

Enterococcus phage SAP6

Enterococcus phage BC-611

Clostridium phage φ8074-B1

Streptococcus phage Sfi19

Streptococcus phage DT1

Streptococcus phage Abc2

ICTV Family
Myoviridae
Podoviridae
Siphoviridae
Unassigned

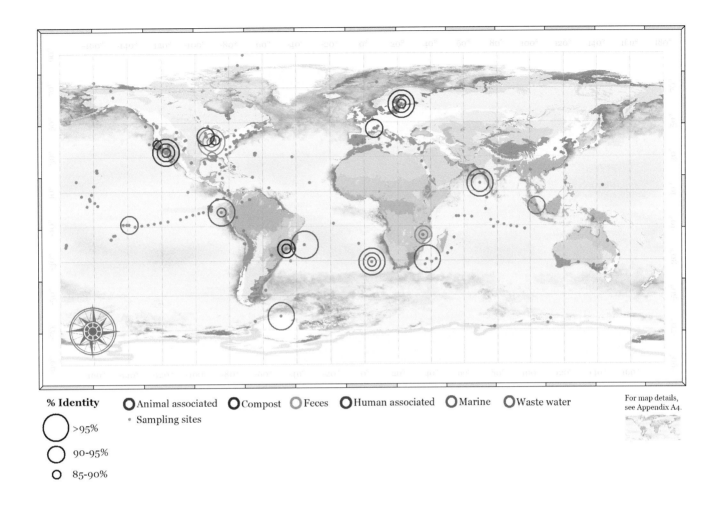

For map details,
see Appendix A4.

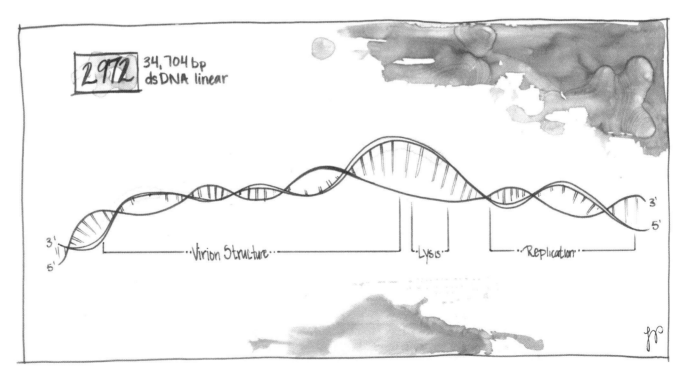

Streptococcus Phage 2972

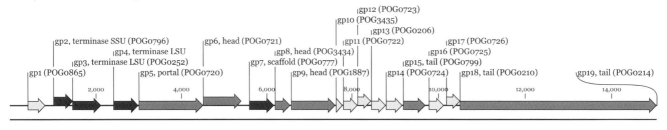

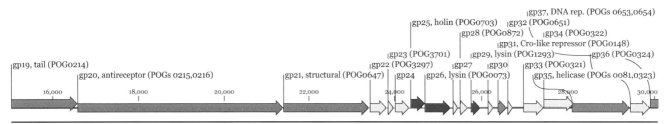

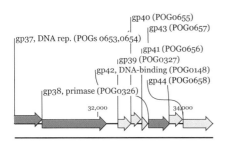

| On the Prowl | Entry | Takeover | Replication Lytic | Replication Lysogenic | Assembly | Escape | Transfer RNA | Non-coding RNA | Multiple | Unknown |

Haven't We Met Before?

Merry Youle

Mise-en-Scène: *Is there an optimal anti-phage defense? Is it better to prevent the wolf from ever getting in the door, or should you let him in, then dice him and eat him for lunch? Host surface modifications to block phage adsorption or genome entry often reduce fitness and also are rapidly countered by the phages. Restriction endonucleases can be dodged by mutation and are, at best, a 'leaky' defense. CRISPRs (see below) offer an adaptive defense that incurs little immediate fitness cost and comes with an intrinsic update mechanism that keeps pace with phage evolution. CRISPRs also defend against other mobile genetic elements, such as plasmids and transposons. To be the ideal anti-phage tactic, CRISPR-associated endonucleases need to discriminate between friend and foe. Some invading phages are temperate and, of those, some will benefit their host as prophages (see page 1-5). A defense that excludes prophages may have long-term fitness costs. While fewer prophages are found in some Bacteria that possess an active CRISPR defense, other discriminating CRISPR systems are tolerant of prophages—until the prophage excises and launches a lytic infection (Goldberg et al. 2014).*

To a 'dairy phage' eyeing a plump *Streptococcus thermophilus*, that bacterium cell represents a packet of nutrients that could be put to good use by being converted into more phage. With that goal in mind, the phage delivers its DNA into the cell. The cell, however, is well prepared. Its surveillance systems are on patrol, keen on converting incoming DNA into nucleotides for synthesizing more bacterial DNA. This particular *S. thermophilus* cell employs a CRISPR guard. This sentinel scans all incoming DNA and compares it with the cell's library of sequences that match previously encountered phages. When there is a match, the guard's associated endonuclease cleaves the DNA, thus callously ending the phage's replication fantasy. CRISPR defenses currently shield 39% of the Bacteria and 88% of the Archaea whose genomes have been sequenced (Van Der Oost et al. 2009).

The spacer library

Key to CRISPR effectiveness is the library of records of previous phage encounters that is handed down to each succeeding generation. Libraries are kept up-to-date with ongoing acquisitions from the current phage community. New entries, termed 'spacers,' are added when the bacterium not only survives the attempted infection but gains a copy of a 'protospacer' segment (26–72 bp long) from the phage's DNA. The original protospacers can be on either strand of the phage genome, in coding or non-coding regions, seem-

ingly from any location provided there is a short protospacer-associated motif (PAM) nearby. The bacterium inserts each new spacer into its own genome at a specific location within its CRISPR/Cas locus. There the spacer joins an array of previously introduced spacers—sometimes one hundred or more—that alternate with identical copies of a palindromic 'repeat' sequence (21-48 bp) that are part of the CRISPR machinery (Deveau, Garneau, Moineau 2010).

Located adjacent to the spacer array are Cas genes that encode the proteins needed to carry out both spacer acquisition and invader destruction. Thus the complete CRISPR/Cas acronym: Clustered Regularly Interspaced Short Palindromic Repeats/ CRISPR-associated proteins. The array is constitutively transcribed, and then the transcripts are processed into short CRISPR RNAs (crRNAs), one per spacer. Each crRNA recognizes and specifically base-pairs with the complementary protospacer sequence in the incoming DNA, then targets it for destruction by the Cas endonucleases.

One or more libraries associated with a set of Cas genes comprise a functional CRISPR locus; both Bacteria and Archaea can encode multiple loci, each with a different library and different Cas genes. A locus can be transferred in toto by horizontal gene transfer (HGT) and immediately put to work by the fortunate recipient. Imagine the benefit to a besieged CRISPR-less bacterium when a functional, preprogrammed CRISPR/Cas cassette arrives!

Defeated?

Will the spread of CRISPR/Cas technology condemn our dairy phage to homelessness? Fear not. Phages have numerous ways to evade or counter even this sophisticated weapon. One tactic takes advantage of the specificity of some bacterial CRISPR systems that attack only if the phage DNA contains a perfect match to an archived spacer *and* also has the correct PAM nearby. Here, all a bacteriophage needs in order to escape is a single base substitution in the protospacer or PAM (Deveau et al. 2008). Other bacterial CRISPRs and most archaeal CRISPRs will ignore a considerable number of mismatches (Manica et al. 2013). Still, given phage abundances and the typical occurrence of one spontaneous mutation in every 300-400 phage genomes (Drake et al. 1998), such escape mutants arise frequently in a phage population. If the host has archived multiple different spacers that match the phage, the frequency of escapees is greatly reduced, but it is not reduced to zero. An escapee phage will prosper and multiply, and its progeny will inherit access to this host—but only until the host acquires a new spacer from some other location in that phage genome.

You can watch this rapid point-counterpoint in the dairy phage, *Streptococcus thermophilus* bacteriophage 2972 (Sun et al. 2013). Researchers introduced wild-type phage 2972 into a population of sensitive *S. thermophilus*, let them fight it out for one week, then examined the survivors. All the surviving Bacteria carried the same newly-acquired spacer targeting 2972, evidence that they all descended from one cell—likely the first one—that managed to acquire a defensive spacer during the slaughter. Likewise, after one week there were none of the original wild-type phage remaining. Selection pressure from the CRISPR/Cas defense was so strong that the current phage lineages all derived from the phage with the first escape mutation. In these phage-host contests, the advantage perpetually see-saws from one side to the other

with no decisive winner. Sensitive strains on either side die out, only to be replaced by a resistant variant. The same dynamic game is being played in your gut (Minot et al. 2013).

Counterattack

But why should a smart phage settle for playing this never-ending game of evasion? Why not knock out some critical component of the CRISPR mechanism and have done with it? Some temperate phages of *Pseudomonas aeruginosa* do exactly that (Bondy-Denomy et al. 2013). Collectively these phages encode at least five different proteins that can each disable CRISPR defenses. Suddenly it is apparent that PAMs and protospacers alone do not determine phage destiny. These phage counterattacks might be one of the forces that has driven the evolution of diverse CRISPR systems and that has prompted some hosts to employ more than one.

Knocking out your host's CRISPR equipment is not the perfect strategy as it would also benefit the other phage types competing for the same host. 'Knowing' this, some archaeal viruses avail themselves of a less charitable tactic: they redirect the host's CRISPRs to actively target competing virus types while ignoring the virus's own genome. One such virus is the fusiform SMV1 (*Sulfolobus* monocaudavirus 1) that dwells in acidic hot springs along with its extremophilic *Sulfolobus* hosts. Another fusiform archaeal virus, STSV1 (*Sulfolobus tengchongensis* spindle-shaped virus 1), infects the same species. During co-infection, SMV1 triggers immediate hyperactive acquisition of spacers by the host but only spacers from STSV1 (Erdmann, Le Moine Bauer, Garrett 2014). The net result? Immunity for SMV1 and death to STSV1.

A co-opted CRISPR

By redirecting its host's CRISPR defense, SMV1 gains some CRISPR clout without encoding its own weaponry. As you would expect, some phages have gone a step farther and have brought a CRISPR defense in house. By encoding their own CRISPR machinery, they can attack their enemies without needing host coop-

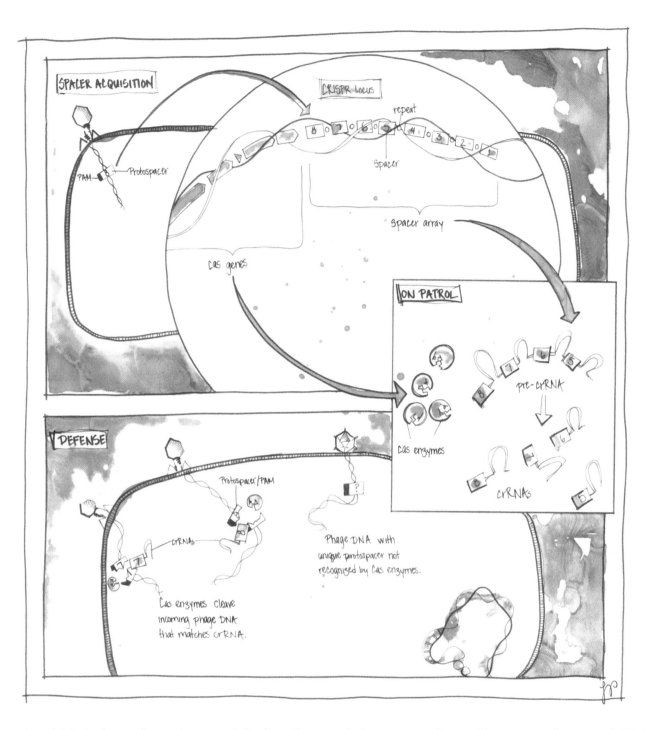

eration. This independence is essential when the phage's nemesis is allied with its host against the phage. For example, some strains of *Vibrio cholerae* O1 carry an 18 kbp chromosomal island (PLE) that protects the cell from infection by many vibriophages. When one of these phages launches an infection, the PLE excises from the host chromosome, replicates, and in the process also blocks phage replication. The story has a different ending when the infecting phage is vibriophage ICP1. Armed with its own active CRISPR

defense, complete with spacers that match PLE sequences, ICP1 successfully replicates in these hosts (Seed et al. 2013). This stratagem puts vibriophages such as ICP1 on our side, as their killing of *V. cholerae* has been implicated in the decline of the semiannual cholera epidemics in Bangladesh and vicinity (Seed et al. 2011).

Using the archive

Any CRISPR spacer array represents a record of past phage encounters and primes the host for rapid response if reinfected—the hallmark of adaptive immunity. Spacer libraries are continually updated by new acquisitions, and the most recently acquired spacers are transcribed at a higher rate. Array size is kept in check by the deletion of older spacers, likely the result of **homologous recombination** between repeat sequences. This loss of old spacers may be no loss at all as these spacers are apt to target only a memory of a phage. Likely by now the targeted phage, evolving rapidly as phages do, will have altered either the protospacer or PAM.

Phage researchers also put these CRISPR archives to work for them when analyzing metagenomes from environmental samples (Andersson, Banfield 2008). Identifying metagenomic reads and scaffolds as phage-derived is difficult because most phage genes are not yet in the databases. However, spacer sequences in a metagenome, nested as they are between two repeats, can be readily discerned. Since the spacer is likely of phage origin, its sequence can be added to the database of known phage genes, thus enabling assignment of a phage origin to otherwise unidentifiable metagenomic scaffolds. Further refinements of this approach have revealed many 'host' metabolic genes to be encoded by phage communities. Moreover, when sequences in viral metagenomes match spacers in known bacterial or archaeal genomes, the source phages can be linked to their hosts (Anderson, Brazelton, Baross 2011).

A spacer array is a 'fossil record' of past phage encounters by the host lineage. The relative position of spacers in the array embodies the history of their sequential acquisition—an open book waiting to be read by phage archaeologists.

Cited references

Anderson, RE, WJ Brazelton, JA Baross. 2011. Using CRISPRs as a metagenomic tool to identify microbial hosts of a diffuse flow hydrothermal vent viral assemblage. FEMS Microbiol Ecol 77:120-133.

Andersson, AF, JF Banfield. 2008. Virus population dynamics and acquired virus resistance in natural microbial communities. Science 320:1047-1050.

Bondy-Denomy, J, A Pawluk, KL Maxwell, AR Davidson. 2013. Bacteriophage genes that inactivate the CRISPR/Cas bacterial immune system. Nature 493:429-432.

Deveau, H, R Barrangou, JE Garneau, J Labonté, C Fremaux, P Boyaval, DA Romero, P Horvath, S Moineau. 2008. Phage response to CRISPR-encoded resistance in *Streptococcus thermophilus*. J Bacteriol 190:1390-1400.

Deveau, H, JE Garneau, S Moineau. 2010. CRISPR/Cas system and its role in phage-bacteria interactions. Annu Rev Microbiol 64:475-493.

Drake, JW, B Charlesworth, D Charlesworth, JF Crow. 1998. Rates of spontaneous mutation. Genetics 148:1667-1686.

Erdmann, S, S Le Moine Bauer, RA Garrett. 2014. Inter-viral conflicts that exploit host CRISPR immune systems of *Sulfolobus*. Mol Microbiol 91:900-917.

Goldberg, GW, W Jiang, D Bikard, LA Marraffini. 2014. Conditional tolerance of temperate phages via transcription-dependent CRISPR-Cas targeting. Nature: 10.1038/nature13637

Manica, A, Z Zebec, J Steinkellner, C Schleper. 2013. Unexpectedly broad target recognition of the CRISPR-mediated virus defence system in the archaeon *Sulfolobus solfataricus*. Nucleic Acids Res 41:10509-10517.

Minot, S, A Bryson, C Chehoud, GD Wu, JD Lewis, FD Bushman. 2013. Rapid evolution of the human gut virome. Proc Natl Acad Sci USA 110:12450-12455.

Seed, KD, KL Bodi, AM Kropinski, H-W Ackermann, SB Calderwood, F Qadri, A Camilli. 2011. Evidence of a dominant lineage of *Vibrio cholerae*-specific lytic bacteriophages shed by cholera patients over a 10-year period in Dhaka, Bangladesh. mBio 2:e00334-00310.

Seed, KD, DW Lazinski, SB Calderwood, A Camilli. 2013. A bacteriophage encodes its own CRISPR/Cas adaptive response to evade host innate immunity. Nature 494:489-491.

Sun, CL, R Barrangou, BC Thomas, P Horvath, C Fremaux, JF Banfield. 2013. Phage mutations in response to CRISPR diversification in a bacterial population. Environ Microbiol 15:463-470.

Van Der Oost, J, MM Jore, ER Westra, M Lundgren, SJ Brouns. 2009. CRISPR-based adaptive and heritable immunity in prokaryotes. Trends Biochem Sci 34:401-407.

Recommended reviews

Deveau, H, JE Garneau, S Moineau. 2010. CRISPR/Cas system and its role in phage-bacteria interactions. Annu Rev Microbiol 64:475-493.

Westra, ER, DC Swarts, RH Staals, MM Jore, SJ Brouns, J van der Oost. 2012. The CRISPRs, they are a-changin': How prokaryotes generate adaptive immunity. Annu Rev Genet 46:311-339.

A Lifetime of Phage Explorations

by Elizabeth Martin Kutter[†]

Abstract: *Bacteriophage T4 has been my scientific partner and friend since 1963, when I began to work on the transition from host to phage metabolism, taking advantage of T4's substitution of HMdC for cytosine in its DNA and of the extensive genetic and molecular studies of T2 and T4 begun in the 1940s. It was with T2 and/or T4 that the beautiful structure of tailed bacterial viruses was first imaged in the EM, DNA was shown to be the genetic material, and the role of messenger RNA in protein synthesis was deciphered. A detailed genetic map of T4, developed with the aid of biochemists, molecular biologists and geneticists, cemented the focus on T4 as a key model system, with conditional-lethal "amber" mutants available in all essential genes. The Cold Spring Harbor tradition of active collaboration among phage researchers was further propagated by the informal Evergreen International T4 Meetings which began in 1975 and led to rapid advances that were documented in the 1983 and 1994 ASM books on T4, which grew out of those meetings and of our analysis of the T4 genome. With uniquely well-established genetics and physiological studies, T4-like phages remain a rich resource for interpreting genomes of other phages and for studies of phage takeover of bacterial metabolism. During a 1990 USSR stay related to the T4 genome project, I was amazed to discover the longstanding phage therapy work centered in Georgia, and the Evergreen International T4 meetings soon morphed into the ongoing, more general biennial Evergreen International Phage Meetings that facilitate worldwide collaborations exploring all aspects of phage ecology and molecular biology, as well as their technological and therapeutic applications. T4 and its relatives continue to play key roles in marine ecology, therapeutic applications, molecular manipulations, and even studies of potentially useful direct interactions between phage and the mammalian immune and mucous membrane systems. Here, I have been invited to share a rather personal view of those adventures.*

[†]Faculty emerita and Head, Evergreen Lab of Phage Biology, The Evergreen State College, Olympia, WA
Email: kutterb@evergreen.edu
Websites: www.evergreen.edu/phage and www.phagebiotics.org

I was first introduced to phages in 1962 while in my first year of graduate school. As an undergrad, I had thrived in my mathematics major at the University of Washington, but then a friend talked me into taking a zoology class taught by a marvelous teacher, someone who had also started as a mathematician and spoke with a beautiful, soft Scottish accent. That course made me decide that the questions at last becoming approachable in biology were far more exciting than those in mathematics, questions like how an egg becomes a human being and how the brain works. But then a summer doing neurobiology research convinced me to look for simpler systems as my husband and I headed off for graduate school.

Not surprisingly, temperate phage λ and its potential lifestyle choices were an intriguing major topic in Alan Campbell's genetics class at the University of Rochester, where I was a graduate student in

Radiation Biology and Biophysics, but it wasn't until that spring that I found my lifelong calling. This came about when John Wiberg visited from MIT to interview for a newly funded biophysics position and the department chairman invited me to go along with them for lunch. I listened in spellbound fascination as John, a nucleic acid biochemist by training, talked about his work with John Buchanan and Salvador Luria identifying the genes and enzymes that T4 uses to substitute 5-hydroxymethylcytosine for cytosine when synthesizing its DNA (see below). I was captivated by how bacteriophage T4 completely took over the host cell and turned it into a phage factory, and by the challenge of figuring out why and how T4 went to so much trouble to substitute this new base. Bale, the department chairman, slipped quietly away after about an hour and a half, while John and I soon agreed that he would come to Rochester and I would be his first graduate student.

Phage and the discovery of the genetic material

I was utterly fascinated by these strange entities— half protein, half DNA—that had been used to show that DNA was the genetic material. In 1940, determining the physical structure of phages had been among the first applications of the recently-developed electron microscope (Luria, Anderson 1942). For most observers, this resolved the ongo-

ing intense argument as to whether phage were indeed viruses or, instead, were some sort of bacterial enzyme—the latter still being firmly stated as fact in a key article at the time disparaging phage therapy (Krueger, Scribner 1941). The intricate pictures of complex phages like T4 from the labs of Eduard Kellenberger, Tom Anderson, and Fred Eiserling captured the imaginations of a new generation of scientists, and continue to do so to

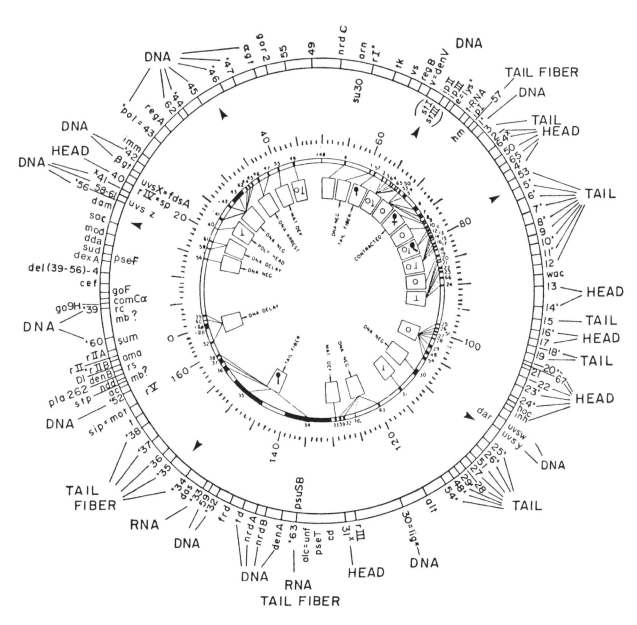

Figure 1. The 1983 genetic map of T4. The central part of this figure is derived from the T4 map developed by Edgar and Epstein (Epstein et al. 1963), based on recombination frequency, with functional gene assignments based on the EM pictures of cells infected by members of their first set of amber mutants. The outer circle, for comparison, is the 1983 version of the genetic map, produced by various members of the phage community when we already had substantial sequence data but still had large gaps, mainly in the non-essential regions, where we still needed to depend on recombination data between widely-spaced genes. It should be noted that the spacing here is still substantially different than that in the map based on the complete sequence, shown in Fig. 2.

this day (Kellenberger 1966). The visions of the T2 particle sitting demurely on the surface of the cell as new phage virions were constructed inside led naturally to the experiment that convinced the last of the stubborn holdouts that DNA was indeed the stuff of which genes are made.

The Hershey-Chase experiment was beautifully simple: either label the phage DNA with ^{32}P or label the phage protein with ^{35}S (Hershey, Chase 1952). Mix the labeled phage with bacterial hosts for a few minutes, throw them in a Waring blender to knock the phage particles off the cells, and then spin out the cells. The protein-labeled capsids stayed in the supernatant, while the phage DNA spun down with the cells—yet each purified cell could still go on to produce a batch of hundreds of new phage after further shaking in fresh medium! This specific approach grew out of a hypothesis developed by Roger Herriot following his discovery that osmotic shock could produce phage 'ghosts' that had lost their DNA but could still kill cells. Herriott wrote Al Hershey that "the virus may act like a little hypodermic needle full of transforming principle; that the virus as such never enters the cell; that only the tail contacts the host and perhaps enzymatically cuts a small hole in the outer membrane and then the nucleic acid of the virus head flows into the cell. If this is so, then... one should be able to get virus formed by the nucleic acid alone ... if one only knew how to get it into cell" (Olby 1974). Al Hershey and Martha Chase then went on to devise the blender experiment, which showed that this was exactly what happened, thereby convincing most skeptics that DNA was indeed the genetic material. Jim Watson, in particular, was really galvanized by this experiment. He was frenziedly looking for ways to determine the structure of DNA that let it perform this wondrous feat as he moved from Salvador Luria's lab to Denmark and thence to Cambridge, where he convinced Crick to join him in that exploration.

Phage and the identification of mRNA: How genes work

A variety of people had been suggesting that RNA somehow plays a role in using the information in the DNA genes to direct the construction of proteins, but efforts to figure out just how that works were confounded because the population of RNA in the bacterial cell seemed to remain unchanged whatever the conditions. A breakthrough came when researchers radioactively labeled the RNA made after T4 infection and observed a pool of RNAs that bound selectively to parts of the phage DNA and not to the host DNA (Brenner, Jacob, Meselson 1961). It turned out that bacteria synthesize so much stable ribosomal RNA (rRNA) that it was completely masking the synthesis of the short-lived templates that we now call messenger RNA (mRNA). Since T4 totally shuts off all host RNA synthesis very soon after infection (*see page 4-7*), it was possible to see and to characterize T4's big population of different mRNAs. Moreover, the rII and lysozyme genes were later used in interesting ways to help sort out the genetic code and the mechanism by which triplets of nucleotides read sequentially from a specific point determine the sequence of amino acids in each particular protein (Sarabhai et al. 1964). The abundant ribosomal RNAs simply serve as components of the assembly plant, the ribosomes, where they provide major parts of both the structure and the complex actively linking successive amino acids.

Genetic mapping of T4

Working with phage back in the 1960s to 1980s largely meant working with T4 or λ. With T4, one could take advantage of the very extensive genetic map recently generated by Dick Epstein, Bob Edgar, and their colleagues (Epstein et al. 1963). A few mutants affecting T4 plaque morphology and/or host range had been isolated and mapped early on, such as those in the rII A and B genes, mutations in either of which yielded phage that couldn't plate on strain CR63 carrying a λ prophage but could make characteristic large plaques on *E. coli* B. These genes were much used to study the triplet nature of the genetic code, but studying those genes didn't tell us much about the nature of the infection process. Then Edgar and Epstein focused in on the concept of conditional lethal mutants and isolated mutants in 50 different genes that impacted growth at high temperature and/

or, surprisingly, in particular hosts (e.g., permitted growth on 'permissive' host *E. coli* CR63 but not on *E. coli* B). It took an enormous number of genetic crosses to figure out the map order of all those genes. They managed to rope a roommate, Harris Bernstein, into playing a major role here. In return, this strange new kind of host-specific mutation bears the name 'amber,' the English word for his family name.

The map, to their surprise, was circular (Fig. 1). Undaunted, they assigned gene numbers starting from an arbitrary origin between the well-characterized genes rIIA and rIIB. Streisinger eventually figured out how the genetic map could be circular even though the virion DNA molecule had to be linear to pass through the phage tail into the host (Streisinger, Edgar, Denhardt 1964). T4's DNA is actually replicated as a long concatemer, eventually involving about 50 copies of the phage DNA. The concatemer is threaded into the procapsid using a 'headful' mechanism to determine when to cut the packaged genome free and thread the newly-formed end into the next waiting head. Doing it this way, each packaged genome is several kb longer than the actual genome, so each headful of DNA starts at a different point on the genome. Apparent circularity results because every gene has neighboring genes on both sides in most of the resultant phage particles. The ends of chromosomes formed this way are always terminally redundant. We now know that many phages use this kind of headful mechanism, while others like T5, T7, and λ make those packaging cuts at a specific place on the DNA, so the chromosomes in their virions all start and end at the same genomic position. This latter group usually have single-stranded ends that also entail some terminal redundancy, providing a handy mechanism for circularizing the genome after arrival in the host cell and thus, in the case of temperate phages, for being able to recombine that circle into the host DNA to form a prophage (*see page 6-5*).

Having been identified by being conditionally lethal, these amber and temperature-sensitive mutations must all be located in genes that are essential

to the phage. To figure out what each of these essential T4 genes was doing required detailed study of a very large number of infections. For an initial screening, Edgar and Epstein took full advantage of the electron microscope to identify which mutants were still able to make at least some phage particle components under non-permissive conditions, and which ones made no detectable phage structures (Fig. 1). Most of the latter were also unable to make phage DNA, and it was these that we would then study enzymatically with an eye particularly for those involved in T4's substitution of a modified base in its DNA. The further very interesting finding was that most of the assembly process could actually take place outside of the cell, as could be seen by combining *in vitro* the lysates from appropriate amber-mutant infections in which only some of the specific components were present (Edgar, Wood 1966).

Phage T4's strange DNA: HMdC rather than C

Before coming to Rochester, John Wiberg had been working with John Buchanan and Salvadore Luria exploring the biosynthesis of T4 DNA during infection of *E. coli*. Seymour Cohen had shown that T4 encoded relevant enzymes that are not provided by the host, enzymes such as hydroxymethylase (HMase), and moreover that the phage used the distinctly different HMdC (5-hydroxymethylcytosine) instead of C in its DNA (Flaks, Cohen 1959). Furthermore, T4 went on to add glucose residues to each of the 5'-OH groups on those HMdCs. For my PhD thesis, I was determined to figure out how and why this phage went to the trouble to do all that, and in the process I would also study the transition from host to viral metabolism that followed phage infection.

In short, two enzymes clearly played key roles in this switch from C to HMdC: HMase (dCMP hydroxymethylase to add the hydroxymethyl group to cytosine) and dCTPase/dCDPase (to deplete the dCTP pool and provide the dCMP). When either or both were inactivated by mutation, cytosine-containing phage DNA was made but then degraded and no progeny phage were produced. At somewhat elevated temperatures, temperature-sensitive dCTPase mutants were able to produce active phage that contained some C in their DNA. When I added a mutation in gene 46, which had been identified as an exonuclease, the new cytosine-containing phage DNA was still degraded, but to operon-sized fragments rather than to acid-soluble form. It was subsequently found that T4 makes a pair of cytosine-specific endonucleases to degrade the host DNA. Endo II was able to nick such DNA at rather rare specific sites, and then endo IV attacked the short single-stranded stretches that developed opposite those nicks (Kutter, Wiberg 1969).

It was years later before we discovered that complete substitution of C for HMdC inT4 DNA had a second lethal effect on the phage: complete blockage of synthesis of the late structural proteins. We also showed that a small early phage protein, gp-alc, blocked elongation during attempted transcription of cytosine-containing DNA. This explained the total inhibition of host transcription after T4 infection, a situation that had been so useful for researchers since it blocked the abundant synthesis of ribosomal RNA that otherwise concealed the mRNA in most systems.

Interestingly, yet another protective effect of glucosylated HMdC in T4 DNA has just been reported. A poster presented at the 2014 Viruses of Microbes meeting in Zurich by Marnix Vlot of Wageningen University, Holland, established that "DNA modifications of bacteriophage T4 inhibit CRISPR-Cas interference." It was awarded the Best Poster prize, with a check for 1,000 Swiss francs. Luckily, we had been able to revive the old 1980s stocks of T4 containing the mutations in HMase, dCTPase, gpAlc, endo II, and endo IV that permit synthesis of viable phage with this phenotype, so he had been able to confirm *in vivo* results that he had first worked out using synthetic templates.

Transition times: a Virginia interlude

The continual novel discoveries and the rapid annual explorations of phage biology advances at the annual Cold Spring Harbor meetings made for

much excitement while I was a grad student. Both of my sons were also born during that era. First-born Bernard went to his first seminar when he was ten days old, having 'attended' the one just before that at minus ten hours, and then spent his early months often 'helping' me from a perch on my back. (John once thanked me for going to so much trouble to provide entertainment for the lab!) By the time Eric came along, I was mainly writing my thesis. As Sig, their dad, finished his astrophysics degree and I finished mine in biophysics, the decision was made that I would take a year off and parent full time while he started his faculty position at the University of Virginia.

After half a year, a welcome change came along. I was invited to give a seminar in the microbiology department. Hearing it, Rolf Benzinger of the biology department became very excited about my thesis work exploring the degradation of host DNA after T4 infection with the initial production of operon-sized pieces. He had worked with Werner Arber and was trying to sort out the mechanisms of restriction enzymes, and this seemed potentially applicable. The research involved unstable enzymes and an unstable assay and went frustratingly slowly, but was promising enough that I soon was awarded my own NSF grant and, two years later, an NIH grant with a very high priority score, all despite my lowly status at UVa. However, when the department received money for four new positions in 1971 and I went to talk to the chairman about applying for one, I was told that if I, as a mother of young children, had the audacity to do so, not only was I guaranteed not to get it, but he would no longer sign for my grant. (I should note that not only were there no women on the biology faculty there in 1971, but women grad students were given teaching assistantships while all the research assistantships went to their male colleagues. Not surprisingly, none of the women had finished their PhDs in recent years.) In some ways, the chairman did me a favor. His outrageous response was enough to incite my husband to leave his tenure-track appointment and we both applied for positions at the new Evergreen State College in Olympia, Washington.

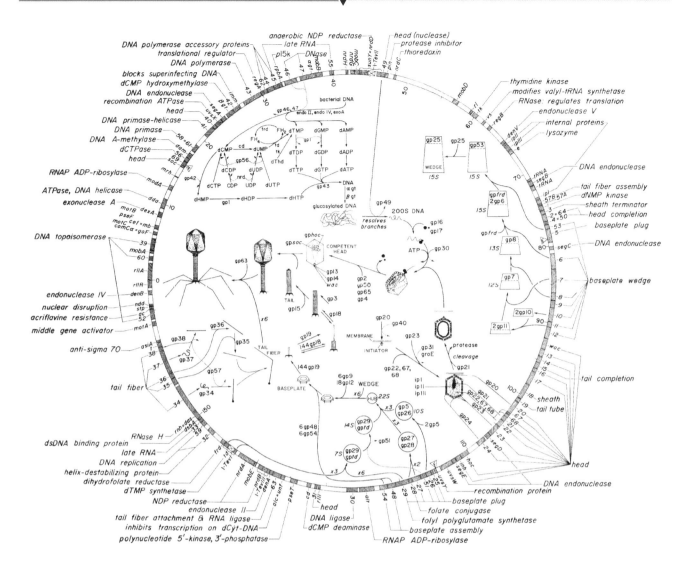

Figure 2. Current T4 genomic map. This map, incorporating detailed illustrations of the pathways of DNA biosynthesis and phage particle construction, has been updated multiple times by Burton Guttman and Elizabeth Kutter from the version originally drawn as the frontispiece for the 1983 book *Bacteriophage T4*. That book, which grew out of the 1981 biennial Evergreen International Phage Meeting and involved 71 members of the T4 community, was the first monograph ever published by ASM press. Only those genes where we have some indication of specific function are shown, but all of the nearly 300 open reading frames determined to probably encode functional products are included in the table presented in the 2003 detailed T4 genomics and gene function paper (Miller et al. 2003). Regions that appear largely empty such as that between *nrdC* and *IPI* are actually densely packed with genes, most of them encoding small 'monkey wrench' proteins (see *also page 4-25*). Credit: Elizabeth Kutter and Burton Guttman.

On to Evergreen

Not only were Sig and I both awarded faculty positions at Evergreen, but I managed to take my NIH grant there with me even though Evergreen was focused on innovative new approaches to undergraduate education rather than on research. As the school was in only its second year, I had to delay the start of that grant by six months until the lab building was finished in January of 1973. Little did Evergreen realize, as the faculty grew from 30 to 60 that year, that they were hiring not one but two

phage biologists. Burt Guttmann, hired as a microbiologist, had spent several exciting years working on phage at Cal Tech with Bob Edgar. The two of us complemented each other well, keeping the lab going strong despite all the demands of collaborative teaching of the full-time integrated programs that took the place of individual courses at this innovative school. This initial NIH grant and an unbroken series of NSF grants until 2000 allowed a range of exciting phage work to flourish there in what was often the only basic research lab at

Evergreen. The excellence of the still-functioning Evergreen Phage Lab depended primarily on intense undergraduate involvement, a series of exceptional post-doctoral fellows and technicians, and periodic involvement of new younger faculty, with key support from amazing people like Phil Harriman at NSF, Bruce Alberts at UCSF and Mike Chamberlin at Berkeley, as well as from the whole wonderful phage community.

Bacteriophage involvement in genetic engineering

In 1975, I was invited to join the NIH Director's Recombinant DNA Advisory Committee right after its initial meeting in Asilomar, CA. I wound up chairing the subcommittee that wrote the first set of safety guidelines, setting a policy that put particular emphasis on broad-based, local biosafety committees and grew out of industrial as well as academic input. While there was much initial concern about the risks, we took pride in the fact that there were no known incidents of harm resulting from work being carried out under those guidelines. While my invitation to participate had specifically acknowledged my coming from a college where "the teaching of the ethics and values of science is a key part of the teaching of science," it also reflected the key roles of phage in the development of genetic engineering. The most obvious aspect then was the use of λ and other phages as vectors. However, some T4 enzymes, including its DNA kinase/phosphatase and blunt-end-joining DNA ligase, soon came to provide vital services in this field. Ironically, cloning of those enzymes for mass production used our cytosine-containing derivative of T4, which I had given away freely to everyone who wanted it.

The Evergreen International T4 Meetings

While I always loved Evergreen, I also felt somewhat scientifically isolated there. Cold Spring Harbor was far away, the phage meetings there had grown quite expensive, and they had become "Phage and Microbial Genetics" meetings, with a strong focus on λ and other temperate phages. The summer of 1975 saw an informal gathering at Evergreen of T4 people who had largely come

west for a California biochemistry meeting. Consensus developed that regular, affordable West Coast phage meetings focused on T4 and other large lytic phages and held at Evergreen would be very useful. The amount of exciting work with T4 was skyrocketing. For the second of these biennial Evergreen T4 meetings, people came from all over the country; for example, Bruce Alberts drove up from the Bay Area with a van full of grad students, postdocs, and Central Valley fruit. The phage field was growing fast. In 1971, Cold Spring Harbor had published the first major bacteriophage monograph, *The Bacteriophage Lambda*. Jim Watson had long been promising to publish a similar monograph on bacteriophage T4, but that kept not happening. By 1981, the attendees at the Evergreen meeting decided to take on that job themselves. Virtually all of the 71 people present agreed to be involved. It turned out that the ASM Press was considering adding a monograph series to their publications. Helen Whitely, the editor in chief there, heard about our plans and invited *Bacteriophage T4* (Mathews et al. 1983) to be their first. These highly collaborative meetings have now morphed into much broader phage meetings; in August of 2013, the 20th biennial Evergreen International Phage Meeting drew 160 participants from 35 countries.

The T4 genome project

By 1983, most of T4's essential genes had been identified and assigned integral gene numbers, and their functions and interrelationships worked out (Fig. 2). Virtually all of them are indeed involved in either phage particle construction or in DNA replication and transcription. For many of the 'nonessential' genes, functions have been identified and serve as the source of the gene names (e.g., ts for thymidylate synthase, tk for thymidine kinase). Our T4 genome project was also well underway by then. During my 1978-79 sabbatical with Bruce Alberts at UCSF, I worked with Pat O'Farrell on a 2D gel approach for ordering DNA restriction fragments. T4's genome size made it a rather ideal test case for this approach— large enough, but still manageable. Further, my cytosine-containing T4 phage derivative could be cut by most six-base cutter restriction enzymes, and we had all that genetic information to help us sort out the details. A number of the essential structural and DNA-related genes had been cloned and used for complementation studies, giving us confirmatory size and genome organization infor-

Figure 3. Host phospholipid synthesis after T4 infection. Although T4 infection inhibits the synthesis of host protein, DNA and RNA, the major host membrane phospholipid, phosphatidyl ethanolamine (PE), continues to be synthesized at the normal rate for the first 20 min after infection. The rate of synthesis actually *doubles* for the first 13 minutes after infection for phosphatidyl glycerol (PG), which makes up 1/3 of the host membrane phospholipid. This helps explain our earlier observations that the optical density of the culture approximately doubles during the first 20 min after T4 infection. This is not just a passive response. As plotted below, infection with T4 strains carrying deletion mutant rIIDD2, which takes out several small genes between *ndd* and *rIIB* including endonuclease *denB* (involved in host DNA breakdown), allows ongoing synthesis of both phospholipids but without the specific stimulation of PG biosynthesis. T4 deletion SaΔ9, which extends further through that 'monkey wrench gene' region to the acridine-resistance gene *ac*, totally eliminates the ability to continue phospholipid synthesis after T4 infection (Harper et al. 1994). Chemical analysis of total phospholipid correlates well with these results, so it is not just a labeling artifact.

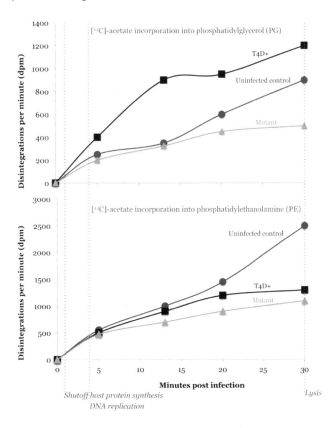

mation. However, it took another decade and the development of new molecular tools to finish sequencing the host-lethal regions and to determine the early and middle-stage promoters.

This genome project involved international collaboration between Gisela Mosig at Vanderbilt, Wolfgang Rüger in Germany, Fumio Arisaka in Japan, and Vadim Mesyanzhinov in Russia, as well as integrating data on specific regions that was collected by other members of the phage community. The communication that required was not then trivial. Assembly of the structural protein regions—sequenced largely by Mesyanzhinov's group in Moscow and Rimas Nivinskas's group in Vilnius—into the genome moved forward when I received a four-month US-USSR Academy of Sciences Exchange Program grant in 1990.

My time in the Soviet Union also aided our ongoing work on the transition from host to viral metabolism in multiple ways. For example, the people in the lab above ours at the Bache Institute of the Soviet Academy of Sciences worked with lipid metabolism, and asked me about what effects T4 infection has on *that*. They then collaborated with us to discover that during the first 15 minutes after infection, while the optical density continues to rise, T4 substantially stimulates synthesis of phosphatidyl glycerol while synthesis of phosphatidyl ethanolamine continues at preinfection levels (Fig. 3). Different small nonessential genes near rIIB control both the continuation of membrane-lipid synthesis long after host RNA and protein synthesis are terminated and the extra stimulation of phosphatidyl glycerol synthesis (Harper et al. 1994).

It was also during this time that I visited the preeminent Soviet RNA polymerase lab and was befriended by Misha Kashlev, then a graduate student there. He later came and helped us figure out how T4 blocks transcription of host DNA. During transcript elongation, the 19 kDa Alc 'monkey wrench' protein interacts with both the RNA polymerase and the non-template strand of the DNA just ahead of the growing transcript, where its recognition of specific C-containing sequences causes termination provided that transcription is proceeding rapidly (Kashlev et al. 1993). That work was done during two summers that he took off from his postdoc with Alex Goldfarb at Columbia and lived in my home, along with his wife and young son, while working in my lab.

T4 turns out to use an additional trick related to nucleotide biosynthesis to make the transition from host to phage metabolism more complete and the T4-infected host cell more efficient. Not only does it encode the dCTPase and HMase that together specify the replacement of dC by HMdC in all of its DNA, but it encodes its own version of most of the other enzymes of nucleotide biosynthesis (Fig. 4). Chris Mathews determined that these T4 enzymes, along with two host enzymes, assemble into a complex that forms a much more efficient production line for nucleotide biosynthesis which is linked directly to the DNA polymerizing enzyme complex. This T4 production line produces twice as much dATP and dTTP as dCTP and dGTP, even when DNA synthesis is blocked by mutation. This implies that the ratios are inherent to the complex itself, not driven by some sort of feedback, and correlates with the fact that T4 DNA is 2/3 AT, rather than ½ AT like its host (Mathews 1993a; Mathews 1993b).

Wolfgang Rüger finally found the secret that let our lab sequence the largely host-lethal, deletable region from 48 to 74 kb on the T4 map. He developed a special trick to identify the immediate-early promoters, which turn out to be exceptionally strong versions of the bacterial housekeeping promoters. Of the 39 immediate early promoters, 15 lie in this particular host-lethal region. What Rüger did was to chop the genome into about 200 bp pieces and clone them just upstream of the β-gal gene fragment that, when active, turns cells blue when IPTG is added. Whereas most of the clones he obtained turned faint blue at most, those containing these promoters turned bright indigo and could easily be isolated and sequenced. As an additional bonus, this gave us the needed informational pedestals for making PCR fragments that spanned the inter-promoter sections without including any of the strong promoters. Bruce

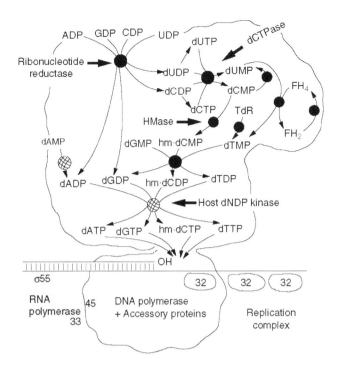

Figure 4. The T4 nucleotide synthetic complex. Over many years, the lab of Chris Mathews determined this model of the interactions in T4's highly efficient nucleotide synthetic complex. It produces nucleotides in the 2:1 AT to GC ratio needed by T4, is capable of processing both NDPs and the dNMPs produced by host DNA breakdown, and interacts directly with the DNA polymerase accessory proteins and with the RNA polymerase containing the unusual T4 gp55 σ factor that is responsible for transcribing T4's late genes. Figure created by Gautam Dutta and Elizabeth Kutter based on the work reported in Mathews 1993a and Mathews 1993b.

Alberts supplied the Evergreen lab with $10,000 worth of the then expensive primers needed to complete the sequencing of that region, a process completed step-by-step by manually producing and arduously reading by hand the necessary Sanger sequencing gels.

By the time we completed analysis of the T4 genome in 1993, it was clear that T4 had far more genes than anyone had imagined—nearly 300, although the precise functions of about 100 (many of them host lethal) are still unknown. The project was completed just in time for the next book that grew out of the T4 meetings, *The Molecular Biology of Bacteriophage T4*, published by ASM in 1994 with over a dozen editors from the T4 community and over 100 authors (Karam et al. 1994). Studies using both molecular tools and an array of experimental approaches to further explore the gene functions and interactions of T4 eventually led to a 70-page 2003 article that continues to be much used (Miller et al. 2003). This has now been complemented by a special 2010 issue of *Virology Journal* that explored further developments in the biochemistry, structural biology, genomics and ecology of T4 and related phages, and later was published as a book (Karam, Miller 2010).

Therapeutic applications of phage

During those four months in the Soviet Union in 1990, I also made my first two trips to Tbilisi, in the Republic of Georgia, where I discovered people who were routinely using phages as antibiotics. Although initially very skeptical of the Georgian therapeutic claims, repeated visits, getting to know the surgeons involved, and viewing my first phage-based surgeries to treat diabetic ulcers in 1995 pushed me to explore further. This was the start of my learning about the first 30 years of phage research, which was of a very different kind, work that had been largely forgotten in the West after Max Delbruck convinced the phage community to focus on just a few phages targeting *E. coli* to better understand the molecular basis of what was going on. This experience also led to substantial further expansions in direction both in the Evergreen meetings and in our own research.

I had much exciting older research to catch up on. At that time I knew only that Frederick Twort (Twort 1915; Twort 1936) and Felix d'Hérelle (d'Hérelle 1917) were jointly credited with the discovery of tiny entities capable of destroying cultures of bacteria. I discovered that Twort was mainly looking for viruses that could infect eukaryotic cells, but also worked with coli-typhoid bacteria and a quite different, larger bacillus from the upper third of the intestine. He reported *a dissolving substance which infected the* (large bacillus) *colonies so rapidly that they were dissolved before attaining a size visible to the eye* which made the large bacillus hard to isolate. It was d'Hérelle who independently discovered viruses of bacteria and gave them the name "bacteriophages" and who played the major role for many years of moving

forward research and practice related to their potential therapeutic applications. Within two years he had carried out trials demonstrating that phages could cure a chicken flock of avian typhosis, went on to successfully treat dysentery in several children, and then turned to learning all he could about the nature and properties of phage before doing further clinical work. The appendix to d'Hérelle's 1938 book, recently translated from the French by his great grandson and Sarah Kuhl (Kuhl, Mazure 2011) reveals a level of understanding of phage and phage therapy truly remarkable for that time. Similarly, with the translation and review of other French work it has become clear that until the 1980s a great deal of successful therapeutic work went on in France tied to the Pasteur Institute, as well as at the production center organized by d'Hérelle. An array of interesting, long-forgotten—but good—early work done in France, the USA, and elsewhere was translated by Sarah Kuhl a few years ago and included in the extensive review by Abedon and colleagues (Abedon et al. 2011), which also includes detailed discussions of the very extensive work in the former Soviet Union and in Poland. The Georgian work from the 1930s up into the 1990s has been made far more accessible through a monograph funded by the UK Global Partnership Program and based on the publications in the library of the Eliava Institute (Chanishvili, 2012), which details extensive use of phage therapy in such areas as surgery and wound treatment, septic infections, dermatology, urology, gynecology, and gastrointestinal infections.

The classic Western work rediscovered and revitalized

In 2000, I was invited to give the keynote speech at an AstraZeneca symposium on Drug Discovery and Development held in Bangalore, India. This event celebrated the formal retirement of V. Ramachandran from the AstroZeneca Research Foundation and the launching of his new Indian phage therapy company, GangaGen, the name reflecting 1894 observations in India by Hankin of the Pasteur Institute, who noted and explored the fact that something in the Ganges could kill the bacteria causing cholera. My lecture was introduced by Gary Schoolnik, leader of medical microbiology at Stanford, who told the following story of how he was drawn into the field of microbiology:

My mother, in 1948, was dying of typhoid fever, before they had antibiotics. My father, a microbiologist at the University of Washington,... read in J. Bacteriology about a Los Angeles scientist who had discovered a phage that killed Salmonella typhi. *He called up this guy, and it was flown up to us in Seattle on a DC-3. My father injected my mother in the hospital with this phage, and the next day she was perfectly well ... That's real infectious disease experimentation: a mix of science, and daring, and desperation.*

We only later learned that there was much data (as discussed above) that laid a solid foundation for this life-saving decision, not just the mouse work of which he was aware. As described by Knouf (Knouf et al. 1946): *One of the most spectacular accomplishments...was the rapidity with which the patient returned to his normal mental outlook. In 24-46 hours, patients who had been comatose and in the 'typhoid state' amazed everyone by cheerful, grateful attitude ... asked for food vociferously.*

As we ponder what we should require today before allowing clinical use of phage therapies that have reportedly saved lives for so many years in Eastern Europe, France, and even the US, it is worth considering earlier clinical experience that was clearly carefully done, even though it did not include the currently expected, enormously expensive, and often ethically challenging double-blind clinical trials.

The early US typhoid clinical work was tracked down by Swiss science writer Thomas Häusler, along with much else of which neither the global phage community nor the modern medical establishment was aware, during a year of broad international research. The resultant book, *Viruses vs. Superbugs* (Häusler 2006) finally made much of these early findings widely available and is a must-read for anyone interested in phage therapy. (I had the good fortune of helping edit both the 2003 German edition and the expanded English

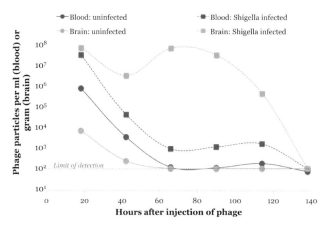

Figure 5. Phage ability to cross and multiply behind the blood-brain barrier given host availability there. This figure is based on the mouse studies of Dubos et al. (Dubos, Straus, Pierce 1943). It provides insight into why phage can often treat infections that are inaccessible to antibiotics. When mice were injected intraperitoneally with 10^9 phage, phage quickly appeared in the bloodstream and a small faction crossed the blood-brain barrier, but they were rapidly cleared. However, when the mice were also injected intracerebrally with *Shigella dysenteriae*, the host of those phage, phage counts in the brain climbed to over 10^9 per gram, the blood level remained somewhat elevated, and 46 of 64 mice survived, compared to only 3/84 in the absence of appropriate viable phage. Once the bacteria cleared, the phage quickly dropped to below detection limits.

translation, learning an enormous amount, and I continually go back to it.)

Among the most remarkable of the early US publications was the key mouse work led by Rene Dubos (Dubos, Straus, Pierce 1943). They found that phage administered intraperitoneally crossed the blood-brain barrier and, if the relevant bacteria were present in the brain, multiplied extensively there to over 10^9 per gram, and in the process rescued most of the mice from the otherwise lethal infection. I have made that data into a figure that I use in every lecture I give because it illustrates so well how a few phage can penetrate protected areas and multiply there to high enough levels to largely destroy the infecting bacteria, thereby saving lives (Fig. 5). This is as true for the osteomyelitis associated with diabetic foot ulcers and for prostatitis as Dubos demonstrated for the blood-brain barrier. Other excellent US and Canadian work focused on treatment of hundreds of patients with staphylococcemia (MacNeal, Frisbee, McRae 1942) and on the successful treatment of typhoid fever patients in hospitals in Los Angeles (Knouf et al. 1946) and Montreal (Desranleau 1949) referred to above. That work halted in the late 1940s when chloramphenicol became readily available for the treatment of typhoid fever.

AstraZeneca had provided me with a round-the-world business class ticket for the India trip, which I used to make very productive visits to the major remaining ongoing centers of phage therapy at the Eliava Institute in Tbilisi and the Hirszfeld Institute of Immunology and Experimental Therapy of the Polish Academy of Sciences in Wroclaw, Poland.

Alfred Gertler, who came to meet me in Tbilisi to be treated with phage, had read about phage therapy and our phage meetings in a NY Times Magazine article, and arranged to come to our 2000 meeting to plea for help. He had severely shattered his ankle while hiking in Costa Rica, and staph still drained out of wounds in both sides of his ankle after four years, one of them spent on IV antibiotics, despite multiple surgeries. As we confirmed, that antibiotic was still perfectly capable of hitting his bacteria *in vitro*, but apparently couldn't reach them well enough in the infected ankle bones to even develop antibiotic resistance, say nothing of curing the infection. He had been told that unless they amputated his foot he would be dead within the year. The phage treatment was very successful, and his story landed him and the Eliava key roles in a 2002 CBS 48 Hours program on antibiotic resistance. He still has his foot and was back playing string bass in his Toronto band (Kutter et al. 2013). Dubow's classic blood-brain barrier paper helps explain how phage therapy can work even in such privileged compartments as infected bone.

This was my first visit to the Hirszfeld Institute. It had long supplied appropriate individual phages for patients in hospitals in the region. I had explored the results of thousands of cases in the 1980s that they published in a series of papers in English, which showed strong indications of success, and the new Institute Director was greatly strengthening their phage focus. Molecular bi-

ologist Malgorzata Lobocka came from Warsaw to meet me there, thus setting up the productive collaboration that has let them sequence and further characterize many of their phages. In 2005, they opened an outpatient Phage Therapy Unit to support research on *the long-term treatment of longstanding chronic infections* under a European Union protocol, *Experimental phage therapy of drug-resistant bacterial infections, including MRSA*. They continue to publish prolifically on all aspects of their work, both basic science and clinical results with detailed lab work (Miedzybrodzki et al. 2012; Kutter et al. 2014).

Phage therapy in the US

An interesting book exploring both the early and more recent Western explorations of phage therapy is *The Forgotten Cure: The Past and Future of Phage Therapy*, by Anna Kuchment (Kuchment 2012). She became fascinated with the field in 2001 while writing a story for *Newsweek*, took a year off to explore intensely, and finally finished her book a decade later as she followed various corporate attempts to reintroduce phage treatments. Both she and Häusler uncovered stories about the early glory days of phage therapy not only in France, Poland, and the Soviet Union, but also in the US where, for example, Western screen idle Tom Mix was rescued from peritonitis following a ruptured appendix by "nature's G-men" and "helpful little bodies", as they were called in a 1934 *Newsweek* article prepared by the phage lab of E. W. Schultz at Stanford (Kuchment, p. 1-4). They also dug deep in the scientific literature from many parts of the world and built a strong case for at least re-exploring phage therapy.

In 1996, entrepreneur Casey Harrington got very excited by the first popular article on phage therapy in recent times, which had just been published in *Discover Magazine*[1]. He called to ask if I would like $5 million to start a phage therapy company. I went so far as to take him to Tbilisi, where he indeed started a company (Georgia Research Incorporated) in cooperation with Nino Chanishvili,

---

[1] http://discovermagazine.com/1996/nov/thegoodvirus918

niece of the director of the Eliava Institute. The rise and fall of that partnership and the goings on that precipitated its disintegration were well documented in the hour-long 1997 BBC *Horizon* program, the *Virus That Cures*[2] and the 1998 follow-up on CBC's *The Nature of Things*[3], both of which helped build rather broad excitement about phage therapy.

About this time, two other phage therapy companies were also launched: Intralytix and Exponential Biotherapies. Both made very promising strides toward phage therapy treatment of vancomycin-resistant *Enterococcus*, using different strategies, but neither was able to get NIH funding for their human studies; though NIH staff people were quite supportive, it seemed as though there were always a few scientists on the study sections who were firmly convinced that phage therapy had long since been proven NOT to work, convincing others, rather than accepting that it just had not yet been proven TO work by Western standards. Despite such setbacks, both companies were moving along well with preparations to expand into clinical trials when the collapse of the dot.com bubble undermined their funding. Intralytix switched to a focus on food safety and agricultural applications, where it has made strong strides, beginning with FDA approval of its product targeting *Listeria* in ready-to-eat meats and cheeses, then progressing to approval of phage against *Salmonella* in collaboration with a major poultry producer.

Exponential Biotherapies carried out a phase I safety trial on ten healthy English subjects in 2000, using a single vancomycin-resistant enterococci (VRE) phage selected to be long-circulating. The results were never published, and still deserve some scrutiny. Former company president Richard Carlton described this IRB-approved trial to me in December, 2012. They had chosen the unconventional but potentially very lucrative route of testing an *intravenous* application for safety and for pharmacokinetics. On days one, three, and nine, 12 healthy volunteers were each injected

with 5×10^6 phage, which is 1000 per ml of blood — about the concentration of VRE seen in septicemia. The expected phage level was seen after the first injection, but after the day 3 injection it was an order of magnitude lower, and only a few score phage per ml were found after the day 9 injection. The conclusion was that the volunteers' immune systems responded by making antibodies that inactivated later doses, making it unlikely that the phage could successfully treat VRE septicemia where multiple doses would be needed, and the company shifted to other pursuits. I include this trial here because it is important to remember that virtually none of the Soviet or more recent Polish and Georgian work and very little earlier work in the West involved intravenous applications, and any efforts to move in that direction call for particularly intense scrutiny.

T4: Where molecular biology and phage therapy converge

Biomedical technology has changed enormously since the early days of phage therapy research, as has our understanding of biological properties of phages and the basic mechanisms of phage-bacterial host interaction. These advances can have a profound impact on the development of safe therapeutic phage preparations possessing optimal efficacy against their specific bacterial hosts. They also provide a footing for designing science-based strategies to integrate phage therapy as a tool in both prevention and treatment of bacterial infections.

Modern phage cocktails use sequenced phages or at least a collection characterized by metagenomic analysis. Infections are also being carried out under conditions more relevant to the real world, such as during anaerobic, stationary-phase, and biofilm growth. Mouse lung infection studies using fluorescently labeled bacteria permit direct visualization of phage effects on the bacterial infection process when added either prophylactically or as a post-infection treatment (Debarbieux et al. 2010). There is a renewed focus on detailed study of the transition from host to phage metabolism in various systems, using new tools to sequence and quantify RNAs (e.g., RNA-Seq).

---

[2] https://www.youtube.com/watch?v=U6sZ7E9Hh-Y
[3] http://www.nytimes.com/movies/movie/262310/The-Nature-of-Things-The-Virus-That-Cures/overview

One of the most interesting and important areas of study is the exploration of direct phage interactions with the mammalian organism, particularly with the immune system, in ways that go beyond the production of anti-phage antibodies. For this, the major studies have been carried out at the Hirszfeld Institute (Dąbrowska et al. 2006; Dabrowska et al. 2007; Gorski et al. 2012), focusing particularly on T4. Since phage can only infect and replicate in their specific host bacteria, much of the rest of the world long discounted the work they were doing on this topic. However, their work and that of others now makes it clear that at least some phages, including T4, do have direct effects on mammalian cells. As K. Dabrowska writes: *Mammals have become "an environment" for enterobacterial phage life cycles. Therefore, it could be expected that bacteriophage adapt to them. GpHoc seems to have significance neither for phage particle structure nor for phage antibacterial activity. ...But the rules of evolution make it improbable that gpHoc really has no function ... More interesting is the eukaryotic origin of gpHoc: a resemblance to immunoglobulin-like proteins that reflects their evolutionary relation. Substantial differences in biological activity between T4 and a mutant that lacks gpHoc were observed in a mammalian system. Hoc protein seems to be one of the molecules predicted to interact with mammalian organisms and/or modulate these interactions* (Dąbrowska et al. 2006).

Barr and colleagues (Barr et al. 2013) have identified a very interesting additional but related role for the T4 Hoc protein: mediating the interaction of the phage with the mucus layer on various animal tissues. The phage helps protect the underlying tissues from bacterial attack while benefiting from a rich source of bacterial hosts.

The most extensive clinical trial to date—the Nestlé study of phage targeting infant diarrhea in Bangladesh—illustrates many of the current challenges, supplies the strongest safety studies to date, and brings us back to the importance of the T4 family of phages in terms of therapeutics as well as molecular biology. Coliform bacteria are responsible for 27% of infant diarrhea in the 3rd world. Immunological-based treatments that

worked for Nestlé against diarrheal viruses were unsuccessful against these bacteria, so Harald Brüssow decided to try phage. We had recently tested nearly 100 T4-related phages from around the world on the broad ECOR collection of *E. coli* and sent him those with the broadest host range to try. A few infected Bangladesh strains, but not as efficiently as he had hoped. He then followed d'Hérelle's method and isolated phage from the stools of infant diarrheal patients at the world's largest pediatric diarrhea center, in Dhaka, using just two host strains: a lab strain and a very widespread and well-studied enteropathogenic *E. coli* (EPEC) strain.

The best choice of host strain for phage selection turned out to be rather surprising. Those isolated from the infant diarrheal patients against the EPEC strain were all siphoviruses, none of which were able to infect more than a few of the Dhaka strains, so they were not candidates for therapeutic use. All of the broad-spectrum phages they obtained were isolated against lab strain K12. These were all related to T4 but varied a lot in range and other properties. Their broad host range reflects the fact that the lab strains have lost the complex O-antigens that protect clinical coliform bacteria against the mammalian immune system; over 150 of these O-antigens have been identified, and at most one of them is produced by each bacterial strain. Phages selected against lab strains that lack these O-antigens must use the highly conserved outer-membrane proteins or the inner stretches of the lipopolysaccharide (LPS) as their receptors. The adhesin that T4 uses to bind to these receptors is located on its long tail fibers, which can penetrate the O-antigen forest to reach those more conserved internal sites. During initial adsorption, these fibers bind the phage to the cell, bending in the process to bring the baseplate down close to the surface, at which point the short tail fibers are deployed to bind irreversibly to the receptor, generally a well-conserved LPS component in these clinical strains. Eventually, Brüssow isolated and characterized nearly 100 T4-related phages, did extensive studies of their genomics, their host ranges on broad sets of pathogenic coli, etc., and determined the properties of some in a mouse-gut model. For their clinical trial, he selected 15 of these phages which between them had a very broad spectrum on Dhaka diarrheal strains. They also carried out metagenomic sequencing on both this phage cocktail and a commercial Russian *E. coli/Proteus* phage cocktail that they used as an extra control.

After over a decade of such work, Nestlé was ready to initiate clinical trials in Dhaka, adding the phage to the routine rehydration fluid. They completed extensive safety studies in adults and older children there, including following a range of physiological parameters. The results of a double-blind trial involving treatment of 100 infants with this cocktail, a placebo or the control Russian *E. coli/Proteus* cocktail are being analyzed. (The original plan called for over 400, but Nestlé apparently ran out of patience and cut back the number before starting to analyze the data.) This very expensive trial has been extremely important as a safety study and for working out various sorts of procedures, whether or not significant efficacy can be seen in this particular set of conditions where the patients were already in the hospital and receiving effective rehydration fluid.

Moving forward

Brüssow (Brüssow 2012), Knoll and Mylonakis (Knoll, Mylonakis 2014), and many others are increasingly exploring what steps are most urgently needed for phage therapy to become a reality in Western medicine. They argue persuasively that industry alone cannot carry out the necessary studies, and suggest that the public health sector must take the lead in the phage therapy field. I strongly agree with Brüssow (Brüssow 2012) that: *Organizations like the US or European CDC should establish phage collections for antibiotic-resistant pathogens which soon are likely to become untreatable in a sizable number of patients. The CDC should characterize, amplify and purify the relevant phages according to safety standards agreed with the FDA… Until the value of PT is established, it is also desirable that…such international organizations supervise the controlled clinical trials…the first trials should target medical priorities. The lower economic prospects in treating these infections will necessitate that the first phase is paid by public money…* It is only once the feasibility, regulatory, and legal parameters have been set that *one might expect private companies to take the lead…for a variety of bacterial infections…we cannot simply wait another 10 years until non-coordinated PT approaches yield reliable answers…Even when it should turn out that PT does not work, knowing this is better than cultivating the illusion that we have a secret weapon in our tool box against antibiotic-resistant infections…*

Given the body of existing documentation of various sorts, broad phage host range, and the small number of phage types involved, phage targeting *S. aureus* to treat diabetic ulcers and MRSA would be one of the most obvious candidates for this approach. (It might help that this would be among the hardest applications to patent.) It is clear that collaboration among governmental entities/health agencies, doctors and patients, academia and industry is crucial. A recent review in *Nature*, encouragingly entitled *Phage Therapy Gets Revitalized* (Reardon 2014), discusses progress in that regard, with emphasis on a project called Phagoburn that is targeting *Pseudomonas* and *E. coli* in burn wounds. This effort involves two companies, a number of institutions, and burn units in seven hospitals in Belgium, France, and Switzerland. The French military, a very active phage-oriented patient support group, and 3.8 million euro in funding from the EU play important roles. Years of work by the P.H.A.G.E. group on regulatory challenges and the experience of a related small-scale phase I clinical trial at the Queen Astrid Military Hospital in Brussels—one of the participants—all help, but progress is still slow. Another key area is the education of doctors and potential patients about the possibilities of phage therapy, as started to happen over a decade ago with the BBC and CBS documentaries, but such exposure breeds frustration and dashed hopes when the only option to obtain phage therapy is to go to Georgia or Poland—further emphasizing the importance of establishing a more clear and rapid path forward for at least one or two clinical applications.

References

Abedon, ST, SJ Kuhl, BG Blasdel, EM Kutter. 2011. Phage treatment of human infections. Bacteriophage 1:66-85.

Barr, JJ, R Auro, M Furlan, KL Whiteson, ML Erb, J Pogliano, A Stotland, R Wolkowicz, AS Cutting, KS Doran, P Salamon, M Youle, F Rohwer. 2013. Bacteriophage adhering to mucus provide a non–host-derived immunity. Proc Natl Acad Sci USA 110:10771-10776.

Brenner, S, F Jacob, M Meselson. 1961. An unstable intermediate carrying information from genes to ribosomes for protein synthesis. Nature 190:576-581.

Brüssow, H. 2012. What is needed for phage therapy to become a reality in Western medicine? Virology 434:138-142.

d'Hérelle, F. 1917. Sur un microbe invisible antagoniste des bacilles dysentériques Comp Rend Acad Sci 165:373-375.

d'Hérelle, F. 1938. *See Kuhl, Mazure.*

Dąbrowska, K, K Świtała-Jeleń, A Opolski, A Górski. 2006. Possible association between phages, Hoc protein, and the immune system. Arch Virol 151:209-215.

Dąbrowska, K, M Zembala, J Boratynski, K Świtała-Jeleń, J Wietrzyk, A Opolski, K Szczaurska, M Kujawa, J Godlewska, A Górski. 2007. Hoc protein regulates the biological effects of T4 phage in mammals. Arch Virol 187:489-498.

Debarbieux, L, D Leduc, D Maura, E Morello, A Criscuolo, O Grossi, V Balloy, L Touqui. 2010. Bacteriophages can treat and prevent *Pseudomonas aeruginosa* lung infections. J Infect Dis 201:1096-1104.

Desranleau, J-M. 1949. Progress in the treatment of typhoid fever with Vi bacteriophages. Can J Public Health:4/3-4/8.

Dubos, RJ, JH Straus, C Pierce. 1943. The multiplication of bacteriophage *in vivo* and its protective effect against an experimental infection with *Shigella dysenteriae*. J Exp Med 78:161.

Edgar, R, W Wood. 1966. Morphogenesis of bacteriophage T4 in extracts of mutant-infected cells. Proc Natl Acad Sci USA 55:498-505.

Epstein, R, A Bolle, CM Steinberg, E Kellenberger, EB De La Tour, R Chevalley, R Edgar, M Susman, G Denhardt, A Lielausis. 1963. Physiological studies of conditional lethal mutants of bacteriophage T4D. Cold Spring Harbor Symp Quant Biol 28:375-394.

Flaks, JG, SS Cohen. 1959. Virus-induced acquisition of metabolic function I. Enzymatic formation of 5-hydroxymethyldeoxycytidylate J Biol Chem 234:1501-1506.

Gorski, A, R Miedzybrodzki, J Borysowski, K Dabrowska, P Wierzbicki, M Ohams, G Korczak-Kowalska, N Olszowska-Zaremba, M Lusiak-Szelachowska, M Klak. 2012. Phage as a modulator of immune responses: Practical implications for phage therapy. Adv Virus Res 83:41-71.

Harper, D, V Eryomin, T White, E Kutter. 1994. Effects of T4 infection on membrane lipid synthesis. In: J Karam, editor. *The Molecular Biology of Bacteriophage T4*: American Society for Microbiology Press. p. 385-390.

Häusler, T. 2006. Viruses vs. Superbugs: A solution to the antibiotics crisis?: Palgrave Macmillan.

Hershey, AD, M Chase. 1952. Independent functions of viral protein and nucleic acid in growth of bacteriophage. J Gen Physiol 36:39-56.

Karam, J, JW Drake, KN Kreuzer, G Mosig, DH Hall, FA Eiserling, LW Black, EK Spicer, E Kutter, K Carlson, editors. 1994. Molecular biology of bacteriophage T4: American Society for Microbiology Press.

Karam, J, E Miller. 2010. Bacteriophage T4 and its Relatives: A Series of Critical Reviews: Biomed Central Ltd. .

Kashlev, M, E Nudler, A Goldfarb, T White, E Kutter. 1993. Bacteriophage T4 Alc protein: A transcription termination factor sensing local modification of DNA. Cell 75:147-154.

Kellenberger, E. 1966. Electron microscopy of developing bacteriophage. In: J Cairns, GS Stent, JD Watson, editors. Phage and the Origins of Molecular Biology: Cold Spring Harbor Laboratory of Quantitative Biology. p. 116-126.

Knoll, BM, E Mylonakis. 2014. Antibacterial bioagents based on principles of bacteriophage biology: An overview. Clin Infect Dis 58:528-534.

Knouf, EG, WE Ward, PA Reichle, A Bower, PM Hamilton. 1946. Treatment of Typhoid Fever with Type Specific Bacteriophage: Preliminary Report. jama 132:134-138.

Krueger, A, EJ Scribner. 1941. The bacteriophage: Its nature and its therapeutic use. JAMA 116:2269-2277.

Kuchment, A. 2012. The Forgotten Cure: The Past and Future of Phage Therapy: Springer.

Kuhl, SJ, H Mazure. d'Hérelle. Preparation of therapeutic bacteriophages, Appendix 1 from: *Le Phénomène de la Guérison dans les maladies infectieuses*: Masson et Cie, 1938, Paris—OCLC 5784382. Bacteriophage 2011; 1:55-65.

Kutter, E, J Borysowski, R Międzybrodzki, A Górski, B Weber-Dąbrowska, M Kutateladze, Z Alavidze, M Goderdzishvili, R Adamia. 2014. Clinical phage therapy. In: J Borysoski, R Miedzybrodzki, A Gorski, editors. *Phage Therapy: Current Research and Applications*: Caister Academic Press. p. 257-288.

Kutter, EM, G Gvasalia, Z Alavidze, E Brewster. 2013. Phage therapy. In: M Grassberger, R Sherman, O Gileva, C Kim, K Mumcupglu, editors. Biotherapy-History, Principles and Practice: Springer Verlag. p. 191-231.

Kutter, EM, JS Wiberg. 1969. Biological effects of substituting cytosine for 5-hydroxymethylcytosine in the deoxyribonucleic acid of bacteriophage T4. J Virol 4:439-453.

Luria, S, TF Anderson. 1942. Identification and characterization of bacteriophage with the electron microscope. Proc Natl Acad Sci USA 28:127-130.

MacNeal, WJ, FC Frisbee, MA McRae. 1942. Staphylococcemia 1931-1940. Five hundred patients. Am J Clin Pathol 12.

Mathews, CK. 1993a. The cell-bag of enzymes or network of channels? J Bacteriol 175:63776381.

Mathews, CK. 1993b. Enzyme organization in DNA precursor biosynthesis. Prog Nucleic Acid Res Mol Biol 44:167-203.

Mathews, CK, E Kutter, G Mosig, P Berget. 1983. *Bacteriophage T4*: American Society for Microbiology Press.

Miedzybrodzki, R, J Borysowski, B Weber-Dabrowska, W Fortuna, S Letkiewicz, K Szufnarowski, Z Pawelczyk, P Rogóz, M Klak, E Wojtasik. 2012. Clinical aspects of phage therapy. Adv Virus Res 83:73-121.

Miller, ES, E Kutter, G Mosig, F Arisaka, T Kunisawa, W Rüger. 2003. Bacteriophage T4 genome. Microbiol Mol Biol Rev 67:86-156.

Olby, RC. 1974. *The Path to the Double Helix: The Discovery of DNA*: Courier Dover Publications.

Reardon, S. 2014. Phage therapy gets revitalized. Nature 510:15-16.

Sarabhai, A, A Stretton, S Brenner, A Bolle. 1964. Co-linearity of the gene with the polypeptide chain. Nature 201:13-17.

Streisinger, G, R Edgar, GH Denhardt. 1964. Chromosome structure in phage T4, I. Circularity of the linkage map. Proc Natl Acad Sci USA 51:775-779.

Twort, FW. 1915. An investigation on the nature of ultra-microscopic viruses. Lancet 186:1241-1243.

Twort, FW. 1936. Further investigations on the nature of ultra-microscopic viruses and their cultivation. J Hyg 36:204-235.

500 mg ml⁻¹

density of dsDNA inside a phage capsid

Panja, D, IJ Molineux. 2010. Dynamics of bacteriophage genome ejection *in vitro* and *in vivo*. Phys Biol 7:045006.

Chapter 5: Replication

STORIES

PERSPECTIVE

    by Graham F. Hatfull

Mycobacteriophage Brujita

a Siphophage that makes its lysis/lysogeny lifestyle decision using only three genes

Genome
dsDNA; linear
47,057 bp
74 predicted ORFs; 0 RNAs

Encapsidation method
Packaging; T = ? capsid

Common host
Mycobacterium smegmatis

Habitat
Water and soil

Lifestyle
Temperate

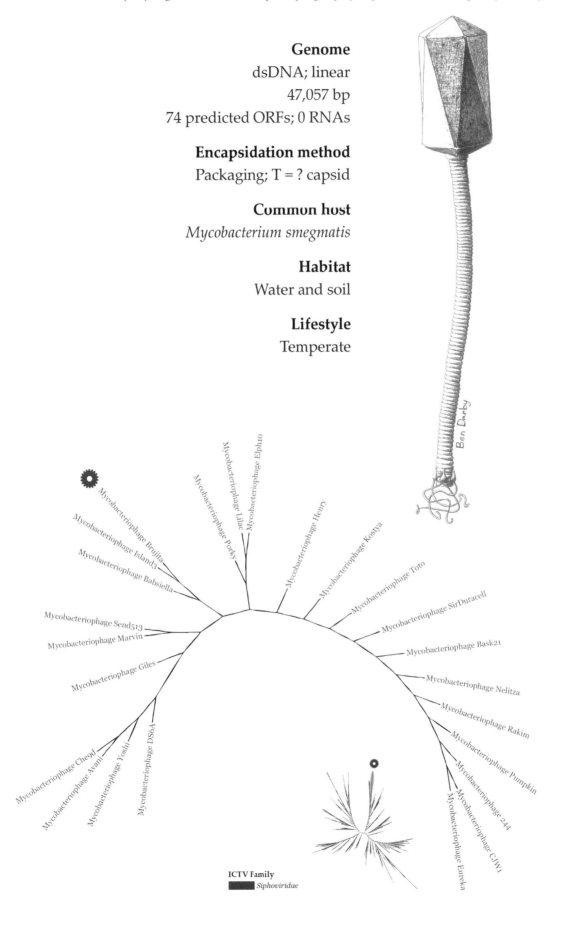

Ben Darby

ICTV Family
Siphoviridae

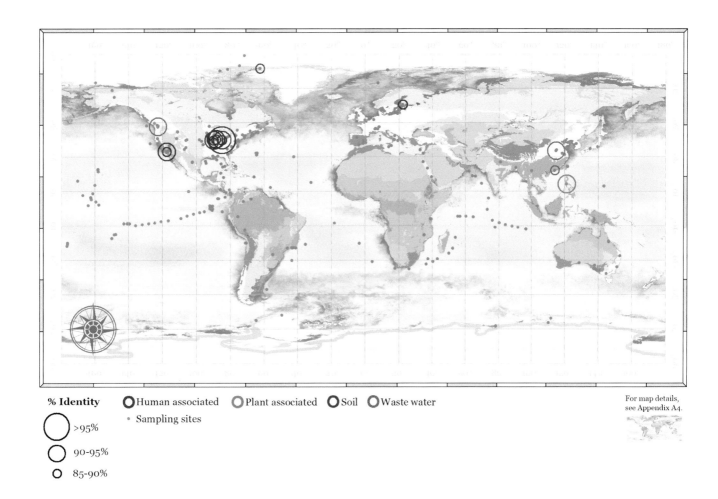

% Identity

○ >95%

○ 90-95%

○ 85-90%

○ Human associated   ○ Plant associated   ○ Soil   ○ Waste water

• Sampling sites

For map details, see Appendix A4.

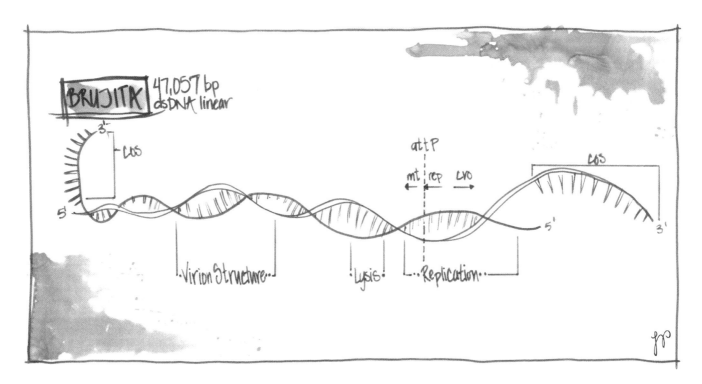

Mycobacteriophage Brujita

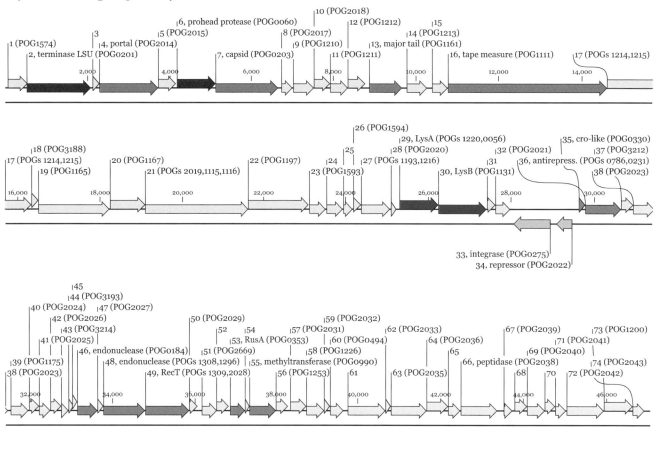

| On the Prowl | Entry | Takeover | Replication Lytic | Replication Lysogenic | Assembly | Escape | Transfer RNA | Non-coding RNA | Multiple | Unknown |

The Simple Switch of Brujita Witch

Heather Maughan

Mise-en-Scène: *Phages often appear the epitome of genomic frugality, having done so much for so long with so little. They wring extra functionality from each kbp by minimizing intergenic regions, including few if any introns, and encoding proficient proteins that are often shorter than their cellular counterparts. In marked contrast, λ, the model Siphophage, makes an extravagant genomic investment in a complex 'genetic switch' to govern its lysis/lysogeny decision. Is this necessary? The functional yet simple integration-dependent mechanism used by some mycobacterial Siphophages as related here proves not. With so many temperate phages that remain to be investigated, likely other strategies will be uncovered and that will, en masse, provide further insights into switch evolution.*

In a cranny within a soil biofilm a *Mycobacterium smegmatis* cell cowers, helpless, as an inescapable mob of Brujita phages approaches. As soon as they are in range, the Brujita gang pounces! Those first on the scene adsorb and enter, the latecomers hot on their heels. The first genomes are delivered and transcription immediately begins. The race is on between two of the first proteins made, the winner to determine whether the phage replicates or collaborates. If the Cro protein wins, the phage goes lytic. If the **integrase** wins, the phage will shelter in the host chromosome, postponing lysis until later.

Brujita is not the only phage faced with this pivotal decision. Upon arrival in a host, every temperate phage chooses lysis or lysogeny (*see page 1-5*). At the molecular level, the phage flips a genetic switch. When 'ON,' the phage replicates posthaste; flip it 'OFF,' and only genes for lysogeny are expressed. To reach its verdict, coliphage λ relies on a jury of at least six different proteins (Oppenheim et al. 2005; Little 2010). The mycobacteriophages have opted for a simpler switch with only three players: integrase, Cro, and a repressor. Together these genes comprise an ~2 kbp locus that is found in diverse mycobacteriophages (Broussard, Hatfull 2013; Broussard et al. 2013).

Placing bets

Which protein—Cro or integrase—will win the race and determine the host's fate? Both proteins begin with a handicap: the transcription rate for the *cro* gene is only half that of *int*; the integrase protein is degraded as soon as it is made, thanks to a C-terminal tag recognized by a cytoplasmic protease. Having active proteases doesn't bode well

for the survival of the host, as lysogeny requires having enough integrase available. Indeed, infection most often leads to lysis (Broussard et al. 2013).

How can integrase ever win? By overwhelming the available proteases, so that a few integrase molecules escape cleavage. Host proteases are constantly patrolling the cell for proteins to destroy. During times of starvation they may be busy rounding up non-essential proteins for amino acid recycling, hence too busy to intercept every integrase. Synthesis of more integrase molecules per cell also improves the odds; thus, a high multiplicity of infection (the mob scene, above) favors the integrase (Broussard et al. 2013). All it takes to flip the switch to lysogeny is enough integrase to successfully integrate just one phage genome.

Switch OFF: The host lives on borrowed time

Keeping one step ahead of the proteases, integrase binds to a specific sequence in the host's genome (*attB*) and to a similar sequence in the phage genome (*attP*), bringing the two genomes together. It then inserts the phage genome into the host chromosome at *attB* via **site-specific recombination**. Immediately, the race is over; the integrase has won. Integration of the phage chromosome as a prophage silences transcription of all the other invading phage genomes, and likewise will block subsequent infections by related phages. Integration also unleashes the repressor, thereby enabling lysogeny. The repressor binds to several regions of the prophage genome and prevents transcription of genes required for lytic replication. Why not bring the repressor into play earlier, before integration, to help in the decision-making?

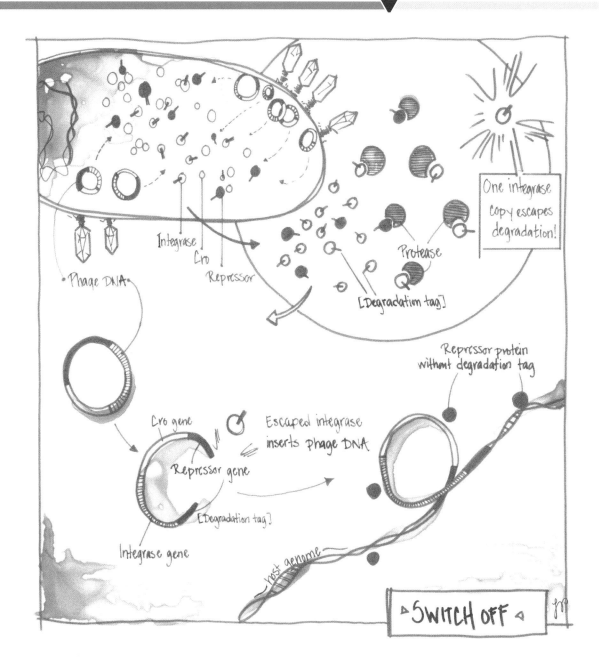

Integrase
Cro
Repressor

• Phage DNA •

One integrase copy escapes degradation!

Protease

[Degradation tag]

Repressor protein without degradation tag

Cro gene

Escaped integrase inserts phage DNA

Repressor gene

[Degradation tag]

Integrase gene

host genome

▲SWITCH OFF◁

Not possible for Brujita because it can make a functional repressor only while integrated as a prophage. Like integrase, Brujita's repressor carries a degradation tag near its C-terminus. But instead of playing the numbers game like integrase, it escapes detection by the protease through integrating and disrupting the tag in the process. Integration splits the circularized phage genome at *attP* to form the linear prophage genome. Since the *attP* site lies within the repressor gene, the repressor gene is divided in two, putting the proteolytic tag at one end of the prophage and the functional part of the gene at the other end (Broussard et al. 2013). Now, the repressor that is made from the functional region need not fear the patrolling proteases.

Switch ON: The host is toast

Repression doesn't last forever. Any integrase around that outwits the protease can act as the excisionase to excise the prophage. The integrase might take advantage of a diminished supply of proteases in a starving host—a time when the phage would be more than happy to abandon its host. Alternatively, more integrase may be made as a result of readthrough of nearby host genes during transcription (Broussard et al. 2013). Or an off-duty repressor may tilt the balance towards lysis by mistakenly allowing transcription of lysis genes.

Typically prophages excise in the presence of DNA damaging agents, but not those of the mycobacte-

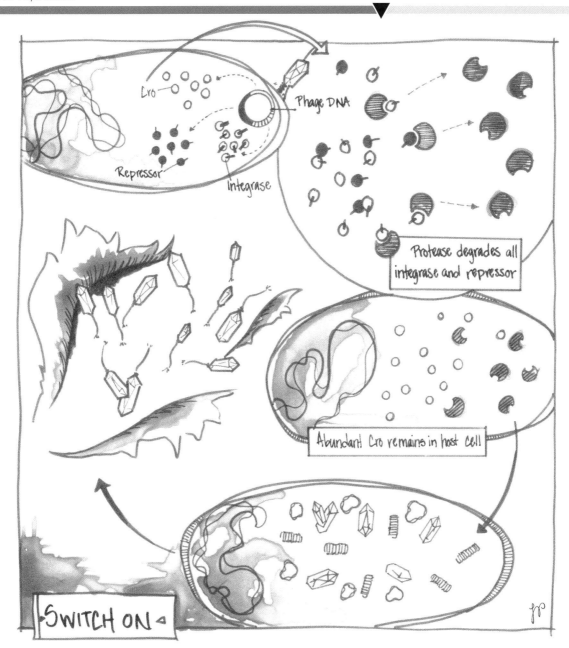

riophages that rely on these simple switches. Although some phages spontaneously turn lytic in cultures of lysogens, the precise mechanism of phage phreedom is not yet known (Broussard et al. 2013). Still, one thing is certain: Brujita's excision will reunite the repressor gene with its proteolytic tag, thus restoring its protease vulnerability and ending its reign of gene silencing. With the repressor destroyed, Cro will slowly accumulate, the phage will replicate, and a bacterial cell will die.

Cited references

Broussard, GW, GF Hatfull. 2013. Evolution of genetic switch complexity. Bacteriophage 3:e24186.

Broussard, GW, LM Oldfield, VM Villanueva, BL Lunt, EE Shine, GF Hatfull. 2013. Integration-dependent bacteriophage immunity provides insights into the evolution of genetic switches. Mol Cell 49:237-248.

Little, JW. 2010. Evolution of complex gene regulatory circuits by addition of refinements. Curr Biol 20:R724-734.

Oppenheim, AB, O Kobiler, J Stavans, DL Court, S Adhya. 2005. Switches in bacteriophage lambda development. Annu Rev Genet 39:409-429.

Recommended review

Oppenheim, AB, O Kobiler, J Stavans, DL Court, S Adhya. 2005. Switches in bacteriophage lambda development. Annu Rev Genet 39:409-429.

Other resources

The Mycobacteriophage Database: http://phagesdb.org/

Synechococcus Phage S-SSM7

a Myophage that encodes a peptide deformylase that preferentially processes a labile photosystem II protein

Genome
dsDNA; linear
232,878 bp
319 predicted ORFs; 5 RNAs

Encapsidation method
Packaging; T = ? capsid

Common host
Synechococcus

Habitat
Marine

Lifestyle
Lytic

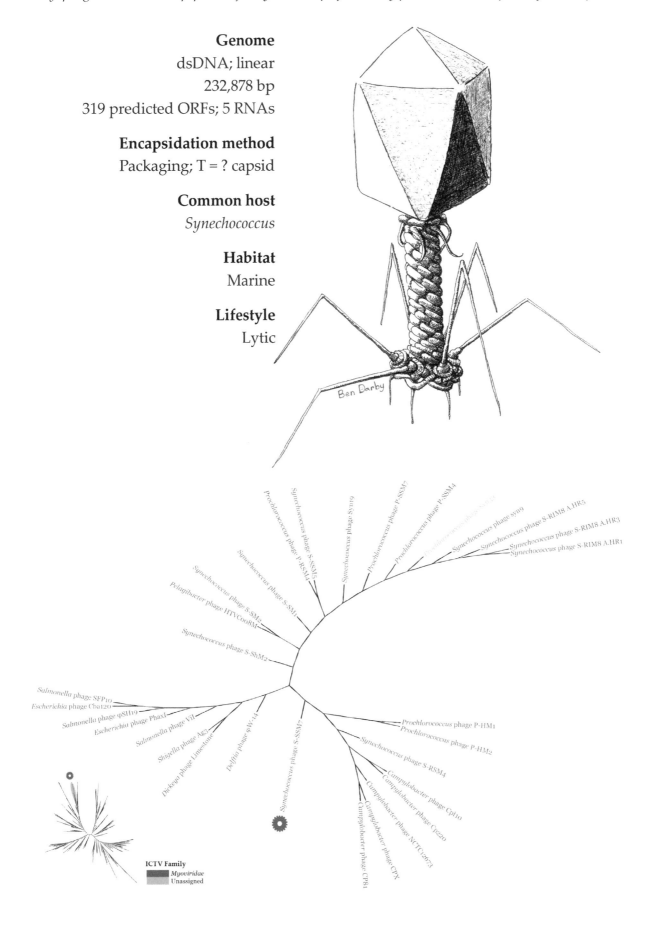

Ben Darby

ICTV Family
Myoviridae
Unassigned

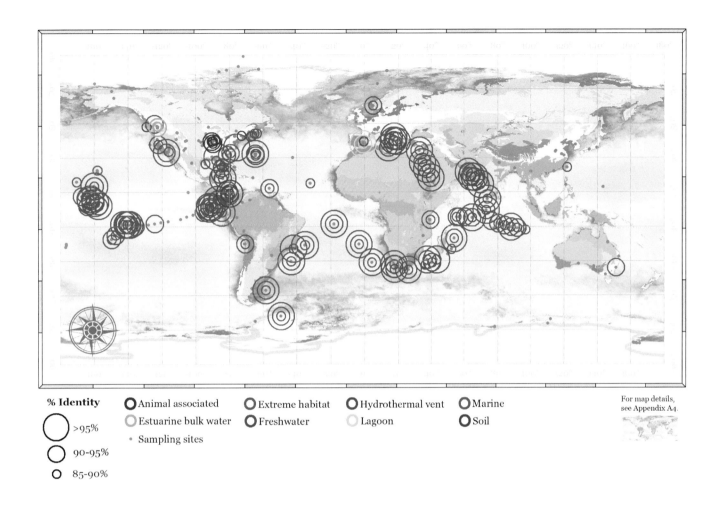

% Identity

- ◯ >95%
- ◯ 90-95%
- ○ 85-90%

- ⬤ Animal associated
- ◯ Estuarine bulk water
- · Sampling sites
- ⬤ Extreme habitat
- ⬤ Freshwater
- ⬤ Hydrothermal vent
- ○ Lagoon
- ⬤ Marine
- ⬤ Soil

For map details,
see Appendix A4.

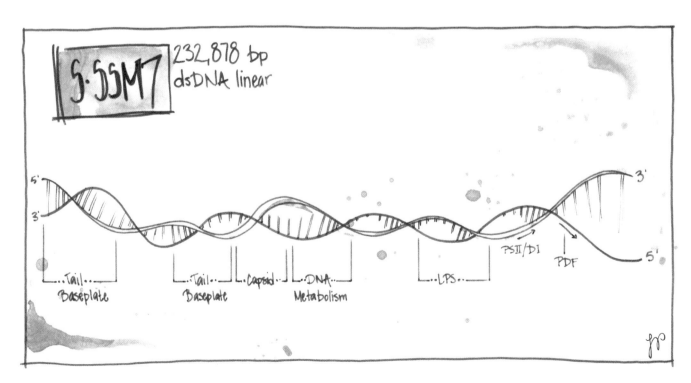

S·SSM7 — 232,878 bp dsDNA linear

Synechococcus Phage S-SSM7

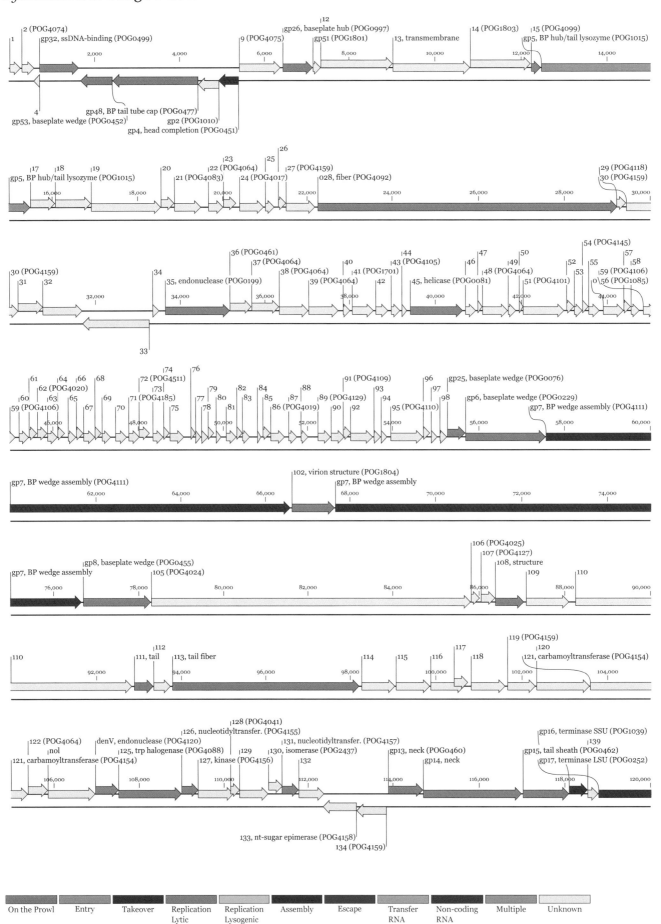

On the Prowl | Entry | Takeover | Replication Lytic | Replication Lysogenic | Assembly | Escape | Transfer RNA | Non-coding RNA | Multiple | Unknown

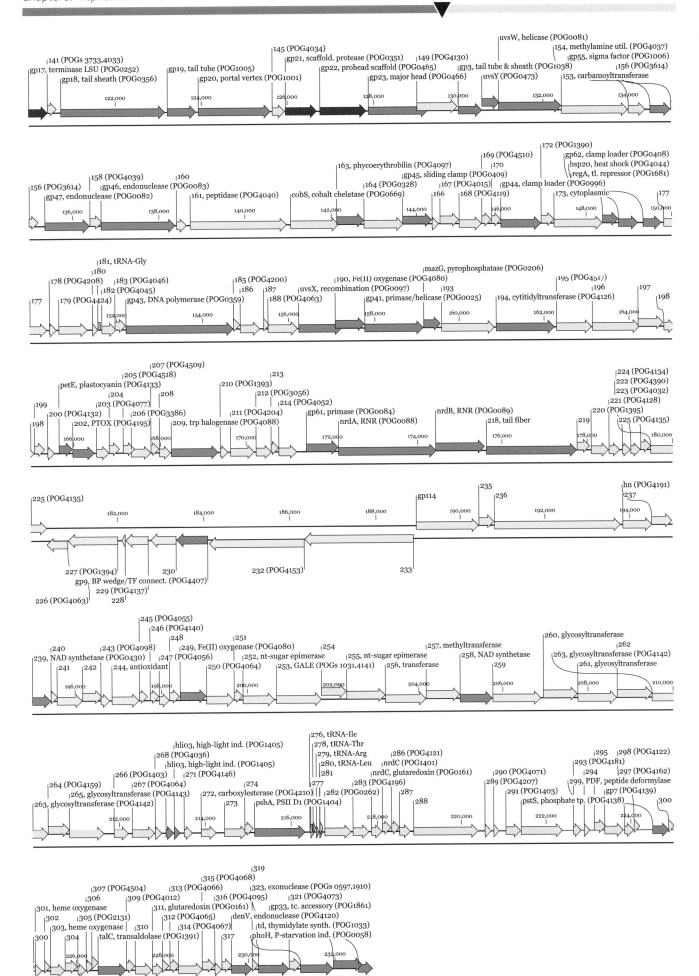

Anything You Can Do, We Can Do Better

Merry Youle

Mise-en-Scène: *Phages are sometimes condemned as thieves for stealing genes from their hosts. Should they be pardoned if they later return an improved version of the stolen goods? Evolution proceeds at a faster pace in phages than in their hosts. The rate of mutation per genome replication in phages with single-stranded RNA genomes is several orders of magnitude greater than for those using double-stranded DNA, but even for the latter the rate is roughly an order of magnitude higher than in Bacteria (Gago et al. 2009). Besides, the 'generation time' is shorter for lytic phages than for their hosts, and while Bacteria double each generation, phages multiply with burst sizes of 25 or more. As a result, phages have the opportunity to test drive more variants more rapidly, under different selection pressures. Should they be condemned or put to work?*

The oligotrophic open oceans, far from coastal nutrient sources, were once thought to be barren expanses supporting only sparse populations of phytoplankton (*see page 2-39*). In actuality they are home to abundant picoplankton, predominantly from two cyanobacterial genera: *Synechococcus* and *Prochlorococcus*. Both genera are rich in diversity with their members adapted to particular depths, temperatures, and latitudes. All Cyanobacteria must cope with the damage inflicted by sunlight on their photosynthetic apparatus. Hardest hit is photosystem II (PSII), particularly the D1/D2 protein heterodimer at its core that is the site of the primary photochemistry. Although many of these Bacteria use protective high-light-inducible proteins to dissipate excess light energy, they still must rapidly recycle and replace their D1/D2 proteins (Millard et al. 2004). When replacement cannot keep pace with destruction, photoinhibition ensues.

Photosynthesis & phage synthesis

Both *Synechococcus* and *Prochlorococcus* are host to numerous cyanophages from all three families within the *Caudovirales* (Myophage, Podophage, Siphophage). For any cyanophage, optimal reproduction depends on sustaining adequate photosynthesis to support replication right up to the time of lysis. This poses a problem. Typical lytic phage strategy is to shutdown host protein synthesis by degrading the host chromosome, blocking host transcription (*see page 4-7*), or other dirty tricks (Koerner, Snustad 1979). However, synthesis of at least the D1 protein is essential to maintain photosynthesis. Without adequate replacement of D1, photosynthesis would decline. The resulting drop in the production of both ATP and reducing power (NADPH) would hamper phage replication (Lindell et al. 2005).

The phage solution? Virtually all of the large, T4-like cyanophages, including phage S-SSM7, carry their own gene for D1 (*psbA*) (Sullivan et al. 2010). So equipped, they have two possible strategies for maintaining adequate D1: they can selectively boost transcription and translation of host-encoded D1 (Clokie et al. 2006) or express their own gene (Lindell et al. 2005). Some use the host gene, some the phage gene, some both.

Post-translation

An effective solution to the D1 problem requires more than generating and then translating abundant D1 gene transcripts. All those nascent polypeptides need to be cotranslationally processed and correctly folded, two tasks that are usually handled by the commandeered host protein translation machinery. That machinery is adequate for the bacterium under normal circumstances, but supporting replication of a large myophage is not the usual situation. Phage replication makes additional demands on these functions at the same time that host macromolecular synthesis is being curtailed.

Once again phage S-SSM7 looks after its own interests by bringing its own gene, in this case a gene encoding a peptide deformylase (PDF) (Frank et

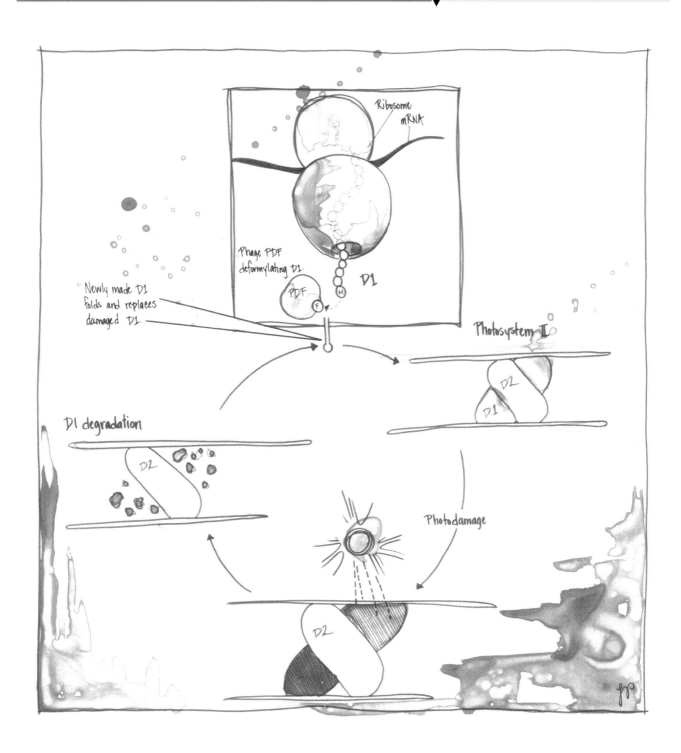

al. 2013). Proteins synthesized by bacterial ribosomes begin with an *N*-formylmethionine at their N-terminus; subsequently the formyl group is removed from at least 98% of them by PDF. Since this excision is essential for correct protein folding and function, sufficient PDF activity is essential for Bacteria. Deformylated D1 is required for an operational PSII in chloroplasts and likely also in Cyanobacteria, the bacterial progenitors of chloroplasts (Hou, Dirk, Williams 2004).

New & improved

Not only does phage S-SSM7 encode its own catalytically active PDF, but it is, from the phage perspective, a superior enzyme. Judged by the amount of substrate each PDF protein can transform per unit time (k_{cat}), the phage PDF performs several-fold better than a cyanobacterial PDF (from *Synechococcus elongatus* PCC 6301; [Frank et al. 2013]). Moreover, a cyanophage and a cyanobacterium have different protein synthesizing priorities. Unlike the cyanophages, the bacterium makes mostly ribosomal proteins. The phage PDF better serves the phage's purposes in that it is optimized to process D1 proteins more efficiently than ribosomal proteins.

Less is more

Based on its amino acid sequence, S-SSM7's PDF is a Type 1B PDF similar to the PDFs from chloroplasts and Cyanobacteria. Its three-dimensional structure, including the active site, very closely matches that of other Type 1B PDFs (such as those from *S. elongatus* and from *Arabidopsis thaliana* chloroplasts) with one marked exception: it lacks the alpha-helical C-terminal domain characteristic of this PDF type (*see page 2-58*). That domain might align the PDF with the ribosome exit portal to ensure access to the formyl group as the new polypeptide emerges (Bingel-

Erlenmeyer et al. 2008). Nonetheless, the phage PDF that lacks that domain deformylates more efficiently. This PDF carried by S-SSM7 is not a rare fluke. PDFs are the most common 'bacterial' genes encoded by marine viruses (Sharon et al. 2011). Like the PDF of S-SSM7, they all lack the C-terminal domain.

The large genomes of cyanomyophages often carry a cluster of various metabolic genes that were originally acquired from their bacterial hosts (Millard et al. 2009). While residing in a phage foster home, these genes continue to evolve, but more rapidly and under different selection pressures than in their original bacterial hosts. Sometimes a new and improved phage version travels back to the bacterial community (Lindell et al. 2004). This occasional exchange between phage and bacterial gene pools can benefit both trading partners.

Cited references

Bingel-Erlenmeyer, R, R Kohler, G Kramer, A Sandikci, S Antolic, T Maier, C Schaffitzel, B Wiedmann, B Bukau, N Ban. 2008. A peptide deformylase–ribosome complex reveals mechanism of nascent chain processing. Nature 452:108-111.

Clokie, MR, J Shan, S Bailey, Y Jia, HM Krisch, S West, NH Mann. 2006. Transcription of a 'photosynthetic' T4-type phage during infection of a marine cyanobacterium. Environ Microbiol 8:827-835.

Frank, JA, D Lorimer, M Youle, P Witte, T Craig, J Abendroth, F Rohwer, RA Edwards, AM Segall, AB Burgin. 2013. Structure and function of a cyanophage-encoded peptide deformylase. ISME J 7:1150-1160.

Gago, S, SF Elena, R Flores, R Sanjuán. 2009. Extremely high mutation rate of a hammerhead viroid. Science 323:1308-1308.

Hou, C-X, LM Dirk, MA Williams. 2004. Inhibition of peptide deformylase in *Nicotiana tabacum* leads to decreased D1 protein accumulation, ultimately resulting in a reduction of photosystem II complexes. Am J Bot 91:1304-1311.

Koerner, JF, DP Snustad. 1979. Shutoff of host macromolecular synthesis after T-even bacteriophage infection. Microbiol Rev 43:199-223.

Lindell, D, JD Jaffe, ZI Johnson, GM Church, SW Chisholm. 2005. Photosynthesis genes in marine viruses yield proteins during host infection. Nature 438:86-89.

Lindell, D, MB Sullivan, ZI Johnson, AC Tolonen, F Rohwer, SW Chisholm. 2004. Transfer of photosynthesis genes to and from *Prochlorococcus* viruses. Proc Natl Acad Sci USA 101:11013-11018.

Millard, A, MRJ Clokie, DA Shub, NH Mann. 2004. Genetic organization of the psbAD region in phages infecting marine *Synechococcus* strains. Proc Natl Acad Sci USA 101:11007-11012.

Millard, AD, K Zwirglmaier, MJ Downey, NH Mann, DJ Scanlan. 2009. Comparative genomics of marine cyanomyoviruses reveals the widespread occurrence of *Synechococcus* host genes localized to a hyperplastic region: Implications for mechanisms of cyanophage evolution. Environ Microbiol 11:2370-2387.

Sharon, I, N Battchikova, EM Aro, C Giglione, T Meinnel, F Glaser, RY Pinter, M Breitbart, F Rohwer, O Béjà. 2011. Comparative metagenomics of microbial traits within oceanic viral communities. ISME J 5:1178-1190.

Sullivan, MB, KH Huang, JC Ignacio-Espinoza, AM Berlin, L Kelly, PR Weigele, AS DeFrancesco, SE Kern, LR Thompson, S Young. 2010. Genomic analysis of oceanic cyanobacterial myoviruses compared with T4-like myoviruses from diverse hosts and environments. Environ Microbiol 12:3035-3056.

Recommended reviews

Kramer, G, D Boehringer, N Ban, B Bukau. 2009. The ribosome as a platform for co-translational processing, folding and targeting of newly synthesized proteins. Nat Struct Mol Biol 16:589-597.

Rohwer, F, RV Thurber. 2009. Viruses manipulate the marine environment. Nature 459:207-212.

4,000,000

families of viral proteins on Earth

Ignacio-Espinoza, JC, SA Solonenko, MB Sullivan. 2013. The global virome: Not as big as we thought? Curr Opin Virol:566-571.

Bacillus Phage PZA

a Podophage that initiates synthesis of its linear DNA genome using a protein primer

Genome
dsDNA; linear
19,366 bp
27 predicted ORFs; 1 RNA

Encapsidation method
Packaging; T = 3 capsid

Common host
Bacillus subtilis

Habitat
Soil

Lifestyle
Lytic

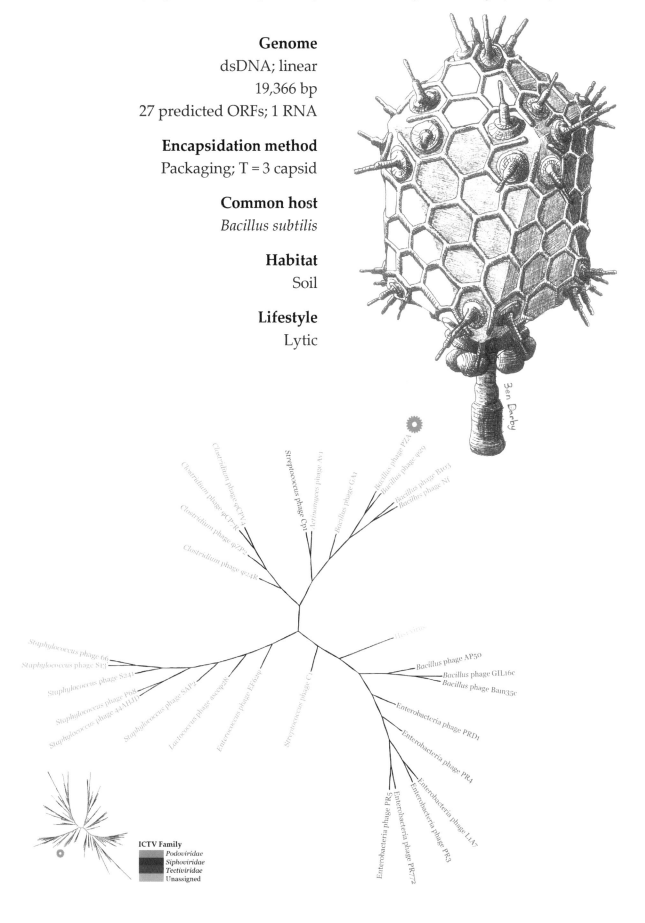

ICTV Family
Podoviridae
Siphoviridae
Tectiviridae
Unassigned

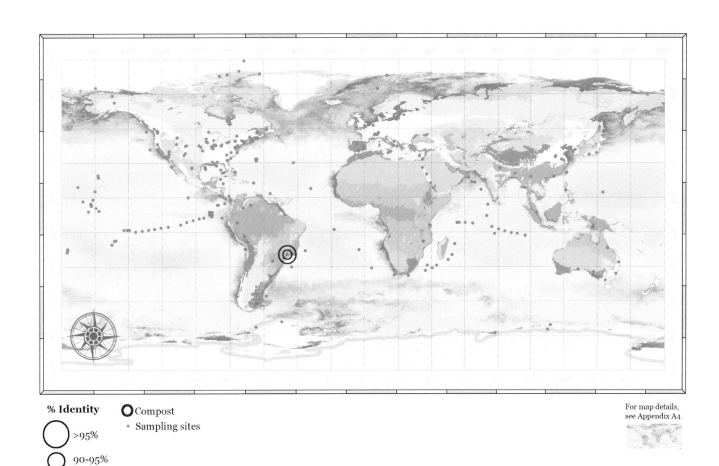

% Identity

◯ >95%

◯ 90-95%

◦ 85-90%

◯ Compost

• Sampling sites

For map details,
see Appendix A4.

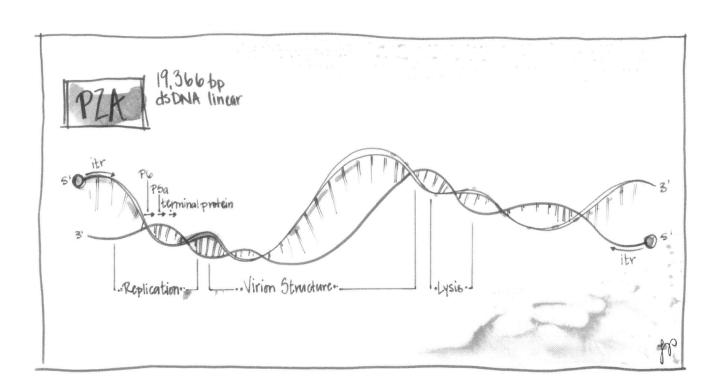

PZA 19,366 bp
dsDNA linear

itr
5'

P6
P5a
terminal protein

3'
3'
itr
5'

Replication   Virion Structure   Lysis

Bacillus Phage PZA

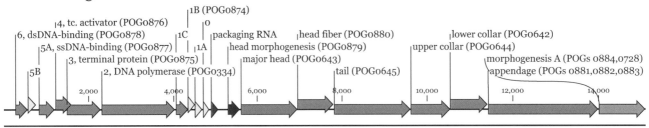

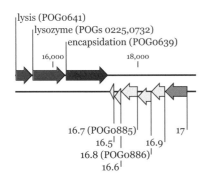

| On the Prowl | Entry | Takeover | Replication Lytic | Replication Lysogenic | Assembly | Escape | Transfer RNA | Non-coding RNA | Multiple | Unknown |

Getting a Handle on Linear Genome Replication

Merry Youle

Mise-en-Scène: *Few are the microbes that live a solitary, self-sufficient existence outside the laboratory; most collaborate or participate in joint projects with others (e.g., biofilms, endosymbioses). The individuals in these co-dependent teams commonly resist our efforts to culture them. To glean information about their lifestyles, we can now scrutinize the genomes of individual cells plucked from the community. Here we rely on the talented DNA polymerase of* Bacillus *phage φ29 to amplify one copy of a genome to yield sufficient DNA for sequencing. This is the enzyme of choice because it possesses exceptionally high fidelity, efficiency, and processivity, the latter demonstrated by its synthesis of long (>10 kb) continuous strands. The field of single cell genomics is indebted to this phage.*

Note: Most of the research related in this story was performed using phage φ29, but PZA uses the same protein-primed replication mechanism and encodes a terminal protein with almost 100% identity to that of φ29.

Any extant life form with a linear, double-stranded DNA (dsDNA) genome has successfully solved a problem—the 'end replication problem.' If left unresolved, a few nucleotides at the 5'-terminus would not be replicated and the genome would shrink each generation. This complication is inherent in the ability of DNA polymerase (DNApol) to synthesize DNA in only one direction: 5' → 3'. Moreover, when beginning a new DNA chain,

DNApol can't position the first nucleotide unless there is something upstream to attach it to. That something is a primer, its function to provide a hydroxyl group to which DNApol can attach the initial nucleotide. Some organisms avoid the new chain issue by using circular chromosomes for replication, while others protect the ends of their linear chromosomes with clusters of repetitive nucleotide sequences known as telomeres.

A diverse group, including some Bacteria, Archaea, **plasmid**s, animal viruses, and phages, have a different handle on this problem—actually two protein handles. They use a protein, rather than RNA, as their primer, with a serine, threonine, or tyrosine residue supplying the priming hydroxyl group. After replication, the primer proteins remain covalently linked to the 5′ terminus at each end of their chromosome—tell-tale evidence of replication by this mechanism. One of phages proficient in this strategy, *Bacillus* phage φ29, has been poked and prodded *in vitro* to reveal its trade secrets, most of which apply not only to its close relatives but to protein-primed DNA replication wherever it is found.

The minimal toolset

φ29's genome is ~19 kbp of linear dsDNA with a 31 kDa parental **terminal protein** (TP) covalently linked to each 5′ end (Salas et al. 1978). φ29's DNApol can replicate this *in vitro* given a supply of TP and dNTPs (Blanco, Salas 1985). No other proteins are required—no helicases, no clamps, etc. DNApol recognizes two origins of replication in this genome, one at each end formed by the short, 6 bp **inverted terminal repeats** (ITRs) (Yoshikawa, Ito 1981).

```
Strand X        3'-TTTCAT ------- ATGAAA-5'-TP
Strand Y     TP-5'-AAAGTA ------- TACTTT-3'
```

From end to end

Two DNApol enzymes independently initiate replication at the two ends; simultaneity is not required. Advancing in opposite directions they semi-conservatively replicate the genome to yield two full-length daughter genomes. To visualize this, consider the DNApol that proceeds from left to right (*see above*) using strand X as its template strand. DNApol's first step is to construct the protein primer. For this, it couples with an unattached TP (termed a priming TP) to form a TP-DNApol heterodimer. Next it covalently links a dAMP to the serine[232] of its TP. This adenine, still attached to TP, will become the first nucleotide at the 5′ end of the new DNA strand.

The TP-DNApol heterodimer loaded with dAMP recognizes the 3′ terminus of strand X as a replication origin. Using strand X as its template, DNApol proceeds to synthesize a new strand Y starting at its 5′ end, displacing the old strand Y in the process. When completed, this yields one full length genome and one displaced strand Y.

Meanwhile another DNApol has likely started replication from the other end of the genome, using strand Y as its template and synthesizing a new strand X beginning with its 5′ terminal nucleotide. If the two working DNApols meet somewhere along the way, each continues along its original template strand. An advancing DNApol seems unperturbed if its template strand has been displaced from the parental dsDNA helix by the other DNApol. It carries on just the same, synthesizing its strand to the very end. When both enzymes have completed their strand, there are two full length genomes, each with a TP linked to each of their 5′ ends.

Working *in vivo*

Of course, the process *in vivo* is a bit more sophisticated from start to finish, as it involves additional helper proteins. First, the origin of replication *in vivo* is a nucleoprotein cluster formed by the binding of numerous copies of the φ29 dsDNA binding protein (p6) to an ITR. This assemblage recruits the priming dAMP-TP-DNApol and also opens the ends of the DNA helix to allow it access (Serrano, Salas, Hermoso 1990). DNApol starts to work here, adding nucleotides as it advances along the template strand. When it has added the tenth nucleotide, it lets go of its TP, leaving the TP covalently bound to the 5′-terminal adenine. DNApol now shifts into elongation mode and rapidly synthesizes the rest of the strand (Méndez, Blanco, Salas 1997). The displaced strand is immediately coated by many copies of φ29's ssDNA binding protein (SSB) that protect it from degradation until its complementary strand is added by the DNApol advancing from the other end (Gutiérrez, Sogo, Salas 1991).

Fidelity

The formation of the protein primer is more error-prone than subsequent steps (Esteban, Salas, Blanco 1993). During priming, DNApol is asked to recognize and attach a dAMP to a protein—not one of its usual tasks. If it errs and links the wrong nucleotide to serine[232] in the primer TP, DNA synthesis could get off to a faulty start. Compounding the difficulty, DNApol can't proofread the initial base. However, φ29 uses another maneuver for quality control: It double-checks the identity of the first nucleotide. Its test equipment consists of the first two bases in the template strand, both Ts. When initiating a new strand, it asks the primer TP-DNApol: do you carry an adenine? It interrogates the bound nucleotide by matching it not to the 3′-terminal T, as you might expect, but rather to the T in the second position (Méndez et al. 1992). If the nucleotide passes this test, then the verified dAMP-TP-DNApol 'slides back' one nucleotide to position that dAMP opposite the terminal T. This step repeats the question: are you an adenine? If some nucleotide other than dAMP were mistakenly linked to the TP, the resultant mismatch would interrupt replication and the replication complex would dissociate.

Versatile handles

Phage φ29 demonstrates that having protein 'handles' at both ends of a linear genome can be useful in other ways, too. For one, when it invades a host, those parental TPs shepherd the genome to the host **nucleoid** (Redrejo-Rodríguez et al. 2013). There the host's RNA polymerase promptly transcribes the early phage genes so infection can proceed. Then, as new copies of φ29 DNApol are made, the TPs also recruit them to the nucleoid to replicate the phage DNA (Kamtekar et al. 2006). When the time comes to fill the waiting **procapsids** with DNA, it is the attached TP that is recognized by the φ29 packaging machinery—a tactic that efficiently packages φ29 DNA and only φ29 DNA (Grimes, Anderson 1989). In short, φ29 and its kin handle their genomes with care.

Cited references

Blanco, L, M Salas. 1985. Replication of phage φ29 DNA with purified terminal protein and DNA polymerase: Synthesis of full-length φ29 DNA. Proc Natl Acad Sci USA 82:6404-6408.

Esteban, J, M Salas, L Blanco. 1993. Fidelity of φ29 DNA polymerase. Comparison between protein-primed initiation and DNA polymerization. J Biol Chem 268:2719-2726.

Grimes, S, D Anderson. 1989. *In vitro* packaging of bacteriophage φ29 DNA restriction fragments and the role of the terminal protein gp3. J Mol Biol 209:91-100.

Gutiérrez, C, J Sogo, M Salas. 1991. Analysis of replicative intermediates produced during bacteriophage φ29 DNA replication *in vitro*. J Mol Biol 222:983-994.

Kamtekar, S, AJ Berman, J Wang, JM Lázaro, M de Vega, L Blanco, M Salas, TA Steitz. 2006. The φ29 DNA polymerase: Protein-primer structure suggests a model for the initiation to elongation transition. EMBO J 25:1335-1343.

Méndez, J, L Blanco, JA Esteban, A Bernad, M Salas. 1992. Initiation of φ29 DNA replication occurs at the second 3′ nucleotide of the linear template: A sliding-back mechanism for protein-primed DNA replication. Proc Natl Acad Sci USA 89:9579-9583.

Méndez, J, L Blanco, M Salas. 1997. Protein-primed DNA replication: A transition between two modes of priming by a unique DNA polymerase. EMBO J 16:2519-2527.

Redrejo-Rodríguez, M, D Muñoz-Espín, I Holguera, M Mencía, M Salas. 2013. Nuclear and nucleoid localization are independently conserved functions in bacteriophage terminal proteins. Mol Microbiol 90:858-868.

Salas, M, RP Mellado, E Viñuela, JM Sogo. 1978. Characterization of a protein covalently linked to the 5′ termini of the DNA of *Bacillus subtilis* phage φ29. J Mol Biol 119:269-291.

Serrano, M, M Salas, JM Hermoso. 1990. A novel nucleoprotein complex at a replication origin. Science 248:1012-1016.

Yoshikawa, H, J Ito. 1981. Terminal proteins and short inverted terminal repeats of the small *Bacillus* bacteriophage genomes. Proc Natl Acad Sci USA 78:2596-2600.

Recommended reviews

de Vega, M, M Salas. Protein-primed replication of bacteriophage φ29 DNA. In: J Kusic-Tisma, editor. *DNA Replication and Related Cellular Processes*: Intech. p. 179-206.

Salas, M. 1991. Protein-priming of DNA replication. Annu Rev Biochem 60:39-71.

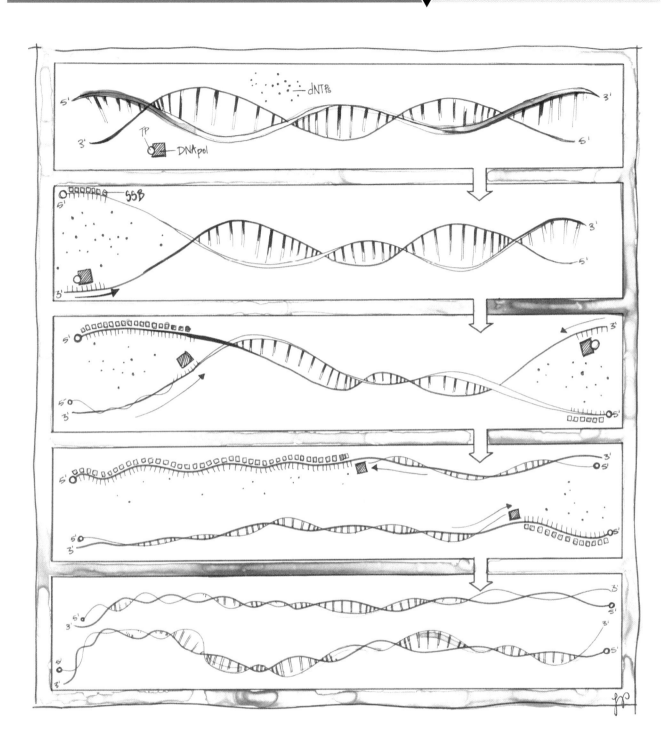

Enterobacteria Phage N15

a Siphophage that, during lysogeny, stably maintains its prophage as a linear plasmid

Genome
dsDNA; linear
46,375 bp
61 predicted ORFs; 0 RNAs

Encapsidation method
Packaging; T = 7 capsid

Common host
Escherichia coli

Habitat
Mammalian intestines & sewage

Lifestyle
Temperate

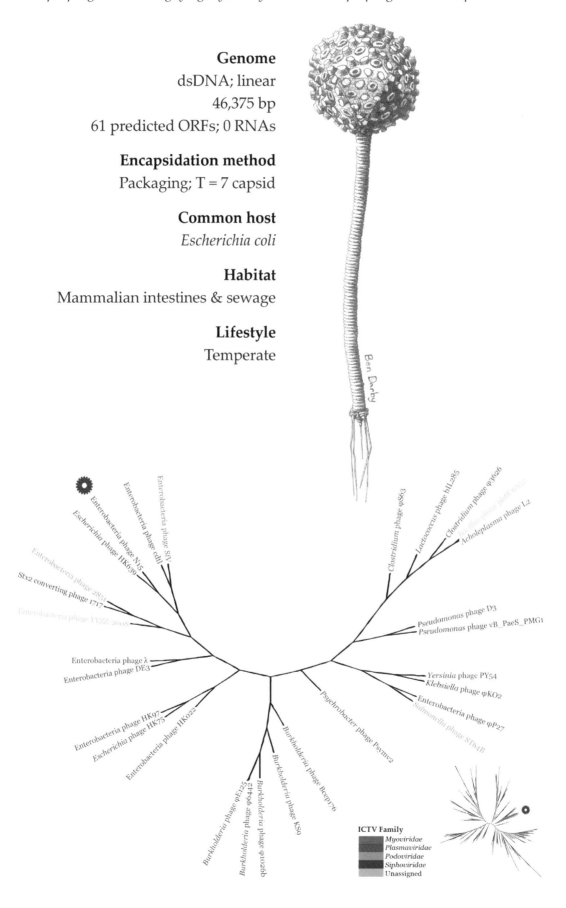

Ben Darby

ICTV Family
Myoviridae
Plasmaviridae
Podoviridae
Siphoviridae
Unassigned

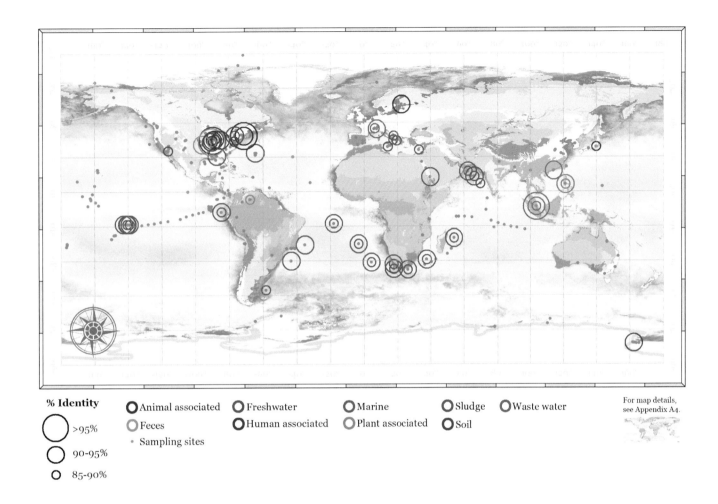

% Identity

◯ >95%

◯ 90-95%

○ 85-90%

◯ Animal associated ◯ Freshwater ◯ Marine ◯ Sludge ◯ Waste water

◯ Feces ◯ Human associated ◯ Plant associated ◯ Soil

• Sampling sites

For map details, see Appendix A4.

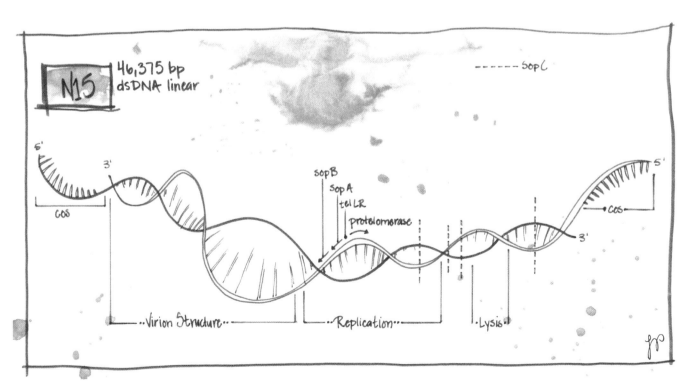

Enterobacteria Phage N15

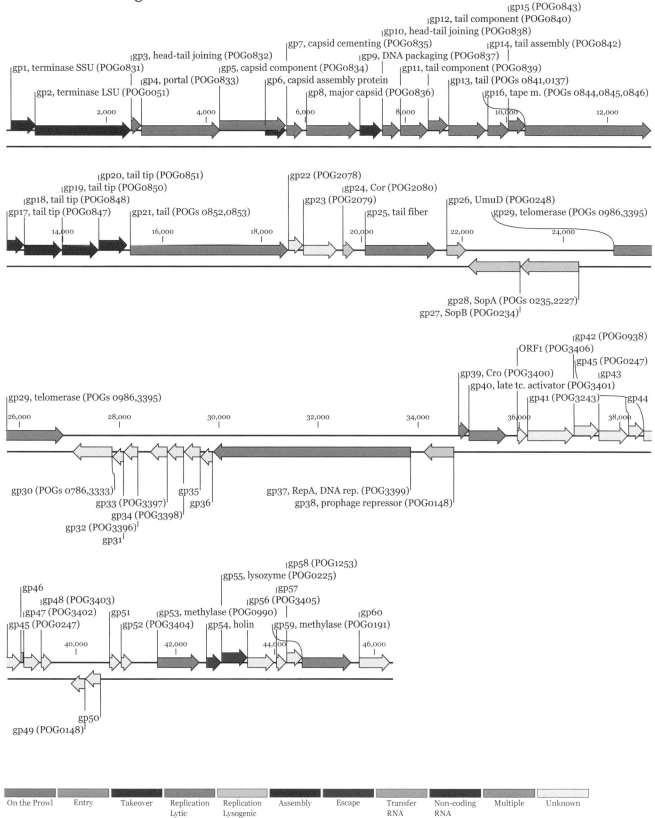

| On the Prowl | Entry | Takeover | Replication Lytic | Replication Lysogenic | Assembly | Escape | Transfer RNA | Non-coding RNA | Multiple | Unknown |

An Independent Prophage

Merry Youle

Mise-en-Scène: *Phages are opportunists that will adopt useful genes from any source: from a host, from other cellular organisms, from other phages, and, as coliphage N15 demonstrates, also from other mobile genetic elements such as plasmids. N15's genome is a particularly striking example of a collage where genes with different evolutionary histories now find themselves side-by-side. Although evolution in the textbooks is pictured as a branching tree with lineages continually bifurcating, during phage evolution branches also anastomose, even branches from different trees.*

At first glance, a phagewatcher might mistake coliphage N15 for the archetypal Siphophage λ. Both phages have similar virion structures and genome lengths, as well as similar **latent periods** and burst sizes when replicating lytically in their host, *E. coli*. A closer look at N15's genome would identify **syntenic homologs** of many of λ's genes for virion structural proteins and lysis tools (Ravin et al. 2000). Their virions both deliver a linear dsDNA genome with 'sticky' **cohesive ends** that anneal to circularize the genome upon arrival inside a host (*see page 6-5*). These sticky overhangs are 12 nts of single-stranded DNA, ten of which match one-to-one between N15 and λ (Rybchin, Svarchevsky 1999). Like λ, N15 is temperate and opts for lysogeny with similar frequency. But once it decides to lysogenize, N15 shows its deviant nature.

Lysogeny styles

Like many temperate phages, λ lysogenizes using **integrase** to insert its genome into a specific location in the host chromosome by **site-specific recombination** (*see page 5-37*). When ensconced there, the λ prophage silences most of its genes, expressing those that suppress lytic replication and protect its home from infection by similar phages. Phage λ then sits back and lets its host replicate the prophage as part of its own chromosome and also see to it that each daughter cell gets a copy.

When N15 lysogenizes, it's a different story. Its prophage does not take up residence in the host chromosome, opting instead to maintain as an independent **plasmid** prophage. This strategy is shared by other λ-like Siphophages (e.g., PY54 of *Yersinia enterocolitica* and φKO2 of *Klebsiella oxytoca*) and by Mu-like Siphophages (e.g., φHAP-1 of *Halomonas aquamarina* and VP882 of *Vibrio para-*

haemolyticus). Soon after arrival in the host, N15's **protelomerase** cuts its circular chromosome at a specific site (*telRL*) to convert the circle into a linear molecule of double-stranded DNA (dsDNA) with *telL* at one end and *telR* at the other—a linear plasmid prophage. Unlike the integrated λ prophage, this plasmid prophage is quite active and transcribes at least 29 of its 60 genes (Ravin et al. 2000). In addition to typical prophage tasks, it carries out some essential plasmid activities, i.e., replication and partitioning into every daughter cell.

A replication scheme

Living the independent life of a plasmid means that an N15 prophage is responsible for its own replication. Being a *linear* prophage plasmid adds an additional challenge—the **end replication problem** (*see page 5-18*). Some phages with linear dsDNA genomes, such as the φ29-like Podophages, use **terminal proteins** to handle this, but not N15. Instead N15 caps its chromosomes using covalently closed hairpin termini known as 'telomeres.' These telomeres differ in structure and mechanism from their eukaryotic namesake, but both solve the same problem. When protelomerase cuts N15's circular dsDNA chromosome, it makes staggered nicks 16 bp apart within a region containing long **inverted repeats** (Rybchin, Svarchevsky 1999). It then shepherds the resulting 16 nt single-stranded overhangs so that they fold back on themselves and anneal to self-complementary sequences (*see the figure*). Protelomerase then ligates the free ends to form a closed hairpin at each end.

Replicating a DNA molecule like this with covalently closed ends calls for specialized replication machinery. N15's replicase, RepA, recognizes an

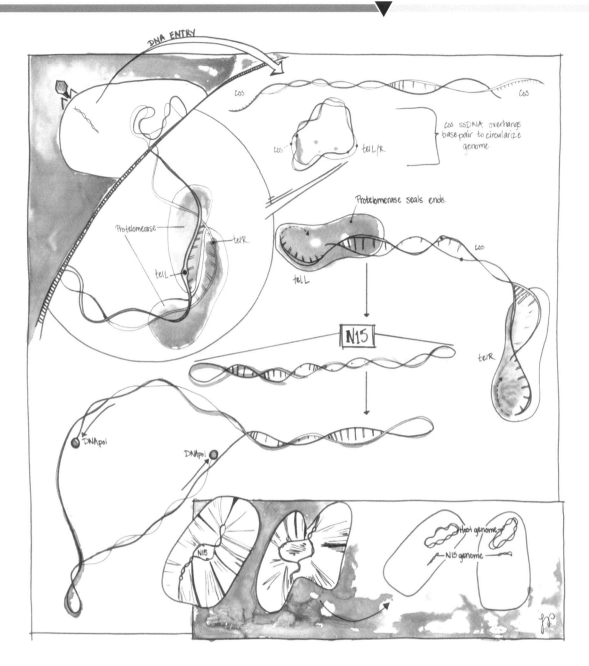

internal replication origin (*ori*) within the linear chromosome. Beginning there it replicates the DNA bi-directionally, converting one circular DNA molecule into two (Ravin 2003). Protelomerase then steps in and separates the topologically interlocked daughter circles by breaking and re-joining the chromosomes.

Sticking around

N15 regulates its replication to maintain ~3-5 prophage plasmids per bacterial chromosome. With so few copies, they cannot be left clustered near the place of their birth lest one of the daughter cells inherit them all. Low-copy-number plasmids, whether linear like N15 or circular like the F plasmid of *E. coli*, actively partition copies to both daughter cells. Since the F plasmid maintains with only 1-2 copies per cell, clearly it must get this right. Its approach is straightforward: it encodes a *sop* operon containing genes for two proteins, SopA and SopB, as well as a centromere-like region, *sopC* (Ogura, Hiraga 1983). SopB binds to *sopC* in the F plasmid to form a nucleoprotein complex. Add both ATP and SopA, and the SopB/*sopC* complex nucleates the polymerization of SopA into dynamic filaments that form a spindle-like structure radiating outward from the complex (Lim, Derman, Pogliano 2005). As a result, whereas these plasmids are centered in newly born daughter cells, they later move to the ¼ and ¾ points along the length of the cell—the mid-points of future daughter cells.

N15 borrowed this technique from bacterial plasmids and encodes its own homologs of SopA and SopB (Ravin et al. 2000). These proteins do the same job for the plasmid prophage as for other plasmids. However, whereas one *sopC* centromere-like locus suffices for a circular plasmid, N15 uses four dispersed along its DNA (Ravin, Lane 1999). Does this system work? In cultures of N15 lysogens, only one daughter cell in ten thousand lacks the N15 prophage (Ravin 2011). To further ensure its maintenance, N15 may also use a plasmid dirty trick—a toxin-antitoxin (TA) module—but it has not yet been proven guilty of murder (Dziewit et al. 2007).

Going lytic

Sometimes, upon arrival, N15 or λ will opt for lytic replication. In that case, λ circularizes and then uses that circle to initiate 5-6 rounds of θ replication thereby providing templates for transcription. Later in the infection it switches to rolling circle (σ) replication to generate linear genomes for packaging in the new virions (*see page 6-5*). You might expect N15 to take the same direct approach and bypass the linear plasmid form altogether, but instead it proceeds the same as for lysogeny. It converts the initial circular chromosome into a linear molecule with hairpin ends, then repeatedly replicates the linear form. Only late in infection does N15 change its routine. At that point, replication of linear chromosomes produces two replication-competent circular chromosomes that each engage in rolling circle replication to yield the linear genomes for packaging (Mardanov, Ravin 2009).

Genomic hodgepodge

Although N15 poses as a lambdoid phage, close scrutiny of its genome discloses a checkered past. In the left arm 25 genes certify its lambdoid ancestry; genes 1-21 correspond, one-to-one, to homologs in λ (Ravin et al. 2000). These genes encode virion structural proteins and proteins that function in assembly and packaging. In contrast, the right arm is a jumbled patchwork of genes from various phages, genes from plasmids, and genes without known homologs. Of the 35 here, only ten have a lambdoid past. Some very promiscuous ancestors must be lurking in N15's family tree.

All of this raises the question how N15 with its cobbled together plasmid phage scheme has managed to keep its toehold in the competition for *E. coli* hosts. Perhaps under some circumstances there is a benefit to having 3-5 copies of its genome transcriptionally active in the lysogen. Perhaps one of the ~20 genes of unknown function gives its host a competitive edge in some environments. Perhaps having its own active replication machinery makes N15 quick on its feet and able to go lytic at the first sign that its host is in trouble. Although employing a plasmid prophage is not the majority's choice, it must work well enough. Phage N15 is still in the game.

Cited references

Dziewit, L, M Jazurek, L Drewniak, J Baj, D Bartosik. 2007. The SXT conjugative element and linear prophage N15 encode toxin-antitoxin-stabilizing systems homologous to the tad-ata module of the *Paracoccus aminophilus* plasmid pAMI2. J Bacteriol 189:1983-1997.

Lim, GE, AI Derman, J Pogliano. 2005. Bacterial DNA segregation by dynamic SopA polymers. Proc Natl Acad Sci USA 102:17658-17663.

Mardanov, AV, NV Ravin. 2009. Conversion of linear DNA with hairpin telomeres into a circular molecule in the course of phage N15 lytic replication. J Mol Biol 391:261-268.

Ogura, T, S Hiraga. 1983. Partition mechanism of F plasmid: Two plasmid gene-encoded products and a cis-acting region are involved in partition. Cell 32:351-360.

Ravin, N, D Lane. 1999. Partition of the linear plasmid N15: Interactions of N15 partition functions with the *sop* locus of the F plasmid. J Bacteriol 181:6898-6906.

Ravin, NV. 2003. Mechanisms of replication and telomere resolution of the linear plasmid prophage N15. FEMS Microbiol Lett 221:1-6.

Ravin, NV. 2011. N15: The linear phage–plasmid. Plasmid 65:102-109.

Ravin, V, N Ravin, S Casjens, ME Ford, GF Hatfull, RW Hendrix. 2000. Genomic sequence and analysis of the atypical temperate bacteriophage N15. J Mol Biol 299:53-73.

Rybchin, VN, AN Svarchevsky. 1999. The plasmid prophage N15: A linear DNA with covalently closed ends. Mol Microbiol 33:895-903.

Recommended reviews

Ravin, NV. 2011. N15: The linear phage–plasmid. Plasmid 65:102-109.
Rybchin, VN, AN Svarchevsky. 1999. The plasmid prophage N15: A linear DNA with covalently closed ends. Mol Microbiol 33:895-903.

Bacillus Phage φ29

a Podophage that, during host sporulation, sequesters its genome in the resistant spore

Genome
dsDNA; linear
19,282 bp
27 predicted ORFs; 1 RNA

Encapsidation method
Packaging; T = 3 capsid

Common host
Bacillus subtilis

Habitat
Soil

Lifestyle
Lytic

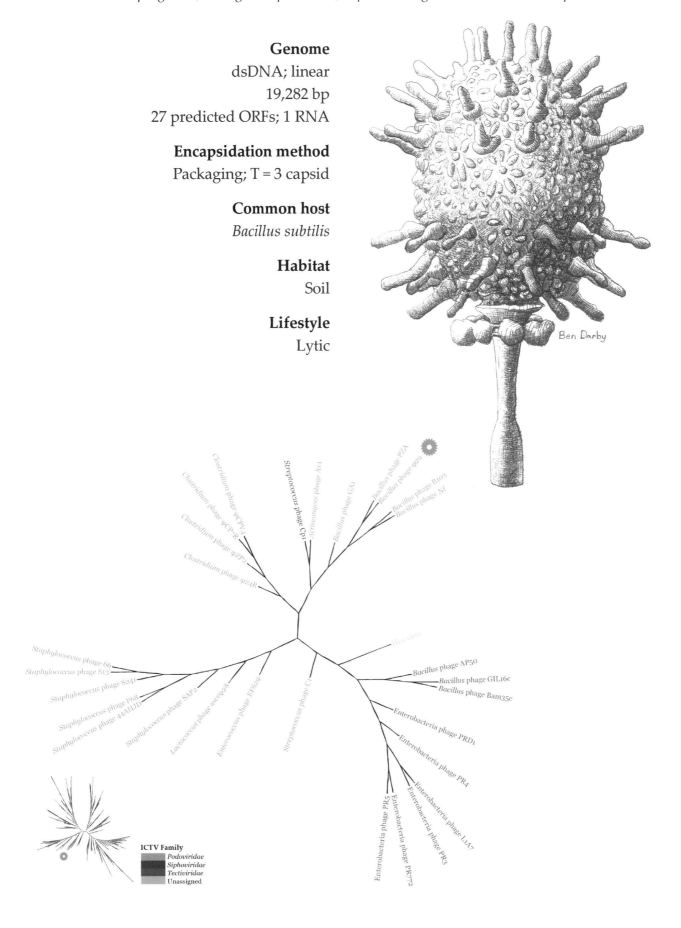

Ben Darby

ICTV Family
Podoviridae
Siphoviridae
Tectiviridae
Unassigned

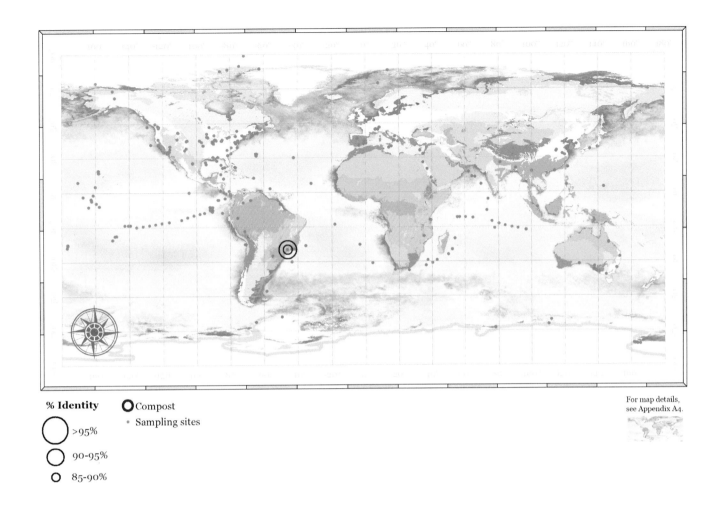

% Identity

○ Compost

• Sampling sites

◯ >95%

◯ 90-95%

○ 85-90%

For map details,
see Appendix A4.

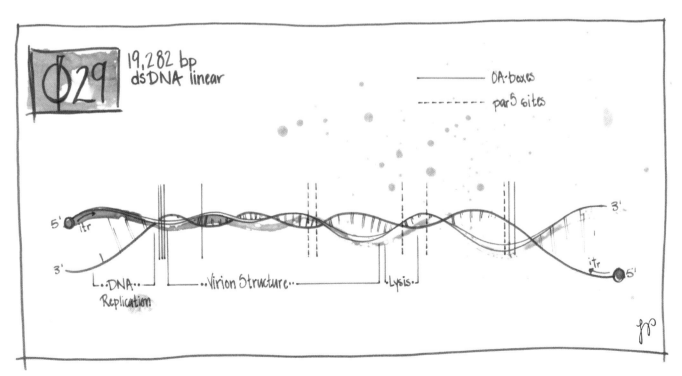

Φ29
19,282 bp
dsDNA linear

——— OA-boxes
- - - - parS sites

5' itr ... 3'
3' ... itr 5'

L..DNA.. Replication   ..Virion Structure..   Lysis.

Bacillus Phage φ29

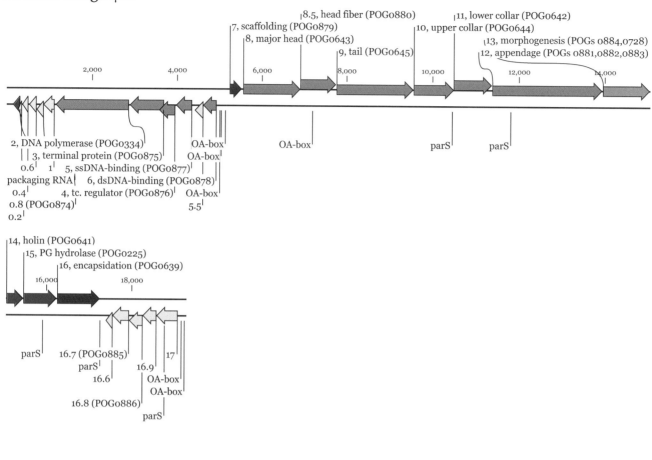

φ29 Hijacks Bomb Shelter for Survival

Heather Maughan

Mise-en-Scène: *When a phage walks about as a virion, it may find itself in harsh conditions that threaten to disrupt its capsid or inactivate its genome. A few phages that infect sporulating Bacteria, such as φ29, have the fortuitous option to hole up somewhere safe and wait for conditions to improve. This doesn't happen by default, but requires skillful manipulation of the host. Even with that, the odds for success may be low.*

Even vicious predators have their vulnerabilities. A phage virion hunting for a host sometimes faces harsh chemicals that threaten to disintegrate its proteinaceous capsid and UV radiation that might irreparably damage its DNA. Although bacterial cells are also susceptible to the effects of toxic or nutrient-depleted environments, some can hunker down and wait for better days. Many members of the bacterial phylum *Firmicutes*, such as *Bacillus subtilis*, are champions of this. They are able to construct a 'bomb shelter' around their genome and remain metabolically inert until the environment improves, even if it takes decades.

That shelter is a spore, a highly resistant and metabolically dormant cell that develops from a vegetative cell during times of stress (Stragier, Losick 1996). In the lab, stressors such as nitrogen starvation and high cell density activate **sporulation**; in nature, the triggers are likely diverse and species-specific. Sporulation is initiated biochemically, but soon morphological changes manifest. An asymmetric cell division yields a small prespore and a larger mother cell. Following chromosome segregation, the mother cell works quickly to engulf the prespore and cover it with a thick, impervious protein armor, while the prespore dehydrates its cytoplasm and coats its DNA with protective proteins. When its work is completed, the mother cell autolyses and releases the spore to face the harsh environment on its own. Little does the spore know that an enemy could be lurking within.

The enemy within

That enemy is φ29, a **lytic** phage that knows how to deal effectively with a sporulating host. When it enters a host cell that is in the early stages of sporulation, it stifles its usual lytic attack, opt-ing instead to become an unobtrusive stowaway. It translocates its genome into the prespore, the two together forming a **virospore**. But this stowaway is treacherous. It keeps quiet as it waits for the good times to return. It watches until increasing nutrients trigger spore germination, and even then it shrewdly delays a bit longer. This allows time for the cell to resume metabolic activity and repair the damage suffered by phage, as well as bacterial, DNA.

Stifling the phage

For φ29 to restrain its killer impulse requires some help from its host. While initiating sporulation, and only then, the host inhibits transcription of early phage genes, thereby blocking phage lytic development. How does the host coordinate this with sporulation? The answer lies with Spo0A, the host's master sporulation regulator. Spo0A initiates spore development by binding to non-coding DNA sequences known as 0A boxes, thereby activating or repressing transcription of specific host genes. The φ29 genome also has some 0A boxes, six of them, all near the promoters of early phage genes. When Spo0A binds to these phage 0A boxes, it represses transcription of phage genes (Meijer et al. 2005), thus silencing phage replication.

Genome entrapment

The next step, getting its genome into the prespore, requires exceptional ingenuity on the part of the phage. φ29 can't take the easy route of integrating its genome into the host chromosome (i.e., lysogenizing its host) because of the structure of its chromosome. The difficulty stems from the **terminal protein** (TP) covalently bound to each 5' end of the phage's linear dsDNA chromosome. These proteins are key to φ29's ability to replicate its linear chromosome, but they prevent lysogeny (*see page 5-18*). Instead, φ29 enlists the services of a host protein (Spo0J) to escort its DNA into the prespore. This protein's usual job is to recognize and bind a specific region on the host chromosome, *parS* (a 16 bp **inverted repeat**), and lead the chromosome to a cell pole. φ29 engages the same polar transport service by including five *parS* sites in its genome.

Which cell pole does φ29's genome move toward? This does matter. Translocation into the prespore spells survival, whereas delivery to the mother cell would be a dead end. Expression of Spo0A by the mother cell would prevent the phage from replicating before the cell autolyses. You might expect the phage chromosome to have a 50% success rate, considering that the host chromosome segregation process delivers

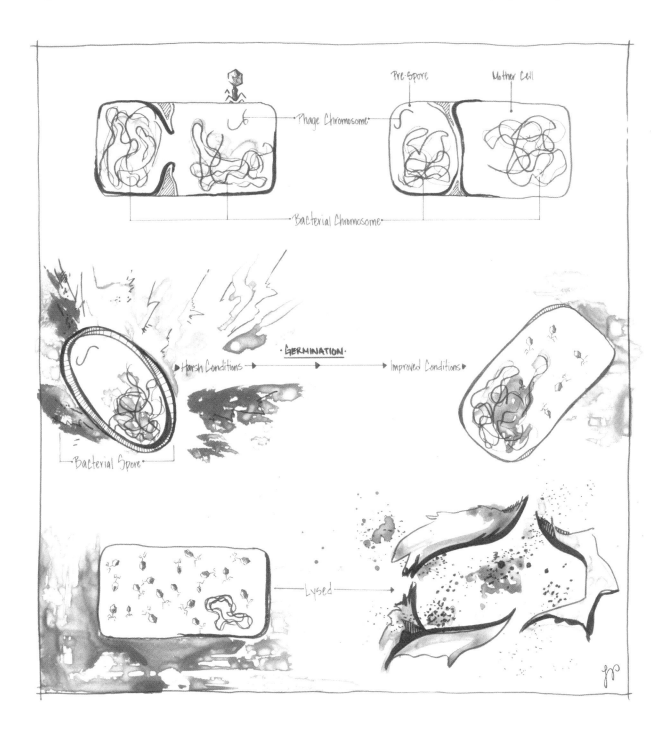

one chromosome to the prespore, the other to the mother cell. But only 20% of spores contain φ29 genomes. Why so few? The smaller size of the prespore makes it less likely to harbor a phage chromosome (Sonenshein 2006).

Betting the odds

The low percentage of spores with φ29 stowaways begs the question: Might it not be better to immediately lyse the sporulating cell despite the unfavorable conditions rather than to take your chances on getting into the bomb shelter? φ29 takes its chances, but a close relative, phage Nf, more often lyses sporulating hosts. The Nf genome reflects its different strategy: there are two *parS* sites for chromosome guidance but only one 0A box. With only a single 0A box, early phage genes are expressed

and Nf proceeds with a lytic infection, thereby condemning the host to death before it completes sporulation (Castilla-Llorente, Meijer, Salas 2009).

Who is smarter, φ29 or Nf? Is lysing a sporulating cell a worthwhile scheme? The dire environmental conditions that induced host sporulation likely mean that a harsh world with few healthy hosts awaits any progeny virions produced. But casting your lot with your host and entering its bomb shelter is a decision that should not be taken lightly. There is no exit until the host decides it is time to germinate. Suppose that meanwhile alternate hosts become available, hosts that can thrive despite the current hard times, hosts that invite fruitful infection, and there you are—trapped, helpless, inside your own virospore. Gamble lost.

Cited references

Castilla-Llorente, V, WJ Meijer, M Salas. 2009. Differential Spo0A-mediated effects on transcription and replication of the related *Bacillus subtilis* phages Nf and φ29 explain their different behaviours *in vivo*. Nucleic Acids Res 37:4955-4964.

Meijer, WJ, V Castilla-Llorente, L Villar, H Murray, J Errington, M Salas. 2005. Molecular basis for the exploitation of spore formation as survival mechanism by virulent phage φ29. EMBO J 24:3647-3657.

Sonenshein, AL. 2006. Bacteriophages: How bacterial spores capture and protect phage DNA. Curr Biol 16:R14-16.

Stragier, P, R Losick. 1996. Molecular genetics of sporulation in *Bacillus subtilis*. Annu Rev Genet 30:297-241.

Recommended reviews

Sonenshein AL. 2006. Bacteriophages: How bacterial spores capture and protect phage DNA. Curr Biol 16:R14-16.

Stragier P, R Losick. 1996. Molecular genetics of sporulation in *Bacillus subtilis*. Annu Rev Genet 30:297-241.

Acholeplasma Phage L2

a Siphophage that can replicate and bud progeny virions or integrate into the host's genome

Genome
dsDNA; circular
11,965 bp
14 predicted ORFs; 0 RNAs

Encapsidation method
Other

Common host
Acholeplasma laidlawii

Habitat
Ubiquitous; free-living or animal-associated

Lifestyle
Non-lytic temperate

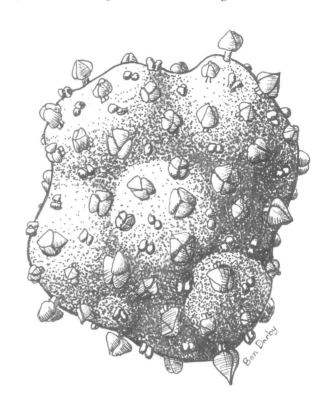

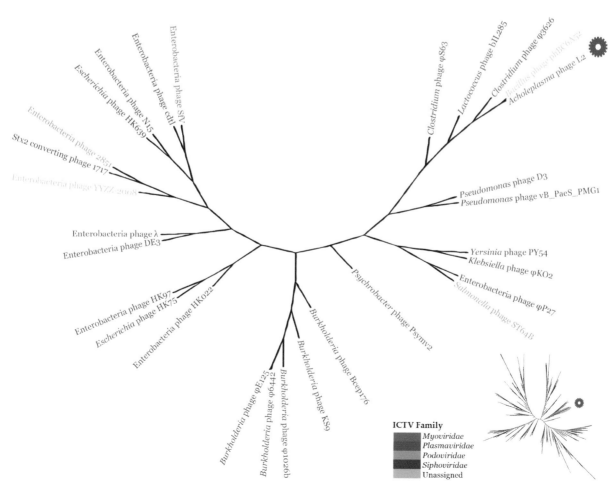

ICTV Family
Myoviridae
Plasmaviridae
Podoviridae
Siphoviridae
Unassigned

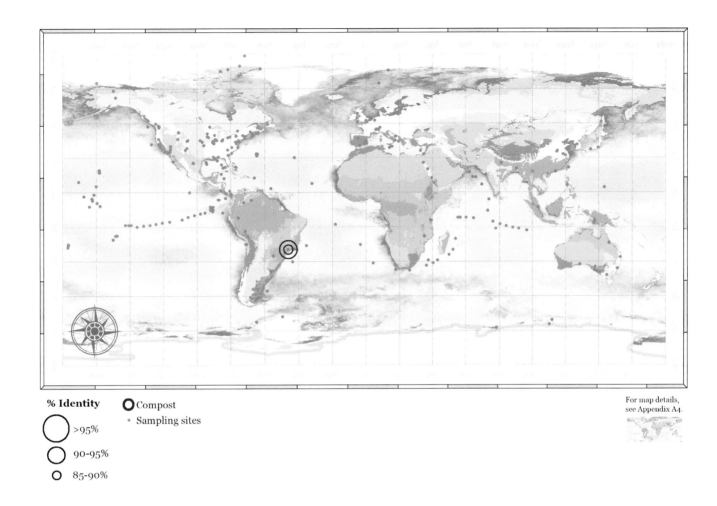

% Identity O Compost
  ○ >95%   • Sampling sites
  ○ 90-95%
  ○ 85-90%

For map details,
see Appendix A4.

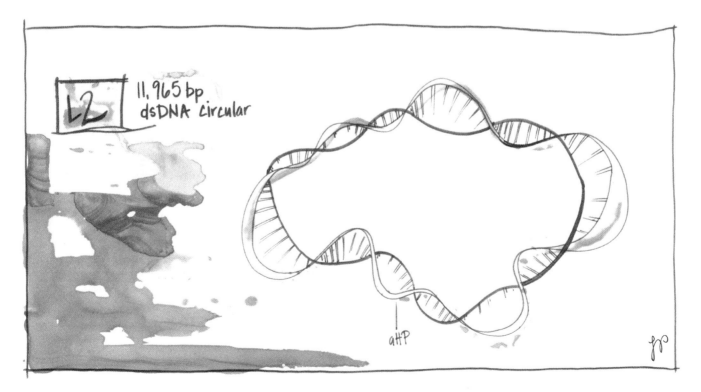

L2 — 11,965 bp
dsDNA circular

gHP

Acholeplasma **Phage L2**

○ Circular genome

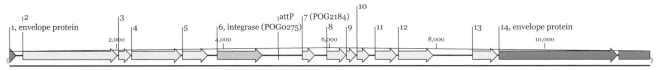

| On the Prowl | Entry | Takeover | Replication Lytic | Replication Lysogenic | Assembly | Escape | Transfer RNA | Non-coding RNA | Multiple | Unknown |

A Phage Infection in Two Acts

Heather Maughan & Merry Youle

Mise-en-Scène: *The term 'virus' defies definition. Nevertheless, one can readily identify a canonical virus. Their hallmark traits include the protein capsid that they synthesize and assemble to encase their genome during intercellular transport. However, there are shrewd entities that behave like viruses but that don't quite conform to this standard, e.g., those that con another phage into synthesizing their capsid components for them (see page 6-18). There are also phages such as* Acholeplasma *phage L2 that, in an apparent adaptation to the surface structure of their Mollicute hosts, eschew a capsid altogether. We still call them viruses.*

The Scene

When under phage invasion, a susceptible bacterium may suffer any of several fates depending on the character of the trespassing phage. If that phage is lytic, the invader quickly takes over without hesitation, redirecting cellular activity to virion production. Before long it discards the host, lysing it to release the progeny virions. A temperate phage behaves differently (*see page 1-5*). Its first task upon arrival in the cytoplasm is to assess the situation and make a management decision: lyse now or later? The kindly filamentous phages dispense with lysis altogether, and instead continually release progeny while the host lives on. Most phages choose one of these three alternatives, but not *Acholeplasma* phage L2. By playing mix and match, it creates a unique lifestyle.

The peculiar structure of L2's host, *Acholeplasma laidlawii*, provides unusual opportunities for a phage. Being a **Mollicute**, *A. laidlawii* foregoes a cell wall and envelopes itself with only a single membrane. L2 has followed suit by spurning the typical protein capsid, instead surrounding its DNA genome with only a protein-rich membrane (Gourlay et al. 1973). During entry, the phage's membrane merges into the host membrane; during exit, each virion acquires its **envelope** by pinching off from the membrane.

Act I

L2 slips into a host smoothly by membrane fusion, then quietly proceeds to replicate profusely. These abundant progeny are released by budding, a process reminiscent of that of the filamentous phages (*see page 7-17*). Although budding averts host lysis,

the host doesn't get off scot-free. To visualize the drain on host energy and resources, consider that the total area of the cell membrane is equivalent to the membrane area used to envelope ~150 virions, yet nearly a thousand virions bud in the first 4-6 h (Putzrath, Maniloff 1977). Exhausted by providing for the exiting horde, the usual 1.5 h host generation time is extended to 4-5 h during this stage—a three-fold increase (Putzrath, Maniloff 1977).

Act II, Scene I

Faced with host fatigue, L2 shifts gears and lysogenizes like a temperate phage by integrating a copy of its genome into the host chromosome (Putzrath, Maniloff 1978; Dybvig, Maniloff 1983). In the process, its **integrase** recognizes a short sequence in the host chromosome (the *attB* site) that is identical to a sequence in the phage genome (the *attP* site), and inserts the phage genome at that location by **site-specific recombination**.

After L2 settles into its new chromosomal home as a prophage, virion release continues at a slower rate for at least another two hours as already accumulated genomes exit. Now behaving like a temperate phage, L2 silences expression of many prophage genes, including those needed for phage replication. It also prevents infection by related phages—a benefit to host and prophage alike (Putzrath, Maniloff 1978; Steinick et al. 1980). As the curtain falls on Scene I, it appears that the host and prophage will live happily ever after.

Act II, Scene II

Scene II opens with the fortunes of the host in decline. A typical prophage in this situation, sensing impending doom, would respond by excising it-self from the chromosome and making a do-or-die attempt to produce progeny before the ship goes down. L2 behaves similarly and resumes repli-cation, but with its own peculiar twist (Putzrath, Maniloff 1978). The genomes of other temperate phages possess an **excisionase** gene, typically lo-cated adjacent to the integrase gene—but not L2. Its genome lacks a recognizable excisionase (Ma-niloff, Kampo, Dascher 1994). Does L2 break out from the chromosome by some other mechanism? Excision has not been explicitly observed, leaving

open the possibility of genome replication from an *in situ* prophage template or perhaps from a ge-nome copy leftover from the reproductive phase and maintained episomally.

Epilogue

L2 is one smart phage. It manages multiple replica-tion strategies with a mere 14 ORFs. By infecting Mollicutes whose naked cell membrane facilitates budding, L2 eliminated the need to encode a holin/ **endolysin** lysis mechanism (*see page 7-5*). Their two-act replication scheme realizes both rapid im-

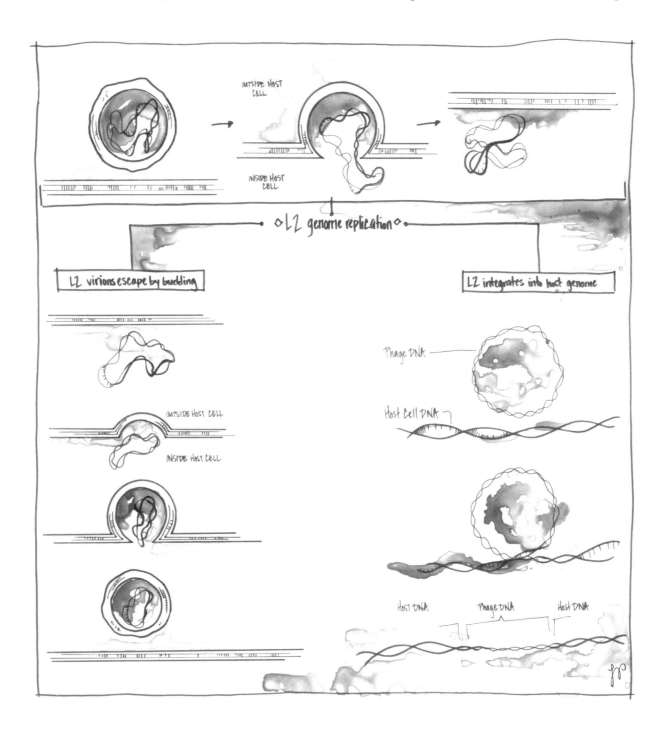

mediate release of abundant progeny by budding *and* the slower, long-term reproduction offered by lysogeny. But is it so smart to package your genome inside an amorphous lipid sac without a protein shell either inside or out? These virions are not only amorphous, but they come in three versions that differ in diameter (74, 88, and 132 nm), genome copy number (1-3), and the stoichiometry of proteins embedded in the membrane (Poddar et al. 1985). Might

this be related to L2's inefficient adsorption (Putzrath, Maniloff 1977; Liss, Heiland 1983)?

Since these eccentric phages have been mostly ignored for decades, every step of their life cycle is enveloped by mystery. It is high time someone—perhaps you—used current methodologies to investigate how these phages live the phage life in such anomalous ways.

Cited references

Dybvig, K, J Maniloff. 1983. Integration and lysogeny by an enveloped mycoplasma virus. J Gen Virol 64 (Pt 8):1781-1785.

Gourlay, RN, DJ Garwes, J Bruce, SG Wyld. 1973. Further studies on the morphology and composition of *Mycoplasmatales* virus-laidlawii 2. J Gen Virol 18:127-133.

Liss, A, RA Heiland. 1983. Characterization of the enveloped plasmavirus MVL2 after propagation on three *Acholeplasma laidlawii* hosts. Arch Virol 75:123-129.

Maniloff, J, GJ Kampo, CC Dascher. 1994. Sequence analysis of a unique temperate phage: Mycoplasma virus L2. Gene 141:1-8.

Poddar, SK, SP Cadden, J Das, J Maniloff. 1985. Heterogeneous progeny viruses are produced by a budding enveloped phage. Intervirology 23:208-221.

Putzrath, RM, J Maniloff. 1977. Growth of an enveloped mycoplasmavirus and establishment of a carrier state. J Virol 22:308-314.

Putzrath, RM, J Maniloff. 1978. Properties of a persistent viral infection: Possible lysogeny by an enveloped nonlytic mycoplasmavirus. J Virol 28:254-261.

Steinick, LE, A Wieslander, KE Johansson, A Liss. 1980. Membrane composition and virus susceptibility of *Acholeplasma laidlawii*. J Bacteriol 143:1200-1207.

Saccharomyces cerevisiae Virus L-A

a member of the Totiviridae that accomplishes a complete life cycle with only two encoded gene products

Genome
dsRNA; linear
L-A: 4,579 bp, 2 ORFs, 0 RNAs
M1: 1,801 bp, 1 ORF, 0 RNAs

Encapsidation method
Co-condensation (?); T = 2
capsid

Common host
Saccharomyces cerevisiae

Habitat
Surfaces of mature fruits and grains

Lifestyle
Non-lytic temperate

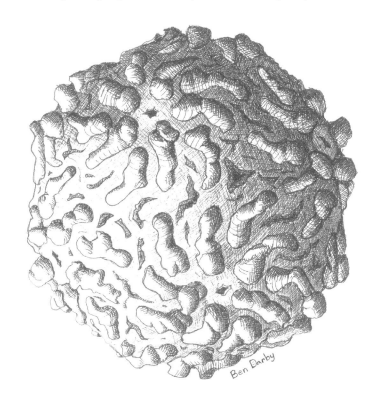

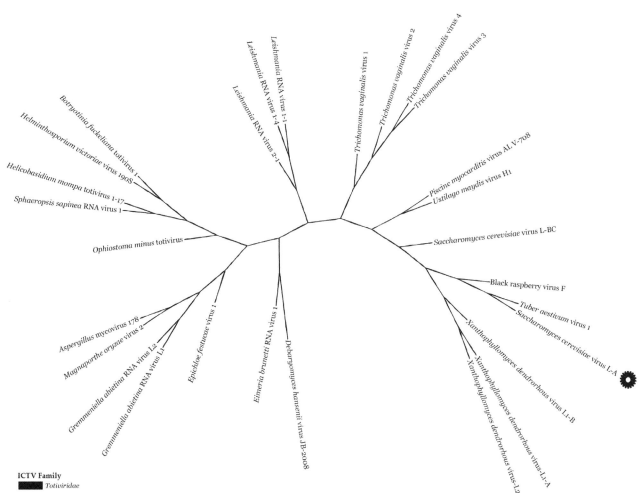

ICTV Family
Totiviridae

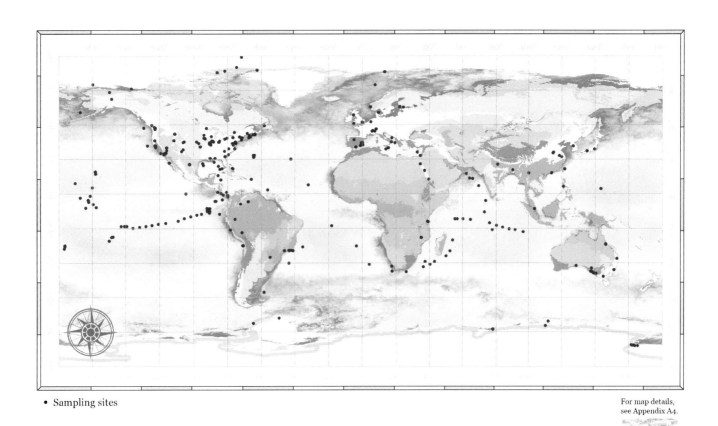

• Sampling sites

For map details,
see Appendix A4.

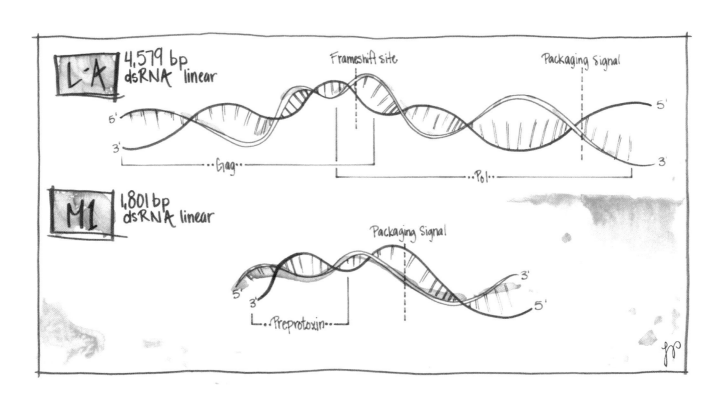

Saccharomyces cerevisiae Virus L-A

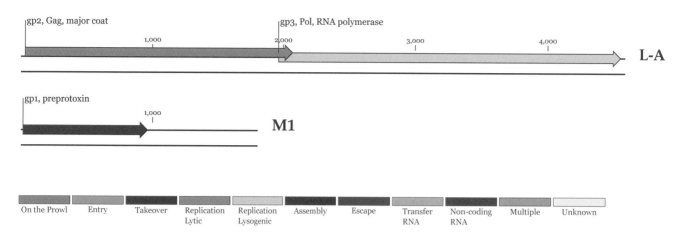

On the Prowl	Entry	Takeover	Replication Lytic	Replication Lysogenic	Assembly	Escape	Transfer RNA	Non-coding RNA	Multiple	Unknown

The Little Phage That Could Kill

Heather Maughan & Merry Youle

Mise-en-Scène: *The question of which cells a phage can infect is not simply a matter of the phage's receptor being present or not. The successful phage must, after adsorption and entry, speak the host's language to redirect its resources to support replication of the phage. The sophistication of this host takeover attests to the lengthy co-evolution of phages with their host. Despite the elegant frugality of the genome of L-A phage, it, too, knows how to use its host well. L-A maintains itself in great numbers in its yeast hosts so skillfully that it does not compromise host growth. L-A also knows the host's life cycle and exploits it to acquire new hosts and spread through the yeast population.*

The genomic thriftiness of small bacteriophages is legendary. Yet it is a phage that infects a microbial eukaryote that captures first prize for frugality. This champion is L-A, a phage that infects the budding yeast, *Saccharomyces cerevisiae*. L-A weighs in with only two genes, but still it manages to replicate and build its 40 nm capsids. Moreover, it spreads through yeast populations without ever venturing outside its fungal host (Wickner 1996).

An armchair traveler

For its propagation, L-A relies on the normal reproduction of its host. Day-to-day, *S. cerevisiae* cells grow and divide as diploid or haploid cells. Each cell division creates two L-A infected cells from one, i.e., vertical transmission of the phage. But the host also provides for horizontal transmission of L-A into new lineages. When yeasts mate, two haploid cells fuse to form a diploid zygote; the zygote later undergoes meiosis to give rise to four haploid cells. When a yeast carrying L-A mates with an uninfected cell, all the progeny will carry L-A. This mechanism enables L-A to spread through the yeast population, albeit slowly. Such sexual adventures are rare as these yeasts mate only once every fifty thousand generations or so (Ruderfer et al. 2006). Both of these modes of transmission are a boon for L-A. It need not risk the hazards of extracellular space travel to search for a new host; the fungus makes new hosts for them. All that is required of L-A is to quietly multiply within the comforts of home. This it does, maintaining a population of approximately a thousand per yeast cell (Wickner 1996).

Two genes, two proteins

L-A's minuscule double-stranded RNA (dsRNA) genome—less than five thousand bp—provides only two genes, but two are sufficient for L-A replication. One gene encodes Gag, the major capsid protein, 60 dimers of which assemble to form each capsid. The second gene, *pol*, encodes the RNA-dependent RNA polymerase (RDRP) capability required to replicate L-A's double-stranded RNA (dsRNA) genome as well as to transcribe viral genes for protein synthesis. One or two copies of the RDRP are assembled into each capsid (Ribas, Wickner 1998).

Since dsRNA is vulnerable to patrolling nucleases, L-A's dsRNA genome spends its entire life hiding inside a capsid in the host cytoplasm. Thus the RDRP has to do all of its work within the confines of a capsid. There, with a dsRNA genome at hand to serve as a template, it transcribes genome-length, **positive-sense** strands that exit the capsid as they are being born. As each nascent transcript lengthens, it sneaks out through one of ~60 holes (Naitow et al. 2002) in the capsid shell. These holes are adequate for the passage of single RNA strands (ssRNA), but not even the largest ones (1.0-1.5 nm diameter) allow the dsRNA genome to escape.

Some of these transcripts function as **polycistronic** messenger RNAs (mRNAs). The *gag* cistron begins at the 5′ end of the transcript, while its 3′ end overlaps 130 bp of the 5′ end of *pol* (Icho, Wickner 1989). During translation, mRNA is a one-way avenue along which ribosomes always travel in the 5′ to 3′ direction. Thus translation begins at the 5′ end of *gag*. Most of the time, when synthesis of the Gag protein is finished, the ribosome is released from the mRNA. But occasionally, when a ribosome reaches the end of *gag*, it slips and moves backward one nucleotide (a -1 **frameshift**), then continues translating as if nothing unseemly had happened. It now continues translating all the way to the 3′ end of the transcript, thereby extending the polypeptide chain to yield a continuous, fused Gag-Pol protein (Dinman, Icho, Wickner 1991). This occurs about once in every 60 times *gag* is translated—a finely tuned 'mistake' that efficiently regulates the relative numbers of Gag and Gag-Pol produced to match what is needed for capsid assembly (Wickner 1996).

Moonlighting proteins

Of necessity, the two proteins encoded by L-A's dsRNA genome are paragons of multi-tasking. For example, capsid protein Gag moonlights as a thief that steals the 'caps' needed to complete L-A's mRNAs. In eukaryotic cells, mRNAs are fitted out at their 5′ end with a 7-methylguanosine cap that protects against exonuclease attack and promotes translation. L-A complies with protocol by using Gag to steal said caps from host mRNAs. Gag, already a structural component of a capsid, swipes a cap from a host mRNA. As the capsid-bound Gag-Pol expels a new transcript 5′ end first, Gag transfers the cap to that end as it snakes out of the capsid (Fujimura, Esteban 2011). This hand-off from Gag to the emerging transcript is precarious as the donor Gag is some distance away from the pore. Once capped, the transcript can be used as mRNA.

Some of the transcripts that exit from the capsid have a different fate. Instead of being capped to serve as mere messengers, these transcripts are destined to become genomes. In their cytoplasmic neighborhood are numerous copies of Gag along with a few Gag-Pol proteins. It is Gag-Pol's turn now to demonstrate multi-tasking by moonlighting as a charismatic organizer. It initiates capsid assembly by latching onto a hairpin structure in the

3′ end of a transcript (Fujimura et al. 1990). Soon nearby Gag proteins assemble into 60 asymmetrical dimers around the ssRNA+Gag-Pol complex (Ribas, Wickner 1998; Naitow et al. 2002; Fujimura, Esteban 2011). The finished capsid shell contains ~120 Gag proteins, 1 or 2 Gag-Pol proteins, and a positive-sense ssRNA transcript (Ribas, Wickner 1998). Gag-Pol promptly changes hats and begins its next task: synthesizing the complementary **negative-sense** strand to complete the encapsidated genome (Wickner 1996).

Even though L-A's proteins perform multiple tasks, there are limits to what one can do with only two genes. Faced with necessity, L-A does not hesitate to exploit three host genes to assist with capsid assembly and virion stability. One host protein (MAK3) acetylates the N-terminus of each Gag monomer—a modification required prior to their self-assembly (Tercero, Wickner 1992; Tercero, Dinman, Wickner 1993). Two other host proteins mysteriously improve virion stability by an unknown mechanism (Wickner 1996).

Killer phage, killer yeast

Not leaving the host can be a double-edged sword. L-A's success depends on the success of its host. The first commandment for such a phage: be not a burden onto your host. L-A complies. A thousand virions can persist in a yeast cell without slowing its growth (Wickner 1996; Schmitt & Breinig 2006). The second commandment: be of use to your host. This L-A does by arming it with a toxin that attacks and kills nearby yeasts if they, too, are not armed with the same weapon. Thus, L-A transforms innocuous yeasts into killers.

L-A's virtuoso multi-tasking notwithstanding, this weaponry called for a third gene to encode the toxin. Given constraints on genome size due to capsid dimensions, L-A's strategy is to encode the toxin gene on a separate satellite dsRNA molecule (e.g., M1). Like the dsRNA encoding Gag and Gag-Pol, the positive-sense strand of this satellite possesses the requisite hairpin structure in its 3′ end and is encapsidated by the same Gag-driven self-assembly mechanism into its own capsids

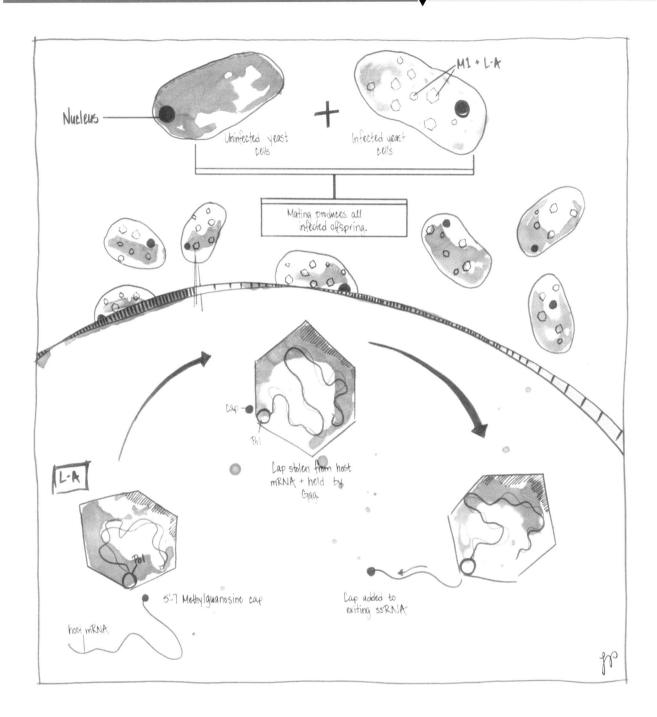

(Fujimura et al. 1990). Being about half the size of the L-A genome, two dsRNA copies of the toxin gene fit in each capsid (Esteban, Wickner 1986; Wickner 1996). As usual, the capsid condenses around a ssRNA+Gag-Pol complex. Gag-Pol synthesizes the complementary strand to complete the genome inside the capsid and begins transcription. But since there is still room inside the capsid, the first positive-sense transcript made remains in the capsid, and it, too, is converted to dsRNA. After that, continued transcription yields transcripts that exit the capsid and are translated

or encapsidated. Even this third gene does a little multi-tasking. It encodes the toxin, while a pre-processed version of the toxin provides immunity to the host (Schmitt, Breinig 2006).

The L-A phages in various killer yeast strains carry one of numerous distinctive toxin genes. Their toxins have diverse modes of action, but all of them interact with receptors in the cell wall of sensitive yeasts. Armed with any one of them, a killer yeast can successfully invade a community of unarmed brethren. When living within a community of cells armed with the same toxin, keeping a killer phage on board becomes essential for yeast survival. In either situation, L-A benefits.

Together we kill

L-A and other Totiviruses that maintain long-term occupancy of their hosts have profited from the opportunity to exchange genes in both directions. By acquiring viral genes, the fungi established their most effective anti-viral defenses (Bruenn 2012). Similarly, the Totiviruses gained the fungal genes that now encode their toxins. Although the cellular functions of those ancestral genes are not known, fungi, like other eukaryotes, often encode secreted toxins. Today's killer phage that reaps the benefits of carrying a toxin is beholden to the yeast for both the original gene and for providing the cellular mechanism for toxin secretion. This little phage can kill, but it cannot do it alone.

Cited references

Bruenn, J. 2012. Genes from double-stranded RNA viruses in the nuclear genomes of fungi. In: G Witzany, editor. *Biocommunication of Fungi*: Springer. p. 71-83.

Dinman, JD, T Icho, RB Wickner. 1991. A -1 ribosomal frameshift in a double-stranded RNA virus of yeast forms a gag-pol fusion protein. Proc Natl Acad Sci USA 88:174-178.

Esteban, R, RB Wickner. 1986. Three different M1 RNA-containing virus-like particle types in *Saccharomyces cerevisiae*: *In vitro* M1 double-stranded RNA synthesis. Mol Cell Biol 6:1552-1561.

Fujimura, T, R Esteban. 2011. Cap-snatching mechanism in yeast L-A double-stranded RNA virus. Proc Natl Acad Sci USA 108:17667-17671.

Fujimura, T, R Esteban, LM Esteban, RB Wickner. 1990. Portable encapsidation signal of the L-A double-stranded RNA virus of *S. cerevisiae*. Cell 62:819-828.

Icho, T, RB Wickner. 1989. The double-stranded RNA genome of yeast virus L-A encodes its own putative RNA polymerase by fusing two open reading frames. J Biol Chem 264:6716-6723.

Naitow, H, J Tang, M Canady, RB Wickner, JE Johnson. 2002. LA virus at 3.4 Å resolution reveals particle architecture and mRNA decapping mechanism. Nat Struct Mol Biol 9:725-728.

Ribas, JC, RB Wickner. 1998. The Gag domain of the Gag-Pol fusion protein directs incorporation into the LA double-stranded RNA viral particles in *Saccharomyces cerevisiae*. J Biol Chem 273:9306-9311.

Ruderfer, DM, SC Pratt, HS Seidel, L Kruglyak. 2006. Population genomic analysis of outcrossing and recombination in yeast. Nat Genet 38:1077-1081.

Schmitt, MJ, F Breinig. 2006. Yeast viral killer toxins: Lethality and self-protection. Nat Rev Microbiol 4:212-221.

Tercero, JC, JD Dinman, RB Wickner. 1993. Yeast MAK3 N-acetyltransferase recognizes the N-terminal four amino acids of the major coat protein (gag) of the L-A double-stranded RNA virus. J Bacteriol 175:3192-3194.

Tercero, JC, RB Wickner. 1992. MAK3 encodes an N-acetyltransferase whose modification of the L-A gag NH2 terminus is necessary for virus particle assembly. J Biol Chem 267:20277-20281.

Wickner, RB. 1996. Double-stranded RNA viruses of *Saccharomyces cerevisiae*. Microbiol Rev 60:250-265.

Recommended review

Wickner, RB. 1996. Double-stranded RNA viruses of *Saccharomyces cerevisiae*. Microbiol Rev 60:250-265.

Unintelligent Design: Mosaic Construction of Bacteriophage Genomes

Graham F. Hatfull[†]

Abstract: *The rapid growth in the numbers of completely sequenced bacteriophage genomes is providing grist for productive comparative analyses that yield insights into both the diversity of extant phages and how they have evolved. The dominant architectural feature of these genomes is the pervasive mosaicism—visible in both nucleic acid and amino acid sequence comparisons—arising from recombination events accumulated over their long evolutionary histories. The size and genetic novelty of the phage population suggests that it harbors the largest unexplored reservoir of sequence information created by Mother Nature, presenting the delightful task of figuring out what these genes do, why they are there, and what ways they may be of use also to us.*

[†]Department of Biological Sciences, University of Pittsburgh, Pittsburgh, PA
Email: gfh@pitt.edu
Website: http://biology.pitt.edu/person/graham-hatfull

During the early studies of bacteriophages following their first reports between 1915 and 1918, there was considerable discourse as to their general nature and existence, but how many different types there are and how they are related was not a question that could be easily addressed (Summers 1999). However, it was surely evident to d'Hérelle in his plaque assay that phages isolated from different sources did not all behave similarly, and there was clearly more than one type. The specificity of phages for certain host strains was also recognized, suggesting that there may be different phages for different bacterial hosts (Summers 1999). It was not until the late 1930's with the development of the electron microscope that bacteriophages could be visualized, and the variety of morphotypes observed further indicated considerable diversity.

In forming the Phage Group in the early 1940's, Max Delbruck certainly appreciated that there were phages of many different types, and although the relationships between these types was unclear, Delbruck recognized the strength in consolidating efforts on a small number of phages (Cairns, Stent, Watson 1992). In the 1945 'Phage Treaty' Delbruck promoted studies on a few select prototype phages and their hosts, which undoubtedly fueled the intensive advances in our understanding of molecular biology over the next two or three decades. Delbruck's push for the Phage Treaty can hardly be faulted, but its successes were doubled-edged in that while great advances were made in understanding biology, the successes were followed by a contraction of phage studies in the years that followed. It would thus be some time before a return to general questions of phage diversity and the issue of how the well-studied prototypes relate to the general phage population. As a colleague once stated, "What we know about phages is an inch wide and a mile deep."

An era of renewed interest in bacteriophage biology began in the 1990's prompted by three developments. First was the advancement of DNA sequencing technologies that permitted relatively cheap and rapid whole genome sequencing of virus-length molecules. Prior to 1990, the only complete genome sequences of phages were of smaller single-stranded DNA (ssDNA) phages (e.g., ϕX174), small RNA phages (e.g., MS2), and the dsDNA tailed phages λ and T7 (Fiers et al. 1976; Sanger et al. 1977; Sanger et al. 1982; Dunn, Studier 1983). But by the year 2000 there were ~30 complete genome sequences (Hendrix et al. 1999), and today there are more than 1,500 and the number is rising rapidly. Second was the determination of the size and dynamics of the phage population and the appreciation that there are perhaps 10^{31} phage particles in the biosphere, participating in 10^{23} infections per second on a global scale (Suttle 2005; Danovaro et al. 2011). The finding that phages represent the majority of all biological entities in the biosphere and play a large role in shaping the geochemical environment meant that they could no longer be overlooked. Third, the growth of antibiotic resistance among bacterial pathogens demanded a search for other antimicrobial strategies and motivated revisiting the prospects of phage therapy, an approach that had been advanced in the former Soviet states but had not generally gained acceptance elsewhere (*see page 4-46*; [(Wittebole, De Roock, Opal 2014]).

Early forays into phage genomics

Our interests in bacteriophage genomics began with the sequencing of mycobacteriophage L5, a project that was started in 1988 but took several years to finish (Hatfull, Sarkis 1993). In this age of next generation sequencing technologies, it is easy to lose sight of the exhilarating sense of achievement that accompanied completion of this project. The original motivation to do this was not rooted in any interest in phage diversity or evolution, but rather was the search for a means of understanding the mechanisms and regulation of site-specific recombination, and ultimately the development of tools for tuberculosis genetics (Hatfull 1994). Nonetheless, it helped to prompt other genome sequencing projects such as HK97 (Juhala et al. 2000). Subsequent genome comparisons sparked our curiosity about their relationships, initiating a long and productive collaboration between the labs of Drs. Hendrix, Casjens, Lawrence, and GFH. In general, there was little nucleotide sequence similarity between phages of phylogenetically distant hosts, thus limiting what could be learned from such DNA comparisons. However, prior work using heteroduplex analysis had established the idea that phage genomes had modular structures (Highton, Whitfield 1975; Highton, Chang, Myers 1990), and a mechanism for targeted recombination at gene boundaries had been proposed (Susskind, Botstein 1978). The availability of additional whole genome sequences facilitated elaboration of these ideas.

All the World's a Phage

Using a dataset of about 30 phage and prophage sequences available at that time, it became apparent that even in the absence of DNA similarities, phages that were evidently distantly related evolutionarily nevertheless sometimes shared common features. Their relatedness might be manifested through common regulatory schemes such as programmed frameshifting of tail assembly chaperones (Xie, Hendrix 1995), or by detectable similarity of the amino acid sequence of the shared gene products (Hendrix et al. 1999). Because only subsets of these features were shared within each group of genomes, their modular structures were apparent. Based on this, we predicted that all dsDNA tailed phage genomes would be found to be architectural mosaics, the result of individual modules being acquired by horizontal genetic exchange, with individual phages having unequal access to a common gene pool. The parameters influencing access to this gene pool remain to be defined, but clearly include host range and its malleability, the numbers of recombination events required to bring DNA segments together, and the requirement to maintain synteny in gene groups with associated functions (such as virion structural genes). We concluded that indeed "All the World's a Phage" but cautioned that "the veracity of this view of bacteriophage population genetics and evolution, and the quantitative nature of the relationships implied, will only be determined, we believe, after substantially more data are determined of sequences and genetic organization of phages and their hosts" (Hendrix et al. 1999).

The blind—and somewhat inept—watchmaker

The advancement of sequencing technologies in the early 2000's spurred a substantial increase in the number of completely sequenced phage genomes and illuminated many facets of this mosaic model. For example, genomic arrangements were found where gene exchange appeared to have resulted from recombination events at gene boundaries or, in some instances, domain boundaries (Pedulla et al. 2003), which could be examined for evidence of conserved boundary sequences serving as hot spots for recombination. Although there is some evidence of such conserved boundary sequences in some genomes (Clark et al. 2001), this does not appear to be a general phenomenon, raising the possibility that the mechanism of recombination is largely illegitimate and not sequence-directed (other perhaps than by a few base pairs). Such events are thus expected to generate genomic trash, destined for the grand garbage piles of biological evolution. However, either extremely rare events, or more likely even more infrequent combinations of events, may on occasion provide a functional advantage, with natural selection favoring the rise of a successful phoenix from the ashes. Presumably this all happens at extremely low frequencies, but given a vast and highly dynamic population—and perhaps 3.5 – 4 billion years—infrequent opportunity would not seem to be an obvious impediment.

This way of generating new viruses may seem dumb, wasteful, and completely thoughtless, but the concept is attractive because it is also highly creative. Illegitimate recombination provides a means of assembling DNA sequences that would not otherwise be adjacent to each other, with the potential not just to place individual genes in disparate genetic contexts, but also to create entirely new genes from unrelated parts. This is not to deny an important and prominent role for homologous recombination in phage evolution, but that provides only new assortments of genes, not new gene boundaries.

Although it is common to think about the participants in these horizontal genetic exchange events as being two phage genomes, this is probably a misplaced concept derived in part from thinking about how phage crosses are performed experimentally. If the events giving rise to mosaicism are not sequence directed, then biology does not 'know' what kind of organism the DNA belongs to, and such recombination is more likely to occur between phage DNA and the host chromosome (which is about 100-fold larger than a phage genome). Many phage genomes contain genes that are otherwise considered to be bacterial genes, and it is plausible that

they have been acquired by this process (Pedulla et al. 2003). Thus the distinction between a 'bacterial' gene and a 'viral' gene becomes quite fuzzy, with the primary criterion being the gene's relative abundance in host versus phage genomes.

It is not clear what specific mechanisms are involved in generating genome mosaicism. In some respects, it is not really necessary to assume any single specific process is involved, as simple 'mistakes' during the routine activities of genome replication or repair could give rise to mosaics. That notwithstanding, there are a variety of enzymatic processes that could be involved, including those of the phage-encoded DNA-pairing recombinases (e.g., Rec T or λ's Red) that are more tolerant of imperfections in homology than host recombination systems and thus that more effectively use shorter segments of homology (Martinsohn, Radman, Petit 2008; De Paepe et al. 2014). But any events that promote breakage and repair are likely to play a role, including transposition and homing-type endonuclease cleavage. Although transposable elements reside in some phage genomes, they are not particularly common, whereas HNH-type nucleases are abundant and are thus prime suspects for stimulating genome rearrangements (Edgell, Gibb, Belfort 2010; Hatfull 2012; Kristensen et al. 2013).

Observing mosaicism: Biding one's time

The evidence for phage genome mosaicism is overwhelming. Still it is helpful to recognize that the genomes we examine today were generated by billions of years of evolution, with most of the actual exchanges having taken place long ago and relatively few having occurred more recently. In general, more recent events are evident when comparing the DNA sequences (Fig. 1), whereas earlier events carry signatures of common ancestry at the amino acid sequence level—nucleotide similarity being long gone (Fig. 2). Inspection of the results of more recent events has shown that large blocks of genes can be exchanged. Such blocks formed in the distant past could be disrupted by subsequent multiple recombination events within that region, with the number of such events accumulating over time (Fig. 1). Eventually large numbers of individual genes emerge as isolated mosaic elements (Figs. 2 and 3). Multitudes of all of these events are apparent using comparative genomic approaches. DNA relationships are easily seen by comparing genomes within mycobacteriophage clusters, as is nicely illustrated by comparing the right ends of genomes within Cluster F (Fig. 1; see below). In contrast, the presence of distantly-related gene homologues that now reside in different genomic contexts can be clearly seen by comparing genomes that are not closely related at the nucleotide sequence level, as illustrated in Figure 2.

Gene acquisition: What & why?

Although DNA rearrangements can involve large sets of genes, there is evidence that these can quickly distill down to smaller sets or single genes that are in rapid flux. This is seen for some of the small genes in phage SPO1 (Stewart et al. 2009) and in comparisons between groups of mycobacteriophages (Hatfull et al. 2010). These small, high flux genes may correspond to a single functional domain (Hatfull et al. 2010). This is consistent with the role of illegitimate recombination, which has no regard for gene boundaries *per se*, and will thus cause reduction in length to the smallest unit of selection, a single protein domain. It is perhaps no surprise then that phage genes are on average only about two-thirds the size of bacterial genes. Furthermore, virion structure and assembly genes are more similar in size to typical host genes (~880 bp in the mycobacteriophages), whereas non-structural genes are small (~470 bp in the mycobacteriophages). The abundance of small genes is evident in Figure 2.

Selection of recombinants for gene function is one obvious driver of productive events. However, phages are unusual in that, for them, genome size is anticipated to also be a major force arbitrating gene acquisition and loss. Because of the constraints imposed by capsid size and DNA packaging densities, there are both minimal and maximal genome sizes for any given capsid structure. Thus, if a genome loses a segment of non-essential genes and its length falls below the minimum required to generate stable particles, then selection

favors acquisition of any DNA segment, without regard for its function. This is expected to have a substantial evolutionary impact, as it provides a bridge to carry acquired genetic information to an environment—in the absence of direct functional selection—where it may subsequently provide a new and advantageous function.

The imbroglios of phage taxonomy

The growing body of phage genomic information and comparative analyses throws some spanners into the works of viral taxonomy. A variety of taxonomic schemes for phages have been proposed, including some based fundamentally on virion morphologies (*see page 1-9*). These are of some use and many distinct morphotypes have been described. The largest and most abundant group is the *Caudovirales* (dsDNA tailed phages) encompassing three main types distinguished by their tails: the *Siphoviridae* (with long, flexible, non-contractile tails), the *Myoviridae* (with contractile tails), and the *Podoviridae* (with short stubby tails). However, phage morphology contains little, if any, phylogenetic information, while constructing taxonomies that accurately reflect phylogeny is confounded by the extensive genetic mosaicism. For example, there are phages with Siphoviral morphology that share more genes with a Myovirus than with any other Siphovirus.

One approach is to compare the collective gene sets of phages and construct a proteomic tree based on shared gene content (*see page 8-8*; Rohwer, Edwards 2002). This successfully groups closely related genomes together, but also generates some strange bedfellows because different parts of each genome can have distinctly different evolutionary histories. An alternative approach is to regard each genome as a composite of separate units, and then to use a non-hierarchical reticulate taxonomy to reflect the multitude of different phylogenies conjoined into each genome (Lawrence, Hatfull, Hendrix 2002). A computational approach has been described to accommodate the reticulate relationships (Lima-Mendez et al. 2008). It should also be noted that the complex genome structures not only confound their taxonomy, but also chal-

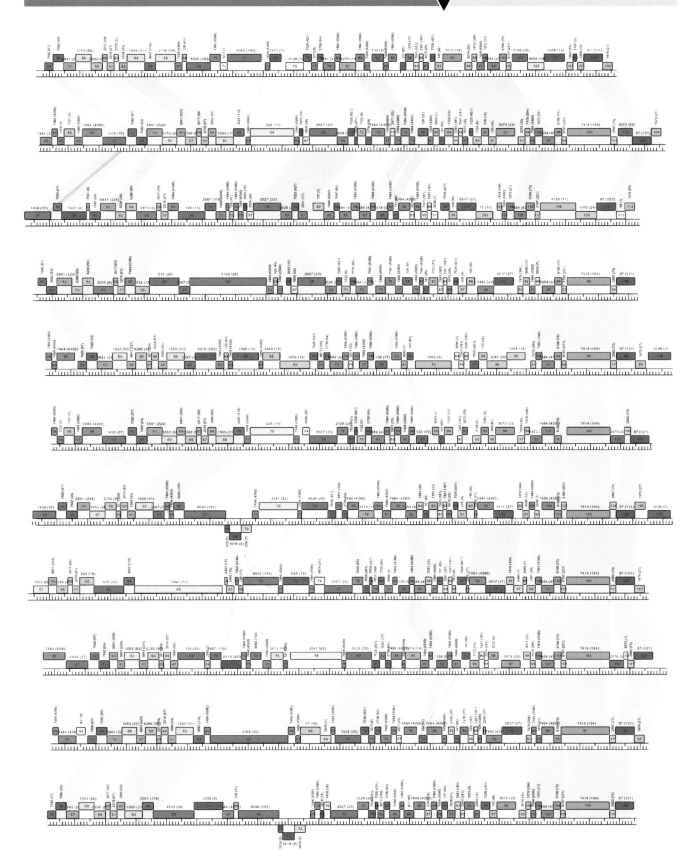

Figure 1. Mosaic relationships among Cluster F phages. The right ends of the genome maps of phages Tweety, PMC, Che8, Velveteen, Llij, Boomer, Shauna1, ShiLan, Wee, Sisi, and Ramsey (top to bottom) are shown, with pairwise nucleotide sequence similarity represented by spectrum colors between them (violet is greatest similarity, red is least similarity). Genes are represented as boxes either above (if transcribed rightwards) or below (if transcribed leftwards) each genome. The relationships between the genomes are highly complex, with segments of similarity dispersed among regions of unrelatedness. Maps were generated using the program Phamerator (Cresawn et al. 2011) and the database 'Mycobacteriophage_627.'

lenge the utility of any Linnaean divisions, including the use of the term 'species' (Lawrence, Hatfull, Hendrix 2002). Because phage genome structures do not accommodate any of the standard definitions of 'species,' we strongly discourage its use for phages. If the term is to be used, it is necessary to provide a definition and an illustration how it accommodates the observed genetic relationships.

The magical mycobacteriophages

In the late 1980's, a time when phage studies had largely fallen out of favor, any new investment in phage biology required an especially compelling rationale. Such justification can be mustered if one chooses to investigate phages whose host has obvious relevance to human health. A convincing reason to study mycobacteriophages—viruses infecting *Mycobacterium tuberculosis, Mycobacterium smegmatis*, and other mycobacterial species—is the preeminent role of mycobacterial diseases. For most of the last thirty years *M. tuberculosis*, the causative agent of human tuberculosis, has rightly claimed to be mankind's deadliest microbial enemy, killing nearly two million people a year by recent estimates (Zumla et al. 2013). Most *M. tuberculosis* infections don't lead immediately to active disease, the more common outcome being establishment of a latent infection. Amazingly, about one-third of the entire world's population is infected with *M. tuberculosis*, with particularly devastating consequences for those also infected by HIV. That tuberculosis is a lesser scourge in the developed world reflects the successful implementation of effective antibiotic regimens, but even so-called 'short-course' therapy requires 3-4 antibiotics for at least six months. Non-compliance with therapy in response to unpleasant side effects has fueled the emergence of antibiotic resistance in the form of multidrug resistant (MDR-TB) and extensively drug resistant (XDR-TB) strains (Abubakar et al. 2013). There is thus a global need for effective vaccines, new drugs, and rapid, cheap diagnostic tests. Enter the mycobacteriophages, whose study promises solutions to these problems by enabling development of advanced methods for genetic manipulation of the mycobacteria, by providing insights into the physiology of *M. tuberculosis*, and

more directly by the deployment of recombinant reporter phages for diagnostics and drug susceptibility testing (Hatfull 2010b; Hatfull 2014a). Phage therapy is also an enticing concept, either in therapeutic or prophylactic context, but remains to be evaluated (Hatfull 2014b).

Our early forays into the genomics of mycobacteriophages tweaked our curiosity in phage diversity

Figure 2. Genome mosaicism visualized by gene product comparisons. Small segments of the Barnyard, Patience, TM4, and Twister genomes are shown to illustrate their mosaic structures. Each gene is shown as a box with its gene number inside, the phamily number above, and the number of phamily members in parentheses. Genes are colored according to their phamily; white boxes are orphams (i.e., there are no relatives in the mycobacteriophage database). Homologues are connected by vertical lines and the percent amino acid sequence identity is shown. Note that the homologues are situated in different genomic contexts with different flanking genes. This is typical of phage genomes and reflects their mosaic nature with segments corresponding to single genes combined in distinctive combinations to generate different genomes.

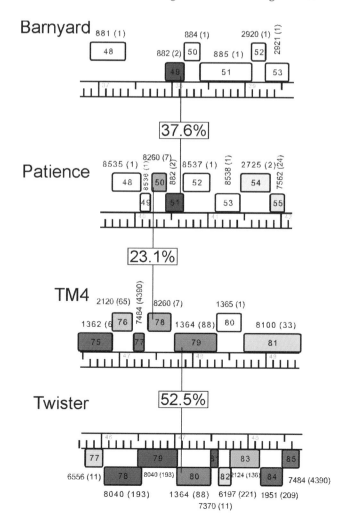

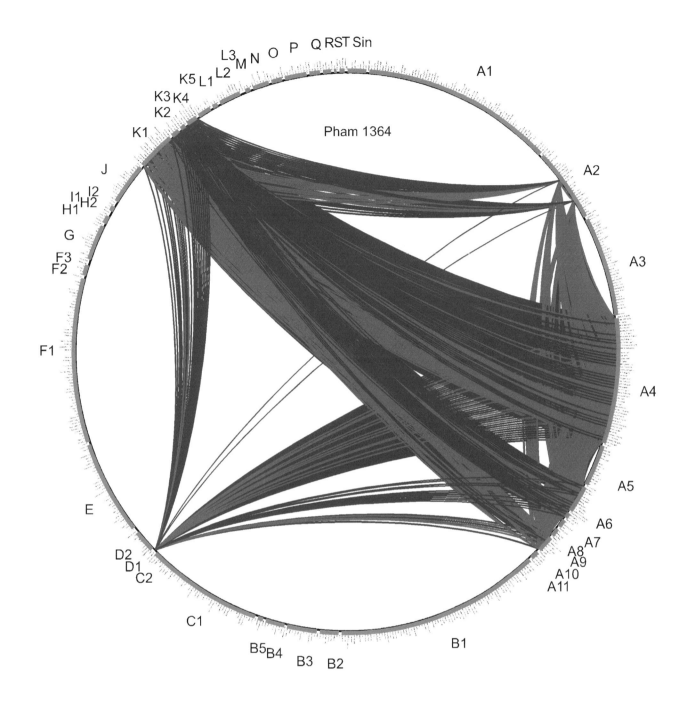

Figure 3. Phamily circle showing the distribution of Pham 1364 members. Phamily circles such as this one show all of the constituent genomes within the database (in this case, the 627 genomes in the Phamerator database Mycobacterio-phage_627) around the circumference of the circle. Genomes are ordered according to cluster and subcluster designations as shown (A to T; Sin: Singletons). Phamilies are generated by pairwise BLASTP and ClustalW comparisons of all predicted gene products, and then grouping together those above threshold values (Cresawn et al. 2011). BLASTP values are shown as blue lines and ClustalW similarities as red lines. Pham 1364 members are broadly distributed within the Cluster K1, K2, K3, K4, A2, A4, A6, A7, A8, A9, A10, and D1 genomes. Pham 1364 gene members in phages TM4 and Twister (*79* and *80* respectively) are illustrated in Figure 2. Comparison of phamily circles of different phams illustrates their different evolutionary histories and thus the mosaic architecture of the genomes currently housing them.

and evolution, but it was striking that the first two mycobacteriophages we sequenced—L5 and D29—were closely related, regardless of their numerous phenotypic differences reported in the literature

(Ford et al. 1998). If we had simply concluded that all mycobacteriophages were that closely related, we may well have stopped at that point. It was a good job we didn't, because as the numbers of sequenced mycobacteriophages continued to grow, the impressive breadth of their diversity became clear. Today, there are a total of 671 completely sequenced mycobacteriophage genomes (www. phagesDB.org), by far the largest group of phages known to infect a single common host strain (*M. smegmatis* mc²155). Not surprisingly, this group of phages provides considerable insight into phage diversity and phage biology in general.

SEA-PHAGES (but nothing to do with the ocean!)

In the race to understand biological diversity, the genome-by-genome exploration of the phage population is more akin to the pace of the tortoise than that of the hare. Although whole genome sequences emerge that are invaluable for elucidating evolutionary mechanisms, the obvious bottleneck is that all those viruses have to come from somewhere. In the absence of any simple high throughput system for recovering individual phages of a specific host, this requires isolation, purification, and analysis of individual viruses from the environment. Fortunately, this matches up beautifully with our interest in identifying projects to engage young high school and undergraduate students in authentic scientific discovery. We developed the Phage Hunters Integrating Research and Education (PHIRE) program at the University of Pittsburgh (Hatfull 2010a), and then extended it nationally with the Howards Hughes Medical Institute (HHMI) Science Education Alliance – Phage Hunters Advancing Genomics and Evolutionary Science (SEA-PHAGES) program (Jordan et al. 2014). The SEA-PHAGES program typically involves freshman undergraduates and is a two-term course in which students isolate new mycobacteriophages from the environment, purify, amplify and name them, isolate their DNA, sequence and annotate their genomes, and then investigate them bioinformatically (Jordan et al. 2014). There are currently 73 participating institutions and in the current academic year (concluding May, 2014), there are over 2,000 students involved. The phage and genome data is coordinated by the phagesDB database and made available through the web site www.phagesDB.org. The PHIRE and SEA-

PHAGES programs clearly have had a huge impact on accumulation of the number of sequenced mycobacteriophage genomes.

The emergence of these programs was something of an unanticipated development. The sequencing of the L5 genome in 1993 had represented a substantial undertaking at the time (prompting a successful application for the Pennsylvania license plate 'PHAGE L5'!) and it seemed as though it might be the only mycobacteriophage we would ever sequence. The subsequent accumulation of sequenced mycobacteriophages has been a remarkable event, and the PHIRE and SEA-PHAGES programs have shown how a research agenda can generate real gains in science education (Jordan et al. 2014). The great unexplored diversity of the phage population piques the interest of young students, and the opportunity to contribute to defining that diversity is compelling. Students also enjoy naming the phage they isolate. Non-systematic names are encouraged to reflect the individual character of viral genomes that are composed of particular combinations of the constituent parts (although veto power is retained to avoid breaching the bounds of decency). Some of our favorite phages are those with names such as 'Corndog', 'Daenerys', and 'Tweety', but the reader is invited to visit phagesDB.org to peruse the full repertoire. A recently purchased new car and the need for a new Pennsylvania license plate prompted a request for the vanity plate 'PHAGES'. This resulted in an inquiry from the Pennsylvania Department of Transportation as to whether this strange term was appropriate for public display. By the end of the phone conversation, the DoT employee was ready to go phage hunting!

Cluster ph***

When the number of sequenced mycobacteriophage genomes had grown to 14, we saw many types that were completely distinct from each other at the nucleotide sequence level (Pedulla et al. 2003). However, as the numbers grew to a total of 30 genomes, it was clear that there were others that are related at the DNA level, some quite closely so (Hatfull et al. 2006). To recognize these closely related phages we decided to group them into 'clusters,' using as our primary criterion that two genomes must share recognizable nucleotide sequence similarity spanning more than 50% of their genomes lengths to be in the same cluster (thus minimizing the chance of phages being assigned to more than one cluster). Nucleotide sequence comparison of some clusters revealed some substructure warranting further division into subclusters (Hatfull et al. 2006). Initially, just six clusters were designated (Clusters A to F), but as the number of sequenced genomes doubled, the number of clusters grew to nine, and there were five 'singleton' phages with no close relatives (Hatfull et al. 2010). The current 671 mycobacteriophage genomes comprise 20 clusters and nine singletons; ten of the clusters are divided into subclusters. These cluster designations generally mirror the relationships displayed by a network analysis using gene content (Hatfull et al. 2006; Hatfull et al. 2010; Hatfull 2014b).

Do mycobacteriophage clusters correspond to discreet population groups?

Comparison of the early collections of mycobacteriophages suggested that most of the clusters were fairly discrete, based on both nucleotide and gene content comparisons. However, as the collection has grown, there are more and more examples challenging the discreteness of the divisions. For example, genomes in the clusters N, I, P, O, and F have a complex set of relationships, in which they share substantial numbers of genes, and some would cluster together given even a modest change in clustering parameters. The simple interpretation is that the clusters may indeed represent predominant groups of phages but that the barriers to exchange are not robust, so that the genomic 'space' between the clusters will be gradually occupied as more and more mycobacteriophage genome sequences are determined. It would seem that maintaining groupings as clusters and subclusters is useful from a pragmatic perspective—especially when discussing relatively large numbers of genomes—but apparently Mother Nature did not intend phage genomes to fit nicely into well-defined square boxes.

Where does all this diversity come from?

The extent of genetic diversity of the mycobacteriophages was somewhat surprising given that they all infect a common host, thus were expected to be in close genetic communication with each other. One obvious interpretation is that they may not actually be in such close genetic contact, an idea that is supported by their substantial variation in GC% content that ranges from 50% to 70% while *M. smegmatis* is 67.3%. One model to account for the variation assumes that phage host range is highly malleable, and thus phages switch or expand host range at reasonably high frequencies (Jacobs-Sera et al. 2012). Provided there are rich diversities of host species available in the source environments of these phages, this model suggests that the phages can jump between phylogenetically-proximal hosts much faster than their genomic signatures can ameliorate to any particular host. Phages within a particular cluster are predicted to have traveled through related hosts and thus to have had access to a genetic pool that is defined by these hosts. Clusters in which there is low variation between the constituent phages may have had to accommodate to more restricted host ranges, whereas those with greater variation represent a much richer collection of individual journeys across the bacterial landscape.

This model makes two clear predictions. First, phages isolated from environments in which there is restricted bacterial diversity should also show low diversity. An example demonstrating this is the limited diversity of phages of *Propionibacterium acnes* isolated from microcomedones (Marinelli et al. 2012), where bacterial diversity is much lower than that in soil or compost, the source of many of the mycobacteriophages. The second prediction is that phages isolated from environments similar to the mycobacteriophage source environments, using different but phylogenetically-proximal hosts, will show similar patterns of diversity and some will possess genetic similarity to the mycobacteriophages. There are hints supporting this from the small numbers of other phages that have been sequenced, but this prediction has yet to be thoroughly tested.

What do all those phage genes do?

We discussed above that the observed patterns of genome mosaicism could arise from non-sequence directed recombination events coupled with selection for gene function. What then, are the functions of all these genes? Comparison of the ~70,000 mycobacteriophage genes described to date sorts them into ~4,000 distinct phamilies of sequences. Searching against public databases shows that about 25% are similar to known proteins, although only about half of these have assigned functions. Thus for any one genome, the virion structural genes can usually be identified (their synteny is especially well-conserved in the *Siphoviridae*) and a smattering of other gene functions can be predicted by homology, but the majority are of unknown function. This general pattern appears to hold for other dsDNA tailed phages, as well. The prevalence of genes of unknown function and the magnitude and diversity of the phage population suggests that bacteriophages represent the largest unexplored reservoir of gene sequences in the biosphere.

Early studies with phage λ showed that significant parts of the genome are not required for lytic growth and can be deleted (a real convenience for using phage λ as a cloning vector). Similarly, the generation of mycobacteriophage shuttle phasmids also relies on the ability to delete segments of phage genomes (Jacobs, Tuckman, Bloom 1987). Systematic deletion of each of the non-structural genes from mycobacteriophage Giles showed that a large proportion of these also are not essential for lytic growth (Dedrick et al. 2013). Preliminary data using ribosomal profiling and mass spectrometry strongly suggests that most of the predicted phage proteins are indeed expressed. So what do they do?

In thinking about this, there are two general themes that we should consider. First is the recognition that co-evolution of host resistance and viral avoidance of host resistance is likely to have dominated microbial evolution for vast numbers of generations. Because many—if not the majority—of phages are temperate, hosts can acquire resistance at a high frequency simply through lysogeny with phages that carry resistance mechanisms. There are several established resistance mechanisms that can be phage-encoded, including restriction, (*see page 4-10*) toxin-antitoxin, and CRISPR systems (*see page 4-33*; [Fineran et al. 2009; Seed et al. 2013]). It seems likely that there are other host resistance mechanisms to be discovered, and these could also be conveyed on phage genomes, along with ways to overcome them. Because such systems are primarily involved in the tug-of-war between hosts and their viruses, most of them will not be required for lytic growth, except perhaps with specific phage-host combinations. The second theme, building on the model described above relating diversity and host range switching, anticipates that genes acquired to provide growth advantages in one host may not be needed during infection of a subsequent host. However, loss of such genes may be slow, so that these genes are simply non-essential in a particular laboratory host. Elucidating the roles of such 'legacy' genes may be especially challenging.

Moving phorward

Our understanding of phage genome architecture has grown substantially over the past twenty years, and some general frameworks for considering phage diversity and evolutionary processes have been established. But it has also become apparent how daunting is the task of achieving a full understanding of phage genomics, gene function, and gene expression and regulation given the size, diversity, and genetic novelty of the phage population. We clearly are nowhere near defining the number of different types of phages or phage genes in the population, or their abundances and distributions. We know little of the rates of horizontal genetic exchange relative to the mutational clock, and defining these may be further complicated by different rates in different types of phages or in different hosts. We have probably only touched upon the variety of processes that influence host range or how it changes. Until we have better grasp on those, it will be a challenge to define how quite different phages have differential access to the larger pool of microbial genes. The next few years of bacteriophage studies promise many new and thrilling discoveries, as the bigger picture of the phage world emerges. A phantastic phuture awaits us.

Acknowledgements

I would like to thank all my terrific colleagues, especially Roger Hendrix and Jeffrey Lawrence who have made phages so much phun to work with, Bill Jacobs who introduced me to the mycobacteriophages and had the vision to see how they could be exploited, and the many lab members who have contributed to this work in numerous ways. Nearly 7,000 undergraduates and over 150 faculty in the SEA-PHAGES program contributed to the current database of mycobacteriophage genome sequences. This was supported by grants from NIH and HHMI.

References

Abubakar, I, M Zignol, D Falzon, et al. 2013. Drug-resistant tuberculosis: Time for visionary political leadership. Lancet Infect Dis 13:529-539.

Cairns, J, GS Stent, JD Watson. 1992. Phage and the origins of molecular biology: Cold Spring Harbor Laboratory Press.

Clark, AJ, W Inwood, T Cloutier, TS Dhillon. 2001. Nucleotide sequence of coliphage HK620 and the evolution of lambdoid phages. J Mol Biol 311:657-679.

Cresawn, SG, M Bogel, N Day, D Jacobs-Sera, RW Hendrix, GF Hatfull. 2011. Phamerator: A bioinformatic tool for comparative bacteriophage genomics. BMC Bioinformatics 12:395.

Danovaro, R, C Corinaldesi, A Dell'anno, JA Fuhrman, JJ Middelburg, RT Noble, CA Suttle. 2011. Marine viruses and global climate change. FEMS Microbiol Rev 35:993-1034.

De Paepe, M, G Hutinet, O Son, J Amarir-Bouhram, S Schbath, MA Petit. 2014. Temperate phages acquire DNA from defective prophages by relaxed homologous recombination: The role of Rad52-like recombinases. PLoS Genet 10:e1004181.

Dedrick, RM, LJ Marinelli, GL Newton, K Pogliano, J Pogliano, GF Hatfull. 2013. Functional requirements for bacteriophage growth: Gene essentiality and expression in mycobacteriophage Giles. Mol Microbiol 88:577-589.

Dunn, JJ, FW Studier. 1983. Complete nucleotide sequence of bacteriophage T7 DNA and the locations of T7 genetic elements. J Mol Biol 166:477-535.

Edgell, DR, EA Gibb, M Belfort. 2010. Mobile DNA elements in T4 and related phages. Virol J 7:290.

Fiers, W, R Contreras, F Duerinck, et al. 1976. Complete nucleotide sequence of bacteriophage MS2 RNA: Primary and secondary structure of the replicase gene. Nature 260:500-507.

Fineran, PC, TR Blower, IJ Foulds, DP Humphreys, KS Lilley, GP Salmond. 2009. The phage abortive infection system, ToxIN, functions as a protein-RNA toxin-antitoxin pair. Proc Natl Acad Sci U S A 106:894-899.

Ford, ME, GJ Sarkis, AE Belanger, RW Hendrix, GF Hatfull. 1998. Genome structure of mycobacteriophage D29: Implications for phage evolution. J Mol Biol 279:143-164.

Hatfull, GF. 1994. Mycobacteriophage L5: A toolbox for tuberculosis. ASM News. 60:255-260.

Hatfull, GF. 2010a. Bacteriophage research: Gateway to learning science. Microbe Wash DC. p. 243-250.

Hatfull, GF. 2010b. Mycobacteriophages: Genes and genomes. Annu Rev Microbiol 64:331-356.

Hatfull, GF. 2012. The secret lives of mycobacteriophages. Adv Virus Res 82:179-288.

Hatfull, GF. 2014a. Molecular genetics of mycobacteriophages. Microbiology Spectrum 2:1-36.

Hatfull, GF. 2014b. Mycobacteriophages: Windows into tuberculosis. PLoS Pathog 10:e1003953.

Hatfull, GF, D Jacobs-Sera, JG Lawrence, et al. 2010. Comparative genomic analysis of 60 mycobacteriophage genomes: Genome clustering, gene acquisition, and gene size. J Mol Biol 397:119-143.

Hatfull, GF, ML Pedulla, D Jacobs-Sera, et al. 2006. Exploring the mycobacteriophage metaproteome: Phage genomics as an educational platform. PLoS Genet 2:e92.

Hatfull, GF, GJ Sarkis. 1993. DNA sequence, structure and gene expression of mycobacteriophage L5: A phage system for mycobacterial genetics. Mol Microbiol 7:395-405.

Hendrix, RW, MC Smith, RN Burns, ME Ford, GF Hatfull. 1999. Evolutionary relationships among diverse bacteriophages and prophages: All the world's a phage. Proc Natl Acad Sci U S A 96:2192-2197.

Highton, PJ, Y Chang, RJ Myers. 1990. Evidence for the exchange of segments between genomes during the evolution of lambdoid bacteriophages. Mol Microbiol 4:1329-1340.

Highton, PJ, M Whitfield. 1975. Similarities between the DNA molecules of bacteriophages 424, lambda, and 21, determined by denaturation and electron microscopy. Virology 63:438-446.

Jacobs-Sera, D, LJ Marinelli, C Bowman, et al. 2012. On the nature of mycobacteriophage diversity and host preference. Virology 434:187-201.

Jacobs, WR, Jr., M Tuckman, BR Bloom. 1987. Introduction of foreign DNA into mycobacteria using a shuttle phasmid. Nature 327:532-535.

Jordan, TC, SH Burnett, S Carson, et al. 2014. A broadly implementable research course in phage discovery and genomics for first-year undergraduate students. MBio 5:e01051-01013.

Juhala, RJ, ME Ford, RL Duda, A Youlton, GF Hatfull, RW Hendrix. 2000. Genomic sequences of bacteriophages HK97 and HK022: Pervasive genetic mosaicism in the lambdoid bacteriophages. J Mol Biol 299:27-51.

Kristensen, DM, AS Waller, T Yamada, P Bork, AR Mushegian, EV Koonin. 2013. Orthologous gene clusters and taxon signature genes for viruses of prokaryotes. J Bacteriol 195:941-950.

Lawrence, JG, GF Hatfull, RW Hendrix. 2002. Imbroglios of viral taxonomy: Genetic exchange and failings of phenetic approaches. J Bacteriol 184:4891-4905.

Lima-Mendez, G, J Van Helden, A Toussaint, R Leplae. 2008. Reticulate representation of evolutionary and functional relationships between phage genomes. Mol Biol Evol 25:762-777.

Marinelli, LJ, S Fitz-Gibbon, C Hayes, et al. 2012. *Propionibacterium acnes* bacteriophages display limited genetic diversity and broad killing activity against bacterial skin isolates. MBio 3(5):e00279-12.

Martinsohn, JT, M Radman, MA Petit. 2008. The lambda Red proteins promote efficient recombination between diverged sequences: Implications for bacteriophage genome mosaicism. PLoS Genet 4:e1000065.

Pedulla, ML, ME Ford, JM Houtz, et al. 2003. Origins of highly mosaic mycobacteriophage genomes. Cell 113:171-182.

Rohwer, F, R Edwards. 2002. The phage proteomic tree: A genome-based taxonomy for phage. J Bacteriol 184:4529-4535.

Sanger, F, GM Air, BG Barrell, NL Brown, AR Coulson, CA Fiddes, CA Hutchison, PM Slocombe, M Smith. 1977. Nucleotide sequence of bacteriophage φX174 DNA. Nature 265:687-695.

Sanger, F, AR Coulson, GF Hong, DF Hill, GB Petersen. 1982. Nucleotide sequence of bacteriophage lambda DNA. J Mol Biol 162:729-773.

Seed, KD, DW Lazinski, SB Calderwood, A Camilli. 2013. A bacteriophage encodes its own CRISPR/Cas adaptive response to evade host innate immunity. Nature 494:489-491.

Stewart, CR, SR Casjens, SG Cresawn, et al. 2009. The genome of *Bacillus subtilis* bacteriophage SPO1. J Mol Biol 388:48-70.

Summers, WC. 1999. *Felix d'Hérelle and the Origins of Molecular Biology*: Yale University Press.

Susskind, MM, D Botstein. 1978. Molecular genetics of bacteriophage P22. Microbiol Rev 42:385-413.

Suttle, CA. 2005. Viruses in the sea. Nature 437:356-361.

Wittebole, X, S De Roock, SM Opal. 2014. A historical overview of bacteriophage therapy as an alternative to antibiotics for the treatment of bacterial pathogens. Virulence 5:226-235.

Xie, Z, RW Hendrix. 1995. Assembly in vitro of bacteriophage HK97 proheads. J Mol Biol 253:74-85.

Zumla, A, A George, V Sharma, N Herbert, I Baroness Masham of. 2013. WHO's 2013 global report on tuberculosis: Successes, threats, and opportunities. Lancet 382:1765-1767.

Chapter 6: Assembly

STORIES

PERSPECTIVE

   by Dennis H. Bamford

Salmonella Phage Gifsy-2

a Siphophage that packages each genome as a linear dsDNA molecule synthesized from a circular template

Genome
dsDNA; linear
45,840 bp
56 predicted ORFs; 0 RNAs

Encapsidation method
Packaging; T = 7 capsid

Common host
Salmonella enterica

Habitat
Animal intestines; aquatic
environments

Lifestyle
Temperate

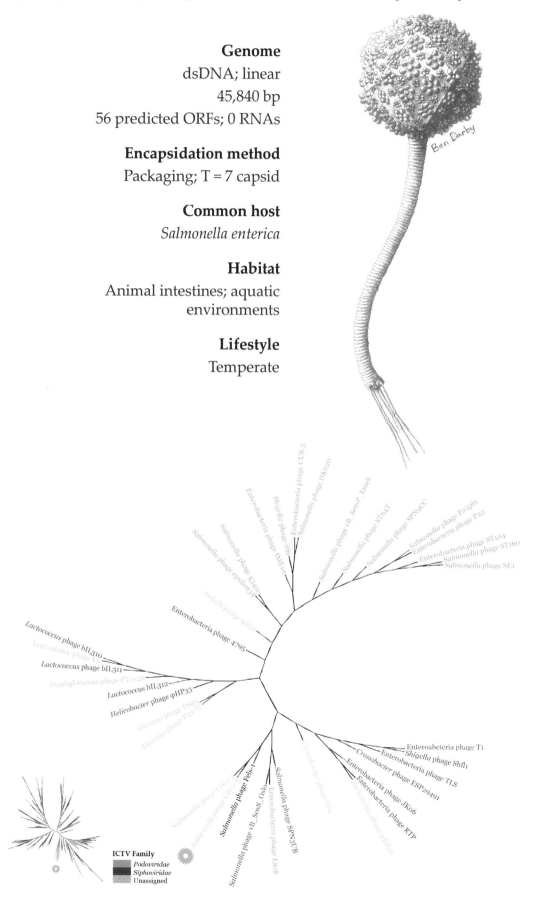

Ben Darby

ICTV Family
Podoviridae
Siphoviridae
Unassigned

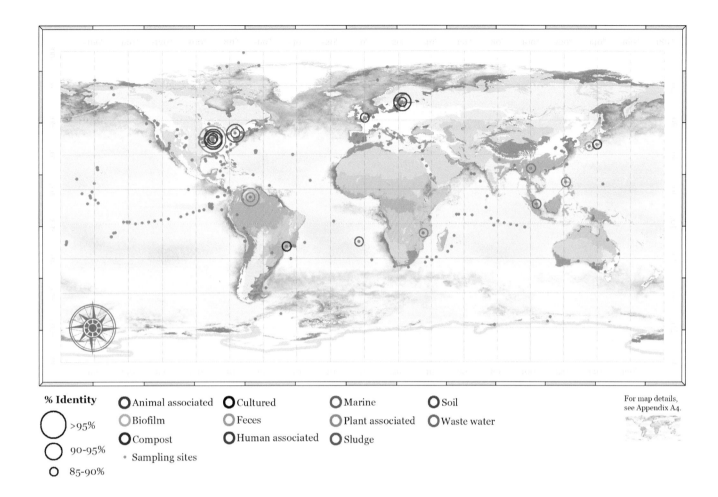

For map details, see Appendix A4.

% Identity

○ >95%

○ 90-95%

○ 85-90%

○ Animal associated   ○ Cultured   ○ Marine   ○ Soil

○ Biofilm   ○ Feces   ○ Plant associated   ○ Waste water

○ Compost   ○ Human associated   ○ Sludge

• Sampling sites

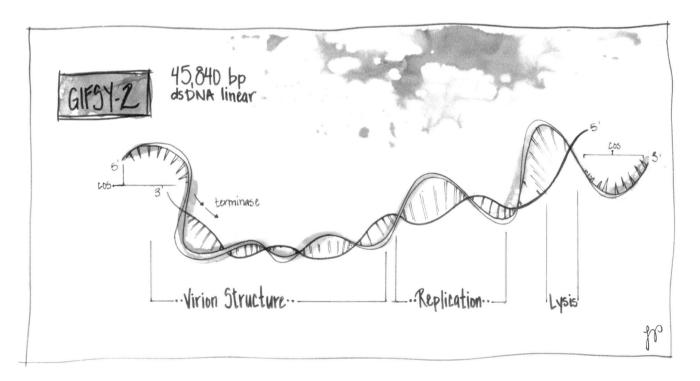

Salmonella **Phage Gifsy-2**

Note: This map shows Gifsy-2's genome as it is structured while integrated as a prophage, not as it would be in a virion (shown in overview on previous page).

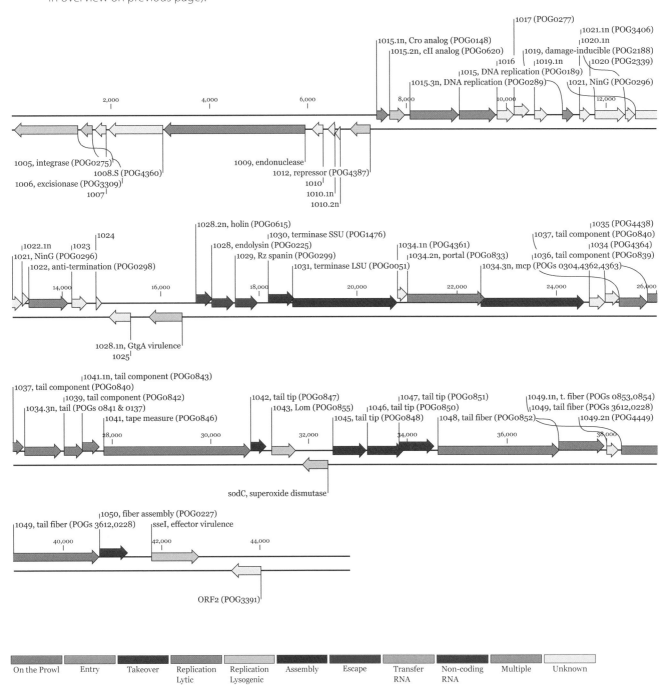

Gifsy-2 Goes Straight

Merry Youle

Mise-en-Scène: *Viruses are chromosomally creative. Their choice of nucleic acid combined with their overall approach to chromosome replication organize them into seven distinct groups under the Baltimore classification system* (see page 1-11; [*Baltimore 1971*])*. Moreover, even within one group there can be numerous variants. Consider the phages in Class I, those with a double-stranded DNA (dsDNA) genome. Some use a linear dsDNA molecule, some a circular one, while others switch between the two when it suits them. Even among the order* Caudovirales, *all of whom package their genomes as a linear dsDNA molecule, one finds a variety of replication strategies. Each method leaves a telltale flag in the termini of the packaged genome, such as 'sticky' ends (Gifsy-2, below), direct terminal repeats (T4), host chromosomal DNA (Mu), or even a terminal protein* (see page 5-18) *covalently attached (PZA).*

Phage Gifsy-2 is a conformist. Like every other tailed phage (order *Caudovirales*), it sends its progeny genomes out into the world packaged inside a capsid as a linear double-stranded DNA (dsDNA) molecule. In addition, Gifsy-2 follows the prevailing *Caudovirales* virion assembly protocol. It assembles a **procapsid**, stuffs its DNA inside, and lastly tacks on a preassembled tail. "And why not?" you might ask. This is the proven method endorsed by upwards of 10^{30} tailed phages (Brüssow, Hendrix 2002). But even within this accepted protocol there is room for variation. Assembly of many copies of the major capsid protein into a procapsid ready to receive the genome typically requires assistance from a scaffolding protein and a maturation protease, the latter usually encoded by separate genes. Not for Gifsy-2. Its multi-domain major capsid protein (mcp) handles all three jobs (Effantin et al. 2010). Genome replication and packaging provide more opportunities for alternative solutions. Here, Gifsy-2 demonstrates one of the frequently used procedures.

What does it cost?

For these phages, genome packaging is a major undertaking that requires specialized equipment. One of the twelve otherwise equivalent vertices of the icosahedral procapsid must be converted into a portal for DNA passage both in and out. For this, Gifsy-2 and others make a dedicated portal protein. During assembly, these proteins position to form a dodecameric ring at the unique vertex (Simpson et al. 2000).

Also essential for packaging is the multifunctional **terminase**, a heterodimer composed of a small and a large subunit. It has responsibility for discriminating between phage and host DNA, and then processing only the former. To begin, the terminase recognizes and binds a specific sequence in the phage DNA, the *cos* site. It cleaves the genome at that location and holds onto one of the resulting ends.

Then, with the phage DNA in tow, it docks at the procapsid portal and forcefully threads the genome into the procapsid. This feat requires brute strength. The negatively-charged DNA helix resists being bent and confined. The terminase motor is convincing, generating a force of up to 60 pN—twenty times the force of the myosin motor (Smith et al. 2001). The terminase prevails and compresses the DNA to quasi-crystalline density. The energy cost of this work is high, demanding hydrolysis of one ATP for each two bps packaged (Guo, Peterson, Anderson 1987). This whole effort would be for naught were the capsids not strong enough to resist these high pressures from within (*see page 6-13*).

Packaging a linear genome in this manner enables the phage to always know which end of its genome will be first to exit the capsid, thus first to enter the host cytoplasm. Based on this certainty, some tailed phages developed sophisticated strategies for genome defense and early transcription upon arrival. Witness Podophage T7 (*see page 3-17*).

The end of linearity

A linear chromosome is fine for packaging through a portal, but linearity creates its own complication. To replicate a linear genome requires solving the **end replication problem** (*see page 5-18*). Gifsy-2's solution? Eliminate the ends by circularizing the genome upon arrival in the host cytoplasm, as do most tailed phages. Thus these phages replicate circular, rather than linear, DNA molecules. Early in an infection Gifsy-2 replicates via bidirectional replication, the common method for duplicating circular chromosomes that is widely used among the Bacteria. The DNA replication machinery assembles at a single origin and then two active replication complexes move away in opposite directions, both replicating the chromosome as they go. (This is called θ replication since the replication intermediate resembles that Greek letter.) Phages similar to Gifsy-2 go through five or six rounds of θ replication first to produce ~50 circular genomes that serve immediately as templates for transcription (Taylor 1995). But Gifsy-2 needs linear, not circular, genomes for packaging in progeny virions.

Going in circles

How to manufacture multiple linear genomes from a circular template? By going in circles. After that initial burst of θ replication, Gifsy-2 diverts a few of those ~50 circles to serve as templates for rolling circle replication. A phage endonuclease makes a single-strand nick in each of these circles at a specific replication origin. The resulting free 3'-end is extended by a DNA polymerase holoenzyme (DNApol). In the process the original complementary strand is continuously displaced starting at the free 5'-end. (This mechanism is called σ replication because the circle, with the tail formed by the displaced strand, resembles that Greek letter.) A second DNApol converts the displaced single-strand to dsDNA by synthesizing a complementary strand as a lagging strand (i.e., as Okazaki fragments that are then joined by DNA ligase). The first DNApol 'rolls' around the circular genome again and again, each revolution extending the tail of the σ by another genome length until that tail is a concatemer of two to five—even as many as ten—dsDNA genomes, all covalently linked head-to-tail.

Cut to length

This string of linear genomes is now sent to the packaging department. Gifsy-2's terminase makes a staggered double-stranded cut at the *cos* site, thereby creating single-stranded overhangs on both of the resulting ends (Casjens et al. 2005). Some phages stagger more than others. In some the two cuts are only seven bp apart, in others they are further removed, up to 19 bp. Thus the overhangs at the DNA ends also range between seven and 19 nt. Similarly, whereas the overhangs are on the 5' strands in Gifsy-2, in some phages they are on the 3' strands instead (*see page 5-2*). After this cleavage, terminase holds onto the end that is to go into the capsid first. It threads that end into the procapsid and continues pumping the DNA in until it encounters the next *cos* site. There it makes a second staggered cut to free the packaged genome from the concatemer, thereby completing that genome. The terminase leaves the portal, the residual concatemer in hand, and proceeds to fill another waiting procapsid in the same fashion. Since this terminase makes these cuts at precisely the same genomic location each time, all the packaged genomes are identical in length and have identical termini (Casjens, Gilcrease 2009).

Full circle

The staggered cuts are key to Gifsy-2's strategy. The overhangs that they create at the two ends are complementary. Upon arrival in the host cytoplasm, these cohesive or 'sticky' ssDNA ends anneal to form a circular structure. DNA ligase then seals the gap in each strand. *Voilà!* A covalently closed circular DNA genome once again. This completes Gifsy-2's genome cycle from circular replicative form to packaged linear chromosome and back to a circle again.

Gifsy-2's way is not the only way for a phage, starting with a concatemer, to package a linear genome capable of recircularizing during infection (Casjens, Gilcrease 2009). Another tactic is for terminase to make an initial blunt cut within a region of the genome, thread the end of the concatemer into a procapsid until the terminase senses that the capsid is 'full' (Tye, Huberman, Botstein 1974; Sun et

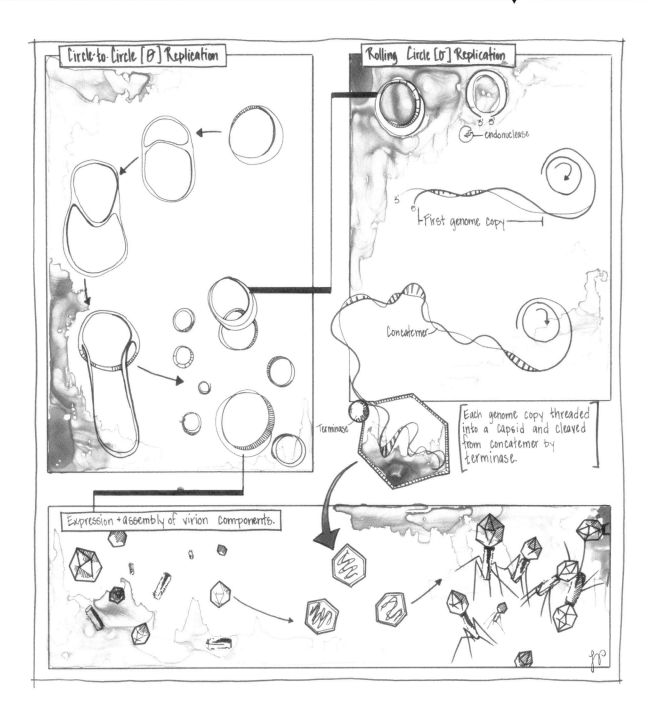

Circle-to-Circle [θ] Replication

Rolling Circle [σ] Replication

endonuclease

First genome copy

Concatemer

Terminase

Each genome copy threaded into a capsid and cleaved from concatemer by terminase.

Expression + assembly of virion components.

al. 2012), then cut the genome free. This 'headful' method typically packages a bit more than a complete genome, i.e., 102-110% of genome length. As a result, the ends of the packaged genomes contain **direct terminal repeats** (DTRs) of varying length. These genomes use the host's **recombination** machinery to circularize upon arrival by **homologous recombination** between the DTRs. This method is imprecise from start to finish. Both the location of the initial cut and the genome length packaged vary; the virions produced carry circularly permuted genomes that carry DTRs but differ slightly in length.

Variations on the theme

Other phages embody different solutions to the end replication problem. One group uses **terminal proteins** (*see page 5-18*) covalently linked to their genome ends, while another replaces genome ends with hairpin 'telomeres' (*see page 5-25*). Temperate phage Mu integrates into the host chromosome and replicates without excision, each copy integrating at a non-specific site. During lytic replication, it generates linear DNA packaging substrates by cleaving the host chromosome some distance away from both ends of its own genome (Bukhari 1976).

When evolution gives the phages a problem to solve, such as a unidirectional DNApol, they respond with multiple innovative solutions. Their ingenuity seems endless. Not surprising, if one extrapolates from the dictum that two heads are better than one.

Cited references

Baltimore, D. 1971. Expression of animal virus genomes. Bacteriol Rev 35:235-241.

Brüssow, H, RW Hendrix. 2002. Phage genomics: Small is beautiful. Cell 108:13-16.

Bukhari, A. 1976. Bacteriophage Mu as a transposition element. Annu Rev Genet 10:389-412.

Casjens, SR, EB Gilcrease. 2009. Determining DNA packaging strategy by analysis of the termini of the chromosomes in tailed-bacteriophage virions. Methods Mol Biol 502:91-111.

Casjens, SR, EB Gilcrease, DA Winn-Stapley, P Schicklmaier, H Schmieger, ML Pedulla, ME Ford, JM Houtz, GF Hatfull, RW Hendrix. 2005. The generalized transducing *Salmonella* bacteriophage ES18: Complete genome sequence and DNA packaging strategy. J Bacteriol 187:1091-1104.

Effantin, G, N Figueroa-Bossi, G Schoehn, L Bossi, JF Conway. 2010. The tripartite capsid gene of Salmonella phage Gifsy-2 yields a capsid assembly pathway engaging features from HK97 and λ. Virology 402:355-365

Guo, P, C Peterson, D Anderson. 1987. Prohead and DNA-gp3-dependent ATPase activity of the DNA packaging protein gp16 of bacteriophage ö29. J Mol Biol 197:229-236.

Simpson, AA, Y Tao, PG Leiman, MO Badasso, Y He, PJ Jardine, NH Olson, MC Morais, S Grimes, DL Anderson. 2000. Structure of the bacteriophage φ29 DNA packaging motor. Nature 408:745-750.

Smith, DE, SJ Tans, SB Smith, S Grimes, DL Anderson, C Bustamante. 2001. The bacteriophage φ29 portal motor can package DNA against a large internal force. Nature 413:748-751.

Sun, S, S Gao, K Kondabagil, Y Xiang, MG Rossmann, VB Rao. 2012. Structure and function of the small terminase component of the DNA packaging machine in T4-like bacteriophages. Proc Natl Acad Sci USA 109:817-822.

Taylor, K. 1995. Replication of coliphage lambda DNA. FEMS Microbiol Rev 17:109-119.

Tye, B-K, JA Huberman, D Botstein. 1974. Non-random circular permutation of phage P22 DNA. J Mol Biol 85:501-527.

Recommended reviews

Casjens, SR, EB Gilcrease. 2009. Determining DNA packaging strategy by analysis of the termini of the chromosomes in tailed-bacteriophage virions. Methods Mol Biol 502:91-111.

Casjens, SR. 2011. The DNA-packaging nanomotor of tailed bacteriophages. Nat Rev Microbiol 9:647-657.

Feiss, M, VB Rao. 2012. The bacteriophage DNA packaging machine. Adv Exp Med Biol 726:489-509.

42,300,000

light years spanned by 10^{31} phage virions placed side-by-side

Calculated assuming 40 nm capsid diameter.

Enterobacteria Phage HK97

a Siphophage that strengthens its capsid by cross-linking all the major capsid proteins into one catenane

Genome
dsDNA; linear
39,732 bp
62 predicted ORFs; 0 RNAs

Encapsidation method
Packaging; T = 7 capsid

Common host
Escherichia coli

Habitat
Mammalian intestines

Lifestyle
Temperate

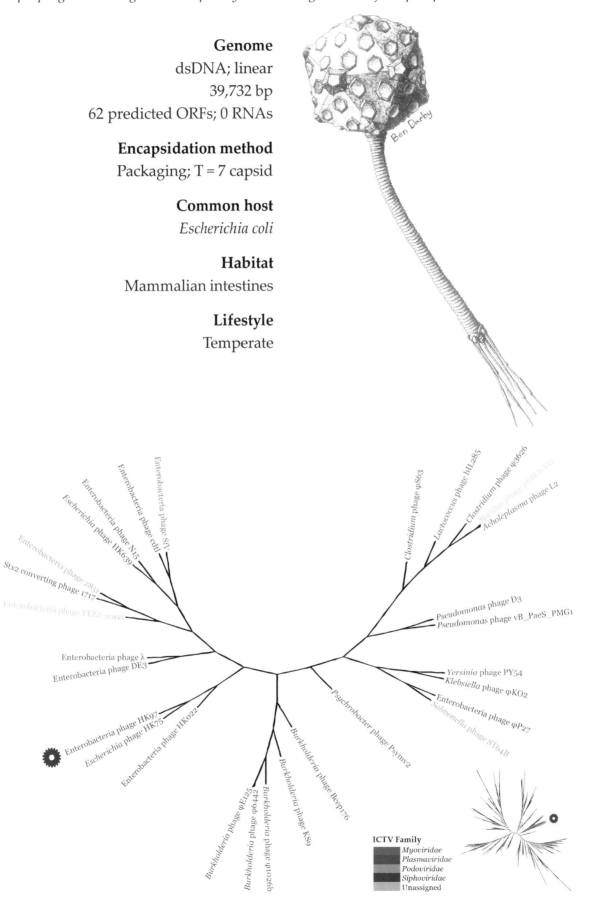

Ben Darby

ICTV Family
Myoviridae
Plasmaviridae
Podoviridae
Siphoviridae
Unassigned

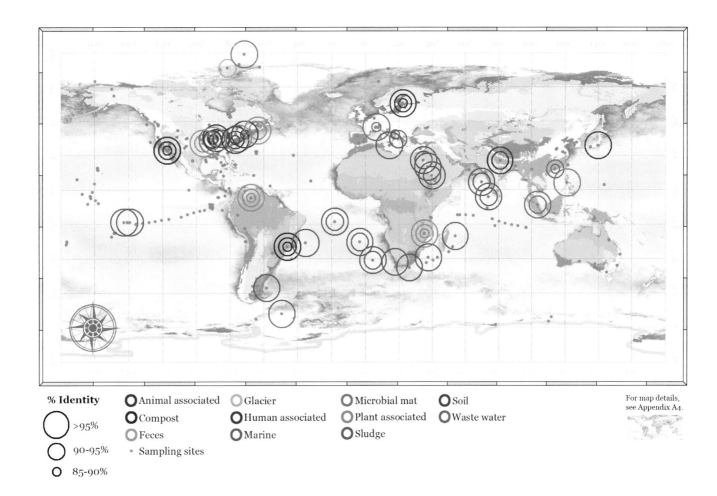

% Identity

- ○ >95%
- ○ 90-95%
- ○ 85-90%

○ Animal associated	○ Glacier	○ Microbial mat	○ Soil
○ Compost	○ Human associated	○ Plant associated	○ Waste water
○ Feces	○ Marine	○ Sludge	
• Sampling sites			

For map details, see Appendix A4.

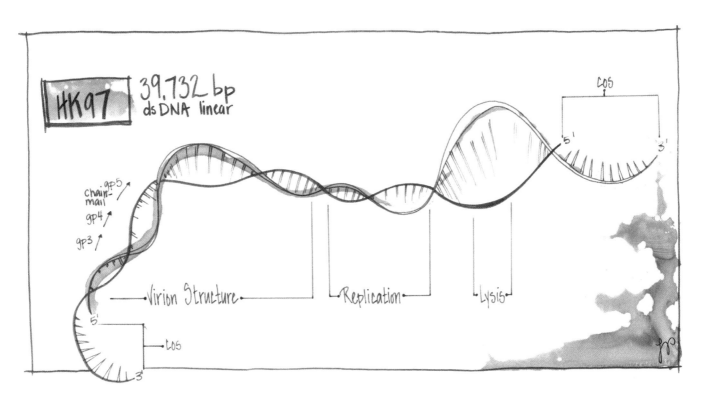

Enterobacteria Phage HK97

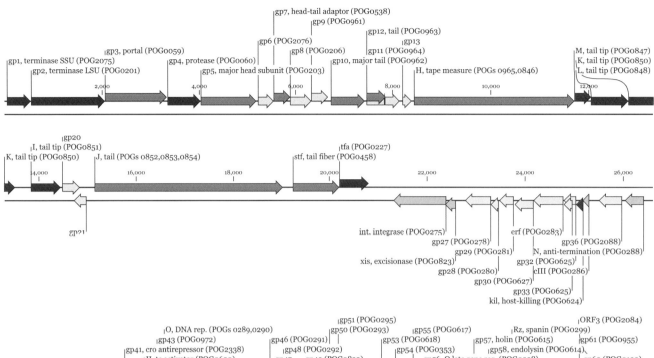

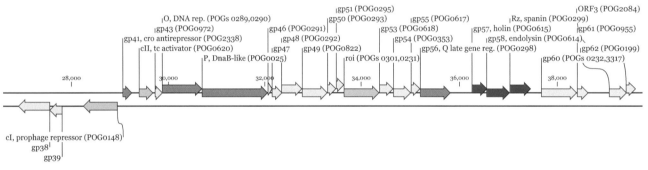

On the Prowl   Entry   Takeover   Replication   Replication   Assembly   Escape   Transfer   Non-coding   Multiple   Unknown
                                   Lytic   Lysogenic   RNA   RNA

Armored Capsids

Merry Youle

Mise-en-Scène: *The rate of phage evolution is rapid. Often protein homologs can be identified by sequence similarity only between recently diverged lineages, leaving more distant relationships unclear. Fortunately, from our perspective, three-dimensional structure and function may be conserved even after sequence similarity has been lost at both the nucleotide and amino acid levels. Application of structural analysis to major capsid proteins has revealed three ancient protein folds. The capsid protein fold first described in Siphophage HK97 was found to be shared by all members of the* Caudovirales. *As the structures of more viral capsid proteins were determined, both the canonical HK97 fold and the fold from PRD1 were found in viruses from all three domains of life, thus sparking added speculation about the early evolution of both viruses and cellular life.*

Lambdoid phage HK97 transports its genome within an inconspicuous, seemingly ordinary Siphophage virion: a flexible, non-contractile, 177 nm-long tail attached to a 55 nm diameter icosahedral capsid (T=7 symmetry). HK97 is also a traditionalist in its overall assembly protocol. Like most members of the *Caudovirales*, it constructs a preliminary **procapsid**, stuffs the genomic DNA inside, and then, as the final step, attaches its tail. However, a closer look at HK97 assembly reveals a most extraordinary twist.

Assembly instructions for HK97

HK97 constructs its capsid from one major capsid protein (gp5) plus a portal protein. Each icosahedral face requires three gp5 hexamers; one gp5 pentamer is needed for each of eleven vertices. The twelfth vertex, the gateway for DNA entry and the attachment site for the tail, is composed of a 12-mer of the portal protein (gp3). Nothing extraordinary so far.

When those parts assemble into a finished capsid, all 415 copies of the major capsid protein are covalently linked into a single **catenane**. Sounds complicated? Only four easy steps are required.

Step 1. For each virion to be produced, synthesize hundreds of copies of the primary building block, the major capsid protein (gp5). The GroEL and GroES chaperonins of your *E. coli* host will properly fold each one for you (Xie, Hendrix 1995).

Step 2. Allow the properly folded gp5 proteins to self-assemble into hexamers and pentamers. For each virion, accumulate 60 hexamers and 11 pentamers. Warning! Do not skip this step. These aggregates, termed **capsomers**, are needed for the next step of capsid shell construction. Only a few mavericks among the phages with double-stranded DNA (dsDNA) genomes, such as P22, can build a capsid directly from individual protein monomers (King et al. 1976).

Step 3. Synthesize and have at hand the other required components: about 50 copies of the protease (gp4) and 12 copies of the portal protein (gp3) for each virion. Don't pay any attention to other phages that say you can't build an icosahedral capsid without a scaffolding protein. There is no rule in The Phage World that can't be broken, or at least bent by a crafty phage.

Step 4. Commingle capsomers, proteases, and portal proteins and then stand back. Inherent within the proteins are all the instructions needed to govern the temporal and spatial sequence of the subsequent steps. (Can you imagine an automobile assembly line that functions similarly?)

Assembly on automatic pilot

Given the opportunity, those capsid components co-assemble to fabricate a dynamic intermediate structure—the roundish, 47 nm-diameter procapsid shell (Procapsid I). The 42 kDa capsid proteins (gp5) form its thick walls while the proteases fill most of the space inside. Eleven of the vertices are occupied by pentamers, the twelfth by the portal complex.

As soon as Procapsid I is assembled, the proteases thin the capsid wall by cleaving the interior-facing N-terminal domain from all of the capsid proteins (Duda et al. 1995). The protein fragments, as well as the proteases themselves, are reduced to small peptides that exit from the shell. This modified structure, Procapsid II, is ready to receive the genome. Similar post-assembly processing of structural proteins is *de rigueur* for many phages (Popa et al. 1991). Such maneuvers during virion maturation dictate the sequence of assembly steps and make the overall process irreversible.

For genome packaging, HK97 uses a two-subunit **terminase**, like countless other phages with dsDNA genomes. As is also common, packaging precipitates reorganization and expansion of the capsid (Jardine, Coombs 1998). Expansion of Procapsid II thins its walls further, approximately doubling its volume. Capsid proteins shift slightly in the process, with compelling effects.

A lysine in each capsid protein moves into close proximity to an asparagine in the neighboring protein. Proximity combined with a favorable microenvironment catalyzes the formation of a covalent amide bond linking each lysine/asparagine pair (Duda et al. 1995). Each of the 415 capsid proteins (gp5) is thus cross-linked to two neighbors: its lysine169 joined to the asparagine356 of one neighbor and its own asparagine356 to the lysine169 of the other. These covalent bonds turn every gp5 pentamer and hexamer into a covalently-linked five- or six-membered ring (Popa et al. 1991).

Chain mail

Although covalent protein cross-linking is common in many organisms, such extensive cross-linking was unheard of in virion assembly prior to its discovery in HK97 (Popa et al. 1991). Moreover, HK97 adds its own extraordinary twist. During capsid expansion but before cross-linking, the ~400 capsid proteins simultaneously shift arm-over-arm to intertwine their polypeptide loops, thereby 'stitching' together adjacent capsomers. The subsequent cross-linking forms closed rings that are topologically interlocked like the links in a chain (Duda 1998). Simultaneous cross-linking over the entire capsid generates one continuous protein catenane. To now disrupt HK97's capsid would require breaking covalent bonds.

Catenanes combine flexibility and strength—thus their historical use in human warfare for chain mail armor and kusari. However, human chain mail was a burdensome external layer. In elegant contrast, phage chain mail is an integral part of a minimal capsid shell. Why might a phage need armor? A capsid shell, composed of a single protein layer only a few nanometers thick, is responsible for protecting the genome against environmental assaults. It also must withstand strong forces exerted by the densely packaged DNA within. To stabilize the capsid against stressors from both within and without, some phages add cementing proteins to the external surface. For example, T4 uses Soc (Qin et al. 2010), P4 uses Psu (*see page 6-18*), and λ attaches gpD (Lander et al. 2008). HK97's capsid chain mail serves the same purpose, but so much more efficiently.

Could λ learn how?

The crystal structure of HK97's capsid protein gp5 is highly similar to that of closely-related phage λ (Lander et al. 2008), yet these two phages employ markedly different strategies for capsid reinforcement: λ adds cementing proteins, HK97 uses chain mail. Comparing their mature capsids reveals that the stabilization locations where λ's cementing proteins attach correspond exactly to the sites of the protein cross-linking in HK97 (Lander et al. 2008). Moreover, expansion of λ's capsid during genome packaging brings into close association the very same protein domains whose juxtaposition triggers the cross-linking in the HK97 capsid. Comparison of their amino acid sequences in this region suggests that two key amino acid substitutions are all that would be required to 'teach' λ the chain mail trick (Lander et al. 2008).

Phage chain mail was first discovered in HK97, and HK97 remains its best-studied practitioner. As would be expected for something so elegant, so efficient, so ingenious, it has now been found in numerous other phages. For example, the majority of the mycobacteriophages also have the knack (Hatfull, Sarkis 1993). Further searching will surely turn up more phage chain mail, but I wonder which phage will be the first one found wearing armor with a distinctly different twist.

Cited references

Duda, RL. 1998. Protein chainmail: Catenated protein in viral capsids. Cell 94:55-60.

Duda, RL, J Hempel, H Michel, J Shabanowitz, D Hunt, RW Hendrix. 1995. Structural transitions during bacteriophage HK97 head assembly. J Mol Biol 247:618-635.

Hatfull, GF, GJ Sarkis. 1993. DNA sequence, structure and gene expression of mycobacteriophage L5: A phage system for mycobacterial genetics. Mol Microbiol 7:395-405.

Jardine, PJ, DH Coombs. 1998. Capsid expansion follows the initiation of DNA packaging in bacteriophage T4. J Mol Biol 284:661-672.

King, J, D Botstein, S Casjens, W Earnshaw, S Harrison, E Lenk. 1976. Structure and assembly of the capsid of bacteriophage P22. Philos Trans R Soc Lond B Biol Sci 276:37-49.

Lander, GC, A Evilevitch, M Jeembaeva, CS Potter, B Carragher, JE Johnson. 2008. Bacteriophage lambda stabilization by auxiliary protein gpD: Timing, location, and mechanism of attachment determined by cryo-EM. Structure 16:1399-1406.

Popa, M, T McKelvey, J Hempel, R Hendrix. 1991. Bacteriophage HK97 structure: Wholesale covalent cross-linking between the major head shell subunits. J Virol 65:3227-3237.

Qin, L, A Fokine, E O'Donnell, VB Rao, MG Rossmann. 2010. Structure of the small outer capsid protein, Soc: A clamp for stabilizing capsids of T4-like phages. J Mol Biol 395:728-741.

Xie, Z, RW Hendrix. 1995. Assembly *in vitro* of bacteriophage HK97 proheads. J Mol Biol 253:74-85.

Recommended review

Aksyuk, AA, MG Rossmann. 2011. Bacteriophage assembly. Viruses 3:172-203.

Enterobacteria Phage P4

a Podophage that compels a 'helper' Myophage to assemble small capsids and fill them with only P4's genome

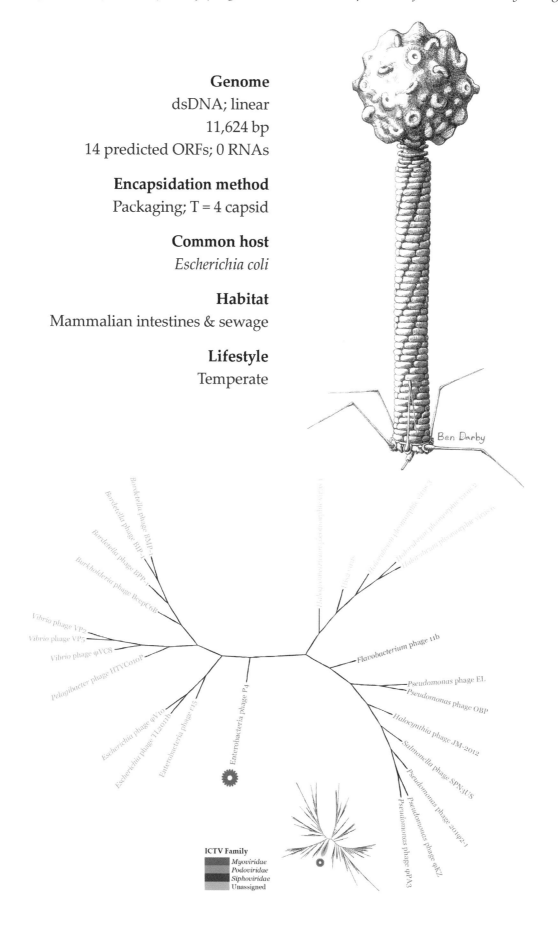

Genome
dsDNA; linear
11,624 bp
14 predicted ORFs; 0 RNAs

Encapsidation method
Packaging; T = 4 capsid

Common host
Escherichia coli

Habitat
Mammalian intestines & sewage

Lifestyle
Temperate

Ben Darby

Bordetella phage BMP-1
Bordetella phage BIP-1
Bordetella phage BPP-1
Burkholderia phage BcepC6B
Vibrio phage VP2
Vibrio phage VP5
Vibrio phage φVC8
Pelagibacter phage HTVC010P
Escherichia phage φV10
Escherichia phage TL2011b
Enterobacteria phage r15
Enterobacteria phage P4
Flavobacterium phage 11b
Pseudomonas phage EL
Pseudomonas phage OBP
Halocynthia phage JM-2012
Salmonella phage SPN3US
Pseudomonas phage 201φ2-1
Pseudomonas phage φPA3
Pseudomonas phage φKZ

ICTV Family
Myoviridae
Podoviridae
Siphoviridae
Unassigned

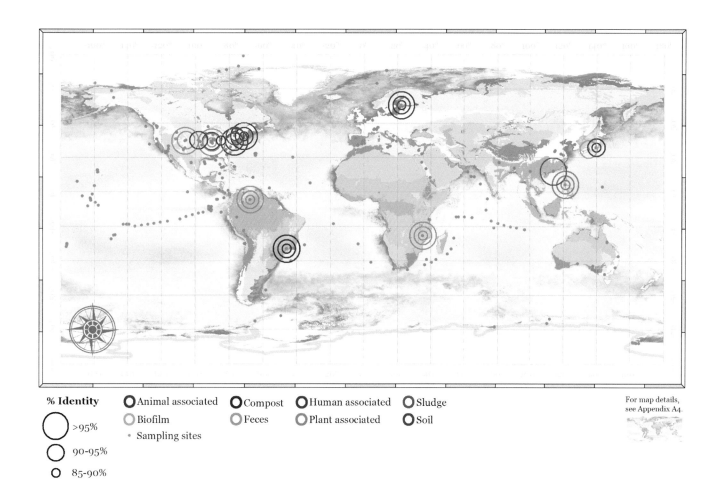

% Identity

⭕ >95%

⭕ 90-95%

○ 85-90%

◯ Animal associated	◯ Compost	◯ Human associated	◯ Sludge
◯ Biofilm	◯ Feces	◯ Plant associated	◯ Soil
• Sampling sites			

For map details, see Appendix A4.

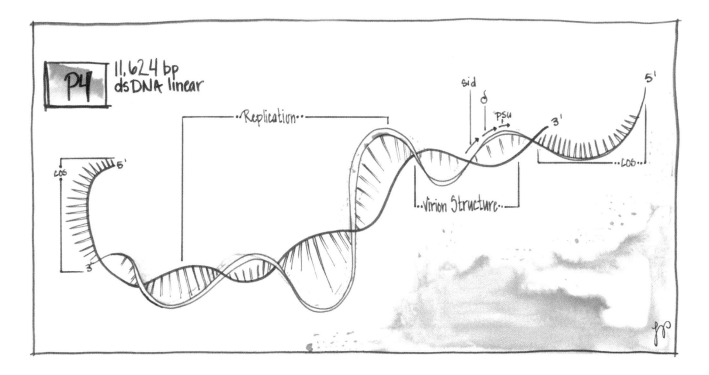

Enterobacteria Phage P4

p1, gop
p2, beta

p13, delta, transactivation
p14, psu, capsid
p12, sid, head size determination

2,000   4,000   6,000   8,000   10,000

p4, integrase (POG0275)
p3, cII (POG0143)
p5
p6, alpha, DNA primase (POG0326)
p7
p10, cI repressor
p8
p11, tc. reg (POGs 1127,2703,3378)
p9, epsilon, helper derepression

On the Prowl	Entry	Takeover	Replication Lytic	Replication Lysogenic	Assembly	Escape	Transfer RNA	Non-coding RNA	Multiple	Unknown

P4: Sophisticated Capsid Thief

Merry Youle

Mise-en-Scène: *Phages exploit the exploitable. Not only do they skillfully take over a host cell, but sometimes they also commandeer resources provided by an exploitable phage. When such opportunistic phages become dependent on a 'helper' phage, they are often called 'defective.' However, since this manipulative strategy works quite well, perhaps such phages should be called skillful or adept, and respected for their ingenious thievery.*

You can't judge a book by its cover, nor can you judge a phage genome by the capsid around it. Some phages have the chutzpah to appropriate their capsids from an unrelated 'helper' phage. Coliphage P4 is one such phage. It not only steals capsid components encoded by its helper, but it then proceeds to assemble them into smaller, P4-sized capsids that are too small to accommodate the helper's larger genome.

P4's game plan

Because it lacks the genes for virion proteins, genome packaging, and host lysis, coliphage P4 can produce infectious virions only with the active assistance of a helper phage. P4 is assured timely assistance when it infects and takes over a P2 lysogen, e.g., an *E. coli* that carries a prophage of a helper phage, P2. Other Enterobacteria with other P2-like resident prophages also serve as well. To show its gratitude, P4 strongly interferes with the helper's own replication.

Enslavement

As soon as P4 enters a suitable lysogen, it replicates its DNA to accumulate more than one hun-

dred copies (Six, Klug 1973). Now what? To proceed with a lytic infection requires proteins that are encoded by the helper prophage. However, prophages are typically quiescent; transcription of most prophage genes is repressed. P4 skillfully activates the needed helper prophage genes without (usually) triggering the prophage to excise from the host chromosome (Six, Lindqvist 1978). Before long, expression is underway of the late genes encoded by both P4 and its helper. Products of both sets of late genes are needed for P4 success: its own set for capsid theft and the helper's set to provide and fill those capsids.

The helper's numerous late genes serve P4 by providing capsids, packaging P4's genome, and ultimately lysing the host. In contrast, P4's own late operon encodes only three genes. Gene products of two are used for capsid theft (Sid and Psu, see below); the third (δ) prompts late gene transcription by both P4 and its helper. The intimate regulatory crosstalk between P4 and its helper that coordinates this activity suggests that these two have been playing this game together for a long, long time. For instance, P4 stimulates expression of a transcription activator encoded by the helper, and this activator, in turn, enhances expression of P4's own late genes (Six, Lindqvist 1978).

Capsid protein diversion

Left to its own devices, the helper phage P2 assembles its structural proteins into a typical Myophage virion the right size for its genome. It constructs a 62 nm diameter **procapsid** from 415 copies of its capsid protein gpN (Dearborn et al. 2012). P4, eyeing those same capsid proteins, redirects them to assemble smaller, P4-sized capsids. Since P4's genome is about one third the size of the helper's (~12 kbp versus ~33 kbp), it can build an adequate 45 nm diameter capsid using just 235 of those same capsid proteins (Dearborn et al. 2012). As an added bonus, P4 has exclusive occupancy of these capsids because the larger helper genome won't fit inside.

Only two P4 genes are required for this build-to-suit alteration. One is known as *sid* (*s*ize *d*etermination). Normally the helper's internal scaffolding protein (gpO) directs construction of the larger capsid, but Sid proteins form an external scaffold that takes control (Shore et al. 1978). As hexamers of gpN are assembled to form the icosahedral shell, Sid trimers bind and tether them, constraining assembly to yield only the smaller capsid (Dearborn et al. 2012).

The packaging department

For packaging, P4 co-opts the helper's machinery, i.e., its **terminase** (*see page 6-5*). This two-component phage enzyme identifies the DNA to be packaged by recognizing a specific packaging sequence (the *cos* **site**) and then translocates the DNA into a waiting procapsid. When unhampered by P4, helper genomes are recognized and packaged by the helper's terminase (Pruss, Wang, Calendar 1975). Although P4 is unrelated to its helper, it has acquired an almost identical *cos* site that it uses to trick the helper's terminase into packaging P4 DNA into either size capsid. P4 mutants that lack Sid assemble only large capsids (see above). In this situation, the terminase packages P4 DNA into the large capsids as best it can, some-times stuffing two or three copies of the smaller P4 genome inside (Shore et al. 1978). Moreover, P4's *almost* identical *cos* site is actually better than the original; the terminase preferentially processes P4 DNA over helper DNA (Bowden, Modrich 1985). This proclivity gives P4 a decided advantage during co-infection when both P4 and its helper are actively replicating in close proximity.

When packaging is completed, Sid's tethering work is finished and it is discarded from the mature capsid. The small capsids being inherently less stable, P4 then adds its own cementing protein (Psu) to the outer surface for reinforcement. The mature virions are released by host lysis, the work of the helper-encoded holin/**endolysin** mechanism (*see page 7-5*). The tally for P4 takeover of a helper lysogen? 100,000 P4 progeny virions for every helper virion produced (Six, Klug 1973).

Questions of identity

That P4 steals the structural components of its virions has led to some taxonomic confusion. As you would expect, its virions look like those of its Myophage helpers. As a result, the **ICTV**, the official arbiters of morphology-based viral taxonomy, assigned P4 to the family *Myoviridae* alongside its helper phage. However, P4 and its helper P2 are not related (Briani et al. 2001). Based on the genome inside the stolen capsid, P4 is a Podophage (*see page 8-14*; [Rohwer, Edwards 2002]).

P4's status as a legitimate phage had already been questioned given that it does not encode a protein capsid, a trait considered a *sine qua non* of a virus. Moreover, it was labeled *defective* when first discovered because it cannot replicate without help from another phage (Six 1963). However, that pejorative label fails to convey the highly-evolved, finely-tuned, integrated interactions between P4 and its helper.

P4 has also blurred the lines of identity by its behavior when infecting a non-lysogen. Here its only viable option is lysogeny and it usually behaves like other temperate phages: it integrates into the host chromosome at a specific location as a prophage. However, about one time in a hundred it goes a different route and maintains itself in the

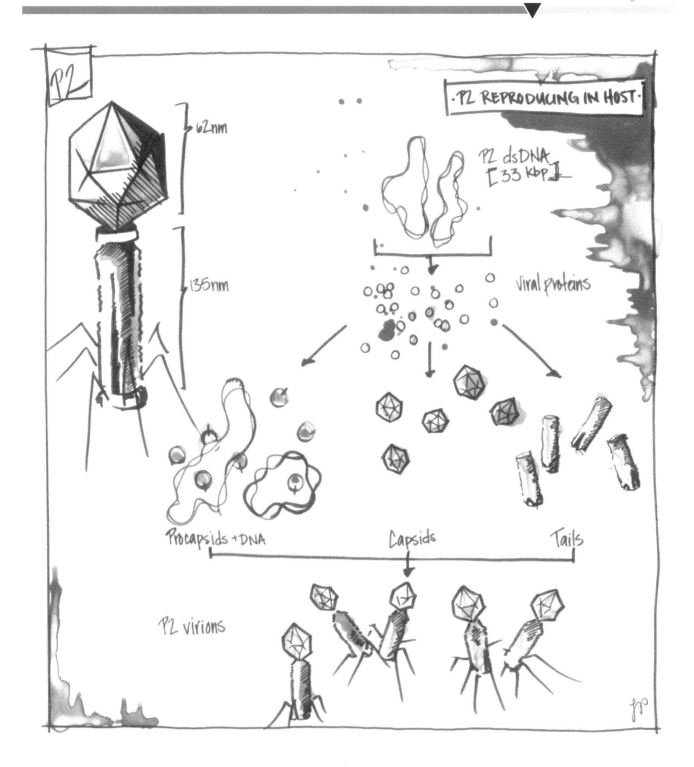

host as a multicopy **plasmid**, 30-50 copies per cell (Briani et al. 2001). This is not phage-like behavior.

If P4 is not a phage, then what sort of a replicon is it? Its plasmid persona suggests it might be an integrative plasmid that has learned the trick of exploiting a helper phage for intercellular transport. However you choose to classify it, it warrants descriptors such as sophisticated, artful, or exploitative, but not defective.

Cited references

Bowden, D, P Modrich. 1985. In vitro maturation of circular bacteriophage P2 DNA. Purification of ter components and characterization of the reaction. J Biol Chem 260:6999-7007.

Briani, F, G Deh, F Forti, D Ghisotti. 2001. The plasmid status of satellite bacteriophage P4. Plasmid 45:1-17.

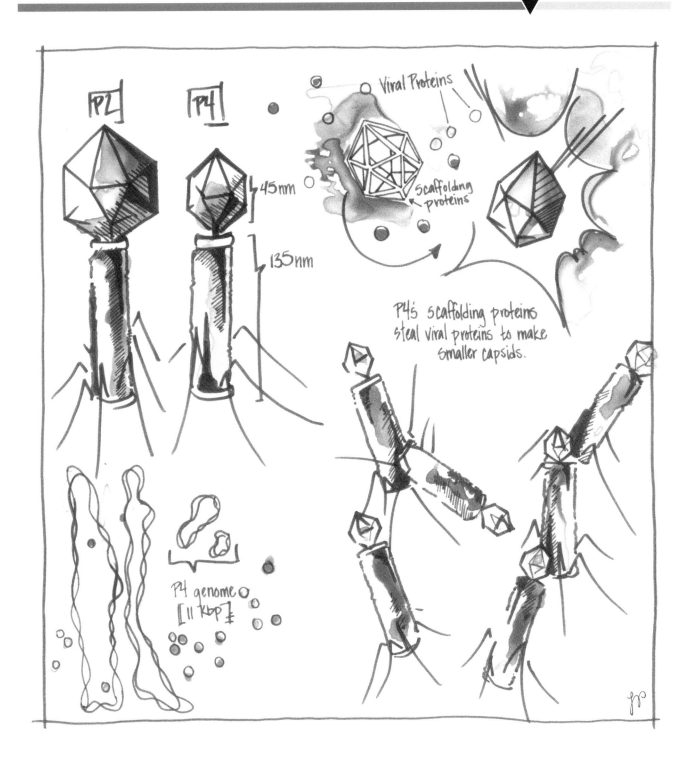

Dearborn, AD, P Laurinmaki, P Chandramouli, CM Rodenburg, S Wang, SJ Butcher, T Dokland. 2012. Structure and size determination of bacteriophage P2 and P4 procapsids: Function of size responsiveness mutations. J Struct Biol 178:215-224.

Pruss, GJ, JC Wang, R Calendar. 1975. *In vitro* packaging of covalently closed circular monomers of bacteriophage DNA. J Mol Biol 98:465-478.

Rohwer, F, R Edwards. 2002. The Phage Proteomic Tree: A genome-based taxonomy for phage. J Bacteriol 184:4529-4535.

Shore, D, G Dehò, J Tsipis, R Goldstein. 1978. Determination of capsid size by satellite bacteriophage P4. Proc Natl Acad Sci USA 75:400-404.

Six, E. 1963. A defective phage depending on phage P2. Bacteriol Proc.: 138.

Six, EW, CAC Klug. 1973. Bacteriophage P4: A satellite virus depending on a helper such as prophage P2. Virology 51:327-344.

Six, EW, BH Lindqvist. 1978. Mutual derepression in the P2-P4 bacteriophage system. Virology 87:217.

Recommended review

Christie, GE, T Dokland. 2012. Pirates of the Caudovirales. Virology 434:210-221.

Pseudomonas Phage φ8

a Cystophage that packages one copy of each of its three chromosomes in each capsid

Genome
dsRNA; linear, segmented
Segment L: 7,051 bp; 7 ORFs; 0 RNAs
Segment M: 4,741 bp; 6 ORFs; 0 RNAs
Segment S: 3,192 bp; 6 ORFs; 0 RNAs

Encapsidation method
Packaging; T = 2 capsid

Common host
Pseudomonas savastanoi pv. phaseolicola

Habitat
Host-associated; plant leaf

Lifestyle
Lytic

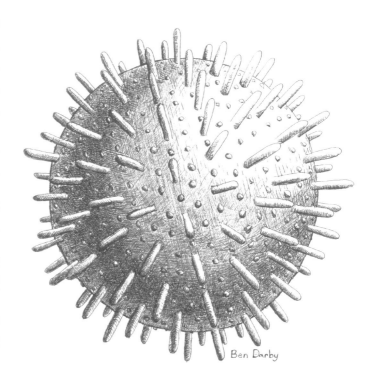

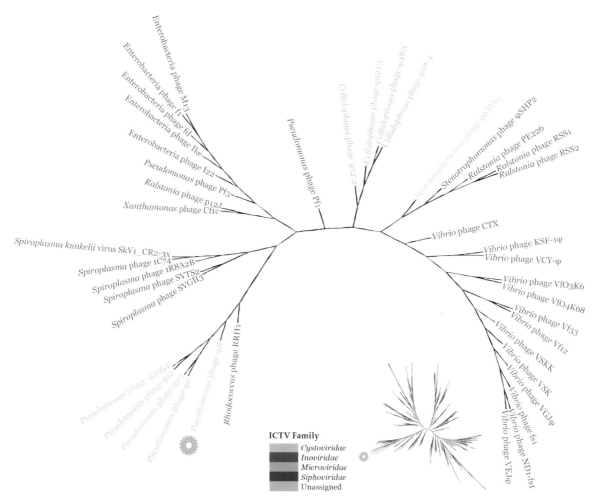

ICTV Family
Cystoviridae
Inoviridae
Microviridae
Siphoviridae
Unassigned

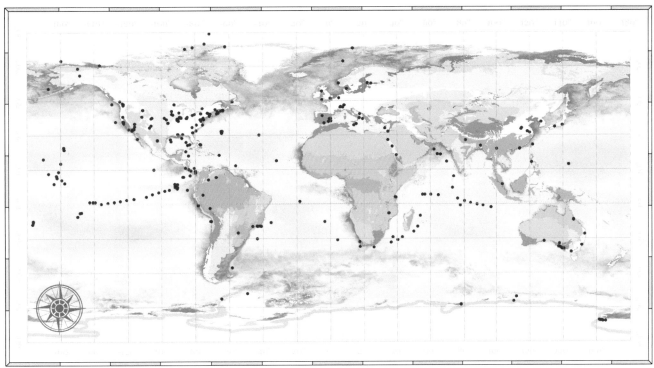

• Sampling sites

For map details,
see Appendix A4.

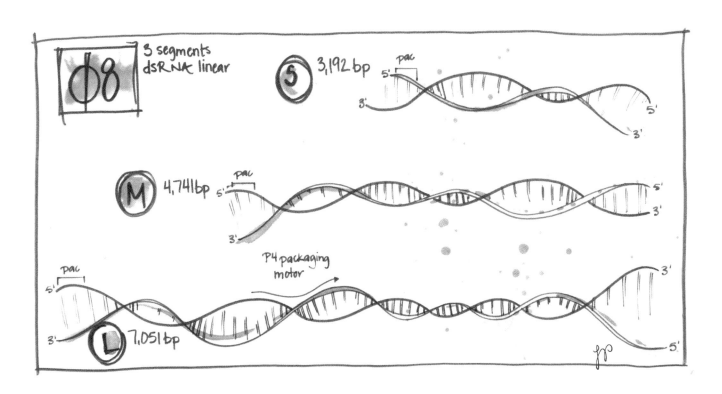

Pseudomonas Phage φ8

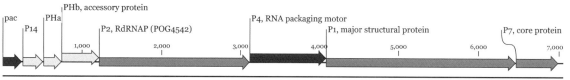

Segment L

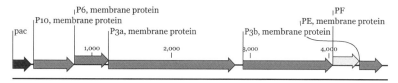

Segment M

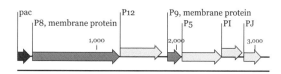

Segment S

On the Prowl   Entry   Takeover   Replication Lytic   Replication Lysogenic   Assembly   Escape   Transfer RNA   Non-coding RNA   Multiple   Unknown

The Great Cystophage Packaging Challenge

Merry Youle

Mise-en-Scène: *The Cystophages are the only double-stranded RNA (dsRNA) phages known to infect Bacteria or Archaea, making them oddities in the phage world. However, they are strikingly similar to a family of vertebrate viruses, the Reoviridae. Both groups have segmented genomes, the Cystophages having three polycistronic segments while some Reoviridae have as many as twelve, all but one of which are monocistronic. Both families have complex icosahedral virions composed of inner and, for some members, outer protein capsids; both enter their host cells through an endocytic process (see page 3-23); both incorporate their RNA-dependent RNA polymerase as a component of the (inner) capsid; in both, transcription and replication take place on the genome in situ within the (inner) capsid. Such similarities tempt one to speculate on an ancient evolutionary link between these families, but comparison of the folds of their major inner capsid proteins has been inconclusive (Veesler, Johnson 2013).*

Note: Most of the research related in this story employed the *in vitro* packaging system developed for the model Cystophage φ6. The structural studies of the packaging ATPase used the P4 protein from the closely related phage φ8 (Huiskonen et al. 2007).

Cystophages boldly defy normal phage conventions. To start with, they use double-stranded RNA (dsRNA) for their genome—the only viruses of Bacteria or Archaea known to choose this option. Moreover, instead of enlisting one chromosome to encode all their genes, Cystophages employ three. Each dsRNA chromosome is termed a segment; all three segments are essential for a successful infection. To qualify as a master Cystophage packer, a phage must package into its virion one, and only one, copy of each of the three segments. Cystophages φ6, φ8, and their kin all qualify.

In-house transcription

By the time an infecting Cystophage genome arrives in the host cytoplasm, it has jettisoned its outer layer(s) (*see page 3-23*). What remains is a efficient, capsid-enclosed replication machine made up of a dsRNA genome and replication enzymes. Cystophage genomes never leave this capsid shelter, a tactic that hides them from host endonucleases that would avidly cleave this dsRNA. Although most phages with dsDNA genomes rely on their hosts to provide at least part of the DNA replication machinery needed, cells don't have enzymes for replicating or transcribing dsRNA. Cystophages fend for themselves by bringing along their own RNA-dependent RNA polymerase (RdRP) as a component of their capsid.

Transcription of the entire genome gets underway soon after this capsid core arrives in the cytoplasm. RdRP transcribes the negative-sense strand in each of the three genome segments (S, small; M, medium; and L, large) to produce equal numbers of full-length, positive-sense S, M, and L transcripts. These transcripts promptly exit from the capsid into the cytoplasm where they act as polycistronic messenger RNAs (mRNAs) for the immediate translation of phage proteins.

Procapsid assembly

The first stage in construction of the complex Cystophage virion is assembly of the dodecahedral procapsids (PCs) that will ultimately become the enveloped core of the next generation virions. The four proteins required are all encoded by segment L (Gottlieb et al. 1990). Each PC assembles from 120 copies of the main structural protein (P1). All twelve vertices are equivalent, all are equipped for genome packaging and transcription. On the exterior face of each vertex sits the RNA packaging motor composed of six copies of the ATPase motor protein (P4) arranged around a central channel. The transcription machinery consists of one copy of RdRP situated on the interior face of each vertex, slightly offset from the packaging channel so as to leave clear passage for the RNAs. In addition, about 30 copies of an auxiliary protein (P7) are also assembled into each PC.

When first assembled, the dodecahedral PC shell is empty and collapsed, appearing somewhat like a deflated basketball that has been punched in at numerous sites. Instead of pointing outward, each vertex is sunken inward. The space inside is small, expandable, and ready to receive the genome segments.

Cystophage packaging basics

Translocating a genome into a procapsid is a skill known to the vast majority of phages, Cystophages included, but Cystophages do this with their own unique (so far) flair, meeting their own particular challenge. For equipment, they have an ATP-fueled packaging motor located at every vertex. These motors package full-length positive-sense transcripts, the same transcripts that can serve as mRNA. Each transcript is actively brought into the PC at one of the vertices, entering through the central channel formed by the hexamers of the packaging motor protein (P4; [Huiskonen et al. 2007]).

The dexterous Cystophages package the S, M, and L strands in a specific order, one copy each, by an elegant and reliable scheme. All of the transcripts contain an essential packaging sequence (*pac*) ~200 nucleotides long located near its 5′ terminus. The sequence of these *pac* regions differs for the L, M, and S transcripts, giving each a markedly different secondary structure and therefore a unique shape. These differences are the cues that orchestrate the packaging process.

Expanding opportunities

While the PC is fully collapsed, exposed on its surface are sites that recognize and bind the *pac* region of S transcripts—only S transcripts. The packaging motor imports the bound S transcript into the empty PC (Qiao et al. 1995). In the process, in order to thread the RNA through a narrow vertex chan-

nel, the packaging motor disrupts the stem loops and other double-stranded regions that give each *pac* its distinctive shape (Huiskonen et al. 2007).

Stuffing the S transcript inside expands the PC slightly (Butcher et al. 1997). In this new conformation, the S binding sites disappear and instead the PC offers binding sites for the M transcript. An M transcript is then packaged, thereby expanding the PC further. Lastly, this expanded PC binds an L strand and pulls it inside the now nearly spherical capsid. This final expansion activates the RdRPs embedded within the capsid to synthesize a complementary negative-sense strand for each of the packaged transcripts, thus completing the dsRNA genome and finishing the conversion of the PC into a core particle (Butcher et al. 1997).

This new core particle might function as a transcription machine, its RdRP producing positive-sense transcripts that exit to serve as mRNA or to be packaged into waiting PCs. Alternatively the core can proceed down the virion assembly pathway: cessation of RdRP activity, the assembly of the outer capsid (protein P8) around the core (in ϕ6 but not ϕ8; Olkkonen, Ojala, Bamford 1991), and addition of the lipid envelope. The mature virion now awaits its release at the time of host lysis.

Using dsRNA for your genome is problematic: host cells are quick to recognize it as foreign and destroy it, and you have to provide your own replication machinery. Having a segmented genome is problematic: each virion must contain an unabridged set of chromosomes. The Cystophages demonstrate that all of this can be done, done even by a phage with a mere dozen genes. Their genome packaging is accurate, precise, and elegant. Their quality control is excellent; nearly every virion that comes off the assembly line is infectious.

Cited references

Butcher, S, T Dokland, P Ojala, D Bamford, S Fuller. 1997. Intermediates in the assembly pathway of the double-stranded RNA virus ϕ6. EMBO J 16:4477-4487.

Gottlieb, P, J Strassman, X Qiao, A Frucht, L Mindich. 1990. *In vitro* replication, packaging, and transcription of the segmented double-stranded RNA genome of bacteriophage ϕ6: Studies with procapsids assembled from plasmid-encoded proteins. J Bacteriol 172:5774-5782.

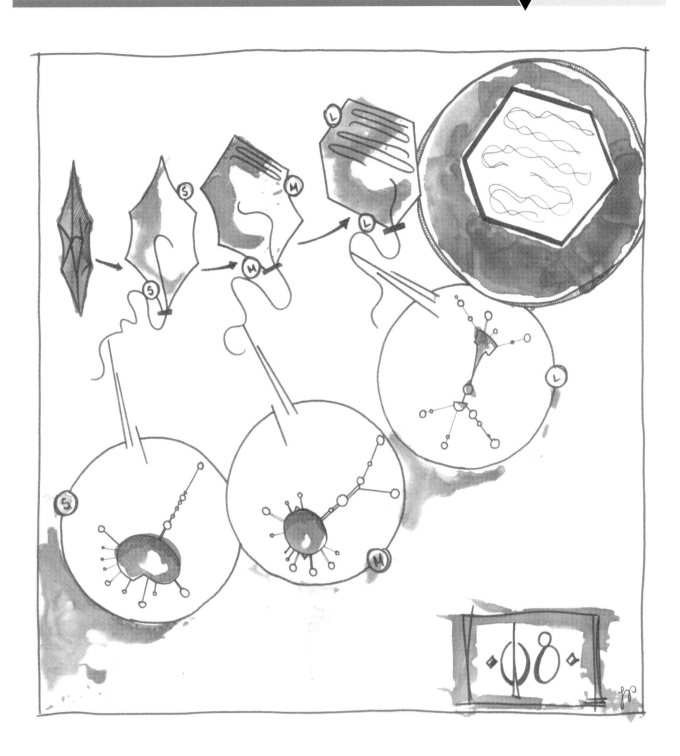

Huiskonen, JT, HT Jäälinoja, JA Briggs, SD Fuller, SJ Butcher. 2007. Structure of a hexameric RNA packaging motor in a viral polymerase complex. J Struct Biol 158:156-164.

Olkkonen, VM, PM Ojala, DH Bamford. 1991. Generation of infectious nucleocapsids by *in vitro* assembly of the shell protein on to the polymerase complex of the dsRNA bacteriophage φ6. J Mol Biol 218:569-581.

Qiao, X, G Casini, J Qiao, L Mindich. 1995. *In vitro* packaging of individual genomic segments of bacteriophage φ6 RNA: Serial dependence relationships. J Virol 69:2926-2931.

Veesler, D, JE Johnson. 2013. Cystovirus maturation at atomic resolution. Structure 21:1266-1268.

Recommended reviews

Mindich, L. 2012. Packaging in dsRNA viruses. Adv Exp Med Biol 726:601-608.

Poranen, MM, MJ Pirttimaa, DH Bamford. 2005. Encapsidation of the segmented double-stranded RNA genome of bacteriophage φ6. In: CE Catalano, editor. *Viral Genome Packaging Machines: Genetics, Structure, and Mechanism*: Springer. p. 117-134.

Acidianus Two-tailed Virus

a Bicaudavirus whose virions complete their assembly after they exit the host cell

Genome
dsDNA; circular
62,730 bp
72 predicted ORFs; 0 RNAs

Encapsidation method
Unknown

Common host
Acidianus convivator

Habitat
Acidic hot springs

Lifestyle
Temperate

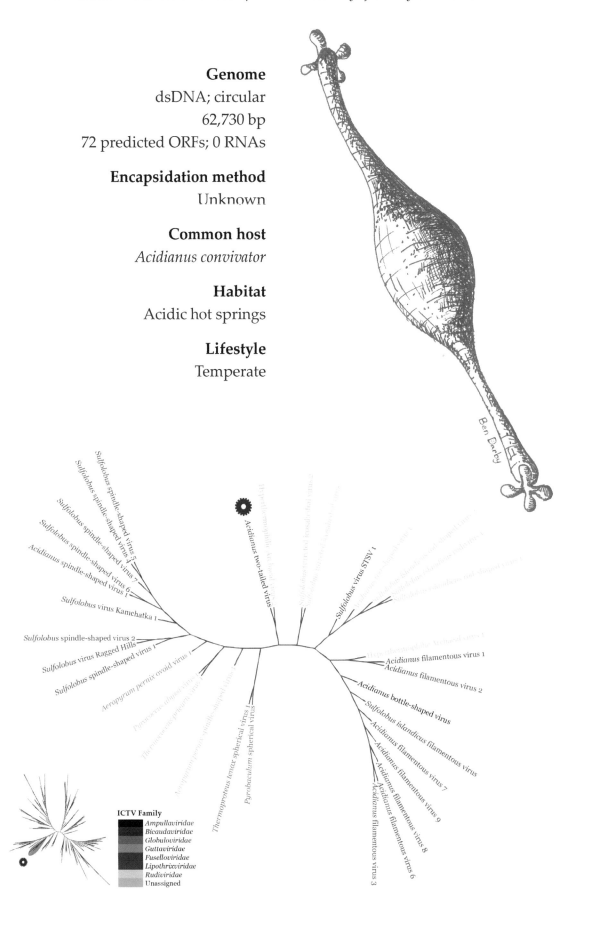

Ben Darby

ICTV Family
Ampullaviridae
Bicaudaviridae
Globuloviridae
Guttaviridae
Fuselloviridae
Lipothrixviridae
Rudiviridae
Unassigned

Sulfolobus spindle-shaped virus 5
Sulfolobus spindle-shaped virus 4
Sulfolobus spindle-shaped virus 6
Sulfolobus spindle-shaped virus 7
Acidianus spindle-shaped virus 1
Sulfolobus virus Kamchatka 1
Sulfolobus spindle-shaped virus 2
Sulfolobus virus Ragged Hills
Sulfolobus spindle-shaped virus 1
Aeropyrum pernix ovoid virus 1
Pyrococcus abyssi virus 1
Thermococcus prieurii virus 1
Aeropyrum pernix spindle-shaped virus 1
Thermoproteus tenax spherical virus 1
Pyrobaculum spherical virus
Acidianus two-tailed virus
Hyperthermophilic Archaeal Virus 2
Sulfolobus turreted icosahedral virus 2
Sulfolobus turreted icosahedral virus 1
Sulfolobus virus STSV1
Sulfolobus islandicus rod-shaped virus 1
Sulfolobus islandicus rod-shaped virus 2
Hyperthermophilic Archaeal Virus 1
Acidianus filamentous virus 1
Acidianus filamentous virus 2
Acidianus bottle-shaped virus
Sulfolobus islandicus filamentous virus
Acidianus filamentous virus 7
Acidianus filamentous virus 9
Acidianus filamentous virus 8
Acidianus filamentous virus 6
Acidianus filamentous virus 3

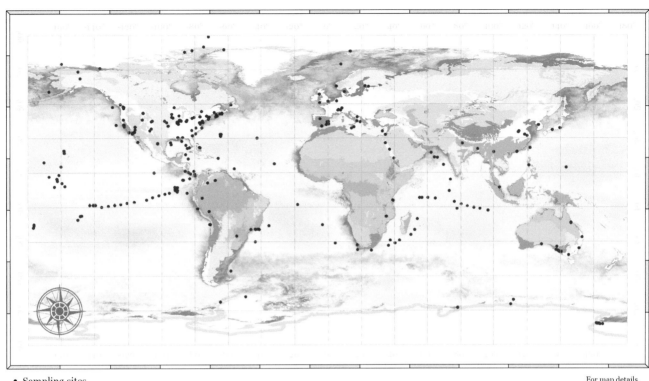

• Sampling sites

For map details,
see Appendix A4.

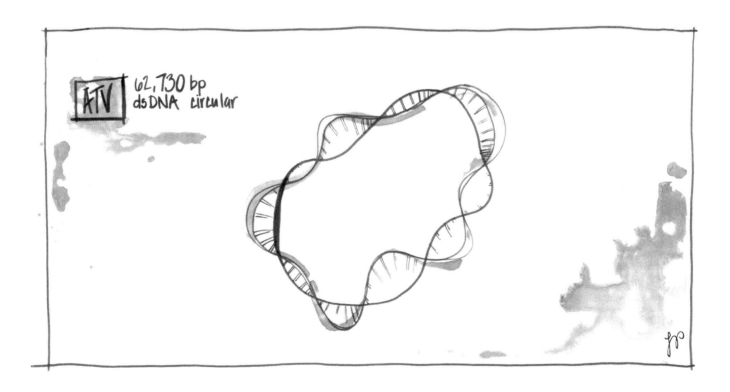

ATV | 62,730 bp
ds DNA circular

Acidianus two-tailed virus

Circular genome

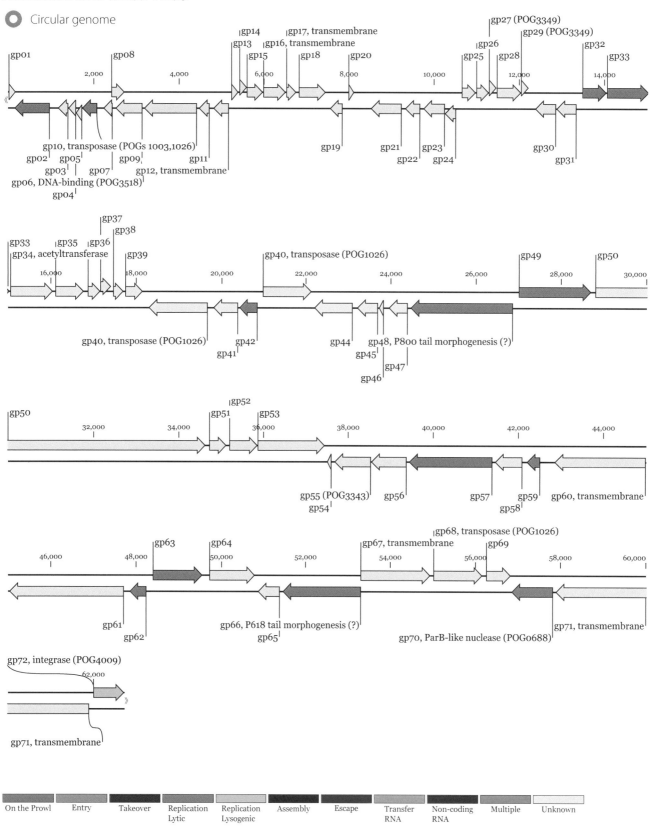

On the Prowl | Entry | Takeover | Replication Lytic | Replication Lysogenic | Assembly | Escape | Transfer RNA | Non-coding RNA | Multiple | Unknown

Virion Grows Long Tails—*Two of Them*—After Leaving Home

Merry Youle

Mise-en-Scène: *By definition, virions are the metabolically inert dispersal form of a phage, but does dramatic restructuring of a virion constitute metabolism? The answer likely depends on your definition of metabolism, but arguably would if the process is driven by energy derived from the hydrolysis of ATP. The argument is weaker if the restructuring is the result of temperature-dependent protein refolding. If the latter, then this morphogenesis may be similar in principle to the familiar structural upheavals triggered by irreversible adsorption to a host. Perhaps the dramatic virion morphogenesis by ATV (see below) is but one point in a continuum that ranges from reversible tail fiber retraction (see page 2-6) to irreversible de novo tail morphogenesis (see page 3-5). If so, one can think of virions as metastable constructs that stably protect the genome within from various environmental insults, yet are primed for precise responses to specific external cues, responses that can include dramatic morphogenesis.*

Every known tailed virion has just one tail. Every one, that is, except those of ATV, the *Acidianus* two-tailed virus. Not only do these virions have two tails, but they actively grow their tails *after* they have left their host cell—no outside energy source or cofactors required. This is extraordinary behavior; except for tail fiber retraction, all other virions observed so far are inert genome packages while in transit.

Extremophilic virion diversity

This eccentric phage resides, along with its archaeal host *Acidianus convivator*, in an acidic hot spring (85–93° C, pH 1.5) at Pozzuoli, Italy—a hellish place (Häring et al. 2005). Such hyperthermophilic environments are home to numerous bizarre viral morphotypes (*see page 7-29*). Among those described by 2003, only 6% were archetypal head-and-tail virions of the *Caudovirales* and about half were filaments or rods. All the rest were surprising structures seen only in archaeal phages, shapes such as spindles (16%), droplets, and bottles (Prangishvili 2003).

ATV virions are part of the spindle-shaped coterie. Their lemon-shaped body averages 243 nm in length and 119 nm across at the center. The two tails triple the average virion length to ~750 nm. Compared to the dimensions of a T4 phage, one of the largest tailed phages, their length is extreme. T4, including its tail and elongated head, measures only ~200 nm.

Survival challenges

Whatever their size and shape, all virions found in these extreme environments face similar survival challenges. A critical one is to find a new host before they are destroyed by the harsh conditions. As soon as progeny virions leave home, the clock starts ticking. Hosts are rare. ATV's apparent strategy is to assemble two long tails that increase its virion length three-fold. As the virion tumbles about, that three-fold increase in length amplifies the swept area nine-fold—significantly increasing the likelihood of colliding with any cells nearby. Is this merely an evolutionary just-so story? That ~50% of the virions in these extreme environments are rods or filaments adds some credibility to this interpretation.

This survival imperative provides a reasonable explanation for why most of the other known phages of hyperthermophilic Crenarchaeota don't lyse their hosts (Peng, Garrett, She 2012). They stay home, living the life of a virocell either as a prophage integrated into the host chromosome or in a non-integrated carrier state. Often such non-lytic lifestyles limit viral replication to a leisurely doubling whenever their host cell divides. However some of these crenarchaeal phages replicate and continuously release virions by an unknown mechanism without killing their host. Some that release their virions by lysis do it in spectacular ways (*see page 7-25*).

Similarly, at 85° C, the host's preferred temperature, new ATV infections lead to stable lysogeny. ATV integrates into the host chromosome and no virions are produced. When the host cell is stressed, for example by a drop in the temperature to 75° C, the ATV prophage excises from the host chromosome, replicates, and converts the virocell into a virion production factory. Host lysis and virion release follow about two days later.

Growing to great lengths

When they first emerge from a host cell, ATV virions are tailless. Tail growth is rapid at 85° C; both tails are fully developed within an hour (Häring et al. 2005). Temperatures below 75° C slow or arrest growth. As the tails lengthen, the virions shrink to about half their original volume (even allowing for the volume of the tails; [Prangishvili et al. 2006]). One cannot help but wonder how ATV virions do this all by themselves, with no outside energy source.

There are two other phages with spindle-shaped virions that assemble a variable length tail: the *Sulfolobus tengchongensis* spindle-shaped viruses (STSV1 and STSV2) from an acidic hot spring in China. They assemble a single tail that measures 130 nm or more. Unlike the case for ATV, their tails are fully developed before the virions are released from the host; no special extracellular acrobatics are required. Although all three phages share 18 genes, some virion structural proteins are exclusive to ATV (Krupovic et al. 2014). Those present in ATV virions, but absent from STSV1 and STSV2 virions, would be suspected of functioning in ATVs unique extracellular morphogenesis. There are four such proteins, two of which warrant particular attention because they contain heptad repeats (i.e., tandem repeats of the same seven amino acid sequence). This protein motif produces the coiled-coil structures used by organisms in all domains of life to build mechanically rigid frameworks such as intermediate filaments (Bagchi et al. 2008).

More evidence? One of ATV's coiled-coil proteins (P800) forms filaments *in vitro* similar to those in the virion tails (Prangishvili et al. 2006). Further, some comparable proteins rearrange their structure by refolding or other processes that would be favored by the higher temperatures in these hot springs (Martin, Gruber, Lupas 2004). Possibly such conformational changes underpin the extracellular growth of ATV's tails.

The other coiled-coil protein (P618) is also guilty by association. It contains an AAA ATPase domain, such as is found in motor proteins. These domains often serve as ATP-requiring chaperones that help assemble or disassemble protein complexes—a potentially useful function when restructuring a virion (Scheele et al. 2011). Virions aren't known to carry ATP, but perhaps ATV is unique in this respect, too. During assembly its large spindles might trap enough ATP from the host's pool to fuel the virion restructuring. This possibility seems worth investigating.

If longer tails improve the chance of finding a host, why aren't the tails even longer? Perhaps tail length is limited by the amount of ATP packaged inside each virion. That some virions might get more than others might explain why some grow longer tails than others. The mechanics of DNA delivery into a host cell through long tail channels might also place a limit on their length.

No other virions are known to alter their structure this markedly outside of their host. Surely this is not a simple thing for a phage to accomplish. However, assembling virions that are ¾ μm long in host cells that are only ~1½ μm in diameter might be downright impossible even for an eccentric archaeal phage.

Cited references

Bagchi, S, H Tomenius, LM Belova, N Ausmees. 2008. Intermediate filament-like proteins in Bacteria and a cytoskeletal function in *Streptomyces*. Mol Microbiol 70:1037-1050.

Häring, M, G Vestergaard, R Rachel, L Chen, RA Garrett, D Prangishvili. 2005. Virology: Independent virus development outside a host. Nature 436:1101-1102.

ACIDIANUS two-tailed virus

~750 nm

243 nm

119 nm

ATVs emerging from the host cell

Krupovic, M, ER Quemin, DH Bamford, P Forterre, D Prangishvili. 2014. Unification of the globally distributed spindle-shaped viruses of the Archaea. J Virol 88:2354-2358.

Martin, J, M Gruber, AN Lupas. 2004. Coiled coils meet the chaperone world. Trends Biochem Sci 29:455-458.

Peng, X, RA Garrett, Q She. 2012. Archaeal viruses—novel, diverse and enigmatic. Sci China Life Sci 55:422-433.

Prangishvili, D. 2003. Evolutionary insights from studies on viruses of hyperthermophilic Archaea. Res Microbiol 154:289-294.

Prangishvili, D, G Vestergaard, M Häring, R Aramayo, T Basta, R Rachel, RA Garrett. 2006. Structural and genomic properties of the hyperthermophilic archaeal virus ATV with an extracellular stage of the reproductive cycle. J Mol Biol 359:1203-1216.

Scheele, U, S Erdmann, EJ Ungewickell, C Felisberto-Rodrigues, M Ortiz-Lombardía, RA Garrett. 2011. Chaperone role for proteins P618 and P892 in the extracellular tail development of *Acidianus* two-tailed virus. J Virol 85:4812-4821.

Recommended review

Prangishvili, D, RA Garrett. 2005. Viruses of hyperthermophilic Crenarchaea. Trends Microbiol 13:535-542.

Acanthamoeba polyphaga Mimivirus

a member of the Mimiviridae that packages its huge genome via an aperture in the center of an icosahedral face

Genome

dsDNA; linear

1,181,549 bp

979 predicted ORFs; 39 RNAs

Encapsidation method

Packaging; T = >972 but <1200 capsid

Common host

Acanthamoeba polyphaga

Habitat

Freshwater; soil; host-associated

Lifestyle

Lytic

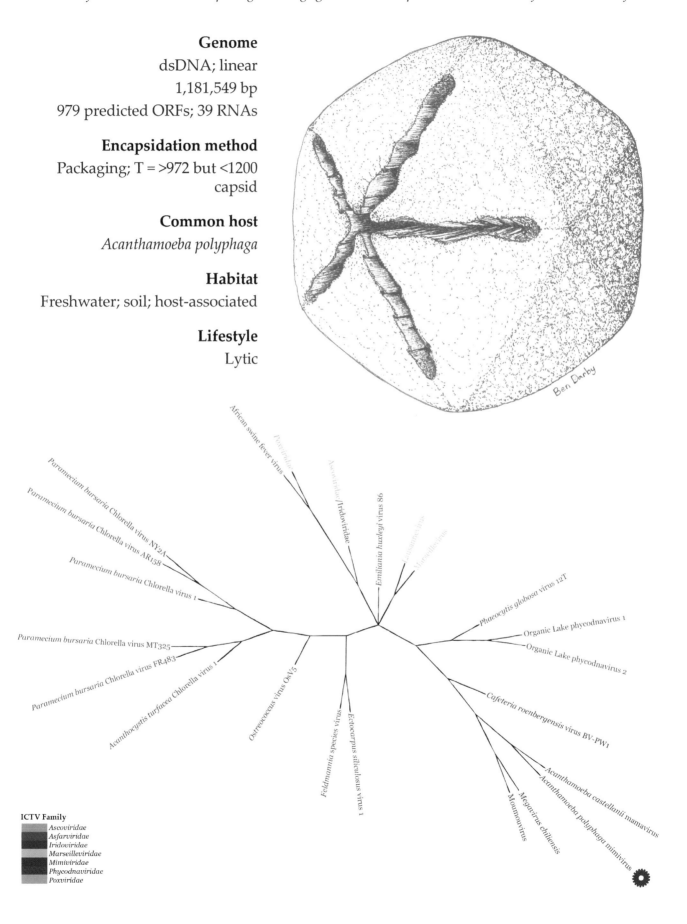

Ben Darby

African swine fever virus

Poxviridae

Ascoviridae/Iridoviridae

Lausannevirus

Marseillevirus

Emiliania huxleyi virus 86

Paramecium bursaria Chlorella virus NY2A

Paramecium bursaria Chlorella virus AR158

Paramecium bursaria Chlorella virus 1

Paramecium bursaria Chlorella virus MT325

Paramecium bursaria Chlorella virus FR483

Acanthocystis turfacea Chlorella virus 1

Ostreococcus virus OsV5

Feldmannia species virus

Ectocarpus siliculosus virus 1

Phaeocystis globosa virus 12T

Organic Lake phycodnavirus 1

Organic Lake phycodnavirus 2

Cafeteria roenbergensis virus BV-PW1

Acanthamoeba castellanii mamavirus

Acanthamoeba polyphaga mimivirus

Megavirus chiliensis

Moumouvirus

ICTV Family

Ascoviridae
Asfarviridae
Iridoviridae
Marseilleviridae
Mimiviridae
Phycodnaviridae
Poxviridae

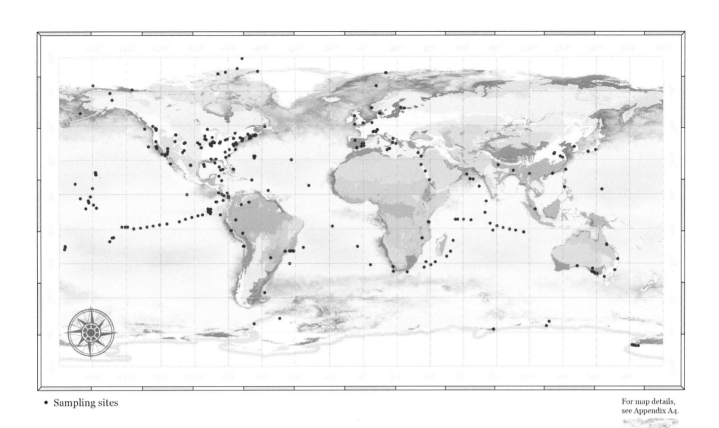

• Sampling sites

For map details,
see Appendix A4.

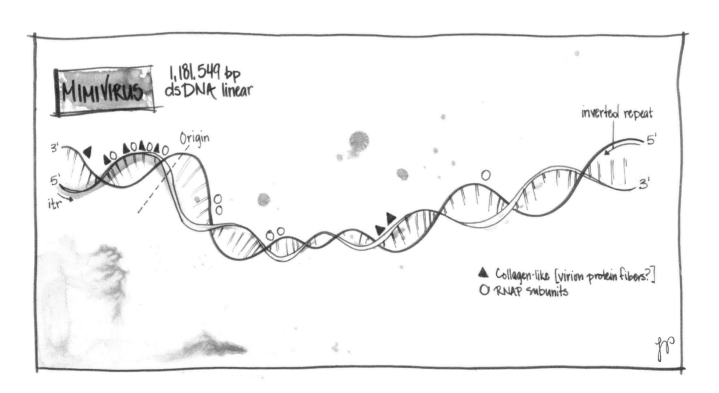

MIMIVIRUS 1,181,549 bp
dsDNA linear

inverted repeat

Origin

3'
5'
itr

5'
3'

▲ Collagen-like [virion protein fibers?]
O RNAP subunits

Acanthamoeba polyphaga Mimivirus

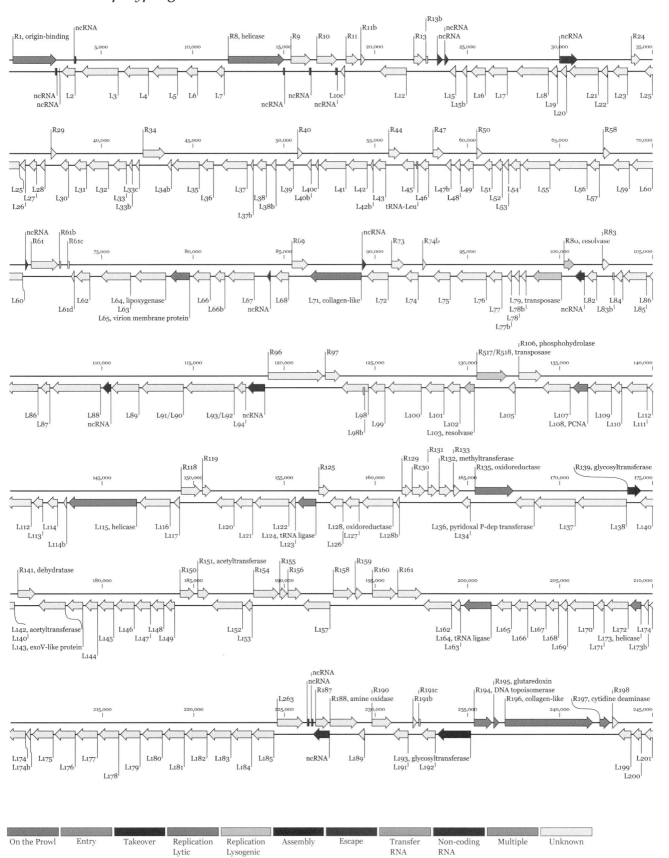

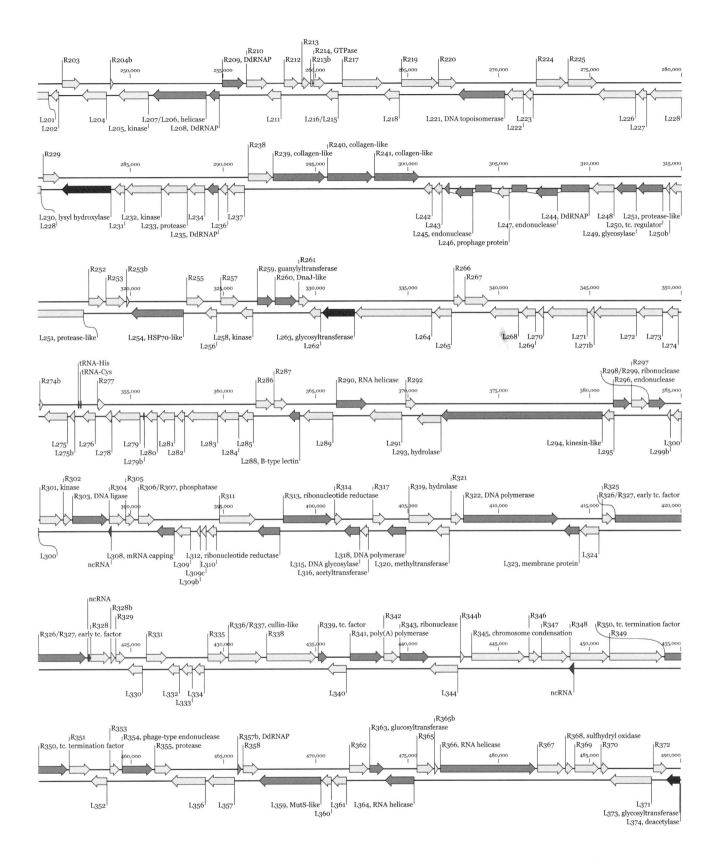

Continued next page

Acanthamoeba polyphaga **Mimivirus** *continued*

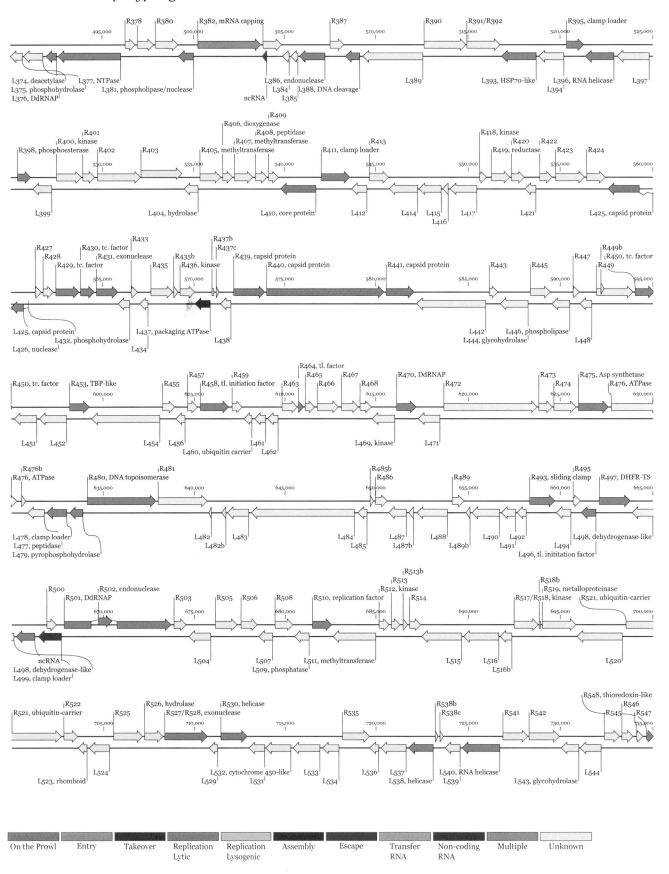

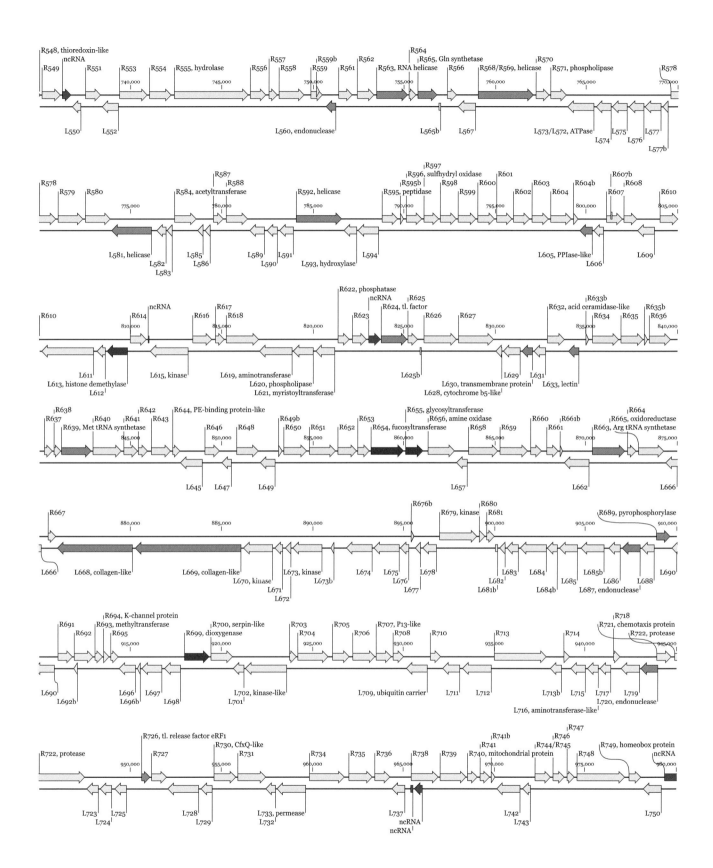

Continued next page

Acanthamoeba polyphaga **Mimivirus** *continued*

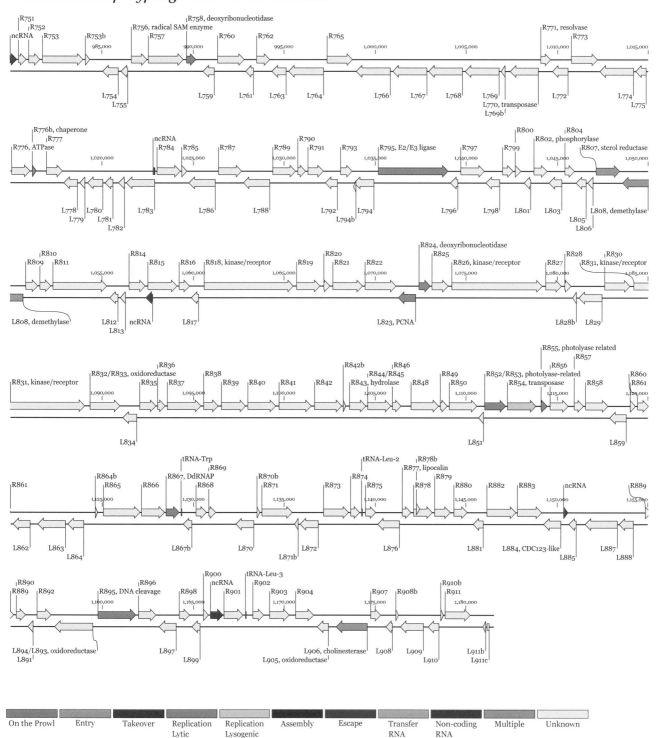

▼

Mimi and the Giant Phage Factory

Merry Youle

Mise-en-Scène: *Although known double-stranded DNA (dsDNA) bacteriophage genomes vary ~45-fold in size, they are all packaged at similar densities within capsids that likewise vary in size. The current record holder, Myophage* Bacillus megaterium *phage G, possesses a genome just under 500 kbp packaged in an icosahedral capsid ~125 nm in diameter. Does 500 kbp represent an inherent limit for bacteriophages? Might an ambitious bacteriophage lineage yet be found to stand tall beside the Mimiviridae? A look at Mimivirus suggests that this would require more than scaling up their capsids and their existing capabilities. Manipulating that much DNA calls for innovations in genome packaging and delivery, as well as in virion assembly and host entry. And besides, bacterial cells do not offer the spacious accommodations and abundant resources of an Amoeba and could scarcely be expected to support replication of such giants.*

Mimivirus ('Mimi') travels luxury class from one host amoeba to the next, its large genome nestled inside a complex virion. These virions are even larger than some bacterial cells—0.75 µm compared to, for example, the mycoplasmas (small **Mollicutes**) that are a mere 0.1 µm across. Mimi's intricate architecture features a lipid membrane, an outer mantle of fibers, and a commodious capsid that affords transport of needed replication machinery along with the genome. Upon arrival in its eukaryote host, Mimi constructs an organized factory for efficient virion manufacture; nearly a thousand come off the assembly line in its 12-14 h replication cycle. Its genome is 1.2 Mbp of DNA, hence larger than those of numerous symbiotic Bacteria and Archaea. While Mimi is in charge, an *Acanthamoeba* that would normally contain about a thousand Mbp (Byers, Hugo, Stewart 1990) supports the synthesis of more than ten thousand Mbp of viral DNA.

The luxury chariot

What sort of a virion can transport and deliver 1.2 Mbp of DNA into an amoeba? An elegant one that is both large and innovative. Mimi is not unique in transporting its genome inside a membranous sac—a tactic used also by some phage families (e.g., *Tectiviridae*, *Corticoviridae*) and by some other eukaryotic phages. However, Mimi's sac also delivers abundant assorted proteins of various sizes and shapes. This lipid pouch is in turn loosely encased by a spacious multi-layered icosahedral protein capsid 0.5 µm across. At one vertex perch-

es a unique proteinaceous 'starfish' structure that will later open into a portal for genome exit. Its five arms radiate out along the icosahedral edges, each one extending almost, but not quite, to the next vertex. As the finishing touch, Mimi covers its surface (except for the starfish) with a thick shag carpet made of thousands of heavily glycosylated protein fibers 125 nm long—an impersonation of a bacterial cell wall. One end of each fiber anchors to the outermost capsid protein lattice, the other sports a globule that makes first contact with a host (Kuznetsov et al. 2010). The net result is a roughly spherical bacterial mimic, 0.75 µm across, that stains Gram-positive. Quite an architectural leap from, for instance, the simple 26 nm icosahedral virion of phage Qβ (*see page 7-8*).

Moving in

Being the right size and suitably camouflaged, Mimi's virions are taken up by the bacterivorous Acanthamoebae via **phagocytosis**. This lands the phage in a host **phagosome**, a most inhospitable compartment that quickly turns acidic and where soon thereafter the intruder is attacked by enzymes. Mimi's virions arrive well prepared. They contain three proteins that protect the phage from the oxidative stress encountered here, while the outermost capsid layer is protected from digestion by its glycosylation. At 2-3 h post infection (PI), the starfish disassembles to allow that vertex to open. The five adjacent icosahedral faces fold back from the vertex to open the 'stargate,' a huge aperture ~400 nm across (Kuznetsov et al. 2010). The viral membrane extrudes through the opening and fuses with the phagosomal membrane to create a massive conduit through which the contents of Mimi's sac exits into the cytoplasm (Suzan-Monti et al. 2007; Zauberman et al. 2008).

Since Mimi replicates in the cytoplasm, it is unable to access the cellular machinery for DNA replication and transcription that resides in the nucleus. Of necessity, it uses some of its numerous genes to encode its own replication machinery. For a quick start, it also arrives carrying 12 proteins needed for transcription, including all five RNA polymerase subunits (Renesto et al. 2006). Other replication factors are among the 114 different protein species stowed in each virion.

The production line

To start protein synthesis rapidly upon arrival, Mimi packages some RNA transcripts in its virion for immediate translation by host ribosomes. Likewise, since it brings along its own RNA polymerase, transcription of viral genes also starts quickly. Intense DNA replication begins as soon as Mimi's genome escapes from the sac. When multiple virions infect a host, each genome establishes its own individual replication center (Mutsafi et al. 2010). By 6 h PI, these replication centers have consolidated to form the core of a single viral factory, complete with an efficient assembly line. DNA replication continues in the central production zone and transcription activity is localized nearby. Successive steps in virion assembly take place as the genome progresses outward from the site of replication. As a result, the factory is organized into concentric zones with particles at the earliest stages closest to the factory center (Mutsafi et al. 2013).

The host provides the membrane needed for the new virions, presumably following instructions from the phage. At 2 h PI, numerous cisternae and vesicles, perhaps derived from rough endoplasmic reticulum, start to form near the nucleus and soon flood the cell. Vesicles fuse into tubules, tubules open into sheets. By 4-5 h PI, hundreds of vesicles have amalgamated and become an inte-

gral part of the viral factory (Mutsafi et al. 2013). Although most icosahedral phages employ temporary protein scaffolding during capsid assembly, Mimi uses a membrane as scaffold and retains that membrane in the mature virion. A new capsid begins when the center of a future protein starfish takes shape on the membrane. Its five arms then quickly grow to a length of ~200 µm (Mutsafi et al. 2013). The rest of the capsid shell fills in around the starfish and the overhanging membrane sheets are trimmed. The surface fibers are added later, in the very last step.

Breaking with tradition

It had been a universal custom, adhered to by all icosahedral phages with a dsDNA genome, that the same vertex serves to package DNA during assembly and to deliver DNA during infection. Now Mimi comes along and defies tradition. Its stargate is a one-way street for genome exiting only. It packages its DNA through a transient 20 nm portal in the center of the icosahedral face directly across the capsid from the starfish (Zauberman et al. 2008). For Mimi, this involves threading a 400 µm long dsDNA molecule into the **procapsid** through a membrane-lined protein shell, then sealing the openings in both layers without a trace (Mutsafi et al. 2013). This is not ordinary viral know-how. Phages such as Mimi may have stolen the mechanism for this from the experts. Bacteria know how to move a lot of DNA through a shrinking membrane portal; this is necessary to ensure a daughter chromosome makes it through the closing septum during cell division (*see page 5-31*). For this they employ various DNA-pumping ATPases that are members of the FtsK-SpoIIIE-HerA superfamily (Errington, Bath, Wu 2001). SpoIIIE even promotes membrane fusion following DNA translocation in Bacteria—a function needed by Mimi, as well (Liu, Dutton, Pogliano 2006). Mimi encodes such an ATPase, as does phage PRD1 with its internal membrane (*see page 3-10*), as do other viruses and phages with lipid membranes (Iyer et al. 2004). These ATPases aren't nearly as powerful as the packaging motors used by tailed phages (*Caudovirales*) (Rao, Feiss 2008), but they're strong enough. Mimi doesn't

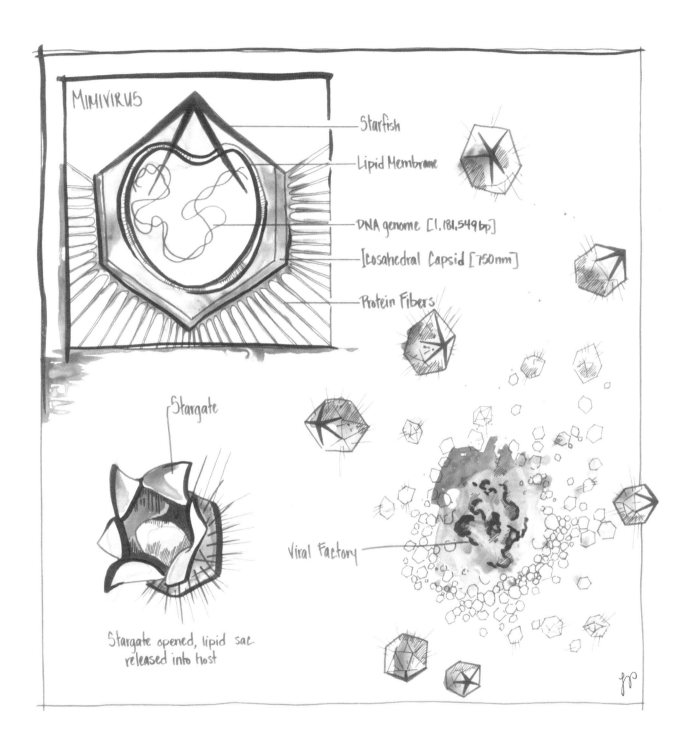

cram its genome into a cramped capsid like those phages, but into spacious quarters where the packaged DNA density is ten-fold less than in the capsids of, for example, Siphophage λ (Kuznetsov et al. 2010).

The future?

Mimi made a big splash when first discovered, proudly winning contests for both the size of its virion and its genomic payload (La Scola et al. 2003). But fame in the viral world is fleeting. Relatives quickly stepped up and established new records: *A. castellanii mamavirus* 1,191,693 bp (Colson et al. 2011) and *Megavirus chilensis* at 1,259,197 bp (Arslan et al. 2011). These were followed by another lineage represented by *Pandoravirus dulcis* and *Pandoravirus salinis* with genomes of at least 1.9 Mbp and 2.8 Mbp, respectively (Philippe et al.

2013). Curiously, so far all of these giants replicate in *Acanthamoeba* hosts. One wonders if that is simply because that's where people are looking for them now, or is something more sinister afoot? An amoeboid conspiracy to achieve world domination with their giant viral henchmen?

Cited references

Arslan, D, M Legendre, V Seltzer, C Abergel, J-M Claverie. 2011. Distant Mimivirus relative with a larger genome highlights the fundamental features of *Megaviridae*. Proc Natl Acad Sci USA 108:17486-17491.

Byers, TJ, ER Hugo, VJ Stewart. 1990. Genes of *Acanthamoeba*: DNA, RNA and protein sequences. J Eukaryot Microbiol 37:17s-25s.

Colson, P, N Yutin, SA Shabalina, C Robert, G Fournous, B La Scola, D Raoult, EV Koonin. 2011. Viruses with more than 1,000 genes: Mamavirus, a new *Acanthamoeba polyphaga* Mimivirus strain, and reannotation of Mimivirus genes. Genome Biol Evol 3:737-742.

Errington, J, J Bath, LJ Wu. 2001. DNA transport in bacteria. Nat Rev Mol Cell Biol 2:538-545.

Iyer, LM, KS Makarova, EV Koonin, L Aravind. 2004. Comparative genomics of the FtsK–HerA superfamily of pumping ATPases: Implications for the origins of chromosome segregation, cell division and viral capsid packaging. Nucleic Acids Res 32:5260-5279.

Kuznetsov, YG, C Xiao, S Sun, D Raoult, M Rossmann, A McPherson. 2010. Atomic force microscopy investigation of the giant Mimivirus. Virology 404:127-137.

La Scola, B, S Audic, C Robert, L Jungang, X de Lamballerie, M Drancourt, R Birtles, J-M Claverie, D Raoult. 2003. A giant virus in amoebae. Science 299:2033-2033.

Legendre, M, S Santini, A Rico, C Abergel, J-M Claverie. 2011. Breaking the 1000-gene barrier for Mimivirus using ultra-deep genome and transcriptome sequencing. Virol J 8:99.

Liu, NJL, RJ Dutton, K Pogliano. 2006. Evidence that the SpoIIIE DNA translocase participates in membrane fusion during cytokinesis and engulfment. Mol Microbiol 59:1097-1113.

Mutsafi, Y, E Shimoni, A Shimon, A Minsky. 2013. Membrane assembly during the infection cycle of the giant Mimivirus. PLoS Pathog 9:e1003367.

Mutsafi, Y, N Zauberman, I Sabanay, A Minsky. 2010. Vaccinia-like cytoplasmic replication of the giant Mimivirus. Proc Natl Acad Sci USA 107:5978-5982.

Philippe, N, M Legendre, G Doutre, Y Couté, O Poirot, M Lescot, D Arslan, V Seltzer, L Bertaux, C Bruley. 2013. Pandoraviruses: Amoeba viruses with genomes up to 2.5 Mb reaching that of parasitic eukaryotes. Science 341:281-286.

Rao, VB, M Feiss. 2008. The bacteriophage DNA packaging motor. Annu Rev Genet 42:647-681.

Renesto, P, C Abergel, P Decloquement, D Moinier, S Azza, H Ogata, P Fourquet, J-P Gorvel, J-M Claverie. 2006. Mimivirus giant particles incorporate a large fraction of anonymous and unique gene products. J Virol 80:11678-11685.

Suzan-Monti, M, B La Scola, L Barrassi, L Espinosa, D Raoult. 2007. Ultrastructural characterization of the giant volcano-like virus factory of *Acanthamoeba polyphaga* Mimivirus. PLoS ONE 2:e328.

Zauberman, N, Y Mutsafi, DB Halevy, E Shimoni, E Klein, C Xiao, S Sun, A Minsky. 2008. Distinct DNA exit and packaging portals in the virus *Acanthamoeba polyphaga* Mimivirus. PLoS Biol 6:e114.

Recommended review

Claverie, J-M. 2006. Viruses take center stage in cellular evolution. Genome Biol 7:110.

Those Magnificent Greasy Phages: A Tribute to Research Collaboration

Dennis H. Bamford[†]

Abstract: *Making a living by studying phages and/or archaeal viruses starts to be an art, particularly in the current funding climate. While managing to do so for the last 40 years, I have experienced that, on many occasions, we have done things not possible using pathogenic viruses. Here I present examples, using lipid-containing bacterial and archaeal viruses, of interesting new insights thus revealed. These include an* in vitro *viral assembly system (phage φ6) as well as atomic-level details of viral genome packaging. Structural data obtained on viruses with a lipid membrane revealed, highly unexpectedly, strong structural similarity between viruses not considered to be related (e.g., adenovirus and phage PRD1). This led to the hypothesis that the total diversity of the viral universe is hugely constrained by the limited protein fold space. Despite the astronomical number of uncharacterized viruses in the biosphere, the great majority of new viral isolates are thus predicted to fall within the already known viral lineages.*

[†]Department of Biosciences and Institute of Biotechnology, University of Helsinki, Finland
Email: dennis.bamford@helsinki.fi
Website: www.helsinki.fi/molecularvirology

I belonged to the fortunate high school student class in Finland that, late Sixties, enjoyed the new edition of our biology textbook by Marja and Veikko Sorsa et al., teachers in the Department of Genetics, University of Helsinki (UH), where the flowers and bees gave way to the central dogma and the establishment of molecular biology in the high school curriculum. However, this paradigm shift was by no means generally observed in many of the schools (and not in a number of university departments, as well). The Department of Genetics established a microbial genetics curriculum in addition to the cyto - and population genetics previously established. After my maturation examination, I managed to get accepted to study genetics and eagerly joined the microbial genetics class. As no serious research in the field was available at our department, the courses were run by senior undergraduate students. I ended up running the microbial genetics advanced class the year after I had been a student. This experience was an eye opener as I certainly had not mastered the subject too well at all. As a consequence many nervous evenings were spent to learn enough not to be immediately spotted as incompetent. In addition Helen Mäkelä (National Public Health Institute, NPHI), when back from the Lederberg laboratory, established microbial genetics to PHI with contacts to Genetics department at the UH.

Those days the Genetics Department was more like a German-type hierarchic organization with a professor (department chair), associate professor, a few lecturers and teaching assistants, but with no well-established research group structure except that Marja Sorsa, when back from UC Davis (USA), assembled a mutagen screening group where I did my master's thesis graduating in 1975. During the last undergraduate year I felt that I was approaching academic status and consequently supposed to do research (how little did I know). The direction was, however, clear and supported by my choice of minor subjects, e.g., chemistry, general microbiology, and biochemistry. My classmates noted, though, that this was not the way to become a qualified biology teacher—the target of the great majority of them—as botanical and zoological studies were required. So the bridges were burned behind me leaving only one way to go. Then what to investigate? The importance of phages in establishing molecular biology was evident, but when considering what was already done with tailed phages, there seemed no space for an ignorant guy from somewhere up North.

At that time our medical school hosted a strong group, led by Kai Simons, studying Semliki Forest virus as a model for membrane biosynthesis. They were later offered the chance to move to EMBL.

Their success encouraged me to join into similar studies. However, it was quickly learned that animal virus work needed infrastructures we certainly could not afford at the Science Faculty. A good idea but not feasible. Having this wish in mind, we learned from the literature that there exist phages with a lipid component and possibly we could afford to study them. The traditional wisdom at the time was that no lipid-containing phages exist (the standard protocol for keeping phage stocks sterile was to add chloroform—not an ideal environment for obtaining membrane-containing phages). Late Sixties the first lipid-containing phage, PM2, was fished out from the Pacific (Espejo, Canelo 1968); then the first dsRNA phage with a membrane envelope, φ6 infecting plant pathogenic *Pseudomonas savastanoi* pv. phaseolicola (previously known as *P. syringae*; Vidaver, Koski, Van Etten 1973), was obtained in the early Seventies. About the same time dsDNA phages PRD1 and PR4, and other related viruses with an internal lipid component, were isolated from several locations round the world (*see page 3-8*); [Olsen, Siak, Gray 1974]). We, meaning me and my fellow student Tapio Palva, currently our genetics professor, decided to ask to obtain φ6 and PR4 from the laboratories where the viruses had been isolated. I found the shipping slip telling that φ6 arrived Helsinki March 6th, 1974.

In addition to Tapio and myself, another fellow student, Kari Lounatmaa who worked at the Department of Electron Microscopy, a university research facility with modern microscopes, went ahead to see whether the viral envelope fused with the host outer membrane—a novelty within the prokaryotic world. Sure enough, our EM studies supported the hypothesis and we pulled together a manuscript that was sent to the *Journal of General Virology*. In a few weeks we got back a letter about the manuscript, thanking us, and asking what possibly was the purpose with it as they did not find any cover letter. Well, the thing was that we did not know that a cover letter was expected. We mailed back that we certainly wanted to submit our work to JGV, where it was published in 1976 (Bamford, Palva, Lounatmaa 1976). Towards the end of the Seventies, we produced a few papers more.

New York, New York

Then something happened. I received an official looking aerogram (those light letters designed for air transportation) in 1979. It was difficult to open and went into several pieces–maybe I was too nervous. Anyway it started by introducing the sender, Dr. Leonard Mindich (Public Health Research Institute, New York; currently at Public Health Research Institute Center, New Jersey Medical School, Rutgers), saying that he had followed our work and proposes that I should come to visit his laboratory with the aim to combine his φ6 genetic system to our electron microscopy techniques. The letter ended with a note that, if interested, the tickets are waiting at the Finnair office (he had obtained NIH support for inviting me). I had been a bit around in Northern Europe but never to US and off I went for a month in Manhattan. Quite a cultural shock, but nevertheless, when the sections were later examined in Helsinki we wrote a paper on the assembly of φ6 (Bamford, Mindich 1980). This turned out to be the last paper in my thesis that I defended late 1980. So, this was my undergraduate and graduate student training, mostly learning by the hard way. In addition, a New York anecdote comes to mind. When Lenny was picking me up from JFK with his rusty Chevrolet Impala, I noticed a not-so-pleasant odor growing stronger. I considered that the car was really rotting badly until I noticed that I had stepped on a juicy dog poop and was spreading it all over the floor. There was I sitting, sweating, and thinking how on earth to open the discussion about the incident. What an introduction of a Finn to New York. Somehow I collected all my courage and pointed down to the floor to show what had happened. I think Lenny welcomed me to New York with laughter in his eyes and I was saved (with some cleaning to do).

There is no such a thing as bad publicity: Introduction to the phage community

The Phage Assembly Meeting was held at Asilomar, California summer 1980. Lenny recommended me to join the conference as its topic was in the heart of our work. I could collect the funds needed to attend. This was my first US meeting which I attended and I was thrilled. I flew over to New York where Lenny joined and we set up to get to Asilomar. The meeting started the same evening we arrived. I selected an aisle seat towards the back of the meeting room just in case I would take a nap due to the long journey (in spite of the interesting talk). The next thing I realized was a lot of noise and me and my chair flat on the floor. The entire audience obviously turned around to find out what was happening. Very sleepy I wished good night to everyone and retired. It was surely noted that there was a delegate also from Finland.

Postdoctoral training 1981-1982

I also got some domestic funding to visit Lenny again as well as an EMBO fellowship to stay in New York as a postdoctoral fellow. These visits opened up my eyes to learn how scientific work was organized in the US—PI-driven, independent research groups and strong funding organizations—a model I immediately copied when back in Finland with subsequent funding from the Academy of Finland. In addition to having a superb time in Lenny's laboratory, I was also socially adopted like a family member. There was always a lot of laughter with Lenny, Lenny's wife Margot, and their three sons. I also adapted to Jewish traditions—a learning journey to a new culture. When looking back, I realize how extremely lucky my study years had been, but how about science? It turned out that we have published close to 20 papers together with Lenny, and several of my students visited Lenny's laboratory. Figure 1 reflects the spirits in Lenny's lab around year 1982.

More learning

There was also another line of learning during the late Seventies. Inspired by the low angle X-ray work done on phage PM2 (Harrison et al. 1971), I asked to collect such data from phage PR4 in the Laboratory of Molecular Biology, Cambridge. I guess with the help of postdoctoral fellows William Earnshaw (currently at the Trust Centre for Cell Biology, University of Edinburgh) and others working at LMB, Sir Aaron Klug agreed for me to use their X-ray facilities. Many films were developed but it took some 25 years before this data was finally utilized in conjunction with the

Figure 1. Snapshots of great moments in the Mindich laboratory around 1982. (A) From the left: Dennis Bamford, Jeffrey Strassman (behind DB), Timothy McGraw, and Leonard Mindich. (B) From the left: Dennis, Jeffrey (behind DB), Timothy, Lenny, and Martin Romantschuk (my first graduate student at the centrifuge).

high-resolution structure determination of phage PRD1 (Bamford et al. 2002; Abrescia et al. 2004; Cockburn et al. 2004) done together with David Stuart's laboratory at the Division of Structural Biology, Oxford University.

From downtown laboratories to the new facilities at the Viikki Campus

Despite the temptation to stay in the US, I was back to Finland in 1983 and started another long journey that ended with the move to new facilities at the Viikki Campus in 1995 (Fig. 2). This strengthened the research infrastructure and brought together the biological community. Our focus was to get deeper mechanistic insights into two lipid-containing bacteriophages (ϕ6 and PRD1) and to bring them out as appreciated model systems for studying virus entry and exit, replication, and virion assembly at medium and high resolution. Successful collaborations have been the necessary and invaluable component of this work. Close to half of our publications have been done as an international collaboration. Listed below are a few examples of the type of work that was carried out to achieve our goals.

During 2000 - 2011 we enjoyed two consecutive six year periods as a National Center of Excellence, the latter one together with Jaana Bamford, Sarah Butcher, and Roman Tuma. We were extremely fortunate that our scientific advisory board included phage wizard Roger Hendrix (Pittsburgh Bacteriophage Institute, University of Pittsburgh, Pennsylvania) for the entire 12 years as well as eminent international members Jack Johnson (Laboratory of Structural Virology at The Scripps Research Institute, La Jolla, California) and Alasdair Steven (Laboratory of Structural Biology, NIH) six years each (Fig. 3). These nominations and my Academy Professor positions (two, each for five years) made it possible to build a strong research unit able to match international challenges.

The complex 66 MDa, membrane-containing virion of PRD1 invited investigation. Its X-ray structure, where the membrane and membrane proteins are visible, was described in two papers in the same issue of *Nature* (Abrescia et al. 2004; Cockburn et al. 2004) and it also provided insights into the size determination of icosahedral viruses. This was a "once in a lifetime" culmination that actually did take more than 15 years to accomplish! There were half a dozen critical determinants that finally allowed the determination of this first, and to my knowledge the only, membrane-containing virion structure with clear signals from the membrane moiety. The critical parameters were that we used receptor binding-deficient particles at high concentration (up to 20 mg/ml). Ultrapure virions were obtained with filter affinity chromatography (MemSep). After purification there was a 30 min window to fix the viruses with very low concentration of glutaraldehyde. The crystallization took place in capillaries so that they were not disturbed when monitoring. Either good looking,

Figure 2. Tilted spirits at the downtown Helsinki laboratory around 1989. Tilted: Dennis Bamford. From the left: Harri Savilahti, Vesa Olkkonen, Martin Romantschuk, and Jarkko Hantula. Behind: Tiina Pakula and additional ladies hiding. (HS, MR, and JH are currently professors, and VO is an institute director.)

shiny crystals or Christmas tree looking, ugly ones were obtained. Surprisingly, the Christmas trees gave the highest resolution. The protocol was a triumph for logistics. Nicola Abrescia (currently a PI at Structural Biology Unit, CIC bioGUNE, CIBERehd, Derio, Spain) landed from Oxford with the last overnight flight. In the meantime the last steps of virus purifications were carried out in Helsinki. I picked him up from the airport to be present when the last affinity chromatography step was done. After an immediate glutaraldehyde addition the capillary crystallization trials were set up. At this time it was past midnight so he got a few hours rest before the 5 am departure to the airport to get to Lyon and further to European troton Radiation Facility at Grenoble to set up data collection with Oxford personnel already present. How many of such trips can a postdoc take before collapse was almost tested.

Let the speed be with us: φ6 and other viruses

φ6, being a dsRNA virus, harbors an RNA-dependent RNA polymerase (RdRp) inside the viral capsid. The polymerase is capable, when using ssRNA as template, to produce a dsRNA molecule (replication) as well as using dsRNA as a template to produce ssRNAs (transcription) by a semiconservative mechanism (*see page 3-23*). Such enzymes had been difficult to produce and purify. Finally, Eugene Makeyev, our graduate student, succeeded and provided Sarah Butcher (then a Marie Curie postdoctoral fellow, now Research Director at the Institute of Biotechnology, UH) with sufficient material to include the polymerase in a crystallization screen. The plates were checked before putting them into the incubator and, hey presto, there were crystals. Our crystallographer colleagues in Oxford (David Stuart's group) were given the alarm and they got the plates in a few days. In a few months, several structures were solved and the paper published in due course (Butcher et al. 2001) with all the amino acids visible in the electron density map and an understanding of the initiation mechanism of RNA synthesis. Sometimes the sun is shining even into the smallest cottage!

This work prompted a number of investigations (e.g., Makeyev, Bamford 2000a; Makeyev, Bamford 2000b; Butcher et al. 2001; Laurila, Makeyev, Bamford 2002; Salgado et al. 2004; Poranen, Koivunen, Bamford 2008; Poranen et al. 2008b; Sarin et al. 2009; Wright et al. 2012) of the polymerase addressing, at the atomic detail, the catalytic mechanisms from preinitiation through initiation and transfer to elongation of this primer-independent polymerase. Interestingly, the closest structural and functional counterpart to the φ6 polymerase

was shown to be the hepatitis C virus polymerase, highlighting the relevance of this information for viral RNA polymerases in general. These results have also led to biotechnical applications (Aalto et al. 2007; Nygårdas et al. 2009; Romanovskaya et al. 2012). This RdRp is now commercially available and used to produce dsRNA for a number of applications, in probing innate immune pathways and RNA interference (RNAi) as examples.

Packing of viral genomes

One of the major steps in virion assembly is genome packaging. In complex icosahedral viruses there is an empty procapsid where the replicated genome is encapsidated with the aid of ATP (NTP) hydrolysis (Catalano et al. 2005). This process is best understood in the tailed dsDNA bacteriophages (*see page 6-5*) and φ6 (*see page 6-25*). Building on the work done in the Mindich laboratory (Gottlieb et al. 1990; Gottlieb et al. 1991; Frilander et al. 1992; Gottlieb et al. 1992), we have obtained mechanistic understanding of the single-stranded RNA (ssRNA) translocation by a φ6 packaging NTPase (Kainov et al. 2003; Mancini et al. 2004a; Mancini et al. 2004b; Lisal et al. 2005; Kainov et al. 2008; El Omari et al. 2013). The packaging motor was structurally related to cellular hexameric helicases, which puts this discovery in a broader perspective. The polymerase and packaging projects were the products of a tight collaboration with David Stuart's re-

Figure 3. The brave Center of Excellence Scientific Advisory Board members. (A) From the left: Dennis Bamford, Jack Johnson and Roger Hendrix photographed around 2001. (B) Roger Hendrix (left) and Alasdair Steven (right) around 2010.

search operation in Oxford, with Jonathan Grimes and Erika Mancini driving the different projects. I quickly counted the number of joint papers published to realize that the number was 36!

Virion assembly

How are biological macromolecular complexes assembled? We have gained insights into such mechanisms by *in vitro* assembly of complex infectious nucleocapsids of phage φ6 using purified protein and nucleic acid constituents (Poranen et al. 2001; Poranen et al. 2008a; Sun, Bamford, Poranen 2012; Sun et al. 2013; Sun, Bamford, Poranen 2014). This was the first time that a complex icosahedral virus was assembled *in vitro*, yielding functional intermediates and end-products with demonstrated infectivity. Remarkably, up to 90% of the precursors were assembled into particles (Sun, Bamford, Poranen 2014).

A more detailed molecular approach to assembly was pursued with George Thomas (Division of Cell Biology and Biophysics, School of Biological Sciences, University of Missouri-Kansas City, Missouri). We employed high-resolution laser Raman spectroscopy to investigate protein conformations and intermolecular interactions in the virions of two phages, φ6 and PRD1, and in their stable subviral particles. The viral complexes were isolated in ultrapure forms, which was essential for the successful application of the Raman spectroscopic probe. While the size and complexity of these supramolecular assemblies ordinarily pose significant challenges, the spectroscopic methods were ultimately successful in revealing details of the viral protein structures and how they are altered step-by-step along the assembly pathways. This collaborative scientific approach led to the publication of 13 primary articles exemplified by the following publications (Bamford et al. 1990; Bamford et al. 1993; Nemecek, Thomas 2009).

In and out of the host cell

Rimantas Daugelavičius (then at Vilnius University, Lithuania, and now at the Department of Biochemistry, Vytautas Magnus University, Lithuania) was one of the few who further de-

veloped electrochemical methods to study bacterial physiology, an important field to supplement more molecularly-oriented studies. When the Soviet Union dissociated, his knowledge and technology were brought to Helsinki and are still operational with a number of students visiting us from Lithuania. Here again our long term cooperation has spawned 20 joint scientific articles on the entry and exit of lipid-containing prokaryotic viruses exemplified by the following publications (Daugelavičius, Bamford, Bamford 1997; Daugelavičius et al. 1997; Poranen et al. 1999; Grahn, Daugelavičius, Bamford 2002a; Grahn, Daugelavičius, Bamford 2002b; Kivelä et al. 2004).

Virion and coat protein structures: Virus evolution

During these years, electron cryomicroscopy was developing fast. It is very suitable for virus-sized particles, especially those with symmetry. Combined with image processing technologies it yields three-dimensional structural models greatly advancing our understanding on virion structures and their dynamics. Stephen Fuller (EMBL, Heidelberg) became our partner to pursue such studies on φ6 and PRD1. Sarah Butcher, after working in Finland as a microbiologist, started as a graduate student with Stephen in 1992 and worked successfully on these viruses (Butcher, Bamford, Fuller 1995; Butcher et al. 1997; Rydman et al. 1999). When structural biology entered Finland in 2001, electron cryomicroscopy was established by purchasing a 200kV instrument as part of the national structural biology package. In the meanwhile, Sarah Butcher had graduated and moved to Helsinki to be the PI responsible for high resolution electron microscopy (currently directing the Structural Biology Program at the Institute of Biotechnology). The work together with Stephen and Sarah, and later with Sarah only, has yielded 10 and 28 primary articles, respectively.

It is not known how virus families are phylogenetically related. Similarly, the origin of viruses is obscure. Using comparative structural analyses and bioinformatics we have observed the same coat protein fold and virion architecture in viruses belonging to a variety of families and infecting hosts from all domains of life (Benson et al. 1999; Abrescia et al. 2008). This led to the hypothesis that all these viruses have a common ancestor dating back to the time before the separation of the current three domains of cellular life (Bamford, Burnett, Stuart 2002; Bamford 2003; Benson et al. 2004; Bamford, Grimes, Stuart 2005; Abrescia et al. 2012). Further support to this relatedness has been obtained by comparing viral packaging ATPases (Strömsten, Bamford, Bamford 2005). When these observations are implemented into virus taxonomy, an extensive revision must be considered (Krupovic, Bamford 2010). Our virus evolution colleagues at Institut Pasteur, Patrick Forterre, David Prangishvili and Mart Krupovic (my previous graduate student), have greatly contributed to the thinking of viral lineages and the consequences it may have to our understanding of the viral universe (Krupovic et al. 2010; Krupovic et al. 2011). This partnership has produced five articles and many more were published with Mart when he was still in Helsinki.

Always adenovirus

Odd things can happen. When attending the first FASEB virus assembly meeting in 1990 at the Vermont Academy (Saxtons River, Vermont), we already had a reasonably good picture of the PRD1 architecture and its biological properties. We also had the methods to obtain large amounts of the major coat protein (MCP) from the virion. Roger Burnett, who worked at the Wistar Institute in Philadelphia (Pennsylvania), proposed to join forces to obtain the high resolution structure of PRD1 MCP. This proposal rose from his wish, after spending a lifetime in studying adenovirus MCP (hexon) and virion structure, to do something very different for a change. After some struggle, as always, I received a phone call one late evening asking me to guess what could be learned from the structure: that the MCP fold of PRD1 was the same as in the adenovirus although there was no detectable sequence similarity between their MCPs (Fig. 4). Roger was doomed; his Midas touch always brought him an adenovirus.

Figure 4. Comparison of the adenovirus (hexon, PDB ID: 1P2Z) and PRD1 (P3, PDB ID: 1HX6) coat proteins (CP). Ribbon diagrams highlight the jelly rolls. Structural data were downloaded from the Protein Data Bank (www.rcsb.org/pdb) and SCOP domains were recolored in Chimera (Pettersen et al. 2004).

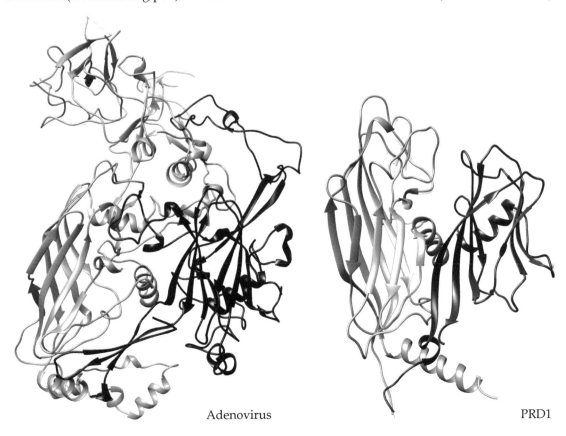

Adenovirus   PRD1

Figure 5. Current senior laboratory members. From the left: Elina Roine, Hanna Oksanen, Janne Ravantti, and Minna Poranen. DB is between Elina and Hanna.

We concluded that this observation was due to divergence and predicted that these viruses have a common ancestor. We also predicted that such an upright double β barrel arrangement in the MCP is shared by many icosahedral viruses, predictions that have proven true (Benson et al. 2004; Abrescia et al. 2012). This architecture applies to viruses infecting cells belonging to all domains of cellular life (Archaea, Bacteria and Eukarya). The paper was published in 1999 (Benson et al. 1999) showing again, as is often the case, that good things take time. This observation was also the initiation to the virus structure-based lineage hypothesis that the enormous virus universe arises from a small number of viral lineages due to the limited protein fold space (Bamford 2003; Benson et al. 2004; Bamford, Grimes, Stuart 2005; Krupovic, Bamford 2008; Krupovic, Bamford 2010; Krupovic, Bamford 2011; Abrescia et al. 2012). We ultimately published a dozen papers together before Roger decided to retire.

The grace of collaborations

Roger Hendrix, when being our Scientific Advisory Board member, could not, in principle, have joint publications with us. However, when the end was approaching, we could not resist the joining of forces to reveal more about tailed phages and archaeal viruses, this resulting in four publications.

Currently we are fortunate to address the issues of asymmetric elements in icosahedral virions with Wah Chiu (Baylor College of Medicine, Structural and Computational Biology and Molecular

Biophysics, Houston, Texas). This is the business end of the virion that has been very difficult to address. However, we now have such information on PRD1 (to be published in due course). In a similar manner Nicola Abrescia (Structural Biology Unit, CIC bioGUNE, CIBERehd, Derio, Spain) and Juha Huiskonen (Structural biology, Oxford University) have been and continue to be involved in structural studies on novel archaeal and bacterial viruses (Aalto et al. 2012; Abrescia et al. 2012; Peralta et al. 2013). Margarita Salas (Centro de Biología Molecular "Severo Ochoa" [CBMSO], Madrid), was supporting us in PRD1 polymerase studies (Caldentey et al. 1992; Caldentey et al. 1993). Nynke Dekker (Kavli Institute of NanoScience, Delft University of Technology) introduced us to single molecule studies on φ6 polymerase (Vilfan et al. 2008), Yi Liu (Department of Physiology, University of Texas Southwestern Medical Center, Dallas, Texas) expanded our interest to *Neurospora crassa* genetics through to polymerase studies (Lee et al. 2009; Lee et al. 2010), and Angela Corcelli (Department of Medical Biochemistry, Medical Biology and Medical Physics, University of Bari) has provided detailed lipid analysis techniques using purified virions without lipid extraction (Vitale et al. 2013). Ilkka Julkkunen (National Institute for Health and Welfare, Helsinki , currently at University of Turku, Finland) and Deying Guo (National Key Laboratory of Virology and Modern Virology Research Center, College of Life Sciences, Wuhan University, China) have been involved in studying innate immunity responses induced by dsRNAs (Jiang et al. 2011). Xiangdong Chen (State Key Laboratory of Virology, College of Life Sciences, Wuhan University, China) revealed with us similarities of a novel archaeal virus with PRD1 (Zhang et al. 2012). Michael Dyall-Smith (Department of Microbiology and Immunology, University of Melbourne, Parkville, Australia) shared archaeal virus SH1 with us to investigate a virus from high salinity environment (Bamford et al. 2005; Porter et al. 2005). Ahron Oren (Department of Plant and Environmental Sciences, The Institute of Life Sciences, and the Moshe Shilo Minerva Center for Marine Biogeochemistry, The Hebrew University of Jerusalem, Israel) contributed archaeal cell knowledge to make order to our archaeal collection (Atanasova et al. 2012). Veijo Hukkanen (University of Turku, Finland) has utilized dsRNA in studying herpes simplex virus (Romanovskaya et al. 2012; Paavilainen et al. 2014).

Currently I have the privilege to have senior members of the laboratory taking charge of scientific projects, teaching, and administrative responsibilities. The major research activities of Minna Poranen focus on the polymerases and virus assembly, while Elina Roine pursues viral genomics, Hanna Oksanen hunts and characterizes novel viruses, and Janne Ravantti provides the computational and evolutionary knowledge (Fig. 5).

Where to go from here?

It gradually became obvious that digging even deeper to atomic detail understanding the viral functions would extend considerably the time needed to accomplish such studies. Time is something that I have started to have a shortage of. On the other hand, I had learned to know Forest Rohwer, a microbial ecologist (Biology Department, San Diego State University, California), who happened to spend the summers collecting coral bacteria in the Caribbean. What a fool I had been spending decades in a smelly microbiology laboratory. Summer of 2006 Michael Rossmann (Department of Biological Sciences, Purdue University, Indiana) organized a meeting on *Structure and Function of Large Molecular Assemblies* in the absolutely beautiful village of Erice, Sicily, where I spoke to support my idea of structure-based viral lineages without being an immediate success. One of the meeting meals was served at the neighboring Trapani Saltern. The hot Mediterranean sun may have affected as I asked my meeting fellows to finish their water/ beer to provide the bottles to carry out an important scientific experiment. We collected both salt and saturated salt water to be brought back home. Sterility was no issue as everything was pickled and only halophilic organisms grew. We isolated both halophilic prokaryotic organisms and their viruses from the Trapani samples with the consequence that a completely novel virion type was discovered. This virus (*Halorubrum* pleomorphic

"I see friends shaking hands, saying, how do you do... ...And I think to myself, what a wonderful world."

Figure 6. You must get intimate with the viruses to reveal their secrets. Dennis communicated this message to the American Society for Virology at their 4th annual meeting (2013) by singing, lyrics provided. (An adaptation of the original figure by Eugene Makeyev.)

virus 1) was a simple membrane vesicle with external spikes and internal protein matrix (Pietilä et al. 2009; Pietilä et al. 2010).

Very peculiarly, the genome of this type of viruses may contain either ssDNA or dsDNA (Roine et al. 2010), and as we learned from further isolates these were globally very common in high salt environments (Pietilä et al. 2012). Encouraged by our findings, we organized several expeditions to further collect specimens from high salt environments populated mainly by Archaea. The spatial sampling locations included the Mediterranean, Eilat, Thailand, and other fine surroundings with the bias to natural beauty (Atanasova et al. 2012). A temporal collection was sampled in Thailand (Atanasova et al., submitted) close by where my daughter happened to live those days—what a revenge to Forest. The only complaint from the laboratory I heard was mutterings about why the boss always got to participate in the expeditions. Well, I think the answer is easy: because he is the boss. Currently we have published some 20 papers on halophilic viruses using them as a test case to support our hypothesis about the low number of structure-based viral lineages. We have also learned a lot about pickled life in general. Sampling of the virosphere has become a rich source of knowledge that has yielded important information and extends to deeper studies of the novel lipid-containing prokaryotic viruses isolated so far and those to be discovered. The journey from φ6 and PRD1 through PM2 and thence to all the viruses in the world has created my close relations to these little creatures (Fig. 6).

Acknowledgements

This article would not have been done without the help of Dr. Katri Eskelin and Dr. Merry Youle. I have enjoyed wonderful collaborations and friendships through the world of science research, probably the only profession that is truly global and international. I have been fortunate to enjoy Academy of Finland support throughout these years. Current grant numbers are 1256518, 1255342, 1271413, and 1255342.

References

Aalto, AP, D Bitto, JJ Ravantti, DH Bamford, JT Huiskonen, HM Oksanen. 2012. Snapshot of virus evolution in hypersaline environments from the characterization of a membrane-containing *Salisaeta* icosahedral phage 1. Proc Natl Acad Sci USA 109:7079-7084.

Aalto, AP, LP Sarin, AA van Dijk, M Saarma, MM Poranen, U Arumäe, DH Bamford. 2007. Large-scale production of dsRNA and siRNA pools for RNA interference utilizing bacteriophage φ6 RNA-dependent RNA polymerase. RNA 13:422-429.

Abrescia, NG, DH Bamford, JM Grimes, DI Stuart. 2012. Structure unifies the viral universe. Annu Rev Biochem 81:795-822.

Abrescia, NG, JJ Cockburn, JM Grimes, GC Sutton, JM Diprose, SJ Butcher, SD Fuller, C San Martin, RM Burnett, DI Stuart, DH Bamford, JK Bamford. 2004. Insights into assembly from structural analysis of bacteriophage PRD1. Nature 432:68-74.

Abrescia, NG, JM Grimes, HM Kivelä, R Assenberg, GC Sutton, SJ Butcher, JK Bamford, DH Bamford, DI Stuart. 2008. Insights into virus evolution and membrane biogenesis from the structure of the marine lipid-containing bacteriophage PM2. Mol Cell 31:749-761.

Atanasova, NS, E Roine, A Oren, DH Bamford, HM Oksanen. 2012. Global network of specific virus-host interactions in hypersaline environments. Environ Microbiol 14:426-440.

Bamford, DH. 2003. Do viruses form lineages across different domains of life? Res Microbiol 154:231-236.

Bamford, DH, JK Bamford, SA Towse, GJ Thomas, Jr. 1990. Structural study of the lipid-containing bacteriophage PRD1 and its capsid and DNA components by laser Raman spectroscopy. Biochemistry 29:5982-5987.

Bamford, DH, RM Burnett, DI Stuart. 2002. Evolution of viral structure. Theor Popul Biol 61:461-470.

Bamford, DH, JM Grimes, DI Stuart. 2005. What does structure tell us about virus evolution? Curr Opin Struct Biol 15:655-663.

Bamford, DH, L Mindich. 1980. Electron microscopy of cells infected with nonsense mutants of bacteriophage φ6. Virology 107:222-228.

Bamford, DH, ET Palva, K Lounatmaa. 1976. Ultrastructure and life cycle of the lipid-containing bacteriophage φ6. J Gen Virol 32:249-259.

Bamford, DH, JJ Ravantti, G Ronnholm, S Laurinavičius, P Kukkaro, M Dyall-Smith, P Somerharju, N Kalkkinen, JK Bamford. 2005. Constituents of SH1, a novel lipid-containing virus infecting the halophilic euryarchaeon *Haloarcula hispanica*. J Virol 79:9097-9107.

Bamford, JK, DH Bamford, T Li, GJ Thomas, Jr. 1993. Structural studies of the enveloped dsRNA bacteriophage φ6 of *Pseudomonas syringae* by Raman spectroscopy. II. Nucleocapsid structure and thermostability of the virion, nucleocapsid and polymerase complex. J Mol Biol 230:473-482.

Bamford, JK, JJ Cockburn, J Diprose, JM Grimes, G Sutton, DI Stuart, DH Bamford. 2002. Diffraction quality crystals of PRD1, a 66-MDa dsDNA virus with an internal membrane. J Struct Biol 139:103-112.

Benson, SD, JK Bamford, DH Bamford, RM Burnett. 1999. Viral evolution revealed by bacteriophage PRD1 and human adenovirus coat protein structures. Cell 98:825-833.

Benson, SD, JK Bamford, DH Bamford, RM Burnett. 2004. Does common architecture reveal a viral lineage spanning all three domains of life? Mol Cell 16:673-685.

Butcher, SJ, DH Bamford, SD Fuller. 1995. DNA packaging orders the membrane of bacteriophage PRD1. EMBO J 14:6078-6086.

Butcher, SJ, T Dokland, PM Ojala, DH Bamford, SD Fuller. 1997. Intermediates in the assembly pathway of the double-stranded RNA virus φ6. EMBO J 16:4477-4487.

Butcher, SJ, JM Grimes, EV Makeyev, DH Bamford, DI Stuart. 2001. A mechanism for initiating RNA-dependent RNA polymerization. Nature 410:235-240.

Caldentey, J, L Blanco, DH Bamford, M Salas. 1993. *In vitro* replication of bacteriophage PRD1 DNA: Characterization of the protein-primed initiation site. Nucleic Acids Res 21:3725-3730.

Caldentey, J, L Blanco, H Savilahti, DH Bamford, M Salas. 1992. *In vitro* replication of bacteriophage PRD1 DNA: Metal activation of protein-primed initiation and DNA elongation. Nucleic Acids Res 20:3971-3976.

Catalano, CE, M Feiss, VB Rao, et al. 2005. Viral Genome Packaging Machines. In: C Catalano, editor. *Viral Genome Packaging Machines: Genetics, Structure, and Mechanism*: Springer US. p. 1-150.

Cockburn, JJ, NG Abrescia, JM Grimes, GC Sutton, JM Diprose, JM Benevides, GJ Thomas, Jr., JK Bamford, DH Bamford, DI Stuart. 2004. Membrane structure and interactions with protein and DNA in bacteriophage PRD1. Nature 432:122-125.

Daugelavičius, R, JK Bamford, DH Bamford. 1997. Changes in host cell energetics in response to bacteriophage PRD1 DNA entry. J Bacteriol 179:5203-5210.

Daugelavičius, R, JK Bamford, AM Grahn, E Lanka, DH Bamford. 1997. The IncP plasmid-encoded cell envelope-associated DNA transfer complex increases cell permeability. J Bacteriol 179:5195-5202.

El Omari, K, C Meier, D Kainov, G Sutton, JM Grimes, MM Poranen, DH Bamford, R Tuma, DI Stuart, EJ Mancini. 2013. Tracking in atomic detail the functional specializations in viral RecA helicases that occur during evolution. Nucleic Acids Res 41:9396-9410.

Espejo, RT, ES Canelo. 1968. Properties of bacteriophage PM2: A lipid-containing bacterial virus. Virology 34:738-747.

Frilander, M, P Gottlieb, J Strassman, DH Bamford, L Mindich. 1992. Dependence of minus-strand synthesis on complete genomic packaging in the double-stranded RNA bacteriophage φ6. J Virol 66:5013-5017.

Gottlieb, P, J Strassman, A Frucht, XY Qiao, L Mindich. 1991. *In vitro* packaging of the bacteriophage φ6 ssRNA genomic precursors. Virology 181:589-594.

Gottlieb, P, J Strassman, X Qiao, M Frilander, A Frucht, L Mindich. 1992. *In vitro* packaging and replication of individual genomic segments of bacteriophage φ6 RNA. J Virol 66:2611-2616.

Gottlieb, P, J Strassman, XY Qiao, A Frucht, L Mindich. 1990. *In vitro* replication, packaging, and transcription of the segmented double-stranded RNA genome of bacteriophage φ6: Studies with procapsids assembled from plasmid-encoded proteins. J Bacteriol 172:5774-5782.

Grahn, AM, R Daugelavičius, DH Bamford. 2002a. Sequential model of phage PRD1 DNA delivery: Active involvement of the viral membrane. Mol Microbiol 46:1199-1209.

Grahn, AM, R Daugelavičius, DH Bamford. 2002b. The small viral membrane-associated protein P32 is involved in bacteriophage PRD1 DNA entry. J Virol 76:4866-4872.

Harrison, SC, DL Caspar, RD Camerini-Otero, RM Franklin. 1971. Lipid and protein arrangement in bacteriophage PM2. Nat New Biol 229:197-201.

Jiang, M, P Osterlund, LP Sarin, MM Poranen, DH Bamford, D Guo, I Julkunen. 2011. Innate immune responses in human monocyte-derived dendritic cells are highly dependent on the size and the 5' phosphorylation of RNA molecules. J Immunol 187:1713-1721.

Kainov, DE, EJ Mancini, J Telenius, J Lisal, JM Grimes, DH Bamford, DI Stuart, R Tuma. 2008. Structural basis of mechanochemical coupling in a hexameric molecular motor. J Biol Chem 283:3607-3617.

Kainov, DE, M Pirttimaa, R Tuma, SJ Butcher, GJ Thomas, Jr., DH Bamford, EV Makeyev. 2003. RNA packaging device of double-stranded RNA bacteriophages, possibly as simple as hexamer of P4 protein. J Biol Chem 278:48084-48091.

Kivelä, HM, R Daugelavičius, RH Hankkio, JK Bamford, DH Bamford. 2004. Penetration of membrane-containing double-stranded-DNA bacteriophage PM2 into *Pseudoalteromonas* hosts. J Bacteriol 186:5342-5354.

Krupovic, M, DH Bamford. 2008. Virus evolution: How far does the double beta-barrel viral lineage extend? Nat Rev Microbiol 6:941-948.

Krupovic, M, DH Bamford. 2010. Order to the viral universe. J Virol 84:12476-12479.

Krupovic, M, DH Bamford. 2011. Double-stranded DNA viruses: 20 families and only five different architectural principles for virion assembly. Curr Opin Virol 1:118-124.

Krupovic, M, S Gribaldo, DH Bamford, P Forterre. 2010. The evolutionary history of archaeal MCM helicases: A case study of vertical evolution combined with hitchhiking of mobile genetic elements. Mol Biol Evol 27:2716-2732.

Krupovic, M, D Prangishvili, RW Hendrix, DH Bamford. 2011. Genomics of bacterial and archaeal viruses: Dynamics within the prokaryotic virosphere. Microbiol Mol Biol Rev 75:610-635.

Laurila, MR, EV Makeyev, DH Bamford. 2002. Bacteriophage φ6 RNA-dependent RNA polymerase: Molecular details of initiating nucleic acid synthesis without primer. J Biol Chem 277:17117-17124.

Lee, HC, AP Aalto, Q Yang, SS Chang, G Huang, D Fisher, J Cha, MM Poranen, DH Bamford, Y Liu. 2010. The DNA/RNA-dependent RNA polymerase QDE-1 generates aberrant RNA and dsRNA for RNAi in a process requiring replication protein A and a DNA helicase. PLoS Biol 8(10): e1000496.

Lee, HC, SS Chang, S Choudhary, AP Aalto, M Maiti, DH Bamford, Y Liu. 2009. qiRNA is a new type of small interfering RNA induced by DNA damage. Nature 459:274-277.

Lisal, J, TT Lam, DE Kainov, MR Emmett, AG Marshall, R Tuma. 2005. Functional visualization of viral molecular motor by hydrogen-deuterium exchange reveals transient states. Nat Struct Mol Biol 12:460-466.

Makeyev, EV, DH Bamford. 2000a. The polymerase subunit of a dsRNA virus plays a central role in the regulation of viral RNA metabolism. EMBO J 19:6275-6284.

Makeyev, EV, DH Bamford. 2000b. Replicase activity of purified recombinant protein P2 of double-stranded RNA bacteriophage φ6. EMBO J 19:124-133.

Mancini, EJ, DE Kainov, JM Grimes, R Tuma, DH Bamford, DI Stuart. 2004a. Atomic snapshots of an RNA packaging motor reveal conformational changes linking ATP hydrolysis to RNA translocation. Cell 118:743-755.

Mancini, EJ, DE Kainov, H Wei, P Gottlieb, R Tuma, DH Bamford, DI Stuart, JM Grimes. 2004b. Production, crystallization and preliminary X-ray crystallographic studies of the bacteriophage φ12 packaging motor. Acta Crystallogr D Biol Crystallogr 60:588-590.

Nemecek, D, GJ Thomas. 2009. Raman spectroscopy in virus structure analysis. In: HG Bohr, editor. *Handbook of Molecular Biophysics: Methods and Applications*. Wiley-VCH. p. 417-456.

Nygårdas, M, T Vuorinen, AP Aalto, DH Bamford, V Hukkanen. 2009. Inhibition of coxsackievirus B3 and related enteroviruses by antiviral short interfering RNA pools produced using φ6 RNA-dependent RNA polymerase. J Gen Virol 90:2468-2473.

Olsen, RH, JS Siak, RH Gray. 1974. Characteristics of PRD1, a plasmid-dependent broad host range DNA bacteriophage. J Gen Virol 14:689-699.

Paavilainen, H, A Romanovskaya, M Nygårdas, DH Bamford, MM Poranen, V Hukkanen. 2014. Innate responses to small interfering RNA pools in astrocytic and epithelial cells during herpes simplex virus infection. Innate Imun doi: 10.1177/1753425914537921.

Peralta, B, D Gil-Carton, D Castaño-Diez, A Bertin, C Boulogne, HM Oksanen, DH Bamford, NG Abrescia. 2013. Mechanism of membranous tunnelling nanotube formation in viral genome delivery. PLoS Biol 11:e1001667.

Pettersen, EF, TD Goddard, CC Huang, GS Couch, DM Greenblatt, EC Meng, TE Ferrin. 2004. UCSF Chimera: A visualization system for exploratory research and analysis. J Comput Chem 25:1605-1612.

Pietilä, MK, NS Atanasova, V Manole, L Liljeroos, SJ Butcher, HM Oksanen, DH Bamford. 2012. Virion architecture unifies globally distributed pleolipoviruses infecting halophilic archaea. J Virol 86:5067-5079.

Pietilä, MK, S Laurinavičius, J Sund, E Roine, DH Bamford. 2010. The single-stranded DNA genome of novel archaeal virus *Halorubrum* pleomorfic virus 1 is enclosed in the envelope decorated with glycoprotein spikes. J Virol 84:788-798.

Pietilä, MK, E Roine, L Paulin, N Kalkkinen, DH Bamford. 2009. An ssDNA virus infecting archaea: A new lineage of viruses with a membrane envelope. Mol Microbiol 72:307-319.

Poranen, MM, SJ Butcher, VM Simonov, P Laurinmäki, DH Bamford. 2008a. Roles of the minor capsid protein P7 in the assembly and replication of double-stranded RNA bacteriophage φ6. J Mol Biol 383:529-538.

Poranen, MM, R Daugelavičius, PM Ojala, MW Hess, DH Bamford. 1999. A novel virus-host cell membrane interaction: Membrane voltage-dependent endocytic-like entry of bacteriophage φ6 nucleocapsid. J Cell Biol 147:671-682.

Poranen, MM, MR Koivunen, DH Bamford. 2008. Nontemplated terminal nucleotidyltransferase activity of double-stranded RNA bacteriophage φ6 RNA-dependent RNA polymerase. J Virol 82:9254-9264.

Poranen, MM, AO Paatero, R Tuma, DH Bamford. 2001. Self-assembly of a viral molecular machine from purified protein and RNA constituents. Mol Cell 7:845-854.

Poranen, MM, PS Salgado, MR Koivunen, S Wright, DH Bamford, DI Stuart, JM Grimes. 2008b. Structural explanation for the role of Mn^{2+} in the activity of φ6 RNA-dependent RNA polymerase. Nucleic Acids Res 36:6633-6644.

Porter, K, P Kukkaro, JK Bamford, C Bath, HM Kivelä, ML Dyall-Smith, DH Bamford. 2005. SH1: A novel, spherical halovirus isolated from an Australian hypersaline lake. Virology 335:22-33.

Roine, E, P Kukkaro, L Paulin, S Laurinavičius, A Domanska, P Somerharju, DH Bamford. 2010. New, closely related haloarchaeal viral elements with different nucleic acid types. J Virol 84:3682-3689.

Romanovskaya, A, H Paavilainen, M Nygårdas, DH Bamford, V Hukkanen, MM Poranen. 2012. Enzymatically produced pools of canonical and Dicer-substrate siRNA molecules display comparable gene silencing and antiviral activities against herpes simplex virus. PloS One 7:e51019.

Rydman, PS, J Caldentey, SJ Butcher, SD Fuller, T Rutten, DH Bamford. 1999. Bacteriophage PRD1 contains a labile receptor-binding structure at each vertex. J Mol Biol 291:575-587.

Salgado, PS, EV Makeyev, SJ Butcher, DH Bamford, DI Stuart, JM Grimes. 2004. The structural basis for RNA specificity and Ca^{2+} inhibition of an RNA-dependent RNA polymerase. Structure 12:307-316.

Sarin, LP, MM Poranen, NM Lehti, JJ Ravantti, MR Koivunen, AP Aalto, AA van Dijk, DI Stuart, JM Grimes, DH Bamford. 2009. Insights into the pre-initiation events of bacteriophage φ6 RNA-dependent RNA polymerase: Towards the assembly of a productive binary complex. Nucleic Acids Res 37:1182-1192.

Strömsten, NJ, DH Bamford, JK Bamford. 2005. *In vitro* DNA packaging of PRD1: A common mechanism for internal-membrane viruses. J Mol Biol 348:617-629.

Sun, X, DH Bamford, MM Poranen. 2012. Probing, by self-assembly, the number of potential binding sites for minor protein subunits in the procapsid of double-stranded RNA bacteriophage φ6. J Virol 86:12208-12216.

Sun, X, DH Bamford, MM Poranen. 2014. Electrostatic interactions drive the self-assembly and the transcription activity of the *Pseudomonas* phage φ6 procapsid. J Virol 88:7712-7116.

Sun, X, MJ Pirttimaa, DH Bamford, MM Poranen. 2013. Rescue of maturation off-pathway products in the assembly of *Pseudomonas* phage φ6. J Virol 87:13279-13286.

Vidaver, AK, RK Koski, JL Van Etten. 1973. Bacteriophage φ6: A lipid-containing virus of *Pseudomonas phaseolicola*. J Virol 11:799-805.

Vilfan, ID, A Candelli, S Hage, AP Aalto, MM Poranen, DH Bamford, NH Dekker. 2008. Reinitiated viral RNA-dependent RNA polymerase resumes replication at a reduced rate. Nucleic Acids Res 36:7059-7067.

Vitale, R, E Roine, DH Bamford, A Corcelli. 2013. Lipid fingerprints of intact viruses by MALDI-TOF/mass spectrometry. Biochim Biophys Acta 1831:872-879.

Wright, S, MM Poranen, DH Bamford, DI Stuart, JM Grimes. 2012. Noncatalytic ions direct the RNA-dependent RNA polymerase of bacterial double-stranded RNA virus φ6 from *de novo* initiation to elongation. J Virol 86:2837-2849.

Zhang, Z, Y Liu, S Wang, et al. 2012. Temperate membrane-containing halophilic archaeal virus SNJ1 has a circular dsDNA genome identical to that of plasmid pHH205. Virology 434:233-241.

Chapter 7: Escape

STORIES

PERSPECTIVE

Enterobacteria Phage λ

the Siphophage that is the source of the paradigm for timed host lysis using a holin-endolysin mechanism

Genome

dsDNA; linear

48,502 bp

74 predicted ORFs; 0 RNAs

Encapsidation method

Packaging; T = 7 capsid

Common host

Escherichia coli

Habitat

Mammalian intestines & sewage

Lifestyle

Temperate

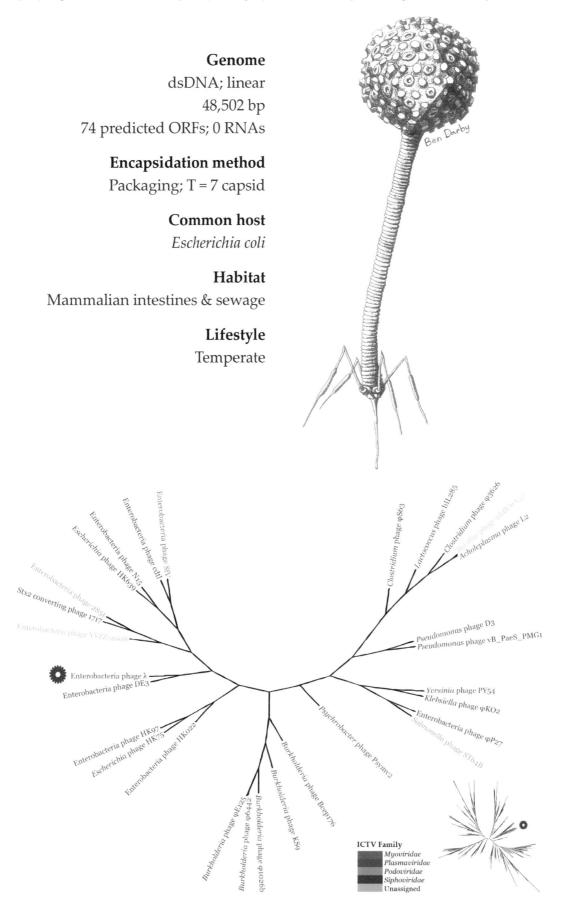

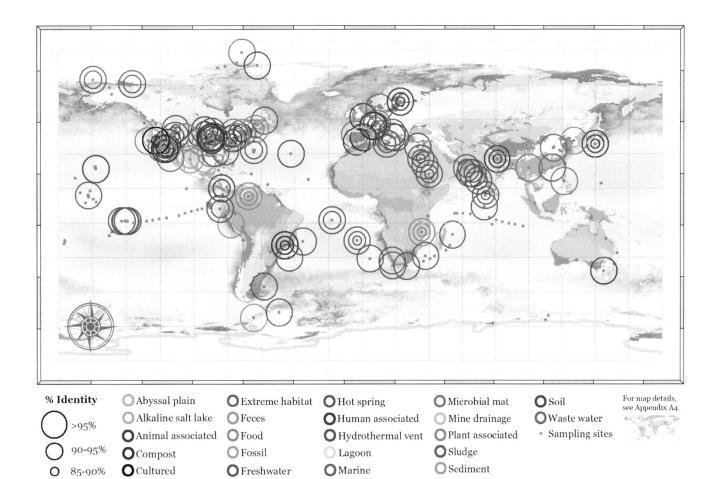

% Identity

- ⬭ >95%
- ◯ 90-95%
- ○ 85-90%

Abyssal plain	Extreme habitat	Hot spring	Microbial mat	Soil
Alkaline salt lake	Feces	Human associated	Mine drainage	Waste water
Animal associated	Food	Hydrothermal vent	Plant associated	• Sampling sites
Compost	Fossil	Lagoon	Sludge	
Cultured	Freshwater	Marine	Sediment	

For map details, see Appendix A4.

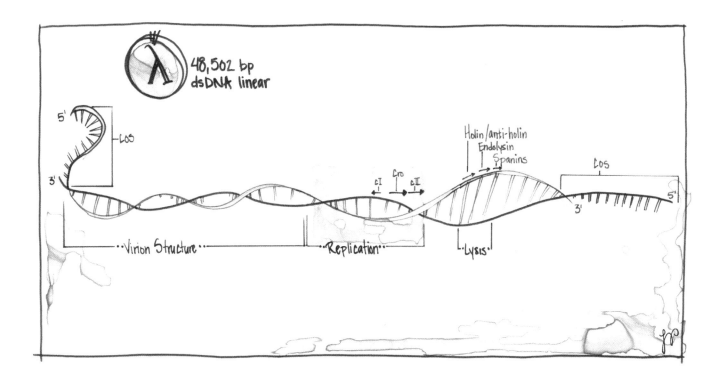

48,502 bp
dsDNA linear

5'

cos

3'

Holin/anti-holin
Endolysin
Spanins

cos

cI   Cro   cII

3'   5'

···Virion Structure···   ···Replication···   Lysis

Enterobacteria Phage λ

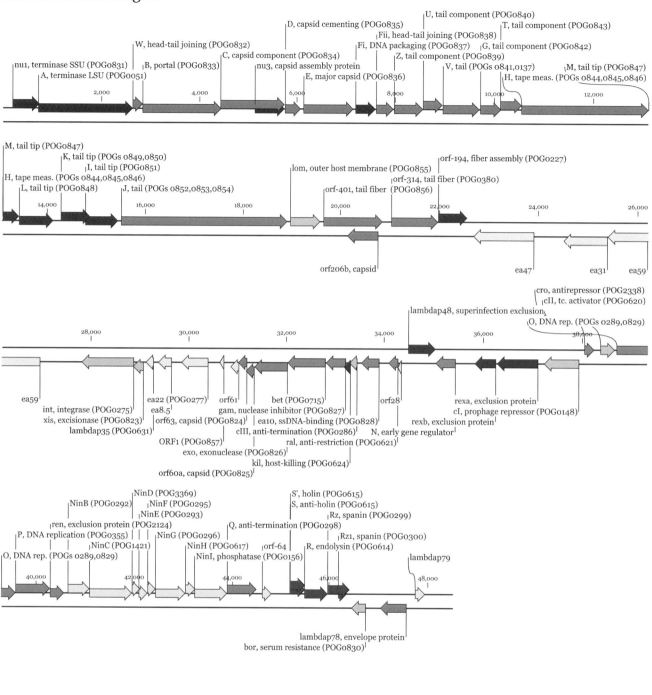

λ Blows Up Host and Splits

Heather Maughan & Merry Youle

Mise-en-Scène: *Every known bacteriophage with a double-stranded genome, either RNA or DNA, employs a holin-endolysin mechanism for host lysis (Young 2002). The holins in this team create lethal lesions in the cell membrane, thereby determining the time of lysis. Although all holins carry out the same functions, they are exceedingly diverse, comprising >30 unrelated protein families. That these families differ not only in amino acid sequence but also in their transmembrane topology demonstrates that there are numerous ways to make a protein that can function as a holin, and phages have found many of them.*

Every second of every day, 10^{24} Bacteria are blown up by phages (Hendrix 2010). There is no place on Earth where you can escape their screams as some emanate from your own gut. Death comes suddenly. Gaping holes open in the cell membrane. Out rush enzymes waiting to attack the cell wall. Proteins spanning the **envelope** demolish the outer membrane. This violence clears the way for the progeny **virions** to escape into the world to wreak similar havoc in new hosts.

Rafts of death

Late in infection, oblivious of its impending doom, *E. coli* swims merrily along while the enemy inside proceeds with its deadly work (Grundling, Manson, Young 2001). While λ virions assemble, the lysis machinery quietly positions itself. *E. coli* presents the typical three-layered cell envelope of Gram-negative Bacteria: cell membrane, **peptidoglycan** layer, and outer membrane. A phage needs to disrupt all three if its progeny virions are to escape. λ carries out a three-step death plan. First come the holes in the cell membrane, each up to a micrometer across, formed by the holin proteins (Dewey et al. 2010). Holin action is tempered by anti-holins until the moment of lysis. Phage λ's holin (S105) contains 105 amino acids that form three α-helical transmembrane domains (TMDs). The TMDs embed in the cell membrane leaving the N-terminus poking into the **periplasm** and the C-terminus facing the cytoplasm. One TMD of each holin interacts with a TMD of a neighboring holin to form dimers. At this stage, the mobile holin dimers—the agents of future destruction—integrate in the cell membrane and accumulate as harmless dispersed dimers that affect neither membrane integrity or cell energetics.

Bacterial life goes on as usual as the lysis timer counts down to zero (White et al. 2011). When the concentration of holin dimers in the cell membrane reaches the critical threshold, then, in less than a minute, dimers aggregate excluding membrane lipids to form large, immobile 'death rafts' (White et al. 2011). The local collapse of the membrane potential in these lipid-depleted regions triggers further aggregation and the formation of a huge hole (>300 nm). Through these gateways stream the waiting **endolysins** that quickly dismantle the second barrier—the peptidoglycan layer. With the peptidoglycan gone, proteins Rz and Rz1 assemble into the spanins that stretch across the cell envelope to disrupt the final barrier, the outer membrane (Berry et al. 2012). The net result? Catastrophic lysis less than a minute after the first hole formed (Grundling, Manson, Young 2001).

Enter the anti-holin

A λ phage equipped with a holin is quite capable of lysing its host at the appropriate time. So why does λ employ an anti-holin, as well? Holin/anti-holin teamwork fine-tunes the timing and ensures sudden, rapid lysis. The genomic cost is minimal as both are encoded by the same S gene, with the anti-holin (S107) being extended by two additional amino acids at its N-terminus. Those two amino acids prevent the first TMD from embedding in the membrane, leaving it in the cytoplasm (Blasi et al. 1990; White et al. 2010). Nevertheless, using its second TMD each anti-holin can form a heterodimer with a neighboring holin in the cell membrane.

As the infection proceeds, both holin homodimers and holin/anti-holin heterodimers accumulate in the membrane. Lysis depends upon the accumulation of sufficient holin homodimers; the anti-holins impede triggering by sequestering some of the holins in inactive heterodimers. Since approximately one anti-holin is made for every two holins, about half of the holins are tied up this way, thereby delaying raft formation. But once enough holin homodimers have accumulated to form the first raft, and thus the first hole, membrane potential collapses, setting off a rapid chain reaction. Now the first TMD of the anti-holins can enter the

membrane. Suddenly all the anti-holins function as holins and all the heterodimers function as homodimers, delivering a swift coup de grâce.

The adaptable timer

The holin is the lysis timer, set to trigger sudden lysis at a particular time. That time is not dictated by the depletion of cellular resources; when lysis is blocked, λ-infected *E. coli* can continue to grow for hours past the usual lysis time and support the

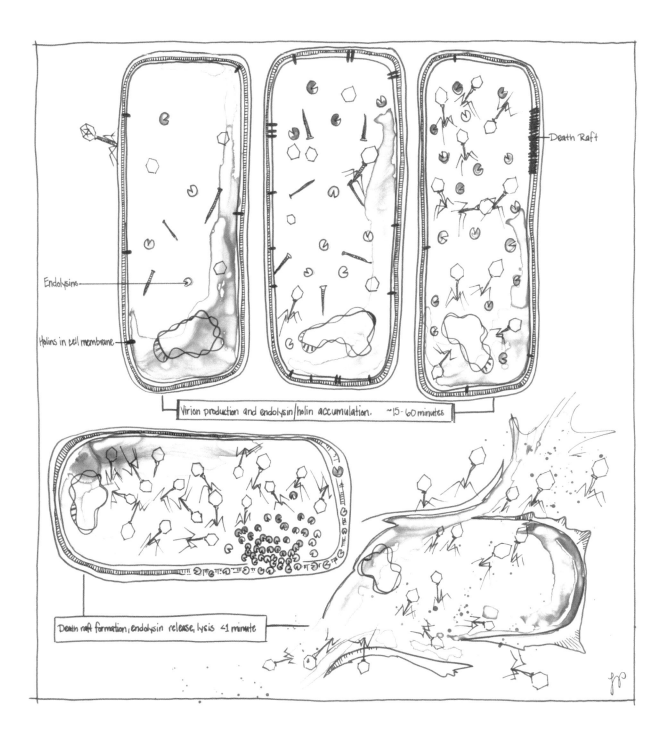

production of at least ten-fold more virions (Reader, Siminovitch 1971; Grundling, Manson, Young 2001). When should λ end an infection and split? It's a numbers game. Lyse later and burst size will be larger but many hosts will escape infection. Lyse earlier and burst size will be smaller, but those virions could already be infecting new hosts, thus increasing the total yield exponentially. As an example, suppose one phage lyses its host after 50 min to release 100 progeny, while a variant phage delays lysis to 3 h and increases its burst size tenfold to 1,000. Suppose further that currently hosts are abundant and half of the progeny successfully launch their own productive infections within 10 min. In this case, the 50 min phage wins. After 3 h, when the late lysis choice has yielded 1,000 progeny, earlier lysis with repeated rounds of replication would have produced $2.5×10^5$ progeny.

In this competitive phage-eat-bacterium world, the phage that wins is the one that lyses at the optimal time to yield the most progeny given the current host abundance. (Even phage Qβ that encodes a single lysis protein knows how to win this game; *see page 7-11*.) Molded by natural selection, λ's holin timer rapidly adapts. A single missense mutation can shorten λ's **latent period** to ~20 minutes, lengthen it to forever, or set it to any time in between (Wang, Smith, Young 2000; Wang 2006). The perpetual fluctuations in host abundance or other limiting factors will favor first one holin mutant, and then another whose lysis time is better suited to the present circumstances.

Quick to adapt, precise in execution, λ remains a world class killer. Listen! Can you hear the doomed Bacteria screaming in *your* gut?

Cited references

Berry, J, M Rajaure, T Pang, R Young. 2012. The spanin complex is essential for lambda lysis. J Bacteriol 194:5667-5674.

Blasi, U, CY Chang, MT Zagotta, KB Nam, R Young. 1990. The lethal lambda S gene encodes its own inhibitor. EMBO J 9:981-989.

Dewey, JS, CG Savva, RL White, S Vitha, A Holzenburg, R Young. 2010. Micron-scale holes terminate the phage infection cycle. Proc Natl Acad Sci USA 107:2219-2223.

Grundling, A, MD Manson, R Young. 2001. Holins kill without warning. Proc Natl Acad Sci USA 98:9348-9352.

Hendrix, RW. 2010. Recoding in bacteriophages. In: JF Atkins, RF Gesteland, editors. *Recoding: Expansion of Decoding Rules Enriches Gene Expression*: Springer. p. 249-258.

Reader, RW, L Siminovitch. 1971. Lysis defective mutants of bacteriophage lambda: Genetics and physiology of S cistron mutants. Virology 43:607-622.

Wang, IN. 2006. Lysis timing and bacteriophage fitness. Genetics 172:17-26.

Wang, IN, DL Smith, R Young. 2000. Holins: The protein clocks of bacteriophage infections. Annu Rev Microbiol 54:799-825.

White, R, S Chiba, T Pang, JS Dewey, CG Savva, A Holzenburg, K Pogliano, R Young. 2011. Holin triggering in real time. Proc Natl Acad Sci USA 108:798-803.

White, R, TA Tran, CA Dankenbring, J Deaton, R Young. 2010. The N-terminal transmembrane domain of lambda S is required for holin but not antiholin function. J Bacteriol 192:725-733.

Young, R. 2002. Bacteriophage holins: Deadly diversity. J Mol Microbiol Biotechnol 4:21-36.

Recommended reviews

Wang, IN, DL Smith, R Young. 2000. Holins: The protein clocks of bacteriophage infections. Annu Rev Microbiol 54:799-825.

Young, R. 2013. Phage lysis: Do we have the hole story yet? Curr Opin Microbiol 16:790-797.

Enterobacteria Phage Qβ

a Leviphage that uses a single multitasking protein to lyse its host by inhibiting peptidoglycan synthesis

Genome
ssRNA; linear
4,215 nt
4 predicted ORFs; 0 RNAs

Encapsidation method
Co-condensation; T = 3
capsid

Common host
Escherichia coli

Habitat
Mammalian intestines

Lifestyle
Lytic

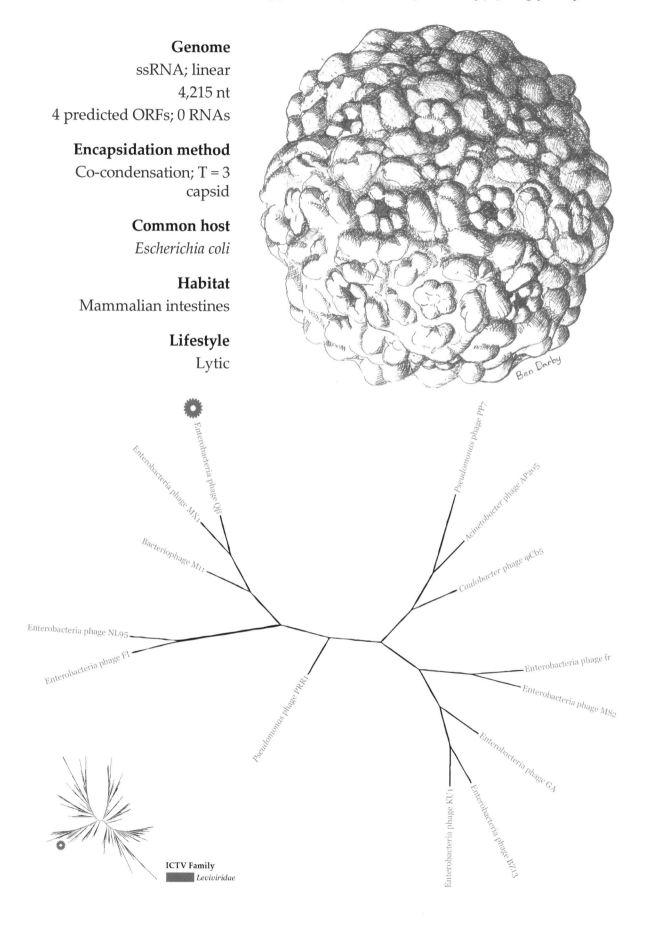

Ben Darby

Enterobacteria phage Qβ

Enterobacteria phage MX1

Bacteriophage M11

Enterobacteria phage NL95

Enterobacteria phage FI

Pseudomonas phage PRR1

Enterobacteria phage KU1

Enterobacteria phage BZ13

Enterobacteria phage GA

Enterobacteria phage MS2

Enterobacteria phage fr

Caulobacter phage φCb5

Acinetobacter phage AP205

Pseudomonas phage PP7

ICTV Family
Leviviridae

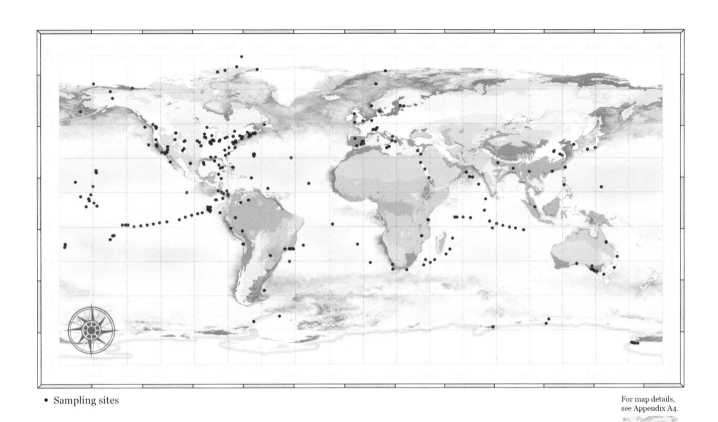

- Sampling sites

For map details,
see Appendix A4.

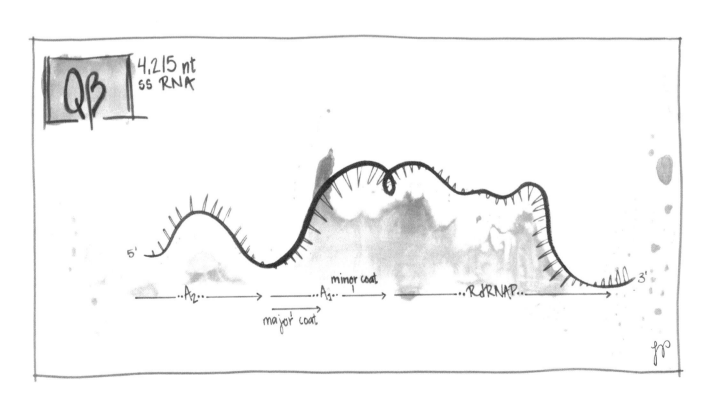

Qβ   4,215 nt
ss RNA

5'   ..A2..   ..A1..   minor coat   ..RdRNAP..   3'
      major coat

Enterobacteria Phage Qβ

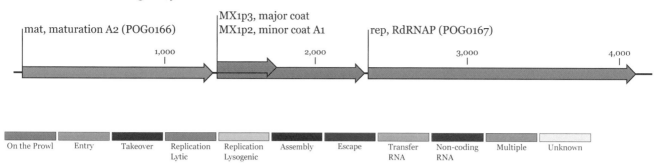

| On the Prowl | Entry | Takeover | Replication Lytic | Replication Lysogenic | Assembly | Escape | Transfer RNA | Non-coding RNA | Multiple | Unknown |

Expert Lysis Timing on a Tight Genomic Budget

Merry Youle

Mise-en-Scène: *As we desperately look for new ways to combat antibiotic-resistant bacterial pathogens, some people have sought to collaborate with the phages—the preeminent experts in bacterial killing. Employing the whole phage to do the killing has already proven applicable in some situations. Alternatively, we can freely steal the phages's anti-bacterial weapons, such as lysins, and wield them ourselves. The 'protein antibiotics' made by phages such as Qβ and φX174 are particularly attractive candidates for this. Not only do they target bacterial peptidoglycan metabolism, but these enzyme inhibitors provide novel approaches for antibiotic development.*

Are you a frugal phage shopping for the ideal tool for host lysis? The holin/**endolysin** system, popular with the *Caudovirales*, is hard to beat (*see page 7-5*). There death is sudden; lysis timing is precise, yet adaptable. But the genomic price is high, too high for some phages. It requires encoding two proteins—an enzyme to digest the **peptidoglycan** cell wall and a holin enabler—and often more. For phage Qβ, having a mere three ORFs in its genomic budget, encoding two proteins to handle host lysis would be preposterous; even the cost of one dedicated to that function would be prohibitive. Instead Qβ relies on one multitasking protein to handle lysis and several other chores.

The mechanism of lysis

Bacteria are full of themselves, so much so that they are hypertonic relative to their environment. If their cell wall is compromised, they burst open due to osmotic forces. As they grow, they maintain cell wall strength by continually intercalating new peptidoglycan (PG in the illustration) into the existing meshwork; when they divide, they add new cell wall at the septum separating the daughter cells. β-lactam antibiotics, such as penicillin, kill growing Bacteria by interfering with peptidoglycan cross-linking, thereby undermining newly-formed cell wall. Qβ exploits the same bacterial vulnerability, but does so by inhibiting the enzyme that catalyzes the first step in peptidoglycan synthesis, MurA (Bernhardt et al. 2001). The inhibitor is the Qβ maturation protein (A_2). This 'protein antibiotic' single-handedly inhibits peptidoglycan synthesis in *E. coli* (Reed et al. 2013). Death by cell wall rupture follows ~20 minutes after peptidoglycan synthesis ceases in rapidly growing cells. Non-growing **stationary phase** cells are immune to A_2 and β-lactam antibiotics alike. This is irrelevant to Qβ as its receptor is on the F **pilus**, and *E. coli* does not produce F pili when in stationary phase.

The cost of lysis

Qβ spends 1266 nt—30% of its genome—to 'purchase' its multitasking protein A_2, but that 30% is well spent. A_2 is responsible for several essential functions. One copy is present in each virion, strategically interposed between genome and environment. There it protects the single-stranded genomic RNA from exogenous RNases. To launch an infection, it recognizes and adsorbs to the specific receptor on the host's F pilus, then accompanies the RNA genome into the cytoplasm (Kozak, Nathans 1971). Later it participates in the assembly of progeny virions (Dykeman et al. 2011) and then lyses the host. That's a lot of services from 1266 nucleotides!

The time of lysis

Host lysis follows in the wake of the suppression of peptidoglycan synthesis, and Qβ suppresses peptidoglycan synthesis by inhibiting the host's enzyme MurA. Each molecule of A_2 can inhibit one copy of MurA. Since rapidly growing *E. coli* cells contain ~400 molecules of MurA, Qβ could halt host peptidoglycan synthesis with only ~400 copies of A_2 available (Reed et al. 2013). Cumulative effects or reduced MurA levels during the infection would mean that fewer copies might still do the job. This gives Qβ control over the time of host lysis. The more swiftly it accumulates A_2, the sooner MurA will be inhibited, the sooner the host will be lysed, and the smaller will be the burst size. Qβ's genomic RNA sets the lysis time by controlling the rate of A_2 translation.

How does Qβ regulate the rate of production of A_2? Different quantities of the various phage proteins are needed. For instance, packaging of each new genome into a virion requires 180 copies of the capsid protein but just one copy of A_2. Nevertheless, having only three ORFs, Qβ can't afford to hire a dedicated regulatory protein to govern

translation rates. Instead, this task is carried out by the genomic RNA itself.

During an infection, new **positive-sense** single-stranded RNA (ssRNA) genomes are synthesized that serve also as **polycistronic** messenger RNAs (mRNAs). Since there is one copy of each gene on each mRNA, the challenge here is to translate some ORFs more frequently than others. Because the A_2 gene is located near the end of this genomic RNA (aka mRNA) that is synthesized first, a ribo-

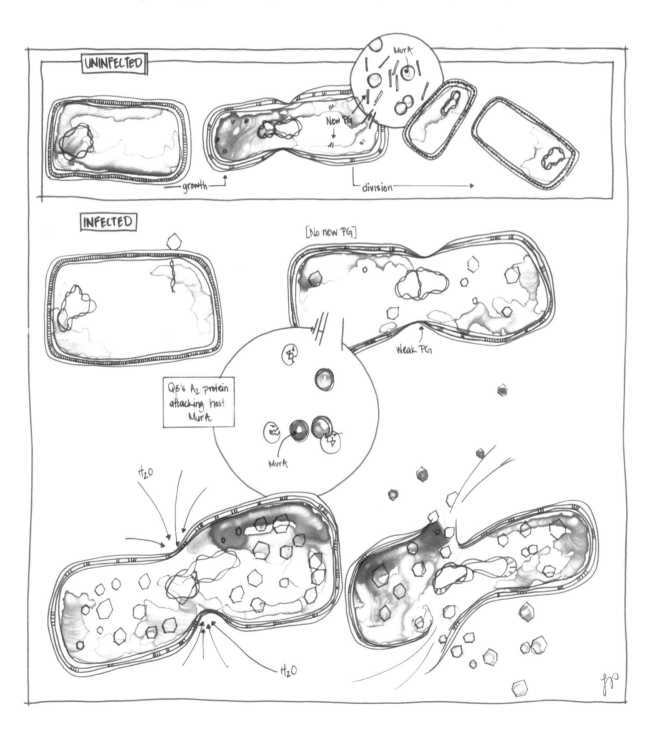

some can bind and initiate translation of A$_2$ early, before the entire RNA chain has been completed. As the nascent Qβ RNA strand grows, it folds into a complex secondary structure that blocks other ribosomes from the A$_2$ initiation site, but allows them access to the start site for the capsid protein further along the mRNA (Beekwilder et al. 1996). This tactic ensures that only one copy of A$_2$—but sometimes two—will be translated from each new genome, yet allows the ribosomes to busily translate the 180 capsid proteins needed for each virion (Hindley, Staples 1969).

The available minority

Significantly, after translation one copy of A$_2$ remains bound to the viral RNA and participates in subsequent virion assembly (Reed et al. 2013). As a result, most of the A$_2$ present in the cell is associated with virions and not free to inhibit MurA. Lysis is the work of those occasional extra copies of A$_2$ that are available for action. Depending on the conditions, Qβ infection can yield a thousand or more progeny virions (García-Villada, Drake 2012), each with one copy of A$_2$ on board. Not surprisingly, by the time peptidoglycan synthesis halts, an infected *E. coli* contains not just 400 A$_2$ proteins, but ~1200 (Reed et al. 2013).

Adapting the time of lysis

The optimal lysis time is continually changing due to fluctuating host abundance and other environmental conditions. Lyse later and the burst size will be larger; lyse earlier and burst size will be smaller. Which is better? When hosts are abundant, earlier lysis can yield more progeny over time. Choose a late lysis time and virions accumulate linearly during the latent period; lyse early, and the first generation virions can already be infecting new hosts, thus multiplying exponentially. Qβ can readily adjust its lysis timer by altering the rate at which inhibiting A$_2$ proteins accumulate. That rate depends directly on how quickly and effectively genome folding blocks ribosome access to the A$_2$ initiation site. A single missense mutation that disrupts intramolecular base pairing in a key position can slow RNA folding or shield that site less effectively. When such a mutant often allows two or three ribosomes to initiate A$_2$ synthesis rather than just one, lysis occurs much sooner and burst size is markedly reduced (Reed et al. 2013).

There you have it: even Qβ's on-the-cheap mechanism can rapidly adapt to optimize lysis time. Impressive!

Cited references

Beekwilder, J, R Nieuwenhuizen, R Poot, Jv Duin. 1996. Secondary structure model for the first three domains of Qβ RNA. Control of A-protein synthesis. J Mol Biol 256:8-19.

Bernhardt, TG, IN Wang, DK Struck, R Young. 2001. A protein antibiotic in the phage Qβ virion: Diversity in lysis targets. Science 292:2326.

Dykeman, E, N Grayson, K Toropova, N Ranson, P Stockley, R Twarock. 2011. Simple rules for efficient assembly predict the layout of a packaged viral RNA. J Mol Biol 408:399-407.

García-Villada, L, JW Drake. 2012. The three faces of riboviral spontaneous mutation: Spectrum, mode of genome replication, and mutation rate. PLoS Genet 8:e1002832.

Hindley, J, D Staples. 1969. Sequence of a ribosome binding site in bacteriophage Qβ-RNA. Nature 224:964-967.

Kozak, M, D Nathans. 1971. Fate of maturation protein during infection by coliphage MS2. Nat New Biol 234:209-211.

Reed, CA, C Langlais, N Wang, R Young. 2013. A2 expression and assembly regulates lysis in Qβ infections. Microbiology 159:507-514.

Weiner, AM, K Weber. 1971. Natural read-through at the UGA termination signal of Qβ coat protein cistron. Nature 234:206-209.

Zheng, Y, DK Struck, R Young. 2009. Purification and functional characterization of φX174 lysis protein E. Biochemistry 48:4999-5006.

Recommended review

Bernhardt, TG, I-N Wang, DK Struck, R Young. 2002. Breaking free: "Protein antibiotics" and phage lysis. Res Microbiol 153:493-501.

Enterobacteria Phage f1

an Inophage whose progeny extrude from the host & use parental coat proteins parked in the membrane

Genome

ssDNA; circular

6,407 nt

10 predicted ORFs; 0 RNAs

Encapsidation method

Co-condensation

Common host

Escherichia coli

Habitat

Mammalian intestines & sewage

Lifestyle

Non-lytic

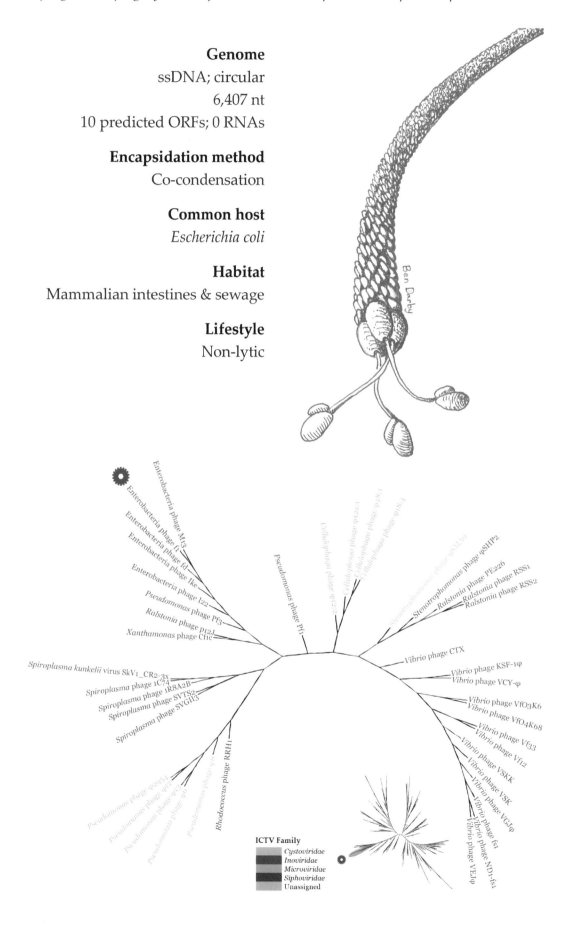

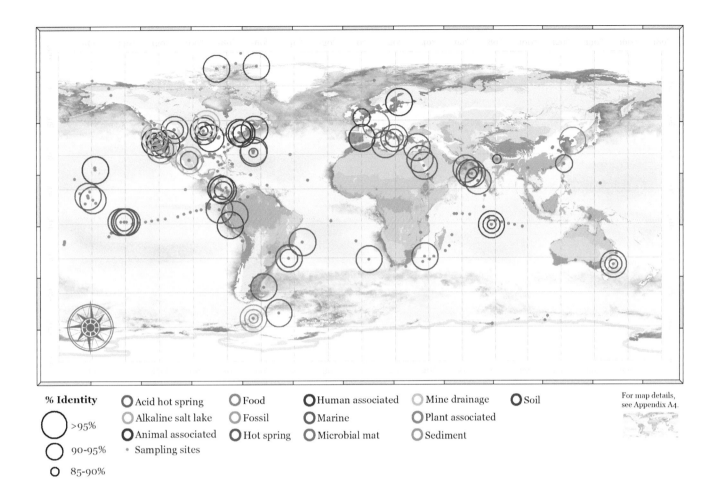

% Identity

O >95%

O 90-95%

o 85-90%

O Acid hot spring
O Alkaline salt lake
O Animal associated
• Sampling sites

O Food
O Fossil
O Hot spring

O Human associated
O Marine
O Microbial mat

O Mine drainage
O Plant associated
O Sediment

O Soil

For map details,
see Appendix A4.

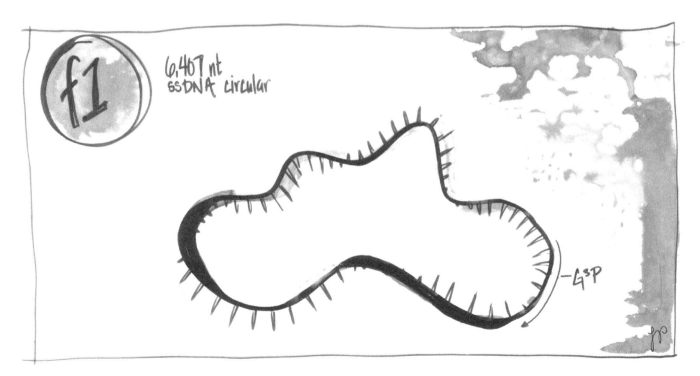

f1

6,407 nt
ssDNA circular

–G³P

Enterobacteria Phage f1

Circular genome

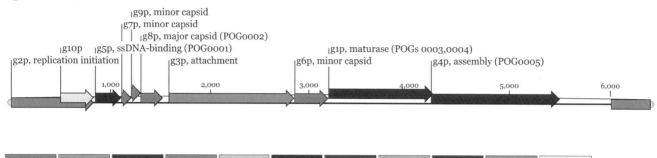

f1: A Pilus Phage

Heather Maughan & Merry Youle

Mise-en-Scène: *The Ff phages are a closely related group of filamentous phages (e.g., f1, fd, and M13) that coexist amiably with their host. They replicate perpetually as **episomes**, and their progeny depart without cell lysis.*

A respectable male *E. coli* moves through the gut, and begins to feel frisky. As its pick-up line, it extends a long conjugative F pilus from its surface to hook up with a female. Thus preoccupied, it is unaware of an f1 phage scouting for **pili** nearby. Once f1 catches a pilus tip, it attaches and holds on for a free ride to its target. As *E. coli* retracts the pilus, it unwittingly escorts f1 to its **secondary receptor** in the **periplasm**. This *E. coli* has been fooled, bringing home an infection instead of a mate.

The single-stranded DNA genome of f1 phages travels inside a skinny cylindrical virion 760-900 nm long and only 4.3-6.3 nm in diameter (Marvin, Hohn 1969). Its simple protein shell is built from 2700 copies of the α-helical major coat protein (g8p) arranged in a closely packed helical array similar to overlapping scales on a fish (Glucksman, Bhattacharjee, Makowski 1992). When entering a host, this frugal phage deposits these coat proteins in the cell membrane (CM) for reuse later to coat progeny as they exit. The two virion ends are adorned with different sets of minor coat proteins, those necessary for entering a host at the 'distal' end and those for exiting at the 'proximal' end. Phage f1 exits quietly without killing its host, using a process that mirrors its entry.

Trailing a pilus to the door

The distal end of the virion sports 3-5 copies of g3p, a multi-tasking protein that contacts the receptors during infection (Gray, Brown, Marvin 1981; Rakonjac et al. 2011) and forms a pore in the CM for genome delivery (Glaser-Wuttke, Keppner, Rasched 1989). Each g3p contains three domains (the N-terminal D1, middle D2, and C-terminal D3) that act in succession during infection as the phage worms its way inside (Marvin 1998).

While f1 is on the prowl in the gut, the g3p N-terminus is exposed to the environment with all three of its domains safely tucked in and held close together, the short flexible linker regions between them forming relaxed loops. D2 acts first by attaching to the tip of a passing pilus (Lubkowski et al. 1999; Deng, Perham 2002). As *E. coli* retracts the pilus, the hitchhiking phage passes through the outer membrane (OM). When D2 grabs the pilus this frees the receptor-binding domain (D1) to dangle with its binding site exposed (Eckert et al. 2005). As the first end of the virion enters the periplasm, f1 peeks under the OM and fishes with D1 for its secondary receptor: the C-terminal domain of TolA, a periplasm-spanning bacterial protein (Holliger, Riechmann 1997; Riechmann, Holliger 1997). D1 binds TolA, which in turn frees D3 and allows it to contact the CM for the next step—DNA entry.

Dissolution on entry

Now f1 is poised to thread its DNA into the cell through a CM pore formed cooperatively by the D3 domains of the multiple g3p proteins at hand (Jakes, Davis, Zinder 1988; Glaser-Wuttke, Keppner, Rasched 1989). Unlike the case for most phages, as f1's genome enters the cell it does not leave its capsid at the door, nor does it bring the capsid along with it into the cell. Instead, as the DNA enters, the capsid disassembles with the assistance of host proteins—the rate-limiting step for phage infection (Click, Webster 1998). Phage f1 stashes monomers of the major coat protein (g8p) and some of the minor capsid proteins in the CM for retrieval and reuse by its progeny as they emerge.

Replication

With its small genome, f1 relies on host proteins for many essential functions including replication. Diverted host enzymes convert the phage's ssDNA into a double-stranded template that serves for

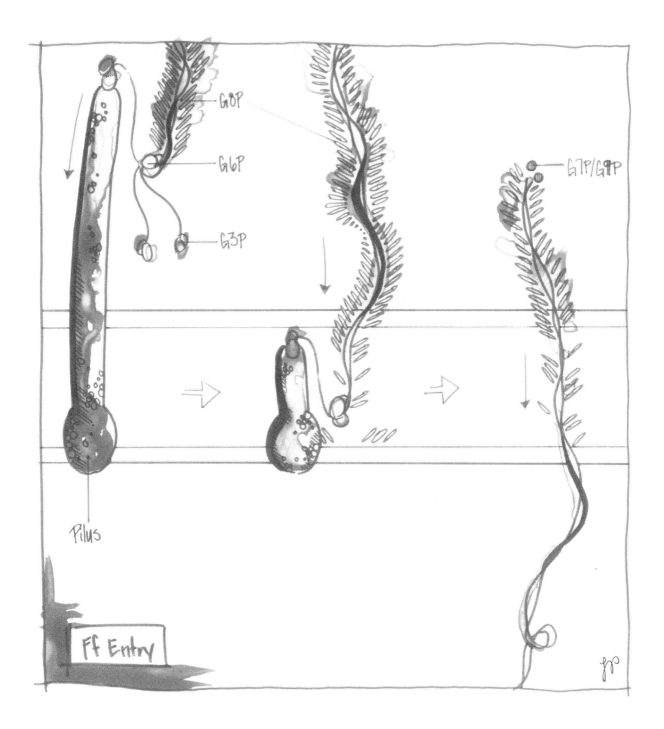

g8p

g6p

g3p

g7p/g9p

Pilus

Ff Entry

both transcription and the synthesis of new single-stranded genomes (Marvin, Hohn 1969; Russel, Linderoth, Sali 1997). Initially f1 produces new genomes at an exponential rate by converting each ss-DNA copy into a double-stranded replicative form. At the same time, f1 actively synthesizes abundant copies of its ssDNA binding protein (g5p), enough copies to soon coat the newly-minted genomes with g5p dimers. Only a short dsDNA hairpin at the proximal end of the genome lacks this interim

protein coat (Russel 1991). This hairpin structure serves as the packaging signal that leads the g5p-coated genomes to the CM for final assembly and export. Phage f1 keeps the replication machinery on task indefinitely to support ongoing continuous phage production, generation after generation. About a thousand progeny phage extrude from each cell each generation, altering membrane properties without bringing significant harm to the accommodating *E. coli* host (Marvin, Hohn 1969).

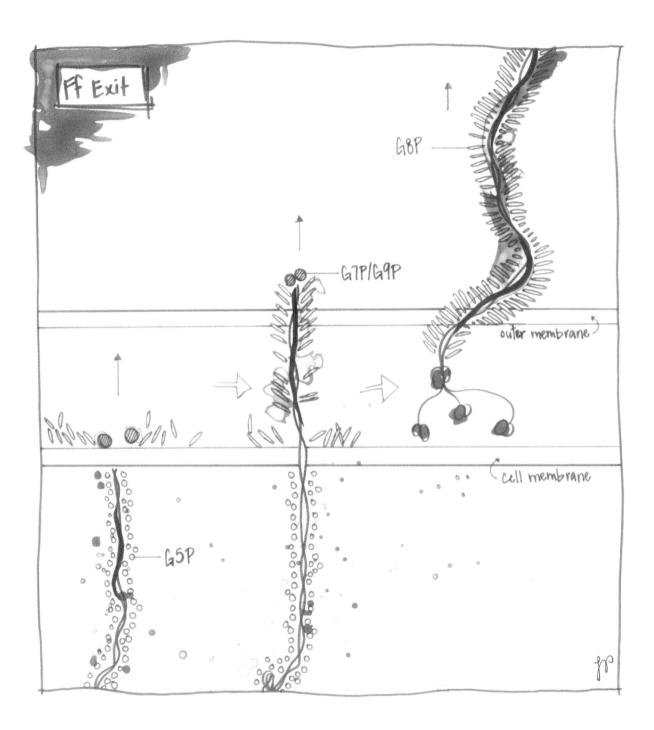

Construction on exit

As f1's cocooned genome approaches the CM with its packaging signal in the lead, it finds the needed virion components waiting as membrane-associated or integral membrane proteins. At the CM, the proximal end acquires its two minor coat proteins (g7p and g9p). Then the DNA passes through the CM, in the process shedding one skin as it acquires another. The g5p dimers are left behind in the cytoplasm, each one replaced by a copy of the major

coat protein g8p. Even though a few of these coat proteins were deposited in the membrane during infection by the parent phage, most were freshly made and anchored in the CM in anticipation. As the extruded proximal end navigates through the periplasm, it identifies its escape hatch in the OM. Although most filamentous phages exit through a borrowed host secretion channel, the f1 phages encode an efficient one of their own, their g4p secretin (Marciano, Russel, Simon 1999). With an average diameter of 14 nm, these secretins are an open road to freedom (Nickerson et al. 2012).

Since the extruding phage is more than 700 nm long, the leading proximal end clears the ~24 nm host envelope long before the trailing end has reached the CM. When the trailing end of the genome finally arrives there, the virion is 'pinched off' by the addition of the two minor coat proteins unique to this end (g6p and g3p). Protein g3p is crucial here (Rakonjac, Model 1998). Without its participation in terminating and releasing each virion, multiple virions fuse end to end to yield a 'polyphage' that looks suspiciously like a pilus.

Kin?

Some intriguing parallels between pili and filamentous phages hint at an evolutionary link between them (Bradley 1967; Rakonjac, Model 1998). The architecture of both includes a hollow cylinder composed of hundreds (if a phage) or thousands (if a pilus) of copies of a small protein arranged in a helical array. These composite structures disassemble to monomers that are inserted into the CM, where they sit tight until called upon to re-emerge and construct a new pilus or phage. To extrude a filamentous phage, membrane-embedded coat protein monomers are recruited and added one by one to the helical shell surrounding the ssDNA genome as it extrudes from the cell through a secretin OM pore. When the end of the DNA is reached, the structure is cut free from the cell as a completed virion that sets out to seek its fortune in the world. A growing pilus, likewise, extends through a secretin pore by the addition of protein monomers at its base. Pilus extension, however, is followed by retraction, the reverse process in which the pilus disassembles at its base and the proteins return to the membrane. If indeed these mechanisms share a common evolutionary history, which came first—the pilus or the phage?

Cited references

Bradley, DE. 1967. Ultrastructure of bacteriophage and bacteriocins. Bacteriol Rev 31:230-314.

Click, EM, RE Webster. 1998. The TolQRA proteins are required for membrane insertion of the major capsid protein of the filamentous phage f1 during infection. J Bacteriol 180:1723-1728.

Deng, LW, RN Perham. 2002. Delineating the site of interaction on the pIII protein of filamentous bacteriophage fd with the F-pilus of *Escherichia coli*. J Mol Biol 319:603-614.

Eckert, B, A Martin, J Balbach, FX Schmid. 2005. Prolyl isomerization as a molecular timer in phage infection. Nat Struct Mol Biol 12:619-623.

Glaser-Wuttke, G, J Keppner, I Rasched. 1989. Pore-forming properties of the adsorption protein of filamentous phage fd. Biochim Biophys Acta 985:239-247.

Glucksman, MJ, S Bhattacharjee, L Makowski. 1992. Three-dimensional structure of a cloning vector. X-ray diffraction studies of filamentous bacteriophage M13 at 7 Å resolution. J Mol Biol 226:455-470.

Gray, CW, RS Brown, DA Marvin. 1981. Adsorption complex of filamentous fd virus. J Mol Biol 146:621-627.

Holliger, P, L Riechmann. 1997. A conserved infection pathway for filamentous bacteriophages is suggested by the structure of the membrane penetration domain of the minor coat protein g3p from phage fd. Structure 5:265-275.

Jakes, K, N Davis, N Zinder. 1988. A hybrid toxin from bacteriophage f1 attachment protein and colicin E3 has altered cell receptor specificity. J Bacteriol 170:4231-4238.

Lubkowski, J, F Hennecke, A Pluckthun, A Wlodawer. 1999. Filamentous phage infection: Crystal structure of g3p in complex with its coreceptor, the C-terminal domain of TolA. Structure 7:711-722.

Marciano, DK, M Russel, SM Simon. 1999. An aqueous channel for filamentous phage export. Science 284:1516-1519.

Marvin, D, B Hohn. 1969. Filamentous bacterial viruses. Bacteriol Rev 33:172-209.

Marvin, DA. 1998. Filamentous phage structure, infection and assembly. Curr Opin Struct Biol 8:150-158.

Nickerson, NN, SS Abby, EP Rocha, M Chami, AP Pugsley. 2012. A single amino acid substitution changes the self-assembly status of a type IV piliation secretin. J Bacteriol 194:4951-4958.

Rakonjac, J, NJ Bennett, J Spagnuolo, D Gagic, M Russel. 2011. Filamentous bacteriophage: Biology, phage display and nanotechnology applications. Curr Issues Mol Biol 13:51-76.

Rakonjac, J, P Model. 1998. Roles of pIII in filamentous phage assembly. J Mol Biol 282:25-41.

Riechmann, L, P Holliger. 1997. The C-terminal domain of TolA is the coreceptor for filamentous phage infection of *E. coli*. Cell 90:351-360.

Russel, M. 1991. Filamentous phage assembly. Mol Microbiol 5:1607-1613.

Russel, M, NA Linderoth, A Sali. 1997. Filamentous phage assembly: Variation on a protein export theme. Gene 192:23-32.

Recommended reviews

Marvin, D, B Hohn. 1969. Filamentous bacterial viruses. Bacteriol Rev 33:172-209.

Marvin, D, M Symmons, S Straus. 2014. Structure and assembly of filamentous bacteriophages. Prog Biophys Mol Biol 114:80-122.

Rakonjac, J, NJ Bennett, J Spagnuolo, D Gagic, M Russel. 2011. Filamentous bacteriophage: Biology, phage display and nanotechnology applications. Curr Issues MolBiol 13:51-76.

Sulfolobus Turreted Icosahedral Virus (STIV)

a Fusellovirus that exits its host through phage-constructed seven-sided pyramids

Genome
dsDNA; circular
17,663 bp
36 predicted ORFs; 0 RNAs

Encapsidation method
Packaging; T = 31 capsid

Common host
Sulfolobus solfataricus

Habitat
Acidic hot springs

Lifestyle
Lytic

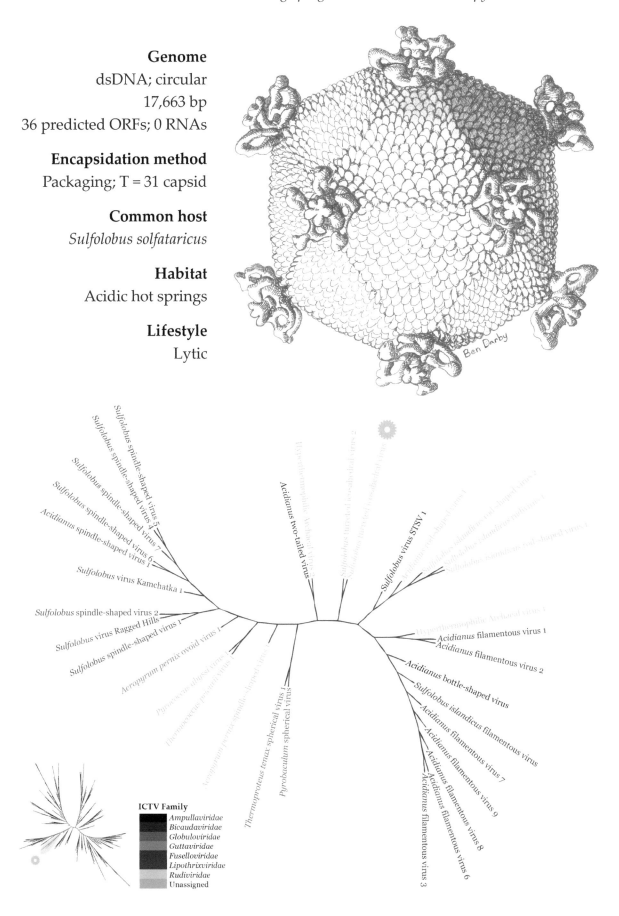

Ben Darby

ICTV Family
Ampullaviridae
Bicaudaviridae
Globuloviridae
Guttaviridae
Fuselloviridae
Lipothrixviridae
Rudiviridae
Unassigned

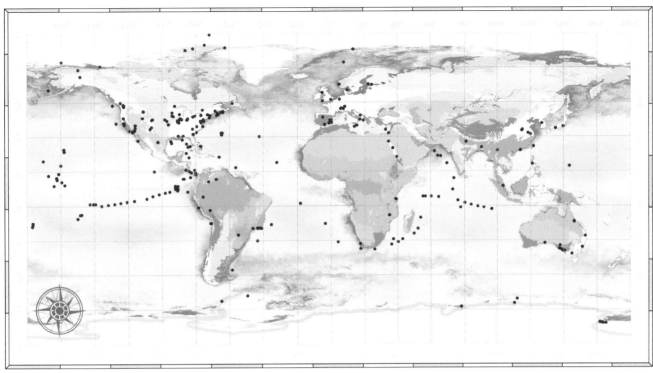

• Sampling sites

For map details,
see Appendix A4.

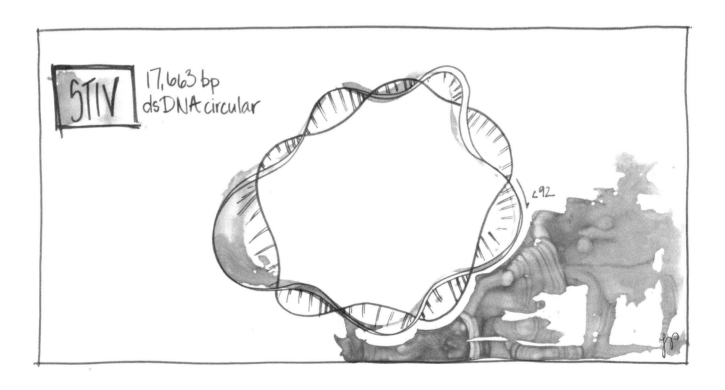

STIV 17,663 bp
dsDNA circular

c92

Sulfolobus Turreted Icosahedral Virus

 Circular genome

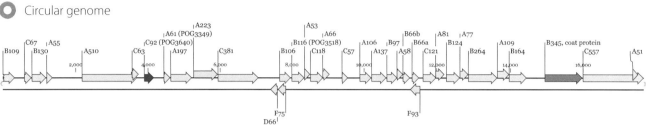

The Mysterious Pyramids of *Sulfolobus*

Merry Youle

Mise-en-Scène: *Do you yearn for uncharted continents to explore, alien terrains where you could discover bizarre life forms and learn their secrets? The vast dark matter of the phage world offers many such opportunities, but no environment serves up more weird critters in greater variety than the bubbling acid baths that are home to extremophilic Crenarchaeota. This group is comprised of ten divergent families, some with virion morphologies never seen elsewhere: droplets, spindles, bottles, and various filaments and rods. These families typically share few homologous genes between them even when they share the same habitat, the same hot spring—even the same host (Prangishvili 2013). Usually less that 10% of their genes have homologs in any public database; some have no known homologs at all. Here indeed are terrae incognitae awaiting reconnaissance by intrepid extremothermophiles* (see page 7-30).

To end an infection, most phages punch holes—large (*see page 7-5*) or small—in the host **envelope**, crude and unimaginative perforations that get the job done and let the waiting virions exit. Not so for *Sulfolobus* turreted icosahedral virus (STIV). This phage is an architect and a craftsman, with a distinctive flair. It crafts precise seven-sided pyramids each of which opens to form an escape hatch (Brumfield et al. 2009; Fulton et al. 2009).

This unusual strategy may reflect an adaptation of this lytic phage to its environment and to the structure of its host, the crenarchaeon *Sulfolobus solfataricus*. Phage and host dwell together in the acidic hot springs of Yellowstone National Park (USA)—from their perspective, a hospitable environment (80° C, pH ~3) that advantageously deters competition from the mesophilic riffraff. STIV infection proceeds at a leisurely pace (Brumfield et al. 2009). Transcription of its early genes is underway by 8 h post infection (PI), followed by transcription of genes for virion structural proteins by 16 h PI (Ortmann et al. 2008). By 24 h PI, several pyramids under construction are evident on each cell. Virion assembly follows soon thereafter.

Lethal pyramids

Each pyramid starts small, a mere angular protrusion of the cell membrane that pushes aside the host's protective outer surface protein layer (S-layer). S-layers composed of an ordered lattice of many copies of a single glycosylated protein shield many Bacteria and Archaea. In *Sulfolobus*, the layer

is firmly anchored to the cell membrane, thereby boosting the physical and chemical robustness of these archaeal **hyperthermophiles**. As assembled virions accumulate inside the small host cell, the pyramids grow to heights of more than 100 nm. By 32 h PI, many waiting virions—50, 100, 150, or more—are crowded together within the pyramids like so many hexagonal close-packed spheres. Here they wait, trapped inside by the pyramid walls. Lysis occurs between 32 and 40 h PI when the pyramids 'open' wide. The 'glue' bonding each triangular face to its two neighbors lets go and the former pyramid collapses into a cluster of loose flaps connected to the cell only at their base. The virions are free, the host dead.

Pyramid construction

What are these pyramids made of? STIV starts with the materials available, i.e., the membrane of the host cell. This is not the usual bacterial phospholipid construct, but rather an archaeal membrane built from various cyclic tetraether lipids (Maaty et al. 2006). STIV adds a protein to the membrane to restructure and thicken it, making of it a suitable material for building pyramids with sharply defined facets. This architect discards the five-fold or six-fold symmetry prevalent in virion structure in favor of heptagonal (septagonal) constructs, each formed by seven triangular faces that rise from a heptagonal portal at the base. All it takes to grow these pyramids on the surface of uninfected *Sulfolobus* cells is the expression of one STIV-encoded protein (c92) (Snyder et al. 2011).

Proprietary technology?

STIV is not the only phage with this knack. A phage that infects *Sulfolobus islandicus* in Iceland, *Sulfolobus islandicus* rod-shaped virus 2 (SIRV2), is also a pyramid builder (Bize et al. 2009). SIRV2 infection proceeds through the same stages, leading to the accumulation of densely packed virions inside the pyramids and their subsequent release. These two phages differ markedly in both genome structure and virion morphology. STIV packages its circular genome inside a complex,

turreted, membrane-containing icosahedral capsid (Fu, Johnson 2012); SIRV2 assembles many copies of a single DNA-binding protein around its linear DNA genome to produce a stiff rod with three short fibers added at each end (Prangishvili et al. 1999). Nevertheless, analysis of their genomes places them as neighbors within the Fuselloviridi in the PPT. They have one key gene in common: SIRV's p98 is a **homolog** of STIV's pyramid protein c92. The amino acid sequences of these two proteins are 55.4% identical. When

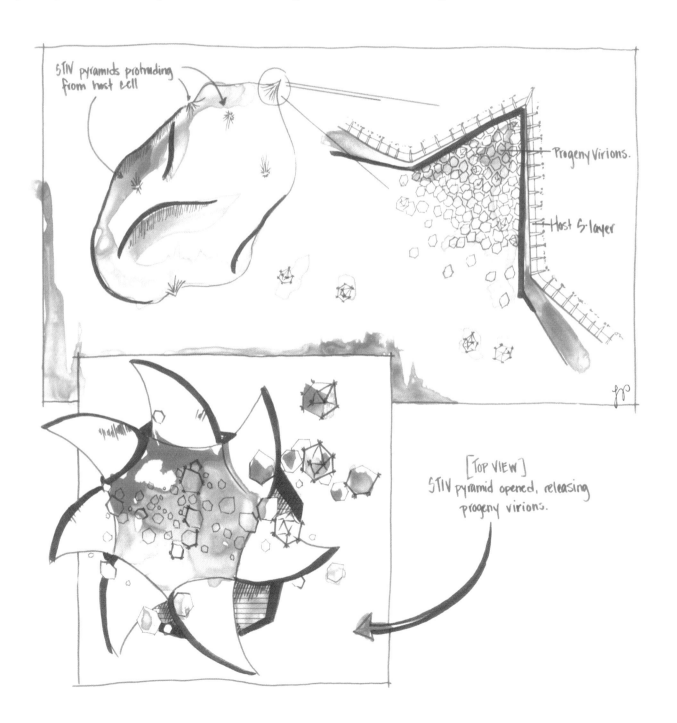

STIV pyramids protruding from host cell

Progeny Virions.

Host S-layer

[TOP VIEW]
STIV pyramid opened, releasing progeny virions.

expressed in an uninfected *Sulfolobus* cell, either-can form pyramids structurally indistinguishable from those made by the phage. Further, p98 did so also when a researcher tested it in *E. coli*, despite the different lipid composition of bacterial and archaeal membranes (Quax et al. 2011).

Are these two phages just the tip of a pyramidal iceberg? Searching for homologs of c92 or p98 in the ~45 other sequenced archaeal phage genomes suggests not. Even STIV's close relative, STIV2, does not have a related gene (Happonen et al. 2010). Likewise, among three sequenced close relatives of SIRV2, only two have a homolog (Quax et al. 2010). All of this leaves us wondering why the close relatives of STIV and SIRV2 aren't also pyramid architects, given that they infect related hosts under similar conditions. Observations suggest that they, like most phages infecting hyperthermophilic Crenarchaeota, forego host lysis and

instead take refuge as prophages or **plasmids** in their host. As such they persist, generation after generation, continuously assembling and gently releasing a few virions (Happonen et al. 2010) to potentially infect new hosts. Not a bad strategy when the outdoors is so inhospitable. But this in turn leaves us wondering why STIV and SIRV2 have opted to do otherwise.

Many other questions remain. How does the structure of the pyramid proteins (c92 and p98) facilitate their assembly into these geometrical forms? Once initiated, how does a pyramid grow in size? How can the same protein (p98) build pyramids both Bacteria and Archaea given the markedly different composition of their membranes? What mechanism triggers pyramid opening and virion release? Do all pyramids on a cell open in unison? All these mysteries, and more, await sleuthing.

Cited references

Bize, A, EA Karlsson, K Ekefjärd, TEF Quax, M Pina, MC Prevost, P Forterre, O Tenaillon, R Bernander, D Prangishvili. 2009. A unique virus release mechanism in the Archaea. Proc Natl Acad Sci USA 106:11306-11311.

Brumfield, SK, AC Ortmann, V Ruigrok, P Suci, T Douglas, MJ Young. 2009. Particle assembly and ultrastructural features associated with replication of the lytic archaeal virus *Sulfolobus* turreted icosahedral virus. J Virol 83:5964-5970.

Fu, C-y, JE Johnson. 2012. Structure and cell biology of archaeal virus STIV. Curr Opin Virol 2:122-127.

Fulton, J, B Bothner, M Lawrence, J Johnson, T Douglas, M Young. 2009. Genetics, biochemistry and structure of the archaeal virus STIV. Biochem Soc Trans 37:114-117.

Happonen, LJ, P Redder, X Peng, LJ Reigstad, D Prangishvili, SJ Butcher. 2010. Familial relationships in hyperthermo-and acidophilic archaeal viruses. J Virol 84:4747-4754.

Maaty, WS, AC Ortmann, M Dlakić, K Schulstad, JK Hilmer, L Liepold, B Weidenheft, R Khayat, T Douglas, MJ Young. 2006. Characterization of the archaeal thermophile *Sulfolobus* turreted icosahedral virus validates an evolutionary link among double-stranded DNA viruses from all domains of life. J Virol 80:7625-7635.

Ortmann, AC, SK Brumfield, J Walther, K McInnerney, SJ Brouns, HJ van de Werken, B Bothner, T Douglas, J van de Oost, MJ Young. 2008. Transcriptome analysis of infection of the archaeon *Sulfolobus solfataricus* with *Sulfolobus* turreted icosahedral virus. J Virol 82:4874-4883.

Prangishvili, D. 2013. The wonderful world of archaeal viruses. Annu Rev Microbiol 67:565-585.

Prangishvili, D, HP Arnold, D Götz, U Ziese, I Holz, JK Kristjansson, W Zillig. 1999. A novel virus family, the *Rudiviridae*: Structure, virus-host interactions and genome variability of the *Sulfolobus* viruses SIRV1 and SIRV2. Genetics 152:1387-1396.

Quax, TEF, M Krupovic, S Lucas, P Forterre, D Prangishvili. 2010. The *Sulfolobus* rod-shaped virus 2 encodes a prominent structural component of the unique virion release system in Archaea. Virology 404:1-4.

Quax, TEF, S Lucas, J Reimann, G Pehau-Arnaudet, MC Prevost, P Forterre, SV Albers, D Prangishvili. 2011. Simple and elegant design of a virion egress structure in Archaea. Proc Natl Acad Sci USA 108:3354-3359.

Snyder, JC, SK Brumfield, N Peng, Q She, MJ Young. 2011. *Sulfolobus* turreted icosahedral virus c92 protein responsible for the formation of pyramid-like cellular lysis structures. J Virol 85:6287-6292.

Recommended reviews

Prangishvili, D. 2013. The wonderful world of archaeal viruses. Annu Rev Microbiol 67:565-585.

Prangishvili, D, RA Garrett. 2005. Viruses of hyperthermophilic Crenarchaea. Trends Microbiol 13:535-542.

Into the Devil's Kitchen:
A Personal History of Archaeal Viruses

by Kenneth Stedman[†]

Abstract: *Archaeal viruses are eccentric in both their virion structures and their genomes (and in their selection of researchers allowed to study them). Even the arguably best-researched archaeal virus, the lemon-shaped Fusellovirus SSV1, is replete with unsolved mysteries. My virus hunting career began at the side of Wolfram Zillig, the pioneer in the field, and over the last 20 years, together with other researchers, we have discovered many viruses of extremophilic Archaea. While more undoubtedly remain to be found, the field is poised to move from the discovery of new viruses to the exploration of the unique replication and host interactions of these fascinating nanobes.*

[†]Department of Biology and Center for Life in Extreme Environments, Portland State University, Portland, OR
Email: kstedman@pdx.edu
Website: http://web.pdx.edu/~kstedman/

Prologue

It was mid-September of 2003. I was hunting for viruses in a solfataric field in Lassen Volcanic National Park with a new Ph.D. student, Adam Clore, and an undergraduate student, Random Diessner. We had tortuously made our way around the moonscape-like environment to get near to a promising bubbling murky spring (Fig. 1). Now it was time to go in to "Devil's Kitchen," one of the main hydrothermal areas of the park. I carefully led my students across the fragile ground towards a promising spring, only to have my boot break through the thin crust of soil into the boiling acidic mud beneath.

Act 1. Archaeal viruses: Extremely different

Why was I endangering myself and my students to collect a small amount of hot, acidic, muddy water? The danger was real. The namesake of a nearby thermal area, Mr. Bumpass, lost both of his legs after falling into some of this stuff. We were hunting new viruses that infect *Sulfolobus* and its relatives. Members of the *Sulfolobales* are among some of the first-discovered and best-studied Archaea. The crenarchaeon *Sulfolobus* thrives in boiling acidic springs at 80° C and at pH 3 or even lower—quite remarkable in itself. The viruses that infect these thermoacidophiles are even more extraordinary with their unique shapes and genomes. They are so divergent, in fact, that an unprecedented ten new virus families were proposed

Figure 1. Overview of Devil's Kitchen, Lassen Volcanic National Park. September 2003. The sampled spring of interest indicated with an orange arrow. Photo credit: K. Stedman.

to accommodate them (Prangishvili 2013). Their virions offer an incredibly diverse assortment of shapes (Fig. 2). The relatively rare types with the familiar icosahedral capsid architecture include the *Sulfolobus* turreted icosahedral virus (STIV) that I discovered in Yellowstone National Park (Rice et al. 2004). Some, such as the aptly-named *Sulfolobus islandicus* rod-shaped virus (SIRV), are indeed rod-shaped (Prangishvili et al. 1999). These are outdone by those with filamentous virions whose length is twice the diameter of the cells that they infect; some of these, the *Sulfolobus islandicus* filamentous virus (SIFV), for instance, have nano-sized claw-like structures at their termini (Arnold et al. 2000). There are also amazing bottle-shaped virions such as ABV (Haring et al. 2005a). The majority of archaeal viruses have spindle or lemon-shaped virions of varying sizes, with or without long slender tails. One of these, the *Acidianus* two-tailed virus (ATV), 'grows' tails after exiting its host provided it is at the usual hot spring temperature (*see page 6-31*; Haring et al. 2005b). David Prangishvili has recently written an excellent review of this bizarre world of archaeal viruses (Prangishvili 2013). Here I will feature the *Sulfolobus* spindle-shaped viruses (SSVs), aka the Fuselloviruses, that are the main focus of my research group (Stedman, Prangishvili, Zillig 2006).

Act 2. SSV1: A lemon full of puzzles

The best-studied Fusellovirus is SSV1. Its genome is unique (Fig. 3); only one of the 35 open reading frames (ORFs), or putative genes, is clearly homologous to sequences found in any other viral or cellular genome (Palm et al. 1991). This one, the SSV1 integrase gene, is homologous to the well-studied integrase of phage λ, but possesses a few quirks of its own. First, the attachment site that is cleaved when the viral genome integrates into the host genome lies within the integrase gene itself. Thus, during integration, the integrase gene is disrupted and presumably inactivated (Reiter, Palm 1990). Another intriguing aspect concerns the structure and activity of the tetrameric functional form of the integrase. That the four monomers act in trans (Letzelter, Duguet, Serre 2004; Eilers, Young, Lawrence 2012) makes SSV1's integrase more similar to the eukaryotic flp-like recombinases than to λ integrase. Moreover, the SSV1 integrase gene is not essential for viral reproduction; if it is deleted, viral infection appears to proceed normally. However, the integrase gene must play some as yet unknown obscure role as viruses lacking this gene are at a competitive disadvantage relative to the wild-type (Clore, Stedman 2007). In contrast to phage λ, SSV1 does not have a lytic replication phase, but releases its virions, apparently without

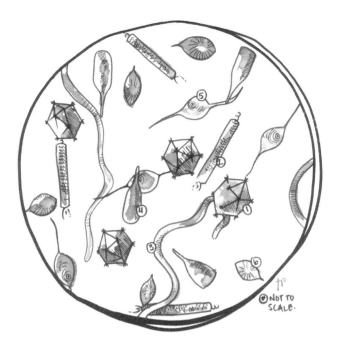

Figure 2. Novel virion shapes in archaeal viruses as drawn based on photographs or diagrams in the cited references. Virion #1: *Sulfolobus* turreted icosahedral virus, STIV (Rice et al. 2004); virion #2: *Sulfolobus islandicus* rod-shaped virus, SIRV (Prangishvili et al. 1999); virion #3: *Sulfolobus islandicus* filamentous virus, SIFV (Arnold et al. 2000); virion #4: *Acidianus* bottle-shaped virus, ABV (Häring et al. 2005a); virion #5: *Acidianus* two-tailed virus, ATV (Häring et al. 2005b); virion #6: *Sulfolobus* spindle-shaped virus, SSV (Martin et al. 1984; Schleper, Kubo, Zillig 1992).

host lysis, by budding at the cellular membrane. Nevertheless, as is the case for phage λ, UV irradiation can induce increased SSV1 virion production up to 100-fold (Martin et al. 1984; Schleper, Kubo, Zillig 1992). The molecular mechanism of this induction is not clear, that of the assembly of the SSV1 virion even less so.

Other SSV1 genes whose function is known include those that encode the three virion structural proteins (VP1, VP2, and VP3). The VP1 protein is the major capsid protein and VP3 the minor capsid protein; together they make up the majority of the proteins in the distinctive, lemon-shaped capsid (Reiter et al. 1987a). The location of VP1 and VP3 in the capsid is not clear. There are many more copies of VP1 than VP3, indicating that the latter may be concentrated at the termini of the particle or in locations of pentagonal symmetry. Preliminary single-particle cryo-EM reconstruction data suggested the presence of some hexagonally arranged capsomers, but no overall icosahedral symmetry. Interestingly, VP1 and VP3 have very similar C-terminal amino acid sequences. Both genes have the identical 61 bp direct repeat at one end (Palm et al. 1991). Another oddity: the SSV1 genome is packaged as a circle of positively supercoiled double stranded DNA (dsDNA; [Nadal et al. 1986]). Positive supercoiling probably makes the SSV1 DNA more stable to thermal denaturation at the high temperatures (80° C) in which its *Sulfolobus* host thrives. The VP2 protein appears to be a non-specific DNA-binding protein, presumably required for this genome packaging (Reiter 1985). Surprisingly, when we deleted the VP2 gene, the virus appears to function normally with no apparent loss of stability or infectivity (Iverson, Stedman 2012). It is hard to imagine that the SSV1 genome does not need to be protein-bound to withstand the 80° C temperatures. Presumably there is a non-orthologous host protein that binds to and packages the

DNA. There are a number of small DNA binding proteins in *Sulfolobus* that may serve this role, but whether or not that is the case remains to be determined. We are actively trying to determine both the role of VP2 in the wild type virus and what allows the mutant virus lacking VP2 to survive.

We, as well as others, have tried to elucidate the function of the other 31 genes using genetic, comparative genomic, biochemical, and structural approaches. My own genetic studies with SSV1 began when I was a postdoc in Wolfram Zillig's lab at the Max Planck Institute for Biochemistry in Martinsried, Germany, in the late 1990s. Here I must digress and tell you about what it was like working with Hr. Prof. Dr. Wolfram Zillig.

Act 3. Wolfram Zillig: Extreme virus hunter

Herr Zillig had started his scientific career working on silkworm pupae, but then quickly moved to viruses. His classic studies of tobacco mosaic virus (TMV) showed that the nucleic acid of TMV, and by extension of all viruses, was required for replication, thus reversing the protein-centric dogma at the time (Schramm, Schumacher, Zillig 1955). Subsequently he extended his work to φX174 (Rueckert, Zillig 1962). That was followed by his investigation of the DNA-dependent RNA polymerase (RNAP) of *E. coli*, particularly as modified by phage T4 during infection (Walter, Seifert, Zillig 1968), work that made him well-known in the field of bacterial transcription.

Very soon after Carl Woese had proposed that Archaea were a fundamentally different group of organisms from the Bacteria (and the Eukarya), Herr Zillig became one of the 'generals of Woese's Army' with his group providing strong supporting evidence for the unique nature of Archaea. In the late 1970s he showed that the RNAPs in Archaea were fundamentally different from those of Bacteria (Zillig, Stetter, Tobien 1978; Stetter, Zillig 1979; Zillig, Stetter, Janekovic 1979). Moreover, he and his group showed that archaeal promoters are much more similar to eukaryotic promoters than to bacterial ones (Huet et al. 1983; Reiter, Palm, Zillig 1988). My original reason for wanting

Figure 3. Genome map of SSV1. ORFs and genes are shown as arrows (Palm et al. 1991) with black arrow points indicating the direct repeat in the VP1 and VP3 genes. Red ORFs do not tolerate insertions or deletions (Stedman et al. 1999; Iverson, Stedman 2012), whereas green ORFs do (Stedman et al. 1999; Clore, Stedman 2007; Iverson, Stedman 2012), and the tolerance of the gray ORFs to insertion or deletion is unknown. Thin black arrows inside the genome circle indicate transcripts (Reiter et al. 1987b; Fröls et al. 2007; Fusco et al. 2013). Credit: Kenneth Stedman.

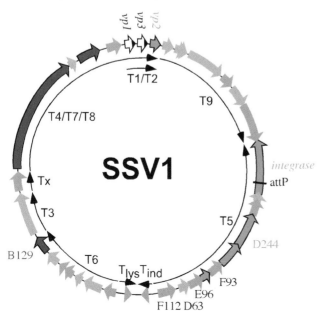

Figure 4. Herr Zillig collecting samples in Yellowstone National Park, summer 2001.
Photo credit: K. Stedman.

to work in the Zillig group was to pursue studies of archaeal transcriptional regulation.

Also in the late 1970s, together with Karl Stetter, Herr Zillig had started to isolate new Archaea from extreme environments, the hotter and more extreme the better, partly because he was unable to obtain specimens from other researchers (Zillig et al. 1980; Zillig, Tu, Holz 1981). Characterizing the RNAPs from these new organisms was only the beginning. He realized that isolating viruses of the Archaea would be critical for understanding their biology. He set about to do that, discovering φH, a tailed virus of extremely halophilic Archaea (Schnabel et al. 1982), and soon thereafter the SSV1 virus. SSV1 was first found as a plasmid in *Sulfolobus shibatae* (Yeats, McWilliam, Zillig 1982). It was only by later electron microscope observations following UV irradiation that this extrachromosomal DNA was shown to be packaged in a virion (Martin et al. 1984). The discovery of an uninfected host (and a laboratory mix-up) allowed the determina-

tion that SSV1 was infectious and thus a true virus (Schleper, Kubo, Zillig 1992). Also, to the best of my knowledge, Herr Zillig wrote the first review of archaeal viruses (Zillig et al. 1986).

Herr Zillig was a character, to say the least. I often described him as a toddler trapped in a 70 year-old body. When I first interviewed about joining the lab, he spent more than half an hour complaining about the German system that forced him at age 68 to retire to an emeritus position, sweetened only by his choice of either a secretary and an office, or a lab. His choice was clear, but then he had to learn how to type instead of dictate, and how to use e-mail (at age 68). After my interview it was clear that I absolutely had to work with this "Naturkraft," as many of his contemporaries and colleagues described him. When I arrived in his lab to begin my postdoc, instead of talking about my proposed research project, he spent another half hour relating how he had shot himself through the hand while spearfishing in Baja

California and was very lucky to have missed all of the bones, ligaments, and nerves. Most importantly, he explained to this newcomer, I was to use the pronoun "Sie," the formal form of address in German, when speaking with him instead of the informal "du" (he would do the same with me). At first, I thought this somewhat odd, but later found it to be extremely useful when we were having arguments (he really liked to argue). By this use of the formal address a certain amount of respect was present in the 'conversation.'

In a less confrontational moment, Herr Zillig told me that scientists were either *Jäger* or *Sammler*, either hunters or gatherers. He was most certainly one of the former, I hope to be considered one, as well. He liked nothing better than to hunt down new viruses and new hosts (Fig. 4). In the middle of a hydrogen sulfide-spewing, bubbling, thermal field, he would whip out a box of pH paper

strips from his customized scientific fishing vest, often managing to spill all but one, but then triumphantly brandish that one to test the pH of a forbidding gray bubbling mud pot, often near the heart of an active volcano. In 2001, we had the opportunity to collect samples at the island of Vulcano, just off the coast of Sicily (a place that all thermophile researchers should visit). Vulcano is the site of Vulcan's forge in Roman and Greek legend. Currently a steaming crater with multiple fumaroles and beautiful crystals of elemental sulfur is but a short climb from the main port, while the real action for thermophile researchers is at the beach, Baia di Levante. In addition to the tourists and locals soaking in a large mud bath, there are many areas of boiling water just off the coast or on the beach. You have to be careful where you walk lest you burn your feet on a superheated steam vent. The famous strains *Pyrococcus furiosus* and *Thermotoga maritima* were both isolated from this beach or nearby.

After arguing for what felt to me like ages with a local Vulcano fisherman, with me frantically trying to translate using my mediocre Italian, Herr Zillig convinced him (for a fee) to sail us around the island to look for places where bubbles of superheated gasses—many of them highly toxic—were coming up from the bottom of the Tyrrhenian Sea. When we saw bubbles, he would throw himself overboard with a 50 mL syringe body and free dive to the bottom to collect as many samples as possible. Unfortunately, no new viruses or organisms were isolated from any of those samples, but I collected wonderful memories (and had some excellent food and wine). Sonja Albers, Patrick Forterre, David Prangishvili, and Christa Schleper have written a very nice retrospective about Wolfram Zillig's accomplishments and his interactions with Carl Woese (Albers et al. 2013).

Act 4. SSV1: The awesome power of genetics

Although I joined the Zillig lab to work on transcriptional regulation in Archaea, it was soon clear that he was much more interested in new viruses. I, in turn, realized that one of the major bottle-

Figure 5. Conservation of SSV1 ORFs. ORFs in the SSV1 genome are color-coded with white denoting ORFs with no homologs in other SSV genomes, black denoting ORFs with homologs in all others, and rainbow coloring denoting intermediate degrees of conservation. The direct repeat in the universally-conserved VP1 and VP3 genes is indicated by the white arrow points. Genome sources: various SSVs (Iverson, Stedman 2012); SSV5, SSV6, SSV7, and ASV (Redder et al. 2009); SMF-1 (Servin-Garcidueñas et al. 2013). Thin black arrows inside the genome circle indicate transcripts (Reiter et al. 1987; Frols et al. 2007; Fusco et al. 2013). Credit: Kenneth Stedman.

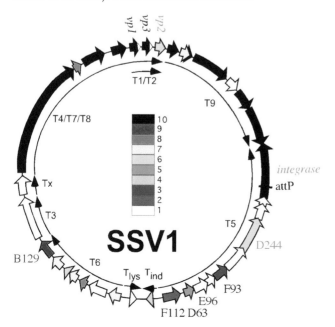

necks to further research in Archaea was the lack of genetic tools. Having been a graduate student of the late Sydney Kustu, working on transcriptional regulation in *Salmonella typhimurium* (Klose et al. 1994), I had experienced firsthand the amazing power of biological systems where one could combine genetics with biochemistry. Biochemistry was well-established in thermophilic Archaea in the late 1990s, but genetics was not. What better tool to use to remedy this deficiency than SSV1? After discovering this virus, the Zillig lab had proceeded to characterize it. Its rather small (15 kbp) genome had been sequenced and was known to be a well-behaved circle of dsDNA. The question looming was where could one change the SSV1 genome without completely disrupting viral replication. To make a long story short, after trying for 18 months to outsmart the virus by inserting an *E. coli* plasmid into various open reading frames in the viral genome, I wised up and let the virus tell me where insertions would and would not be tolerated. I did a partial digestion of the SSV1 genome and randomly inserted an *E. coli* plasmid. I then selected for virus genomes that had taken up the *E. coli* plasmid by antibiotic selection in *E. coli*. Finally, I selected for functional constructs by transforming the mixture into *Sulfolobus* and screening to see which cells were productively infected. Eventually I succeeded in making an *E. coli-Sulfolobus* shuttle vector from the SSV1 genome, i.e., a replicon that functions as a plasmid in *E. coli* and as a virus in *Sulfolobus*, and in the process discovered where the SSV1 genome could tolerate insertions (Stedman et al. 1999). Moreover, this showed that, somewhat surprisingly, the SSV1 genome could tolerate a three kbp insertion and still be packaged and function normally. Since then, we and others have developed this vector further and have used it for gene expression studies in *Sulfolobus* (Jonuscheit et al. 2003; Albers et al. 2006; DeYoung et al. 2011).

In parallel with the screen for functional virus, I collected the virus constructs that had *E. coli* plasmid insertions but were not functional. These mutants identified regions of the SSV1 genome that could not tolerate insertions without impairing

essential virus activities. Since this screen was not complete, i.e., we did not have insertions in every SSV1 open reading frame, we are now systematically disrupting each of the ORFs in the SSV1 genome in turn. For this we developed a long-inverse PCR technique in which we amplify the entire SSV1 genome except for the gene that we want to disrupt, then ligate the ends together to reform the circular chromosome. We started with the viral integrase gene since that was the one gene in the SSV1 genome that was clearly homologous to genes in other viruses. Since SSV1 was known to integrate soon after infection (Schleper, Kubo, Zillig 1992), we were surprised (as described above) to find that without this gene SSV1 appeared to be fully functional, with the exception of genome integration (Clore, Stedman 2007). Since then we have used this technique to delete three other ORFs in turn: VP2, D244, and B129. Deletion of the VP2 gene yields a functional virus, but ORF B129 that encodes a DNA binding protein appears to be essential. ORF D244 served up surprises. Its deletion generated a mutant with a hyper-virulent phenotype that strongly retards growth of its *Sulfolobus* host, whereas infection by wild-type SSV1 causes only a very minor growth defect, if any at all (Iverson, Stedman 2012). SSV-K, aka SSV9, a related virus that I isolated from the Kamchatka Peninsula in far Eastern Russia, lacks that ORF (Wiedenheft et al. 2004), is hyper-virulent, and also has a very wide host range (Ceballos et al. 2012). The protein encoded by a homologous ORF in a Yellowstone SSV-isolate appears to be a nuclease (Menon et al. 2010). Why deletion of a nuclease should increase the growth impairment of the host during viral infection is not clear.

Act 5. SSV1: What do those genes do?

Concurrently with the genetics work and in the footsteps, almost literally, of Herr Zillig, we pursued some comparative genetic studies. Herr Zillig had observed that about 5% of enrichment cultures from Icelandic hot springs with temperatures above 70° C and a pH of less than 4 contained SSV-like viruses (Zillig et al. 1998). So it was off to the hot springs, including Devil's Kitchen in Lassen Volcanic National Park. I helped sequence and analyze one of these new viruses from Iceland, SSV2 (Stedman et al. 2003). Unexpectedly the SSV2 isolate also contained a satellite plasmid-virus hybrid, pSSVx (Arnold et al. 1999). I also collected and sequenced new SSVs from Kamchatka and Yellowstone National Park (Wiedenheft et al. 2004). All SSV viruses that I worked with were isolated on *S. solfataricus* strain P2, but some have different host ranges, and some were isolated on other strains (Redder et al. 2009; Ceballos et al. 2012). Between Herr Zillig, my colleagues Rachel Whitaker, Roger Garrett, Christa Schleper, and David Prangishvili, their co-workers, and myself, we have isolated about 20 new SSV viruses and proviruses and have sequenced their ~15 kbp genomes (Held, Whitaker 2009; Redder et al. 2009). Notably, even though genome synteny is conserved, their genomes are only ~50% identical at the nucleotide level and only ~50% of their ORFs are conserved (Fig. 5). Conserved genes are apt to be critical for SSV function, while those not conserved may reflect specialization among these different viruses. Consistent with those ideas, the VP2 gene that we were able to knock out of SSV1 is not well-conserved (Iverson, Stedman 2012), whereas the B129 ORF that appears to be essential in SSV1 is well-conserved.

Most of the biochemical research to investigate the functions of SSV1 proteins was done by Herr Zillig's co-workers. Georgi Muskhelishvili showed that the viral integrase has the expected nuclease and ligase activity (Muskhelishvili, Palm, Zillig 1993). This work has been followed up by Marie Claude Serre and colleagues in Paris (Letzelter, Duguet, Serre 2004). Wolf-Deiter Reiter showed that the SSV1 VP2 protein binds to DNA non-specifically (Reiter 1985) and we have similar unpublished results with recombinant SSV1 VP2 protein. Strangely, no high-affinity site-specific DNA binding proteins, other than the integrase, have been identified and characterized in SSV1. This is despite bioinformatic prediction of three ribbon-helix-helix putative DNA-binding proteins, three ORFs with putative zinc-fingers, (including SSV1 ORF B129 that appears to be essential for virus replication), and one ORF with a predicted helix-turn-helix DNA-binding motif (Lawrence et al. 2009; Fusco et al. 2013). Martin Lawrence and colleagues at Montana State University have determined high-resolution structures for the products of five SSV1 ORFs, some of which posses non-specific DNA-binding activity (Lawrence et al. 2009). These proteins probably need other partners or as yet untested conditions to bind DNA specifically.

Act 6. STIV: A voyage back in time?

Martin Lawrence and colleagues have also been determining structures for another virus that I discovered in Yellowstone while looking for SSVs. (Lawrence et al. 2009; Veesler et al. 2013). My discovery had come a few weeks after collecting samples from an acidic hot spring in the Yellowstone backcountry (See Stedman, Porter, Dyall-Smith 2010 for more information on how to isolate archaeal viruses). Now I was at the TEM screening through *Sulfolobus* cultures derived from single colonies isolated from a resulting enrichment culture. I was looking for some of my beloved SSVs, when all of a sudden an icosahedral particle practically jumped off the phosphor screen and into my lap. That virus was STIV (Fig. 2; *see page 7-22*). I thought to myself that I was probably the first person in the world to have ever seen this virus or anything like it. Quite an epiphany.

STIV is fascinating for a number of reasons. It is the first archaeal virus described that has a relatively 'normal' icosahedral capsid morphology—a particularly striking finding given the array of weird morphologies seen in the previously known viruses of thermoacidophilic Archaea. Having a symmetrical virion allowed Jack Johnson and coworkers to rapidly determine its 3D structure using cryo-electron microscopy (Fig. 2; Rice et al. 2004]). Looking at the whole virion, one can't help but admire the attractive knobs at the twelve vertices (the five-fold axes of symmetry), but the most surprising finding awaited closer examination. Based on the cryo-EM structure and confirmed by subsequent X-ray crystallography, the STIV major capsid protein was found to be practically identical in structure to the major capsid proteins of the bacterial virus PRD1 and the algal virus PBCV-1

(Rice et al. 2004; Khayat et al. 2005). This was unexpected given that there is no detectable amino acid sequence similarity between these proteins.

How might the structural similarity of these capsid proteins have arisen? We have several hypotheses. Convergent evolution is one possibility, but this would require that the same outcome arose independently in three extremely diverse lineages: Bacteria, Archaea and Eukarya. Moreover, clearly other solutions to the problem of how to make an icosahedral virion are available, solutions that are used by other viruses. On the other hand, horizontal gene transfer between viruses infecting different domains could yield the observed similarity, or similarly ancestors of these viruses might have switched host range across domains. Both of these seem unlikely due to the markedly different mechanisms of gene regulation in the three domains. We think that the most parsimonious argument is that all three viruses—STIV, PRD1, and PBCV-1—are descended from a common ancestral virus existing billions of years ago, and predating the divergence of the three domains, whose major capsid protein fold was very similar to that found in these three extant viruses (Rice et al. 2004). Until we have a reliable time machine available, it will be challenging to determine which of these explanations is correct.

Adding a bit more mystery to the story, both STIV with its icosahedral virion and the structurally and genetically divergent Rudivirus SIRV (*Sulfolobus islandicus* rod-shaped virus; Fig. 2) use a unique virion exit mechanism (*see page 7-25*). During infection, the virus directs the formation of heptagonal pyramidal structures in the host membrane built from multiple copies of a single viral protein (Bize et al. 2009; Brumfield et al. 2009). After progeny virions have accumulated in the host, these pyramids open, forming flower-shaped portals through which the virions escape. There is only one gene that is conserved between the two viruses, and that is the gene encoding the protein involved in pyramid formation. Expression of this protein is necessary and sufficient for pyramid formation—even when expressed heterologously

Figure 6. Spring KS1 in Devil's Kitchen, Lassen Volcanic National Park (orange arrow). September 2003. Photo credit: K. Stedman. (Editor's note: You're even crazier than I thought to walk on this stuff!)

in *E. coli* or *Saccharomyces cerevisiae* (Quax et al. 2011)! These archaeal viruses with their unique virions undoubtedly contain many more surprises awaiting our discovery.

Act 7. Beyond the Devil's Kitchen

The diverse virion morphologies of viruses of the thermoacidophilic Crenarchaea also set them apart from most of the other known viruses of Archaea. Most of those others studied to date, particularly those from halophilic and methanogenic Archaea, have the typical head-and-tail virion morphology. This may, however, reflect our biased searching, since the discovery of the unusual virion morphologies first by Wolfram Zillig and then by David Prangishvili and others came much later (Stedman, Prangishvili, Zillig 2006). In the last five years, now that we are looking for them, a number of odd-shaped virions have been found in environments dominated by Archaea, such as hypersaline lakes (e.g. Sime-Ngando et al. 2011). The recent wide survey of halophilic environments by Dennis Bamford's group netted them a large number of new halophilic viruses (Atanasova et al. 2012), including the first archaeal virus with a single-stranded DNA (ssDNA) genome, the *Halorubrum* pleomorphic virus-1 (HRPV-1; [Pietilä et al. 2009]). This virus seems to represent a continuum between plasmids and viruses, as both related plasmids and viruses with double stranded genomes have also been found (Roine et al. 2010). Viruses of extremely halophilic Archaea also include a spindle-shaped Fusellovirus, a virion morphology seen so far in a diversity of extreme environments, but only in viruses of Archaea (Bath, Dyall-Smith 1998). Whether the spindle-shape will be found to be unique to archaeal viruses, or used also by other viruses found in extreme environments is presently an open question (Krupovic et al. 2014).

Another archaeal virus with ssDNA was recently discovered that infects a thermophilic archaeon, the *Aeropyrum* coil-shaped virus (ACV; [Mochizuki et al. 2012]). ACV has an astonishingly large 25 kb genome—twice the size of the largest ssDNA genome previously known. It also adds yet another unusual virion structure to the world of archaeal viruses, this one being a hollow cylinder formed by a coiled nucleoprotein.

The anaerobic environments inhabited by methanogenic Archaea are, in their own way, an extreme environment, and the viruses that infect these Archaea represent a vast terra incognita. So far, a strange spindle-shaped virion has been reported in a culture of the anaerobic archaeon *Methanococcus voltae* (Wood, Whitman, Konisky 1989), and putative proviruses have been found in the genome of *M. voltae* and other archaeal methanogens (Krupovic, Bamford 2008). The little-known viruses of the archaeal methanogens that inhabit animal rumens and other environments warrant special attention as they are potentially useful biocontrol agents to reduce methane emissions— a possibility of increasing importance given the

major contribution of atmospheric methane as a greenhouse gas.

But not all Archaea are extremophiles. Others appear to constitute a large fraction of the microbial biomass in the world's oceans (Karner, DeLong, Karl 2001). To date, none of their viruses have been isolated—but not for want of trying.

Act 8. The next (careful) steps on the road ahead

Above and beyond finding new archaeal viruses, what else lies ahead in the field of archaeal virus research? Staring into my crystal ball, I can see breakthroughs coming soon in several areas, and undoubtedly there are others hidden by the swirling steamy acidic mists. Virus-host interactions have been understudied to date, with the notable exception of STIV and SIRV (Brumfield et al. 2009; Quax et al. 2013) for which microarrays, proteomics and RNA-Seq have been performed. As usual with these "omics" techniques, this work raised more questions to which it cannot provide answers. One fascinating outcome of the SIRV work is that transcription of the host CRISPR/Cas genes is very strongly induced upon virus infection (Quax et al. 2013). As of yet, no archaeal virus receptors have been conclusively identified. Mechanisms of virus entry are also totally obscure. Genetic analyses of the genetically tractable viruses, particularly SSV1, have already served up some surprising developments and can be guaranteed to yield many more (unpublished results).

Elucidating how the viruses of the thermoacidophilic Archaea assemble their unique virions is likely to take a bit longer to do. We now have a few of the tools needed to study the molecular basis of their assembly. The first step will be to determine both their virion and their protein structures at high resolution, something that has been done only for STIV to date. As a bonus, this structural information may provide insight into the factors influencing thermal and acid stability of their virions and as well as the stability of proteins and macromolecular structures in general. Much novelty remains to be discovered at this level, too, as evidenced by our observation that the main structural protein of the SSV1 virion appears to have a new fold (M.C. Morais and K.M. Stedman, unpublished).

Epilogue

When I embarked on that hunt for more SSVs at Devil's Kitchen on that September afternoon in 2003, I went prepared. Herr Zillig had taught me well. Don't even think about going into a thermal area without rubber boots, or, better yet, the hip waders that many of my colleagues wear. Being sensibly attired on that afternoon, I pulled my somewhat warmed foot (with boot) back out of the hole and proceeded on to the spring that had lured us into this solfataric field. I collected a sample of its bubbling water, a sample that we would later find contained a new SSV virus—a prize well worth this slight misadventure. We named that virus SSV-L for Lassen Volcanic National Park (K.M. Stedman, unpublished). And the hole that my boot made, which persists to this day, my students called spring KS1 (Fig. 6). We have not sampled KS1 for new viruses yet. To do so, maybe we should use drones—but then we'd be missing half the fun and most of the adrenaline kick.

References

Albers, S-V, P Forterre, D Prangishvili, C Schleper. 2013. The legacy of Carl Woese and Wolfram Zillig: From phylogeny to landmark discoveries. Nat Rev Microbiol 11:713-719.

Albers, SV, M Jonuscheit, S Dinkelaker, T Urich, A Kletzin, R Tampé, AJM Driessen, C Schleper. 2006. Production of recombinant and tagged proteins in the hyperthermophilic archaeon *Sulfolobus solfataricus*. Appl Environ Microbiol 72:102-111.

Arnold, HP, Q She, H Phan, K Stedman, D Prangishvili, I Holz, JK Kristjansson, R Garrett, W Zillig. 1999. The genetic element pSSVx of the extremely thermophilic crenarchaeon *Sulfolobus* is a hybrid between a plasmid and a virus. Mol Microbiol 34:217-226.

Arnold, HP, W Zillig, U Ziese, et al. 2000. A novel lipothrixvirus, SIFV, of the extremely thermophilic crenarchaeon *Sulfolobus*. Virology 267:252-266.

Atanasova, NS, E Roine, A Oren, DH Bamford, HM Oksanen. 2012. Global network of specific virus-host interactions in hypersaline environments. Environ Microbiol 14:426-440.

Bath, C, ML Dyall-Smith. 1998. His1, an archaeal virus of the Fuselloviridae family that infects *Haloarcula hispanica*. J Virol 72:9392-9395.

Bize, A, EA Karlsson, K Ekefjärd, TEF Quax, M Pina, MC Prevost, P Forterre, O Tenaillon, R Bernander, D Prangishvili. 2009. A unique virus release mechanism in the Archaea. Proc Natl Acad Sci USA 106:11306-11311.

Brumfield, SK, AC Ortmann, V Ruigrok, P Suci, T Douglas, MJ Young. 2009. Particle assembly and ultrastructural features associated with replication of the lytic archaeal virus *Sulfolobus* turreted icosahedral virus. J Virol 83:5964-5970.

Ceballos, RM, CD Marceau, JO Marceau, S Morris, AJ Clore, KM Stedman. 2012. Differential virus host-ranges of the Fuselloviridae of hyperthermophilic Archaea: Implications for evolution in extreme environments. Front Microbiol 3:295.

Clore, AJ, KM Stedman. 2007. The SSV1 viral integrase is not essential. Virology 361:103-111.

DeYoung, M, M Thayer, J van der Oost, KM Stedman. 2011. Growth phase-dependent gene regulation *in vivo* in *Sulfolobus solfataricus*. FEMS Microbiol Lett 321:92-99.

Eilers, BJ, MJ Young, CM Lawrence. 2012. The structure of an archaeal viral integrase reveals an evolutionarily conserved catalytic core yet supports a mechanism of DNA cleavage in trans. J Virol 86:8309-8313.

Fröls, S, PMK Gordon, MA Panlilio, C Schleper, CW Sensen. 2007. Elucidating the transcription cycle of the UV-inducible hyperthermophilic archaeal virus SSV1 by DNA microarrays. Virology 365:48-59.

Fusco, S, Q She, S Bartolucci, P Contursi. 2013. T-lys, a newly identified *Sulfolobus* spindle-shaped virus 1 transcript expressed in the lysogenic state, encodes a DNA-binding protein interacting at the promoters of the early genes. J Virol 87:5926-5936.

Häring, M, R Rachel, X Peng, RA Garrett, D Prangishvili. 2005a. Viral diversity in hot springs of Pozzuoli, Italy, and characterization of a unique archaeal virus, acidianus bottle-shaped virus, from a new family, the *Ampullaviridae*. J Virol 79:9904-9911.

Häring, M, G Vestergaard, R Rachel, LM Chen, RA Garrett, D Prangishvili. 2005b. Virology: Independent virus development outside a host. Nature 436:1101-1102.

Held, NL, RJ Whitaker. 2009. Viral biogeography revealed by signatures in *Sulfolobus islandicus* genomes. Environ Microbiol 11:457-466.

Huet, J, R Schnabel, A Sentenac, W Zillig. 1983. Archaebacteria and eukaryotes possess DNA-dependent RNA-polymerases of a common type. EMBO J 2:1291-1294.

Iverson, E, K Stedman. 2012. A genetic study of SSV1, the prototypical fusellovirus. Front Microbiol 3:200.

Jonuscheit, M, E Martusewitsch, KM Stedman, C Schleper. 2003. A reporter gene system for the hyperthermophilic archaeon *Sulfolobus solfataricus* based on a selectable and integrative shuttle vector. Mol Microbiol 48:1241-1252.

Karner, MB, EF DeLong, DM Karl. 2001. Archaeal dominance in the mesopelagic zone of the Pacific Ocean. Nature 409:507-510.

Khayat, R, L Tang, ET Larson, CM Lawrence, M Young, JE Johnson. 2005. Structure of an archaeal virus capsid protein reveals a common ancestry to eukaryotic and bacterial viruses. Proc Natl Acad Sci USA 102:18944-18949.

Klose, KE, AK North, KM Stedman, S Kustu. 1994. The major dimerization determinants of the nitrogen regulatory protein NTRC from enteric bacteria lie in its carboxy-terminal domain. J Mol Biol 241:233-245.

Krupovic, M, DH Bamford. 2008. Archaeal proviruses TKV4 and MVV extend the PRD1-adenovirus lineage to the phylum Euryarchaeota. Virology 375:292-300.

Krupovic, M, ERJ Quemin, DH Bamford, P Forterre, D Prangishvili. 2014. Unification of the globally distributed spindle-shaped viruses of the Archaea. J Virol 88:2354-2358.

Lawrence, CM, S Menon, BJ Eilers, B Bothner, R Khayat, T Douglas, MJ Young. 2009. Structural and functional studies of archaeal viruses. J Biol Chem 284:12599-12603.

Letzelter, C, M Duguet, MC Serre. 2004. Mutational analysis of the archaeal tyrosine recombinase SSV1 integrase suggests a mechanism of DNA cleavage in trans. J Biol Chem 279:28936-28944.

Martin, A, S Yeats, D Janekovic, WD Reiter, W Aicher, W Zillig. 1984. SAV-1, a temperate UV-inducible DNA virus-like particle from the archaebacterium *Sulfolobus-acidocaldarius* isolate B-12. EMBO J 3:2165-2168.

Menon, SK, BJ Eilers, MJ Young, CM Lawrence. 2010. The crystal structure of D212 from *Sulfolobus* spindle-shaped virus ragged hills reveals a new member of the PD-(D/E)XK nuclease superfamily. J Virol 84:5890-5897.

Mochizuki, T, M Krupovic, G Pehau-Arnaudet, Y Sako, P Forterre, D Prangishvili. 2012. Archaeal virus with exceptional virion architecture and the largest single-stranded DNA genome. Proc Natl Acad Sci USA 109:13386-13391.

Muskhelishvili, G, P Palm, W Zillig. 1993. SSV1-encoded site-specific recombination system in *Sulfolobus shibatae*. Mol Gen Genet 237:334-342.

Nadal, M, G Mirambeau, P Forterre, WD Reiter, M Duguet. 1986. Positively supercoiled DNA in a virus-like particle of an archaebacterium. Nature 321:256-258.

Palm, P, C Schleper, B Grampp, S Yeats, P McWilliam, WD Reiter, W Zillig. 1991. Complete nucleotide sequence of the virus SSV1 of the archaebacterium *Sulfolobus shibatae*. Virology 185:242-250.

Pietilä, MK, E Roine, L Paulin, N Kalkkinen, DH Bamford. 2009. An ssDNA virus infecting archaea: A new lineage of viruses with a membrane envelope. Mol Microbiol 72:307-319.

Prangishvili, D. 2013. The wonderful world of archaeal viruses. Annu Rev Microbiol 67:565-585.

Prangishvili, D, HP Arnold, D Götz, U Ziese, I Holz, JK Kristjansson, W Zillig. 1999. A novel virus family, the *Rudiviridae*: Structure, virus-host interactions and genome variability of the *Sulfolobus* viruses SIRV1 and SIRV2. Genetics 152:1387-1396.

Quax, TEF, S Lucas, J Reimann, G Pehau-Arnaudet, M-C Prevost, P Forterre, S-V Albers, D Prangishvili. 2011. Simple and elegant design of a virion egress structure in Archaea. Proc Natl Acad Sci USA 108:3354-3359.

Quax, TEF, M Voet, O Sismeiro, et al. 2013. Massive activation of archaeal defense genes during viral infection. J Virol 87:8419-8428.

Redder, P, X Peng, K Brügger, et al. 2009. Four newly isolated fuselloviruses from extreme geothermal environments reveal unusual morphologies and a possible interviral recombination mechanism. Environ Microbiol 11:2849-2862.

Reiter, W. 1985. Das Virusartige Partikel SSV-1 von *Sulfolobus solfataricus* Isolat B12: UV-Induktion, Reinigung und Characterizierung. Diploma thesis, Eberhard-Karls Universität, Tübingen.

Reiter, WD, P Palm. 1990. Identification and characterization of a defective SSV1 genome integrated into a tRNA gene in the archaebacterium *Sulfolobus* sp. B12. Mol Gen Genet 221:65-71.

Reiter, WD, P Palm, A Henschen, F Lottspeich, W Zillig, B Grampp. 1987a. Identification and characterization of the genes encoding three structural proteins of the *Sulfolobus* virus-like particle SSV1. Mol Gen Genet 206:144-153.

Reiter, WD, P Palm, S Yeats, W Zillig. 1987b. Gene-expression in Archaebacteria–Physical mapping of constitutive and UV-inducible transcripts from the *Sulfolobus* virus-like particle SSV1. Mol Gen Genet 209:270-275.

Reiter, WD, P Palm, W Zillig. 1988. Analysis of transcription in the archaebacterium *Sulfolobus* indicates that archaebacterial promoters are homologous to eukaryotic pol-II promoters. Nucleic Acids Res 16:1-19.

Rice, G, L Tang, K Stedman, F Roberto, J Spuhler, E Gillitzer, JE Johnson, T Douglas, M Young. 2004. The structure of a thermophilic archaeal virus shows a double-stranded DNA viral capsid type that spans all domains of life. Proc Natl Acad Sci USA 101:7716-7720.

Roine, E, P Kukkaro, L Paulin, S Laurinavicius, A Domanska, P Somerharju, DH Bamford. 2010. New, closely related haloarchaeal viral elements with different nucleic acid types. J Virol 84:3682-3689.

Rueckert, RR, W Zillig. 1962. Biosynthesis of viral protein in *Escherichia coli* C *in vivo* following infection with bacteriophage φX174. J Mol Biol 5:1-&.

Schleper, C, K Kubo, W Zillig. 1992. The particle SSV1 from the extremely thermophilic archaeon *Sulfolobus* is a virus: Demonstration of infectivity and of transfection with viral DNA. Proc Natl Acad Sci USA 89:7645-7649.

Schnabel, H, W Zillig, M Pfäffle, R Schnabel, H Michel, H Delius. 1982. *Halobacterium halobium* phage øH. EMBO J 1:87.

Schramm, G, G Schumacher, W Zillig. 1955. Infectious nucleoprotein from tobacco mosaic virus. Nature 175:549-550.

Servin-Garcidueñas, LE, X Peng, RA Garrett, E Martinez-Romero. 2013. Genome sequence of a novel archaeal fusellovirus assembled from the metagenome of a Mexican hot spring. Genome Announc 1:e0016413-e0016413.

Sime-Ngando, T, S Lucas, A Robin, et al. 2011. Diversity of virus-host systems in hypersaline Lake Retba, Senegal. Environ Microbiol 13:1956-1972.

Stedman, KM, K Porter, ML Dyall-Smith. 2010. The isolation of viruses infecting Archaea. In: SW Wilhelm, MG Weinbauer, CA Suttle, editors. *Manual of Aquatic Viral Ecology*: American Society for Limnology and Oceanography. p. 57-64.

Stedman, KM, D Prangishvili, W Zillig. 2006. Viruses of Archaea. In: R Calendar, editor. *The Bacteriophages*: Oxford University Press. p. 499-516.

Stedman, KM, C Schleper, E Rumpf, W Zillig. 1999. Genetic requirements for the function of the archaeal virus SSV1 in *Sulfolobus solfataricus*: Construction and testing of viral shuttle vectors. Genetics 152:1397-1405.

Stedman, KM, QX She, H Phan, HP Arnold, I Holz, RA Garrett, W Zillig. 2003. Relationships between fuselloviruses infecting the extremely thermophilic archaeon *Sulfolobus*: SSV1 and SSV2. Res Microbiol 154:295-302.

Stetter, KO, W Zillig. 1979. Unusual DNA-dependent RNA-polymerases in archaebacteria. Hoppe Seylers Z Physiol Chem 360:381-382.

Veesler, D, T-S Ng, AK Sendamarai, BJ Eilers, CM Lawrence, S-M Lok, MJ Young, JE Johnson, C-Y Fu. 2013. Atomic structure of the 75 MDa extremophile *Sulfolobus* turreted icosahedral virus determined by CryoEM and X-ray crystallography. Proc Natl Acad Sci USA 110:5504-5509.

Walter, G, W Seifert, W Zillig. 1968. Modified DNA-dependent RNA polymerase from *E. coli* infected with bacteriophage T4. Biochem Biophys Res Commun 30:240-&.

Wiedenheft, B, K Stedman, F Roberto, D Willits, AK Gleske, L Zoeller, J Snyder, T Douglas, M Young. 2004. Comparative genomic analysis of hyperthermophilic archaeal *Fuselloviridae* viruses. J Virol 78:1954-1961.

Wood, AG, WB Whitman, J Konisky. 1989. Isolation and characterization of an archaebacterial virus-like particle from *Methanococcus voltae* A3. J Bacteriol 171:93-98.

Yeats, S, P McWilliam, W Zillig. 1982. A plasmid in the archaebacterium *Sulfolobus-acidocaldarius*. EMBO J 1:1035-1038.

Zillig, W, HP Arnold, I Holz, D Prangishvili, A Schweier, K Stedman, Q She, H Phan, R Garrett, JK Kristjansson. 1998. Genetic elements in the extremely thermophilic archaeon *Sulfolobus*. Extremophiles 2:131-140.

Zillig, W, F Gropp, A Henschen, H Neumann, P Palm, WD Reiter, M Rettenberger, H Schnabel, S Yeats. 1986. Archaebacterial virus host systems. Syst Appl Microbiol 7:58-66.

Zillig, W, KO Stetter, D Janekovic. 1979. DNA-dependent RNA polymerase from the archaebacterium *Sulfolobus acidocaldarius*. Eur J Biochem 96:597-604.

Zillig, W, KO Stetter, M Tobien. 1978. DNA-dependent RNA polymerase from *Halobacterium halobium*. Eur J Biochem 91:193-199.

Zillig, W, KO Stetter, S Wunderl, W Schulz, H Priess, I Scholz. 1980. The *Sulfolobus*-"Caldariella" group: Taxonomy on the basis of the structure of DNA-dependent RNA polymerases. Arch Microbiol 125:259-269.

Zillig, W, J Tu, I Holz. 1981. Thermoproteales—A third order of the thermoacidophilic archaebacteria. Nature 293:85-86.

55

different structural proteins in the virion of *Bacillus thuringiensis* phage 0305φ8-36

Thomas, JA, SC Hardies, M Rolando, SJ Hayes, K Lieman, CA Carroll, ST Weintraub, P Serwer. 2007. Complete genomic sequence and mass spectrometric analysis of highly diverse, atypical *Bacillus thuringiensis* phage 0305φ8–36. Virology 368:405-421.

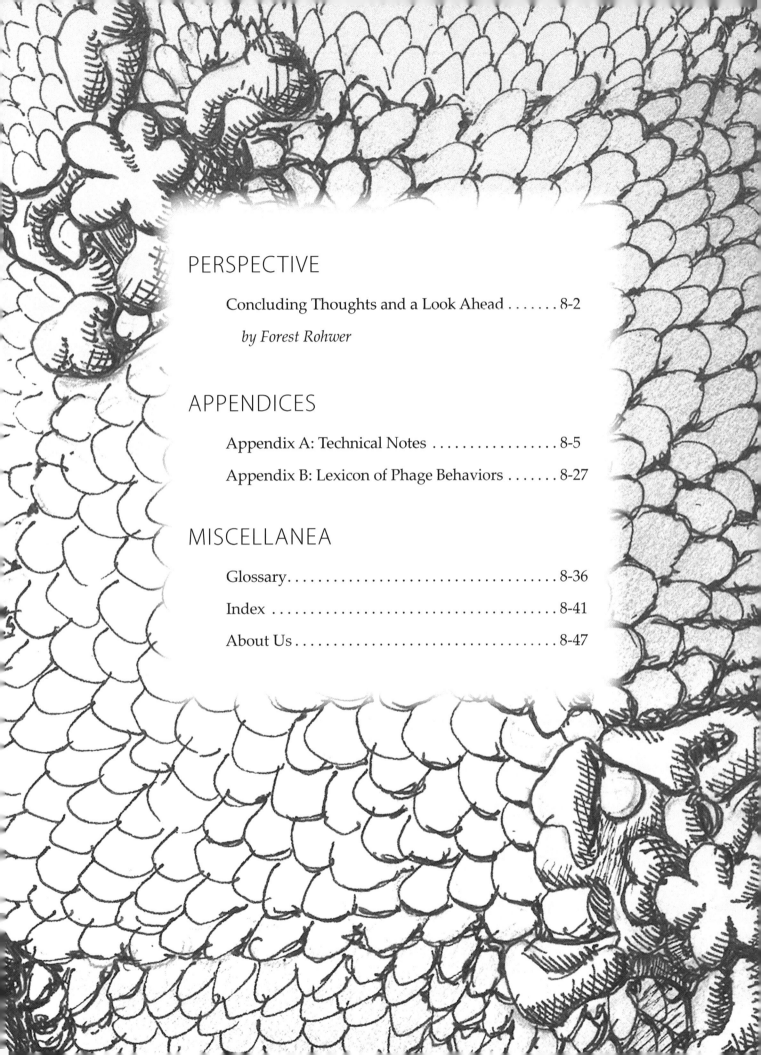

PERSPECTIVE

APPENDICES

MISCELLANEA

Concluding Thoughts and a Look Ahead

by Forest Rohwer

Figure 1. Viral sampling on Hall Island in Franz Josef Land. Cory Richards, risking ignominy and sacrificing comfort, entered the freezing waters, tube in gloved hand, determined to bring back a single precious sample or die.

Where do you sample for phage when in a completely unexplored land? This was the challenge that faced Steve Quistad and myself in Franz Josef Land, the most northern territory in Eurasia. One approach is to just sample everything you see, but most of these samples will just sit in the freezer unloved and unanalyzed. I know, because my lab has several -80 °C freezers housing just these sort of samples. It is important to have some reason to sample a particular site. Then someone may actually take responsibility for analyzing the data.

To avoid this 'unloved samples' phenomenon in FJL, we concentrated on a few key ecological or geological features: places where the glaciers were retreating, blood snow that might be contributing to melting, walrus and polar bear poop, surfaces of sea anemones, etc. However, this approach wasn't taking advantage of the collection potential of tens of other expedition members who were traipsing over this unique land, climbing the cliffs, or diving in the freezing waters (-1.8 °C). So while drinking some of the worst wine in the world, we came up with an idea: let anyone sample something that they found interesting. And since we were in a new land, we let them name their sampling spot. So 50 ml conical tubes went into the hands of Russian guards, photographers, geologists, bird scientists, various hangers-on, and even the loneliest Fish Guy in the world. (The Fish Guy was lonely; there are almost no fish in the freezing waters of FJL.)

Soon we started getting samples back from all over the archipelago. There were samples of ptarmigan poop named "To Be or Not To Be," a mossy loam called "Wegley Island," a fresh water slime named "Willow's Wallow," a weirdly-colored soil dubbed "Paul's Pink," a tuft of polar bear fur christened "Winnie," and many more. All the newly-minted phage hunters conscientiously delivered GPS coordinates and photos of their samples, and used gloves when collecting.

Not to be out done, Cory Richards, one of the National Geographic photographers, decided to sample the phage in one of the very shallow lakes on Hall Island. To avoid freezing to death, he put on his hand-me-down dry suit (complete with customized dishwashing gloves) and waded into the lake, tube in hand. A dry suit works by keeping a cushion of air between you and the freezing environment. This keeps you alive, but also makes you exceedingly buoyant (a you-need-to-carry-40-pounds-of-lead-in-order-to-sink sort of buoyant). Soon Cory's feet were floating as he floundered around trying to paddle to the middle of the lake.

To help, those of us on shore threw small rocks at him in hopes of creating waves to move him along, or maybe just because it was fun to throw rocks at someone you knew. You can get very bored in the frozen North. Soon a breeze joined in and 'sailed' him along the shore. Eventually Cory hit on the idea of floating on his back and 'rowing' with his hands. After only about 45 minutes he managed to make it to the middle of the lake (at least

Figure 2. The infamous viral explorers Steven Quistad (left) & Cory Richards (right) reunited after Cory's successful sampling 50 feet from shore, in a lake that was at least 5 feet deep. One of the more pivotal, emotional moments in Arctic phage exploration.

50 feet from shore), sample, and return to shore where he was warmly received by his fellow explorer pre-Dr. Quistad (see photo). It was almost as emotional as Nansen running into Jackson after the race to the North Pole. The sample, labeled "Cory's Crawl," has sat at -80 °C next to all of the other unloved samples to this day.

We hope that you have enjoyed this survey of phage diversity. Phage exploration is exciting and essentially a completely open field with room for newcomers alongside confirmed devotees. Most of the Franz Josef Land samples will probably never be analyzed, and even for those lucky few that eventually are, we will only scratch the surface of what they have to tell. A look at the maps in this book shows that most of the world's biomes have not been sampled even once. What other field of biology offers so much *terra nova*? What's more, advances in the phage field have moved mainstream science forward time and time again. Fruitful studies can range from the isolation, culturing, and sequencing of new phages by students to some employing the most costly, high-tech methods in biology such as tomography and metabolomics. Anyone can join in and make very significant contributions.

In this book we have introduced a number of innovations that we hope will be sustained by others into the future. The use of a lexicon (detailed in Appendix B) that conveys the active, organic character of phage is important for reminding people that these are the most abundant and successful organisms on the planet. This lexicon helps us to imagine additional phage behaviors that, once imagined, some young, motivated scientist will look for and indeed find. In the future, these behaviors will be linked to specific genes, cellular structures, and environments.

The widespread adoption of a genome-based phage taxonomy is how I envision the future. However, as pointed out by a number of the guest authors, this is by no means a finished discussion. Today's various taxonomical systems will continue to evolve and in so doing may clarify, rather than confound, our ability to talk about phages to one another. Phages, more than other organisms, remind us of the transient nature of a species. Evolution is on-going and Darwin's Doubtful Species concept keeps us on our toes.

The genomic and physical maps included are to give you an idea about what the phages are doing and where they are found. The production of these maps will be automated in the near future and generated for many more phages. The stories and artwork are another matter. Each of these represents much love and care. It is just going to take a lot of excited artists, writers, and scientists working together to produce future versions. My present vision for the next couple of years is to solicit more of these stories and perspectives to fill a second edition of this book. It is particularly important to capture the insights and personal experiences of our oldest generation of phage biologists. Most of them have already retired and several of the greats have already passed away. It would be a great shame if the future generations of phage biologists do not know about Lambda Lunches, the Intergalactic Phage Meetings, etc.

Phage are going to be extremely important in the future of biology. They will be the tools used to engineer the human microbiome, even though the 'microbiomists' are still ignoring them! Not only can the bacteriophage and other microbial viruses tell us something about what makes for biological success, but they have properties that point to currently unknown scientific territory. Phage are essentially different from cells. For cellular life, material substance passes from parent to daughter cells during cellular division; in contrast, it is information, not parental material, that travels from generation to generation during lytic phage replication. Phage are an idea that perpetuates itself in a manner analogous to what humans do through books. Scientifically this is quantified by Information theory. Understanding how phage actually perpetuate as an idea will give us insights into both the natural and human worlds. Phage are also extremely small, so small that they exist below the cutoff between the quantum and classical physical worlds. This observation is both exciting and disconcerting. Biology is grounded in the classical physics of the 1800. It is time to move forward and develop a biology informed by modern physical models of the Universe. And I fully expect phage research to lead the way as it has done for the last 100 years.

Excerpted from a speech given on the occasion of the bicentennial of phage discovery: "Looking back on this occasion, we are reminded that it was the phage therapy technology developed in the 2050s that led to the era of remote manufacturing controlled by Information transfer. In turn, the ability to remotely manufacture by moving Information alone allowed humanity in the early years of the twenty-second century to explore the Universe by propagating biological explorers throughout the Galaxy. Now in 2115, the bicentennial of the discovery of phage, we can see a new era opening where phage can be used to immortalize our very personalities. The next 100 years should be extremely exciting..."

Appendix A: Technical Notes

Appendix A1: Sources for Field Guide Data

Field guide data for each of the featured phages was derived from several sources.

Taxonomic designations for bacteriophages and archaeal viruses come from their proteome-based family assignment as shown in the phage proteomic tree (PPT); those for the eukaryotic viruses are their ICTV assignments.

Genome structure, length, and number of predicted ORFs were taken from the GenBank annotation files used for the genome maps[1].

Encapsidation method was inferred from the cited sources used for the story content. *Packaging* was designated for phages that translocate their genome into a preassembled procapsid; *co-condensation* was assigned to those whose nucleic acid and virion structural proteins assemble together.

Common host and habitat were obtained from the cited story references. Neither are exhaustive lists but rather are only a smattering of the phage's ecology. Typically habitat corresponds to where the host is known to reside.

Although most of the featured phages are known to have a canonical lytic or temperate lifestyle, some are designated as non-lytic (e.g., the budding of virions by enterobacteria phage f1) or non-lytic temperate (e.g., the combined budding and lysogenic strategies of *Acholeplasma* phage L2).

While the book features only 28 of the 1220 phages in the Phage Proteomic Tree, field guide data were collected for all 1220 using automated methods. A Perl script was written to search the National Center for Biotechnology Information (NCBI) nucleotide database[2] to obtain the following information for each phage: name, ICTV taxonomy, 'strandedness' (e.g., dsDNA, ssDNA), genome size (bp or nts), habitat, and location where isolated. The lifestyle of each phage was predicted using PHACTS (McNair et al., 2012).

Reference

McNair, K, BA Bailey, RA Edwards. 2012. PHACTS, a computational approach to classifying the lifestyle of phages. Bioinformatics 28:614-618.

[1] ftp.ncbi.nlm.nih.gov/genomes/Viruses/

[2] http://www.ncbi.nlm.nih.gov/

Appendix A2: Drawing Phage Portraits

A functional field guide uses both words and pictures to identify its subjects. Words can tell the reader about an organism's behavior whereas a picture will show what the organism looks like. The same is true for this field guide, with action stories that relate how a phage behaves and portraits that provide a glimpse of what each phage's virion would look like if caught on camera.

Tactics

Although virions are everywhere, obtaining their pictures is challenging and requires sophisticated microscopy techniques, even the best of which have drawbacks. Many virions have been captured in electron micrographs, but this approach cannot capture their three-dimensional structure and the resolution is inadequate. Both of those shortcomings are addressed by three-dimensional virion reconstructions generated using cryo-electron tomography, but these often cannot capture capsid and tail together, and non-isometric capsids pose a particular challenge. X-ray crystallography provides an even closer look for those virions that can be crystallized, and likewise for the various structural proteins comprising them.

Instead of choosing a single image or even one type of image to represent each of our featured phages, we perused the available images for data to serve as a basis for drawing the virions that they assemble. For some phages, virion structural data were abundant (e.g., T4, T7) whereas for others we only had predictions from a genome sequence (e.g., S-SSM7). Details known for one phage were, when necessary, applied to close relatives (e.g., T4, RB49, RB51, and RB69). We didn't think the phages would mind.

All portraits were based on structural data that were available in April 2014, with a dash of caricature added to fit the playful style of this book. Because virion structures are rapidly being described at finer scales, there will no doubt be structural updates that will fill in the gaps and/or change our understanding of certain virion structures. Thus, these portraits are only snapshots that represent our current interpretation of the data. We have made them as accurate as possible with the data at hand.

Geometry

Often when we think of what virions look like, we think of icosahedral symmetry. Of course not all phages have adopted this geometry for their capsids; notable exceptions include the helical architecture of filamentous phages, pleomorphic virions, and the impressive morphological diversity of Archaeal viruses (*see 7-29*). However, the best-studied phages build their capsids using an icosahedral blueprint. Why might this be such a common strategy? Early on it was proposed that viruses favor this structure because it can be easily constructed using a large number of small, chemically identical subunits (Crick, Watson 1956). Small is the keyword here. To make a fully functional protein coat from one or a few large protein molecules would require more protein coding capacity than allowed by the phage genome to be housed. Phages are thus forced to assemble their capsids from many copies of small proteins encoded by short genes.

The exact way in which these proteins assemble into icosahedral capsids varies between phages and is largely influenced by virion size. Typically one or more proteins assemble to form the protomers that are then assembled into two types of capsomers (pentamers and hexamers). Groups of five protomers assemble into the pentamers, 12 of which will form the vertices of the icosahedron. (For phages with a unique vertex for DNA packaging or delivery, such as

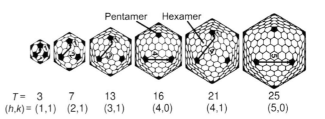

Figure 1. Capsomer assembly into icosahedra of various sizes. Variation in capsid size with increasing T-number. The number of hexamers (white) varies while the number of pentamers (black) stays constant. Arrows indicate the 'walk through' path to calculate h and k. (Mannige, Brooks III 2010)

the tailed phages, only 11 standard pentamers are needed.) Groups of six protomers assemble into the hexamers that will fill each face of the isosahedron. As a group, the phages have exploited the potential for building icosahedral capsids of different sizes by using different numbers of hexamers per face. These different architectures can be described by their triangulation number, or T-number, defined by the following equation: $T = h^2 + hk + k^2$. When moving from one pentamer vertex to an adjacent one, h refers to the number of hexamers 'walked through' in one direction and, if a 90° turn to the left is then required, k refers to the number of hexamers 'walked through' in that direction (Fig. 1). Thus, as a general rule, capsids with higher T-numbers are larger than those with smaller T-numbers (though this can also depend on the size of individual protomers).

We were able to obtain the T-numbers for the majority of our icosahedral featured phages from the literature, though some were assumed based on close relation to a phage with a known T-number. However, there were a few phages whose T-numbers are currently unknown (e.g., S-SSM7). For these we simply illustrated their capsids as icosahedra lacking this detail.

Cited reference

Crick, FH, JD Watson. 1956. Structure of small viruses. Nature 177:473-475.

Mannige, RV, CL Brooks III. 2010. Periodic table of virus capsids: Implications for natural selection and design. PLoS ONE 5:e9423.

Appendix A3: The Phage Proteomic Tree

Historically, when faced with the challenge of classifying phages, the most common approaches attempted to define a hierarchical Linnaean system based on observable traits, primarily virion morphology and genome type (*see 1-10*). The system introduced by the International Committee on the Taxonomy of Viruses (ICTV) in 1971 employs this approach and has been widely adopted. However the committee-based procedures of the ICTV cannot keep up with the escalating pace of phage discovery. Many phages now known from genome data cannot be cultured, thus determination of their morphology by EM is not possible. Most significantly, the extent of phage diversity and the phylogenetic relationships that have been inferred from genome data are simply not detectable when looking at the traits traditionally used for classification by such systems.

Phylogenetic relationships are often inferred by tracing the divergence of one gene (or several genes) from a common ancestor to all of its descendants. The resultant trees intuitively portray their evolutionary relationships. Such inferences are not possible with phages because no gene is shared by all phages. Thus, organizing phages into taxonomically useful groups remains challenging, despite—and partly because of—the amount of genome data now accumulating. The key to resolving this is to develop methods of analysis that infer relatedness from these data. We have used one such approach, one that individually compares all proteins encoded by each phage (i.e., its proteome) to the proteome of every other phage (Rohwer, Edwards 2002). The resultant similarity pairs are used to build a tree showing phage relatedness.

To estimate proteomic similarities, genome sequences from 1,220 phages infecting Bacteria and Archaea were first downloaded from NCBI[1] Genbank (see list below). To ensure the same annotation method was used for each genome, sequences were re-annotated prior to identifying individual

protein similarities. Protein sequences were predicted for each phage genome by searching for potential open reading frames and then translating those into protein sequences. This was done using the PhAnToMe[2] phage annotation pipeline, which predicts genes using a self-trained version of GeneMark[3].

To calculate distances between proteins, each protein sequence was compared with all other protein sequences using BLASTP (Altschul et al. 1997) with an e-value cutoff of 0.1. Pairwise protein comparisons with e-values below this cutoff were then grouped together using single linkage clustering; this resulted in 91,405 groups that each represented a protein family. All protein sequences in each group were aligned using CLUSTALW[4] with default settings, and pairwise distances were calculated between all members of a group using PROTDIST[5] with default parameters. This yielded a distance matrix for each protein family. These 91,405 individual matrices were then merged into a single master matrix. This matrix was created by taking the average distance of all proteins shared between each phage pair, imposing a maximum penalty of 10 for proteins not shared between each phage pair, and correcting for differences in protein length. The information in this single master matrix was utilized to generate the Phage Proteomic Tree using NEIGHBOR[6] with default parameters. This pipeline closely follows that reported for the original PPT (Rohwer, Edwards 2002). Scripts used in this pipeline are available at Sourceforge[7].

Our current tree (*pages 8-10 and 8-11*) includes 1220 phages with sequenced DNA or RNA ge-

---

[1] https://www.ncbi.nlm.nih.gov/

[2] http://www.phantome.org/PhageSeed/Phage.cgi?page=phas

[3] http://opal.biology.gatech.edu/

[4] http://www.clustal.org/clustal2/

[5] http://evolution.genetics.washington.edu/phylip/doc/protdist.html

[6] http://evolution.genetics.washington.edu/phylip/doc/neighbor.html

[7] http://edwards-sdsu.cvs.sourceforge.net/viewvc/edwards-sdsu/bioinformatics/phage_tree/

nomes, both bacteriophages and archaeal viruses. This PPT was drawn as an unrooted cladogram using the Interactive Tree of Life (iTOL) (Letunic, Bork 2007; Letunic, Bork 2011) and further edited in Adobe Photoshop CS5. The tree is unrooted because it is based on shared characters rather than the divergence of a single shared protein. Similarly, the lack of a single common ancestor precludes the usefulness of estimated branch lengths; thus these are not included in the tree graphic.

Phages with similarities cluster together as 'groups' in the PPT. Phages within a group share more of their proteome than phages in different groups. This sharing may be due to horizontal gene transfer among these phages or due to the presence of these genes in an ancestor of that group. The groups (represented by different colored branches) consist of lineages whose distance from each other was less than 0.05. Each group was named after the most commonly occurring ICTV family within the group, as in the original PPT (Rohwer, Edwards 2002). To distinguish between the morphology-based ICTV family classifications and the proteome-based PPT groups, for bacteriophage the suffix -viridae was replaced with -phage (i.e., Siphophage instead of *Siphoviridae*). To make a similar distinction for the archaeal phages, we used the suffix -viridi.

Many ICTV families were split into multiple groupings in the PPT, thus yielding several subgroups with the same family name (e.g., Siphophage). Subgroups were named by adding as a suffix the name of the most studied member, as determined using Google Scholar search results, and all subgroups are listed below. Additionally,

the ICTV family composition of each subgroup was calculated and the results shown as gearshaped 'pie charts' next to the relevant subgroup in the PPT. Individual branches that are not part of any group may be founding members of new groups that will become populated as additional phage genomes are sequenced.

Overall, the PPT groups echo the phage ICTV families. A closer look reveals considerable proteomic diversity within each ICTV-defined family. For instance, the PPT depicts seven groups of Siphophages and six groups of Podophages. Moreover, the majority of PPT-based groups include some phages with differing morphologies. The P2-like Myophage sub-family, for example, includes phages classified by the ICTV as *Myoviridae*, *Cystoviridae*, *Inoviridae*, and *Siphoviridae*. By not relying on morphology as the main criterion for classification, the PPT has provided some phages with a new identity. One phage thus reclassified is phage P4. Because it steals its virion proteins from a helper phage such as P2 that has a typical Myovirus morphology (*see 6-18*), P4 was previously classified as a Myovirus. Its proteome shows its Podophage character.

Located on the field guide page for each featured phage is a regional PPT that displays 30 – 40 of its closest neighbors with each phage name colored according to its ICTV family. These groups were manually chosen by visually selecting a clade of that size that includes the featured phage. Occasionally, the evident clade was significantly smaller (e.g., 15 phages), in which case the small size was retained rather than adding phages from a different clade.

Cited references

Altschul, SF, TL Madden, AA Schäffer, J Zhang, Z Zhang, W Miller, DJ Lipman. 1997. Gapped BLAST and PSI-BLAST: A new generation of protein database search programs. Nucleic Acids Res 25:3389-3402.

Letunic, I, P Bork. 2007. Interactive Tree Of Life (iTOL): An online tool for phylogenetic tree display and annotation. Bioinformatics 23:127-128.

Letunic, I, P Bork. 2011. Interactive Tree Of Life v2: Online annotation and display of phylogenetic trees made easy. Nucleic Acids Res 24:1641-1642.

Rohwer, F, R Edwards. 2002. The phage proteomic tree: a genome-based taxonomy for phage. J Bact 184:4529-4535.

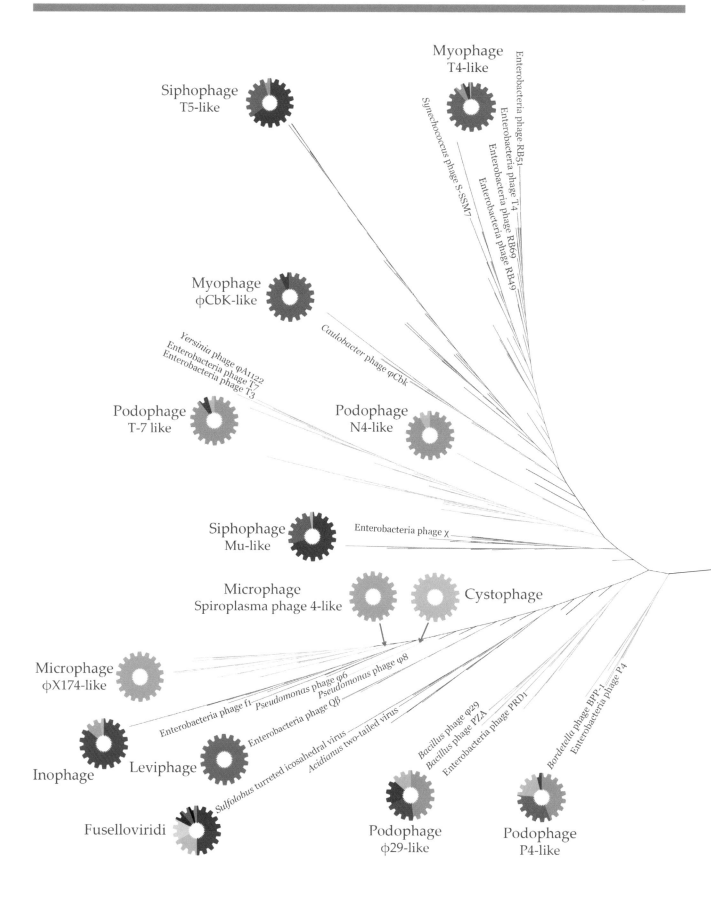

Myophage
T4-like

Enterobacteria phage RB51
Enterobacteria phage T4
Enterobacteria phage RB69
Enterobacteria phage RB49

Synechococcus phage S-SSM7

Siphophage
T5-like

Myophage
φCbK-like

Caulobacter phage φCbk

Yersinia phage φA1122
Enterobacteria phage T7
Enterobacteria phage T3

Podophage
T-7 like

Podophage
N4-like

Siphophage
Mu-like

Enterobacteria phage χ

Microphage
Spiroplasma phage 4-like

Cystophage

Microphage
φX174-like

Enterobacteria phage f1
Pseudomonas phage φ6
Pseudomonas phage φ8
Enterobacteria phage Qβ

Bacillus phage φ29
Bacillus phage PZA
Enterobacteria phage PRD1

Bordetella phage BPP-1
Enterobacteria phage P4

Inophage

Leviphage

Sulfolobus turreted icosahedral virus
Acidianus two-tailed virus

Fuselloviridi

Podophage
φ29-like

Podophage
P4-like

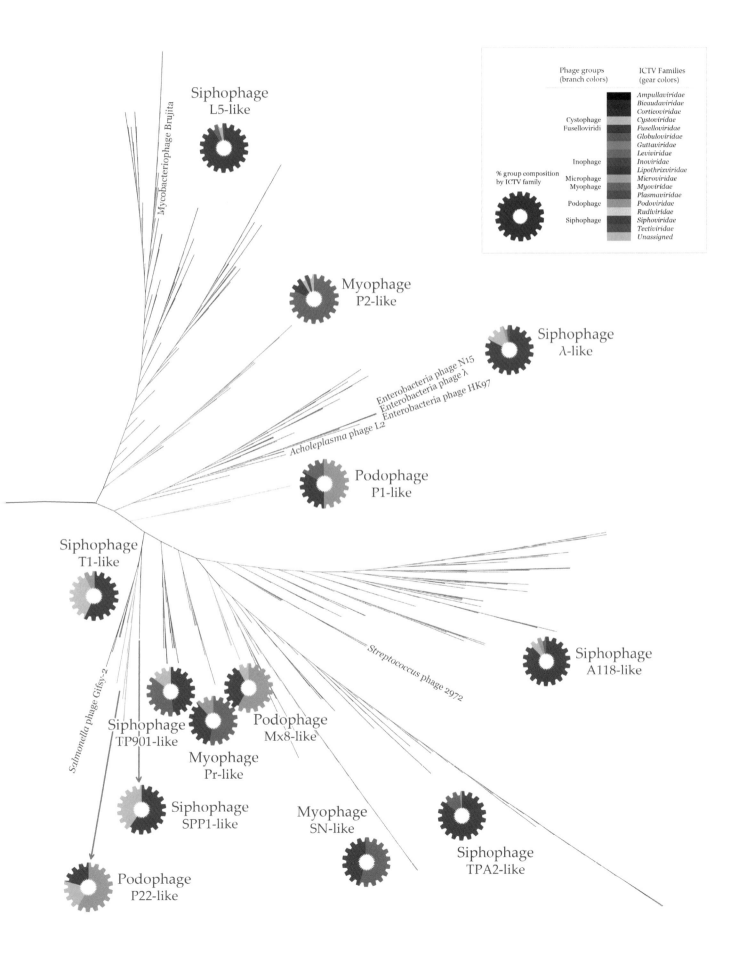

Bacteriophages and Archaeal Viruses in the PPT

Cystophage

Pseudomonas bacteriophage φ13
Pseudomonas phage φ12
Pseudomonas phage φ2954
Pseudomonas phage φ6
Pseudomonas phage φ8
Rhodococcus phage RRH1

Fuselloviridi

Acidianus bottle-shaped virus
Acidianus filamentous virus 1
Acidianus filamentous virus 2
Acidianus filamentous virus 3
Acidianus filamentous virus 6
Acidianus filamentous virus 7
Acidianus filamentous virus 8
Acidianus filamentous virus 9
Acidianus rod-shaped virus 1
Acidianus spindle-shaped virus 1
Acidianus two-tailed virus
Aeropyrum pernix ovoid virus 1
Aeropyrum pernix spindle-shaped virus 1
Hyperthermoφlic Archaeal Virus 1
Hyperthermoφlic Archaeal Virus 2
Pyrobaculum spherical virus
Pyrococcus abyssi virus 1
Sulfolobus islandicus filamentous virus
Sulfolobus islandicus rod-shaped virus 1
Sulfolobus islandicus rod-shaped virus 2
Sulfolobus islandicus rudivirus 1 variant XX
Sulfolobus spindle-shaped virus 1
Sulfolobus spindle-shaped virus 2
Sulfolobus spindle-shaped virus 4
Sulfolobus spindle-shaped virus 5
Sulfolobus spindle-shaped virus 6
Sulfolobus spindle-shaped virus 7
Sulfolobus turreted icosahedral virus
Sulfolobus turreted icosahedral virus 2
Sulfolobus virus Kamchatka 1
Sulfolobus virus Ragged Hills
Sulfolobus virus STSV1
Thermococcus prieurii virus 1
Thermoproteus tenax spherical virus 1

Inophage

Cellulophaga phage φ12:2
Cellulophaga phage φ12a:1
Cellulophaga phage φ18:4
Cellulophaga phage φ48:1
Enterobacteria phage f1
Enterobacteria phage fd
Enterobacteria phage I22
Enterobacteria phage Ike
Enterobacteria phage M13

Pseudomonas phage Pf1
Pseudomonas phage Pf3
Ralstonia phage p12J
Ralstonia phage PE226
Ralstonia phage RSS0
Ralstonia phage RSS1
Spiroplasma kunkelii virus SkV1 CR23x
Spiroplasma phage 1C74
Spiroplasma phage 1R8A2B
Spiroplasma phage SVGII3
Spiroplasma phage SVTS2
Stenotrophomonas phage φSHP2
Stenotrophomonas phage φSMA9
Vibrio phage CTX
Vibrio phage fs1
Vibrio phage KSF1φ
Vibrio phage ND1fs1
Vibrio phage VCYφ
Vibrio phage VEJφ
Vibrio phage Vf12
Vibrio phage Vf33
Vibrio phage VfO3K6
Vibrio phage VfO4K68
Vibrio phage VGJφ
Vibrio phage VSK
Vibrio phage VSKK
Xanthomonas phage Cf1c

Leviphage

Acinetobacter phage AP205
Bacteriophage M11
Caulobacter phage φCb5
Enterobacteria phage BZ13
Enterobacteria phage FI sensu lato
Enterobacteria phage FI sensu lato
Enterobacteria phage fr
Enterobacteria phage GA
Enterobacteria phage KU1
Enterobacteria phage MS2
Enterobacteria phage MX1
Enterobacteria phage NL95
Enterobacteria phage Qβ
Pseudomonas phage PP7
Pseudomonas phage PRR1

Microphage *Spiroplasma* phage 4-like

Bdellovibrio phage φMH2K
Chlamydia phage 2
Chlamydia phage 3
Chlamydia phage 4
Chlamydia phage Chp1
Chlamydia phage CPAR39
Guinea pig *Chlamydia* phage
Microviridae φCA82
Spiroplasma phage 4

Microphage X174-like

Enterobacteria ID2 MoscowID2001
Enterobacteria phage alpha3
Enterobacteria phage G4 sensu lato
Enterobacteria phage ID1
Enterobacteria phage ID11
Enterobacteria phage ID12
Enterobacteria phage ID18 sensu lato
Enterobacteria phage ID21
Enterobacteria phage ID22
Enterobacteria phage ID32
Enterobacteria phage ID34
Enterobacteria phage ID41
Enterobacteria phage ID45
Enterobacteria phage ID52
Enterobacteria phage ID62
Enterobacteria phage ID8
Enterobacteria phage NC1
Enterobacteria phage NC10
Enterobacteria phage NC11
Enterobacteria phage NC13
Enterobacteria phage NC16
Enterobacteria phage NC19
Enterobacteria phage NC2
Enterobacteria phage NC28
Enterobacteria phage NC29
Enterobacteria phage NC3
Enterobacteria phage NC35
Enterobacteria phage NC37
Enterobacteria phage NC41
Enterobacteria phage NC5
Enterobacteria phage NC51
Enterobacteria phage NC56
Enterobacteria phage NC6
Enterobacteria phage NC7
Enterobacteria phage φK
Enterobacteria phage φX174 sensu lato
Enterobacteria phage S13
Enterobacteria phage St1
Enterobacteria phage WA10
Enterobacteria phage WA11
Enterobacteria phage WA13 sensu lato
Enterobacteria phage WA14
Enterobacteria phage WA2
Enterobacteria phage WA3
Enterobacteria phage WA4
Enterobacteria phage WA45
Enterobacteria phage WA5
Enterobacteria phage WA6

Myophage φCbK-like

Caulobacter phage φCbK
Colwellia phage 9A
Cronobacter phage CR3
Enterobacteria phage φ92

Enterobacteria phage vB_EcoM_FV3
Erwinia phage φEa104
Erwinia phage φEa214
Erwinia phage vB_EamMM7
Escherichia phage rv5
Escherichia phage wV8
Microcystis aeruginosa phage MaLMM01
Pseudomonas phage JG004
Pseudomonas phage KPP10
Pseudomonas phage P3 CHA
Pseudomonas phage PAK P1
Pseudomonas phage PAK P3
Pseudomonas phage PaP1
Salmonella phage FelixO1
Salmonella phage PVP SE1
Staphylococcus phage SA1
Vibrio phage ICP1
Vibrio phage ICP1_2001_A
Vibrio phage ICP1_2004_A
Vibrio phage ICP1_2005_A
Vibrio phage ICP1_2006_A
Vibrio phage ICP1_2006_B
Vibrio phage ICP1_2006_C
Vibrio phage ICP1_2006_D

Myophage P2-like

Aeromonas phage φO18P
Archaeal BJ1 virus
Burkholderia phage KL3
Burkholderia phage KS14
Burkholderia phage KS5
Burkholderia phage φ52237
Burkholderia phage φE122
Burkholderia phage φE202
Cronobacter phage ESSI2
Enterobacteria phage 186
Enterobacteria phage P2
Enterobacteria phage PsP3
Enterobacteria phage Wφ
Erwinia phage ENT90
Haemophilus phage HP1
Haemophilus phage HP2
Mannheimia phage φMHaA1
Mannheimia phage φMhaA1BAA410
Mannheimia phage φMhaA1PHL101
Methanobacterium phage ψM2
Methanothermobacter phage ψM100
Natrialba phage φCh1
Pasteurella phage F108
Pseudoalteromonas phage PM2
Pseudomonas phage φCTX
Ralstonia phage RSA1
Salmonella phage Fels2
Salmonella phage RE2010
Vibrio phage fs2
Vibrio phage K139
Vibrio phage kappa
Yersinia phage L413C

Myophage Pr-like

Actinoplanes phage φAsp2
Bdellovibrio phage φ1422
Brucella phage Pr
Brucella phage Tb
Burkholderia phage Bcep1
Burkholderia phage Bcep43
Burkholderia phage Bcep781
Burkholderia phage BcepB1A
Burkholderia phage BcepNY3
Corynebacterium phage BFK20
Listonella phage φHSIC
Nocardia phage NBR1
Persicivirga phage P12024L
Persicivirga phage P12024S
Vibrio phage CPT1
Vibrio phage vB_VchM138
Xanthomonas phage OP2

Myophage SN-like

Burkholderia phage BcepF1
Clavibacter phage CMP1
Cyanophage PSS2
Lactococcus phage 949
Mycobacteriophage Adjutor
Mycobacteriophage Alice
Mycobacteriophage Ava3
Mycobacteriophage Butterscotch
Mycobacteriophage Bxz1
Mycobacteriophage Cali
Mycobacteriophage Catera
Mycobacteriophage Drazdys
Mycobacteriophage ET08
Mycobacteriophage Ghost
Mycobacteriophage Gumball
Mycobacteriophage Konstantine
Mycobacteriophage LinStu
Mycobacteriophage LRRHood
Mycobacteriophage MoMoMixon
Mycobacteriophage Myrna
Mycobacteriophage Nappy
Mycobacteriophage Nova
Mycobacteriophage PBI1
Mycobacteriophage Pio
Mycobacteriophage Pleione
Mycobacteriophage PLot
Mycobacteriophage Predator
Mycobacteriophage Rizal
Mycobacteriophage ScottMcG
Mycobacteriophage Sebata
Mycobacteriophage SirHarley
Mycobacteriophage Spud
Mycobacteriophage Troll4
Mycobacteriophage Wally
Pseudomonas phage 141
Pseudomonas phage F8
Pseudomonas phage JG024
Pseudomonas phage LBL3
Pseudomonas phage LMA2

Pseudomonas phage PaMx13
Pseudomonas phage PB1
Pseudomonas phage SN
Synechococcus phage SCBS2
Synechococcus phage SCBS4

Myophage T4-like

Acinetobacter phage 133
Acinetobacter phage Ac42
Acinetobacter phage Acj61
Acinetobacter phage Acj9
Acinetobacter phage ZZ1
Aeromonas phage 25
Aeromonas phage 31
Aeromonas phage 44RR2.8t
Aeromonas phage 65
Aeromonas phage Aeh1
Aeromonas phage CC2
Aeromonas phage φAS4
Aeromonas phage φAS5
Aeromonas phage PX29
Bacillus phage 0305φ836
Bacillus phage G
Bacillus phage SPBc2
Campylobacter phage CP220
Campylobacter phage CP81
Campylobacter phage CPt10
Campylobacter phage CPX
Campylobacter phage NCTC12673
Cellulophaga phage φ13:1
Cellulophaga phage φ14:2
Cellulophaga phage φ17:2
Cellulophaga phage φ19:2
Cellulophaga phage φ4:1
Cellulophaga phage φST
Deftia phage φW14
Dickeya phage Limestone
Enterobacteria phage AR1
Enterobacteria phage CC31
Enterobacteria phage IME08
Enterobacteria phage JS10
Enterobacteria phage JS98
Enterobacteria phage JSE
Enterobacteria phage φ1
Enterobacteria phage RB14
Enterobacteria phage RB16
Enterobacteria phage RB32
Enterobacteria phage RB43
Enterobacteria phage RB49
Enterobacteria phage RB51
Enterobacteria phage RB69
Enterobacteria phage T2
Enterobacteria phage T4
Enterobacteria phage T4T
Enterobacteria phage vB_EcoMVR7
Enterobacteria phage vB_KleMRaK2
Escherichia phage Cba120
Escherichia phage PhaxI
Escherichia phage wV7

Klebsiella phage KP15
Pelagibacter phage HTVC008M
Planktothrix phage PaVLD
Prochlorococcus phage P-HM1
Prochlorococcus phage P-HM2
Prochlorococcus phage PRSM4
Prochlorococcus phage P-SSM2
Prochlorococcus phage PSSM4
Prochlorococcus phage PSSM7
Prochlorococcus phage Syn1
Prochlorococcus phage Syn33
Roseobacter phage RDJL φ 1
Salmonella phage φSH19
Salmonella phage SFP10
Salmonella phage Vil
Shigella phage Ag3
Shigella phage Shfl2
Shigella phage SP18
Sφhngomonas phage PAU
Synechococcus phage S-CRM01
Synechococcus phage S-PM2
Synechococcus phage SRIM8_A_HR1
Synechococcus phage SRIM8_A_HR3
Synechococcus phage S-RSM4
Synechococcus phage S-ShM2
Synechococcus phage SSM1
Synechococcus phage S-SM2
Synechococcus phage SSSM5
Synechococcus phage S-SSM7
Synechococcus phage Syn19
Synechococcus phage syn9
Synechococcus phage_SRIM8_A_HR5
Thermus phage φYS40
Thermus phage TMA
Vibrio phage KVP40
Vibrio phage φpp2

Podophage Mx8-like

Cellulophaga phage φ10:1
Cellulophaga phage φ13:2
Cellulophaga phage φ18:3
Cellulophaga phage φ19:1
Cellulophaga phage φ19:3
Cellulophaga phage φ38:1
Cellulophaga phage φ40:1
Cellulophaga phage φ46:3
Lactobacillus phage LBR48
Myxococcus phage Mx8
Pseudoalteromonas phage H105/1
Tetrasphaera phage TJE1

Podophage N4-like

Bas Gut Phage
Enterobacter phage EcP1
Erwinia phage vB_EamP_S6
Escherichia phage N4
Escherichia phage vB_EcoP_G7C
Pseudomonas phage 119X
Pseudomonas phage LIT1

Pseudomonas phage LUZ7
Pseudomonas phage PaP2
Silicibacter phage DSS3φ2
Sulfitobacter phage EE36φ1

Podophage P1-like

Bacteroides phage B12414
Bacteroides phage B408
Burkholderia phage Bcep22
Burkholderia phage BcepIL02
Burkholderia phage BcepMigl
Croceibacter phage P2559S
Enterobacteria phage P1
Enterobacteria phage P7
Erwinia phage PEp14
Pseudomonas phage F116
Sinorhizobium phage PBC5
Vibrio phage VvAW1

Podophage P22-like

Enterobacteria phage 4795
Enterobacteria phage CUS3
Enterobacteria phage IME10
Enterobacteria phage P22
Enterobacteria phage ST104
Helicobacter phage φHP33
Lactococcus phage bIL310
Lactococcus phage bIL311
Lactococcus phage bIL312
Leuconostoc phage L5
Salmonella phag SE1
Salmonella phage ε34
Salmonella phage g341c
Salmonella phage HK620
Salmonella phage P22pbi
Salmonella phage SPN9CC
Salmonella phage ST1605
Salmonella phage ST64T
Salmonella phage vB_SemP_Emek
Shigella phage Sf6
Sodalis phage φSG1
Staphylococcus phage PT1028
Thermus phage IN93
Thermus phage P2377

Podophage P4-like

Bordetella phage BIP-1
Bordetella phage BMP-1
Bordetella phage BPP-1
Burkholderia phage BcepC6B
Enterobacteria phage P4
Escherichia phage φV10
Escherichia phage TL2011b
Flavobacterium phage 11b
Halocynthia phage JM2012
Halogeometricum pleomorφc virus 1
Halorubrum pleomorφc virus 2
Halorubrum pleomorφc virus 3
Halorubrum pleomorφc virus 6

His2 virus
Pelagibacter phage HTVC010P
Pseudomonas phage 201φ21
Pseudomonas phage EL
Pseudomonas phage OBP
Pseudomonas phage φKZ
Pseudomonas phage φPA3
Salmonella phage ε15
Salmonella phage SPN3US
Vibrio phage φVC8
Vibrio phage VP2
Vibrio phage VP5

Podophage φ29-like

Actinomyces phage Av1
Bacillus phage AP50
Bacillus phage B103
Bacillus phage Bam35c
Bacillus phage GA1
Bacillus phage GIL16c
Bacillus phage Nf
Bacillus phage φ29
Bacillus phage PZA
Clostridium phage φ24R
Clostridium phage φCP7R
Clostridium phage φCPV4
Clostridium phage φZP2
Enterobacteria phage 933W
Enterobacteria phage L17
Enterobacteria phage Min27
Enterobacteria phage PR3
Enterobacteria phage PR4
Enterobacteria phage PR5
Enterobacteria phage PR772
Enterobacteria phage PRD1
Enterobacteria phage VT2-Sakai
Enterococcus phage EF62φ
Escherichia phage TL2011c
Escherichia Stx1 converting phage
His1 virus
Lactococcus phage asccφ28
Mycoplasma phage P1
Organic Lake virophage
Sputnik virophage
Sputnik virophage 2
Sputnik virophage 3
Staphylococcus phage 44AHJD
Staphylococcus phage 66
Staphylococcus phage P68
Staphylococcus phage S13'DNA
Staphylococcus phage S241 DNA
Staphylococcus phage SAP2
Streptococcus phage C1
Streptococcus phage Cp1
Stx2 converting phage I
Stx2 converting phage II
Stx2 converting phage vB_EcoP_24B
Stx2converting phage 86

Podophage T7-like

Acinetobacter phage φAB1
Aeromonas phage φAS7
Celeribacter phage P12053L
Cyanophage 951510a
Cyanophage NATL1A7
Cyanophage NATL2A133
Cyanophage P60
Cyanophage PSSP2
Enterobacteria phage 13a
Enterobacteria phage 285P
Enterobacteria phage BA14
Enterobacteria phage EcoDS1
Enterobacteria phage K15
Enterobacteria phage K1E
Enterobacteria phage K1F
Enterobacteria phage K30
Enterobacteria phage φeco32
Enterobacteria phage SP6
Enterobacteria phage T3
Enterobacteria phage T7
Erwinia amylovora phage Era103
Erwinia phage φEa100
Erwinia phage φEa1H
Erwinia phage vB_EamPL1
Escherichia phage φKT
Klebsiella phage K11
Klebsiella phage KP32
Klebsiella phage KP34
Kluyvera phage Kvp1
Morganella phage MmP1
Pantoea phage LIMElight
Pantoea phage LIMEzero
Pelagibacter phage HTVC011P
Pelagibacter phage HTVC019P
Phormidium phage PfWMP3
Phormidium phage PfWMP4
Prochlorococcus phage PSSP7
Pseudomonas phage Bf7
Pseudomonas phage gh1
Pseudomonas phage LKA1
Pseudomonas phage LKD16
Pseudomonas phage LUZ19
Pseudomonas phage LUZ24
Pseudomonas phage PA11
Pseudomonas phage PaP3
Pseudomonas phage φ15
Pseudomonas phage φ2
Pseudomonas phage φIBBPF7A
Pseudomonas phage φkF77
Pseudomonas phage φKMV
Pseudomonas phage PT2
Pseudomonas phage PT5
Pseudomonas phage tf
Pseudomonas phage vB_PaeTbilisiM32
Ralstonia phage RSB1
Ralstonia phage RSB2
Roseobacter phage SIO1
Salinivibrio phage CW02

Salmonella phage 711
Salmonella phage φSGJL2
Salmonella phage Vi06
Synechococcus phage Syn5
Vibrio phage ICP2
Vibrio phage ICP2_2006_A
Vibrio phage ICP3
Vibrio phage ICP3_2007_A
Vibrio phage ICP3_2008_A
Vibrio phage ICP3_2009_B
Vibrio phage N4
Vibrio phage VP3
Vibrio phage VP4
Vibrio phage VP93
Vibrio phage VpV262
Xanthomonas phage CP1
Xanthomonas phage OP1
Xanthomonas phage φL7
Xanthomonas phage Xop411
Xanthomonas phage Xp10
Yersinia phage Berlin
Yersinia phage φA1122
Yersinia phage φYeO312
Yersinia phage Yepe2
Yersinia phage Yepφ

Siphophage A118-like

Azospirillum phage Cd
Bacillus phage 250
Bacillus phage BceA1
Bacillus phage BCJA1c
Bacillus phage IEBH
Bacillus phage PBC1
Bacillus phage phBC6A51
Bacillus phage TP21L
Bacillus virus 1
Bacteriophage APSE2
Bacteriophage sk1
Brochothrix phage BL3
Brochothrix phage NF5
Cellulophaga phage φ39 1
Clostridium phage φ8074B1
Clostridium phage φC2
Clostridium phage φCD119
Clostridium phage φCD27
Clostridium phage φCD382
Clostridium phage φCD6356
Clostridium phage φCP13O
Clostridium phage φCP26F
Clostridium phage φCP34O
Clostridium phage φCP39O
Clostridium phage φCP9O
Deepsea thermophilic phage D6E
Endosymbiont phage APSE-1
Enterobacteria phage SSL2009a
Enterococcus phage BC611
Enterococcus phage EFAP1
Enterococcus phage EFRM31
Enterococcus phage φEf11

Enterococcus phage φFL1A
Enterococcus phage φFL1B
Enterococcus phage φFL1C
Enterococcus phage φFL2A
Enterococcus phage φFL2B
Enterococcus phage φFL3A
Enterococcus phage φFL3B
Enterococcus phage φFL4A
Enterococcus phage SAP6
Escherichia phage K1dep 1
Escherichia phage K1dep 4
Escherichia phage K1ind 1
Escherichia phage K1ind 2
Escherichia phage K1ind 3
Geobacillus phage GBSV1
Geobacillus virus E2
Lactobacillus bacteriophage φJL1
Lactobacillus johnsonii prophage Lj771
Lactobacillus phage A2
Lactobacillus phage c5
Lactobacillus phage KC5a
Lactobacillus phage LF1
Lactobacillus phage LLH
Lactobacillus phage LLKu
Lactobacillus phage Lrm1
Lactobacillus phage Lv1
Lactobacillus phage φg1e
Lactobacillus phage φPYB5
Lactobacillus phage Sha1
Lactobacillus prophage Lj928
Lactobacillus prophage Lj965
Lactococcus phage 1706
Lactococcus phage 712
Lactococcus Phage ASCC191
Lactococcus Phage ASCC273
Lactococcus Phage ASCC281
Lactococcus Phage ASCC284
Lactococcus Phage ASCC287
Lactococcus Phage ASCC310
Lactococcus Phage ASCC324
Lactococcus Phage ASCC337
Lactococcus Phage ASCC356
Lactococcus Phage ASCC358
Lactococcus Phage ASCC365
Lactococcus Phage ASCC368
Lactococcus Phage ASCC395
Lactococcus Phage ASCC397
Lactococcus Phage ASCC406
Lactococcus Phage ASCC454
Lactococcus Phage ASCC460
Lactococcus Phage ASCC465
Lactococcus Phage ASCC473
Lactococcus Phage ASCC476
Lactococcus Phage ASCC489
Lactococcus Phage ASCC497
Lactococcus Phage ASCC502
Lactococcus Phage ASCC506
Lactococcus Phage ASCC527
Lactococcus Phage ASCC531

Lactococcus Phage ASCC532
Lactococcus Phage ASCC544
Lactococcus phage blBB29
Lactococcus phage blL170
Lactococcus phage CB13
Lactococcus phage CB14
Lactococcus phage CB19
Lactococcus phage CB20
Lactococcus phage jj50
Lactococcus phage P008
Lactococcus phage SL4
Leuconostoc phage 1A4
Liberibacter phage FP2
Liberibacter phage SC1
Liberibacter phage SC2
Listeria phage A006
Listeria phage A118
Listeria phage A500
Listeria phage P35
Listeria phage P40
Microbacterium phage Min1
Pediococcus phage clP1
Propionibacterium phage PA6
Propionibacterium phage PAD20
Propionibacterium phage PAS50
Pseudomonas phage PAJU2
Rhodococcus phage ReqiDocB7
Rhodococcus phage ReqiPepy6
Rhodococcus s phage ReqiPoco6
Riemerella phage RAP44
Salmonella phage SE2
Salmonella phage SETP3
Salmonella phage SS3e
Salmonella phage vB SenSEnt1
Shigella phage EP23
Sodalis phage SO1
Staphylococcus aureus phage φNM1
Staphylococcus aureus phage φNM2
Staphylococcus aureus phage φNM4
Staphylococcus phage 11
Staphylococcus phage 187
Staphylococcus phage 29
Staphylococcus phage 37
Staphylococcus phage 52A
Staphylococcus phage 53
Staphylococcus phage 55
Staphylococcus phage 69
Staphylococcus phage 71
Staphylococcus phage 80
Staphylococcus phage 80alpha
Staphylococcus phage 85
Staphylococcus phage 88
Staphylococcus phage 92
Staphylococcus phage 96
Staphylococcus phage CNPH82
Staphylococcus phage EW
Staphylococcus phage PH15
Staphylococcus phage φETA
Staphylococcus phage φETA2

Staphylococcus phage φETA3
Staphylococcus phage φMR11
Staphylococcus phage φMR25
Staphylococcus phage φSauSIPLA88
Staphylococcus phage ROSA
Staphylococcus phage SAP26
Staphylococcus phage SpaA1
Staphylococcus phage StB12
Staphylococcus phage StB27
Staphylococcus phage TEM123
Staphylococcus phage vB SepiSφIPLA5
Staphylococcus phage vB SepiSφIPLA7
Staphylococcus phage X2
Streptococcus bacteriophage Sfi11
Streptococcus phage 2972
Streptococcus phage 5093
Streptococcus phage 858
Streptococcus phage Abc2
Streptococcus phage ALQ13 2
Streptococcus phage DT1
Streptococcus phage EJ1
Streptococcus phage MM1
Streptococcus phage MM1 1998
Streptococcus phage O1205
Streptococcus phage P9
Streptococcus phage PH10
Streptococcus phage PH15
Streptococcus phage Sfi19
Streptococcus phage Sfi21
Streptococcus phage SMP
Streptococcus pyogenes phage 315 4
Streptococcus pyogenes phage 315 5
Streptococcus pyogenes phage 315 6
Temperate phage φNIH1 1
Thermoanaerobacterium phage THSA485A
Xylella phage Xfas53

Siphophage D29-like

Corynebacterium phage P1201
Gordonia phage GRU1
Gordonia phage GTE5
Gordonia phage GTE7
Mycobacteriophage 244
Mycobacteriophage Adephagia
Mycobacteriophage Airmid
Mycobacteriophage Alma
Mycobacteriophage Anaya
Mycobacteriophage Angel
Mycobacteriophage Angelica
Mycobacteriophage Ardmore
Mycobacteriophage Astro
Mycobacteriophage Avani
Mycobacteriophage Avrafan
Mycobacteriophage Babsiella
Mycobacteriophage Backyardigan
Mycobacteriophage BAKA
Mycobacteriophage Barnyard
Mycobacteriophage BarrelRoll
Mycobacteriophage Bask21

Mycobacteriophage Batiatus
Mycobacteriophage Benedict
Mycobacteriophage BigNuz
Mycobacteriophage Blue7
Mycobacteriophage Bongo
Mycobacteriophage Boomer
Mycobacteriophage BPs
Mycobacteriophage Brujita
Mycobacteriophage Bxz2
Mycobacteriophage Charlie
Mycobacteriophage Che12
Mycobacteriophage Che8
Mycobacteriophage Che9c
Mycobacteriophage Che9d
Mycobacteriophage CJW1
Mycobacteriophage Corndog
Mycobacteriophage Courthouse
Mycobacteriophage CrimD
Mycobacteriophage D29
Mycobacteriophage DaVinci
Mycobacteriophage DeadP
Mycobacteriophage DLane
Mycobacteriophage DotProduct
Mycobacteriophage Drago
Mycobacteriophage DS6A
Mycobacteriophage Eagle
Mycobacteriophage Elph10
Mycobacteriophage ElTiger69
Mycobacteriophage EricB
Mycobacteriophage Eureka
Mycobacteriophage Faith1
Mycobacteriophage Fezzik
Mycobacteriophage Fionnbharth
Mycobacteriophage Firecracker
Mycobacteriophage Flux
Mycobacteriophage Fruitloop
Mycobacteriophage George
Mycobacteriophage Giles
Mycobacteriophage Gladiator
Mycobacteriophage Gumbie
Mycobacteriophage Halo
Mycobacteriophage Hammer
Mycobacteriophage Hamulus
Mycobacteriophage HelDan
Mycobacteriophage Henry
Mycobacteriophage Hope
Mycobacteriophage Ibhubesi
Mycobacteriophage ICleared
Mycobacteriophage Island3
Mycobacteriophage JAWS
Mycobacteriophage Jebeks
Mycobacteriophage Jeffabunny
Mycobacteriophage JHC117
Mycobacteriophage JoeDirt
Mycobacteriophage Kostya
Mycobacteriophage L5
Mycobacteriophage Larva
Mycobacteriophage LeBron
Mycobacteriophage LHTSCC

Mycobacteriophage Liefie
Mycobacteriophage Lilac
Mycobacteriophage LittleE
Mycobacteriophage Llij
Mycobacteriophage MacnCheese
Mycobacteriophage Marvin
Mycobacteriophage Maverick
Mycobacteriophage Microwolf
Mycobacteriophage Mozy
Mycobacteriophage Mutaforma13
Mycobacteriophage Nelitza
Mycobacteriophage Omega
Mycobacteriophage Optimus
Mycobacteriophage Ovechkin
Mycobacteriophage Pacc40
Mycobacteriophage PackMan
Mycobacteriophage Peaches
Mycobacteriophage Pixie
Mycobacteriophage PMC
Mycobacteriophage Porky
Mycobacteriophage Pukovnik
Mycobacteriophage pumpkin
Mycobacteriophage Rakim
Mycobacteriophage Ramsey
Mycobacteriophage Redi
Mycobacteriophage Redno2
Mycobacteriophage RedRock
Mycobacteriophage Rey
Mycobacteriophage Rockstar
Mycobacteriophage RockyHorror
Mycobacteriophage Rumpelstiltskin
Mycobacteriophage Saal
Mycobacteriophage Saintus
Mycobacteriophage Send513
Mycobacteriophage SG4
Mycobacteriophage Shaka
Mycobacteriophage Shauna1
Mycobacteriophage ShiLan
Mycobacteriophage SirDuracell
Mycobacteriophage Spartacus
Mycobacteriophage SWU1
Mycobacteriophage Theia
Mycobacteriophage Thibault
Mycobacteriophage Tiger
Mycobacteriophage Timshel
Mycobacteriophage TiroTheta9
Mycobacteriophage TM4
Mycobacteriophage Toto
Mycobacteriophage Trixie
Mycobacteriophage Turbido
Mycobacteriophage Tweety
Mycobacteriophage Twister
Mycobacteriophage UPIE
Mycobacteriophage Vix
Mycobacteriophage Wee
Mycobacteriophage Wildcat
Mycobacteriophage Wile
Mycobacteriophage Wonder
Mycobacteriophage Yoshi

Ralstonia phage RSM1
Ralstonia phage RSM3
Rhodococcus phage REQ1
Rhodococcus phage REQ2
Rhodococcus phage REQ3
Rhodococcus phage RER2
Rhodococcus phage RGL3
Saccharomonospora phage PIS 136
Salisaeta icosahedral phage 1
Streptomyces phage mu1/6
Streptomyces phage VWB

Siphophage λ-like

Acholeplasma phage L2
Bac Gamma isolate d'Herelle
Bacillus phage BtCS33
Bacillus phage Cherry
Bacillus phage Fah
Bacillus phage Gamma
Bacillus phage phBC6A52
Bacillus phage φ105
Bacillus phage φS3501
Bacillus phage Wβ
Burkholderia phage Bcep176
Burkholderia phage KS9
Burkholderia phage φ1026b
Burkholderia phage φ6442
Burkholderia phage φE125
Clostridium phage φ3626
Clostridium phage φS63
Clostridium phage φSM101
Enterobacteria phage 2851
Enterobacteria phage cdtI
Enterobacteria phage DE3
Enterobacteria phage HK022
Enterobacteria phage HK97
Enterobacteria phage λ
Enterobacteria phage N15
Enterobacteria phage φP27
Enterobacteria phage SfV
Enterobacteria phage YYZ2008
Escherichia phage HK639
Escherichia phage HK75
Klebsiella phage φKO2
Lactobacillus phage JCL1032
Lactobacillus phage LcNu
Lactobacillus phage φadh
Lactobacillus phage φAT3
Lactococcus phage 4268
Lactococcus phage bIL285
Lactococcus phage bIL286
Lactococcus phage bIL309
Lactococcus phage bIL67
Lactococcus phage BK5T
Lactococcus phage c2
Lactococcus phage Q54
Listeria phage 2389
Listeria phage B025
Pseudomonas phage D3

Pseudomonas phage vB_PaeS_PMG1
Psychrobacter phage Psymv2
Rhizobium phage 163
Rhodobacter phage RcapNL
Salmonella phage ST64B
Staphylococcus aureus phage φ 13
Staphylococcus phage 2638A
Staphylococcus phage 3A
Staphylococcus phage 42E
Staphylococcus phage 47
Staphylococcus phage 77
Staphylococcus phage P954
Staphylococcus phage φ 12
Staphylococcus phage φ2958PVL
Staphylococcus phage φ5967PVL
Staphylococcus phage φ7247PVL
Staphylococcus phage φN315
Staphylococcus phage φNM3
Staphylococcus phage φPVL108
Staphylococcus phage φPVLCN125
Staphylococcus phage φSauSIPLA35
Staphylococcus phage φSLT
Staphylococcus phage PVL
Staphylococcus phage SMSAP5
Staphylococcus phage StB20
Staphylococcus phage tp3101
Staphylococcus phage tp3102
Staphylococcus phage tp3103
Staphylococcus prophage φPV83
Streptococcus phage 7201
Streptococcus phage M102
Streptococcus phage φ3396
Streptococcus phage YMC2011
Streptococcus pyogenes phage 315 1
Streptococcus pyogenes phage 315 2
Streptomyces phage φBT1
Streptomyces phage φC31
Stx2converting phage 1717
Yersinia phage PY54

Siphophage Mu-like

Burkholderia phage AH2
Burkholderia phage BcepMu
Burkholderia phage BcepNazgul
Burkholderia phage KL1
Burkholderia phage KS10
Burkholderia phage φE255
Enterobacter phage Enc34
Enterobacteria phage Chi
Enterobacteria phage Mu
Escherichia phage D108
Escherichia phage vB_EcoM_ECO123010
Halomonas phage φHAP1
Phage φJL001
Pseudomonas phage 73
Pseudomonas phage B3
Pseudomonas phage D3112
Pseudomonas phage DMS3
Pseudomonas phage F_HA0480sp

Pseudomonas phage JBD26
Pseudomonas phage LPB1
Pseudomonas phage M6
Pseudomonas phage MP1412
Pseudomonas phage MP22
Pseudomonas phage MP29
Pseudomonas phage MP38
Pseudomonas phage MP42
Pseudomonas phage PA1/KOR/2010
Pseudomonas phage PaMx25
Pseudomonas phage PaMx73
Pseudomonas phage vB_PaeKakheti25
Pseudomonas phage YuA
Rhodobacter phage RcapMu
Stenotrophomonas phage S1
Synechococcus phage SCBS1
Synechococcus phage SCBS3
Vibrio phage SSP002
Vibrio phage VHML
Vibrio phage VP58.5
Vibrio phage VP882
Xanthomonas phage Xp15

Siphophage SPP1-like

Bacillus phage SPP1
Marinomonas phage P12026
Mycoplasma phage MAV1
Mycoplasma phage φMFV1
Pseudomonas phage F10
Pseudomonas phage φ297

Siphophage T1-like

Cronobacter phage ES2
Cronobacter phage ESP29491
Enterobacteria phage
Enterobacteria phage JK06
Enterobacteria phage RTP
Enterobacteria phage T1
Enterobacteria phage TLS
Escherichia phage φEB49
Salmonella phage Fels 1
Salmonella phage Gifsy-1
Salmonella phage Gifsy-2
Salmonella phage SPN3UB
Salmonella phage vB_SosS_Oslo
Shigella phage Shfl1

Siphophage T5-like

Bacillus phage B4
Bacillus phage BCP78
Bacillus phage SPO1
Bacillus phage W Ph
Brochothrix phage A9
Clostridium phage φCTP1
Enterobacteria phage EPS7
Enterobacteria phage SPC35
Enterobacteria phage T5
Enterococcus phage φEF24C

Enterococcus phage φEF24CP2
Escherichia phage bV_EcoS_AKFV33
Gordonia phage GTE2
Lactobacillus phage Lb3381
Lactobacillus phage LP65
Lactococcus phage KSY1
Listeria phage A511
Listeria phage P100
Mycobacteriophage Switzer
Mycobacteriophage Aeneas
Mycobacteriophage Bethlehem
Mycobacteriophage BillKnuckles
Mycobacteriophage BPBiebs31
Mycobacteriophage Bruns
Mycobacteriophage BxB1
Mycobacteriophage DD5
Mycobacteriophage Doom
Mycobacteriophage Dreamboat
Mycobacteriophage Jasper
Mycobacteriophage JC27
Mycobacteriophage KBG
Mycobacteriophage KSSJEB
Mycobacteriophage Kugel
Mycobacteriophage Lesedi
Mycobacteriophage Lockley
Mycobacteriophage MrGordo
Mycobacteriophage Museum
Mycobacteriophage Nepal
Mycobacteriophage Pari
Mycobacteriophage Perseus
Mycobacteriophage SkiPole
Mycobacteriophage Solon
Mycobacteriophage U2
Mycobacteriophage Violet
Pseudomonas phage Lu11
Puniceispirillum phage HMO2011
Ralstonia phage RSL1
Staphylococcus phage A5W
Staphylococcus phage G1
Staphylococcus phage ISP
Staphylococcus phage K
Staphylococcus phage Sb1
Staphylococcus phage Twort
Streptococcus phage Dp1
Streptomyces phage φSASD1
Thermus phage P2345
Thermus phage P7426
Vibrio phage 1
Vibrio phage pVp1
Vibrio phage SIO2

Siphophage TP901-1-like

Acinetobacter phage AB1
Acinetobacter phage AP22
Aeromonas phage vB_AsaM56
Aggregatibacter phage S1249
Clostridium phage cst
Clostridium phage D1873

Enterobacteria phage φEcoMGJ1
Erwinia phage φEt88
Erwinia phage vB_EamMY2
Haemophilus phage Aaφ23
Iodobacteriophage φPLPE
Lactococcus phage P335
Lactococcus phage φLC3
Lactococcus phage φsmq86
Lactococcus phage r1t
Lactococcus phage TP901-1
Lactococcus phage Tuc2009
Lactococcus phage ul36
Lactococcus phage ul36.k1
Lactococcus phage ul36.k1t1
Lactococcus phage ul36.t1
Lactococcus phage ul36.t1k1
Listeria phage B054
Pectobacterium phage ZF40
Streptococcus phage SM1
Streptococcus pyogenes phage 315 3
Xanthomonas phage vB_XveM_DIBBI

Siphophage TPA2-like

Burkholderia phage BcepGomr
Cellulophaga phage φ12:1
Cellulophaga phage φ12:3
Cellulophaga phage φ17:1
Cellulophaga phage φ18:1
Cellulophaga phage φ18:2
Cellulophaga phage φ3:1
Cellulophaga phage φ38:2
Cellulophaga phage φ3ST:2
Cellulophaga phage φ46:1
Cellulophaga phage φ47:1
Cellulophaga phage φSM
Lactococcus phage P087
Mycobacteriophage Kamiyu
Mycobacteriophage Stinger
Mycobacteriophage ABU
Mycobacteriophage Acadian
Mycobacteriophage Akoma
Mycobacteriophage AnnaL29
Mycobacteriophage Arbiter
Mycobacteriophage Ares
Mycobacteriophage Athena
Mycobacteriophage Chah
Mycobacteriophage ChrisnMich
Mycobacteriophage Colbert
Mycobacteriophage Cooper
Mycobacteriophage Daisy
Mycobacteriophage Dori
Mycobacteriophage Fang
Mycobacteriophage Frederick
Mycobacteriophage Gadjet
Mycobacteriophage Harvey
Mycobacteriophage Hedgerow
Mycobacteriophage Hertubise
Mycobacteriophage IsaacEli

Mycobacteriophage JacAttac	Mycobacteriophage Rosebush	Unassigned
Mycobacteriophage Kikipoo	Mycobacteriophage Scoot17C	*Acholeplasma* phage MVL-1
Mycobacteriophage KLucky39	Mycobacteriophage Serendipity	*Bdellovibrio* phage φ1402
Mycobacteriophage Morgushi	Mycobacteriophage Suffolk	*Campylobacter* phage vB_CcoM_IBB35
Mycobacteriophage Murdoc	Mycobacteriophage TallGrassMM	*Cellulophaga* phage φ48:2
Mycobacteriophage Nigel	Mycobacteriophage Thora	Cyanophage S-TIM5
Mycobacteriophage Oline	Mycobacteriophage ThreeOh3D2	*Haloarcula hispanica* icosahedral virus 2
Mycobacteriophage Oosterbaan	Mycobacteriophage UncleHowie	*Haloarcula* phage SH1
Mycobacteriophage Orion	Mycobacteriophage Vista	*Halorubrum* phage HF2
Mycobacteriophage OSmaximus	Mycobacteriophage Vortex	Halovirus HF1
Mycobacteriophage PG1	Mycobacteriophage Yahalom	*Nitrososphaera* phage Pro-NVie1
Mycobacteriophage Phaedrus	Mycobacteriophage Yoshand	*Propionibacterium* phage B5
Mycobacteriophage φpps	Mycobacteriophage Zemanar	*Pseudomonas* phage φ_Pto6g
Mycobacteriophage Phlyer	*Rhodococcus* phage ReqiPine5	*Rhodothermus* phage RM378
Mycobacteriophage Pipefish	*Salmonella* phage Vi IIE1	*Thermoproteus* tenax virus 1
Mycobacteriophage Puhltonio	*Thalassomonas* phage	*Yersinia* phage φR137
Mycobacteriophage Qyrzula	*Tsukamurella* phage TPA2	

Eukaryotic Viruses

L-A neighbors[1]

Aspergillus mycovirus 178
Black raspberry virus F
Botryotinia fuckeliana totivirus 1
Debaryomyces hansenii virus JB-2008
Eimeria brunetti RNA virus 1
Epichloe festucae virus 1
Gremmeniella abietina RNA virus L1
Gremmeniella abietina RNA virus L2
Helicobasidium mompa totivirus 1-17
Helminthosporium victoriae virus 190S
Leishmania RNA virus 1 – 1
Leishmania RNA virus 1 – 4
Leishmania RNA virus 2 – 1
Magnaporthe oryzae virus 2
Ophiostoma minus totivirus
Piscine myocarditis virus AL V-708
Saccharomyces cerevisiae virus L-A
Saccharomyces cerevisiae virus L-BC
Sphaeropsis sapinea RNA virus 1
Trichomonas vaginalis virus 1
Trichomonas vaginalis virus 2
Trichomonas vaginalis virus 3
Trichomonas vaginalis virus 4
Tuber aestivum virus 1
Ustilago maydis virus H1
Xanthophyllomyces dendrorhous L1-A
Xanthophyllomyces dendrorhous L1-B
Xanthophyllomyces dendrorhous L2

Mimivirus neighbors[2]

Acanthamoeba castellanii mamavirus
Acanthamoeba polyphaga mimivirus
Acanthocystis turfacea Chlorella virus 1
African swine fever virus
Ascoviridae/Iridoviridae
Cafeteria roenbergensis virus BV-PW1
Ectocarpus siliculosus virus 1
Emiliania huxleyi virus 86
Feldmannia species virus
Lausannevirus
Marseillevirus
Megavirus chiliensis
Moumouvirus
Organic Lake phycodnavirus 1
Organic Lake phycodnavirus 2
Ostreococcus virus OsV5
Paramecium bursaria Chlorella virus 1
Paramecium bursaria Chlorella virus AR158
Paramecium bursaria Chlorella virus FR483
Paramecium bursaria Chlorella virus MT325
Paramecium bursaria Chlorella virus NY2A
Phaeocystis globosa virus 12 T
Poxviridae

---

[1] Evolutionary relationships based on Fig. 5 in: Baeza, M, N Bravo, M Sanhueza, O Flores, P Villarreal, V Cifuentes. 2012. Molecular characterization of totiviruses in *Xanthophyllomyces dendrorhous*. Virol J 9:140.

[2] Evolutionary relationships based on Fig. 3 in: Yutin, N, P Colson, D Raoult, EV Koonin. 2013. Mimiviridae: Clusters of orthologous genes, reconstruction of gene repertoire evolution and proposed expansion of the giant virus family. Virol J 10:106.

Appendix A4: Global Maps

Mapping the global distribution of a particular phage is challenging. If we want to know where a featured phage such as T4 hangs out, we cannot simply use binoculars to observe it in the wild. Instead we isolate and sequence fragments of DNA from an environment (i.e., prepare a metagenome or virome) and scrutinize them closely for evidence of T4's presence. This involves comparing the nucleotide sequence of each fragment against all the sequences in a collection of known phage genomes, including T4, by attempting to align them, nucleotide for nucleotide. We don't expect to find perfect matches because phages evolve rapidly, more rapidly than cellular organisms. To claim T4 was at that location, we require instead that a fragment must be a closer *match* to T4's genome than it is to any other sequenced phage genome, and this match must be a good enough match. Because the vast majority of the Earth has not been sampled for metagenomic sequencing, the distribution of each phage thus observed is likely a gross underestimate of its true worldwide distribution.

Finding a good match

In search of good matches, each phage genome sequence in the Phage SEED[1] was compared with metagenome data obtained from MG-RAST (Meyer et al. 2008) and the Tara Oceans project (Karsenti et al. 2011). A total of 1152 metagenomes with available geographic coordinates were downloaded from MG-RAST[2], along with the pertinent metadata: location, biome, feature, material, and package. An additional 80 metagenomes from the Tara

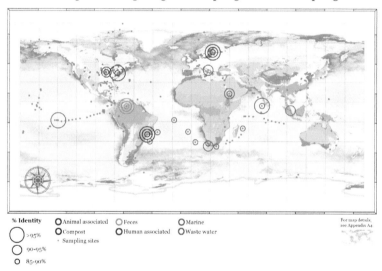

Figure 1. A sample global map showing the percent identity and the material sampled for 'sightings' of one phage, in this case phage RB69.

Oceans project and their associated metadata were graciously provided by Matthew Sullivan. Sequence reads from each metagenome were mapped to the database of phage genome sequences using BLASTN, which has been shown to have superior recruitment rates (Niu et al. 2011).

For a phage to be considered present at a particular location, metagenomic reads mapped to a region of its genome (i.e., a 'hit') were required to align for at least 50 base pairs with >85% nucleotide identity. Previous studies have shown that this is an acceptable threshold for avoiding spurious matches during viral fragment recruitment (Symonds, Griffin, Breitbart 2009; Mizuno et al. 2013a; Mizuno et al. 2013b). Hits meeting those requirements were retained. A metagenomic fragment might have hits to more than one phage genome. For each fragment, the best hit was identified by having the highest alignment quality (bit score) and the lowest expectation that the match is due to chance alone (E-value) (Madden, 2002). This best hit was considered to identify the source of the metagenome fragment, and thus that phage was deemed to be present at that metagenome sampling site.

---

[1] http://www.phantome.org/PhageSeed/seedviewer.cgi
[2] metagenomics.anl.gov/

Figure 2. Biomes shown on the global maps.

Figure 3. Ocean chlorophyll levels shown on the global maps.

From matches to maps

The best-hit results for each phage were organized into a list of metagenomes that then served as the 'sightings' indicated on the global map for that phage (circles in sample map below). Some phages did not show any significant matches, indicating they were undetectable or not present in the sampled environments. The identical 'sightings' shown for three phages (PZA, φ29, and L2) were verified by manually checking the metagenome matches.

Each phage's list of metagenome sightings was then imported into ArcMap 10.2 along with the 1232 metagenome sampling locations. All sampled locations were included on each map, thus showing where each phage could have been detected but was not (Fig. 1).

Interpreting the maps

Colored rings. For sampling locations with significant matches to a phage, the colored rings around each sampling location convey both the material sampled and the percent identity.

Material sampled: Ring color indicates the type of material sampled for each metagenome (e.g., freshwater, feces). These data were obtained from the MG-RAST metadata designated as feature, material, and/or package. For the Tara Oceans project, all material sampled was marine.

Percent identity: Ring size indicates the percent identity of the matches between the metagenomic fragments and the phage genome. A minimum 85% identity was required (see above). These reads were binned into three categories based on their alignment identities: 85% to <90%, ≥90 to <95%, and ≥95%. Each category corresponds to a ring size, as indicated in the map legend. For interpretation, see below.

Map colors. The colors of the land in these maps (Fig. 2) correspond to regions that have similar environmental conditions, habitat structure, and patterns of biological complexity, and that support communities with similar guild structures and species adaptations (Olson, Dinerstein 1998). These major habitat types are very similar to biomes (Olson, Dinerstein 1998), and were evaluated using multiple maps such as floristic and zoogeographic provinces, distribution of plants and animals, and biogeographic realms (Olson et al. 2001). The major habitat type map was downloaded from The Nature Conservancy's GIS Data webpage[3].

The ocean coloration depicts chlorophyll concentration (Fig. 3). This map was obtained from NASA's Ocean Color webpage, using the Level 3 data browser[4]. The data depicted were derived from images acquired by the Aqua MODIS satellite by measuring the reflectance of particular

---

[3] http://maps.tnc.org/gis_data.html
[4] http://oceancolor.gsfc.nasa.gov/

spectral bands. The 2013 annual composite map at 4 km resolution was chosen for this project because its global coverage of chlorophyll data is one of the most comprehensive, and it is freely and easily accessible. The original data in HDF format was converted to GeoTiff using SeaDAS open source image analysis software[5] before importation into ArcMap.

Percent identity and phage identity

One criterion used to classify a metagenome fragment as a significant match was the percent identity of its alignment to a known phage genome. Percent identity is an approximate measure of the relatedness between these aligned sequences. Any ancestral gene that is evolving in two descendant lineages will increasingly diverge, that is the percent nucleotide identity shared between those two gene sequences will decrease through time. Thus, the greater the identity, the more closely related (i.e., more recently diverged) are the phages carrying those gene sequences. Due to the rapid evolution of phages, detectable sequence similarity between phage genomes generally indicates recent divergence and thus close relatedness.

However, time is not the only factor determining the rate of sequence divergence. Due to natural selection, different genes in a phage genome evolve at different rates and thus would give differing relatedness signals based on this criterion. Some genes tolerate more sequence modification without loss of essential function, some less, and for yet others sequence change can be under strong positive selection to keep the phage one step ahead in the arms race with its host. Despite such confounding factors, we can say that a particular phage, or a close relative, was present in the sampled environment if the metagenome from that environment includes a genome fragment that is highly similar to a genomic region of that phage.

Matches with greater than 95% identity indicate that the metagenome included DNA from the featured phage or a very close relative. Fragments aligning with lower percent identities may represent matches to quickly evolving genome regions of the same phage or a very close relative. Alternatively, they may originate from a phage that is more distantly related to the featured phage but still shows a relatively high signal of relatedness.

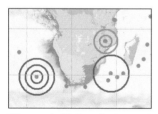

Figure 4. Colored rings at sampling sites.

At some geographic sites, different fragments of sampled DNA match a featured phage genome with varying identities. For example, at an ocean site west of South Africa, metagenomic fragments matched phage 2972 at all three identity levels (Fig. 4). What does this mean? It could be that a collection of 2972's relatives was sampled and/or that sampled regions of the 2972 genome are evolving at different rates in these phage populations.

Caveats? In some cases a featured phage such as T4 may be falsely identified in an environment if it recently exchanged a gene with an unknown phage. That gene would be associated with only T4 in the genome databases although it actually resides in *both* T4 and an unknown phage. Finding a fragment of that gene in a sampled environment we would conclude that T4 was present, but we might be wrong.

Conversely, a phage may also go undetected when it was present. This 'missing phage' may have been too rare to be included in the sample, or it could have been sampled but not se-

---

[5] http://seadas.gsfc.nasa.gov/

quenced. The RNA that some phages use for their genomes requires a sufficiently different protocol for isolation and characterization that it is often overlooked by phage-hunting safaris.

Cited references

Karsenti, E, SG Acinas, P Bork, C Bowler, C De Vargas, J Raes, M Sullivan, D Arendt, F Benzoni, J-M Claverie. 2011. A holistic approach to marine eco-systems biology. PLoS Biology 9:e1001177.

Madden T. 2002. The BLAST Sequence Analysis Tool [Updated 2003 Aug 13]. In: McEntyre J, Ostell J, editors. *The NCBI Handbook* [Internet]: National Center for Biotechnology Information (US). Chapter 16. Available online[6].

Meyer, F, D Paarmann, M D'Souza, R Olson, EM Glass, M Kubal, T Paczian, A Rodriguez, R Stevens, A Wilke. 2008. The metagenomics RAST server–a public resource for the automatic phylogenetic and functional analysis of metagenomes. BMC bioinformatics 9:386.

Mizuno, CM, F Rodriguez-Valera, I Garcia-Heredia, A-B Martin-Cuadrado, R Ghai. 2013a. Reconstruction of novel cyanobacterial siphovirus genomes from Mediterranean metagenomic fosmids. Appl Environ Microbiol 79:688-695.

Mizuno, CM, F Rodriguez-Valera, NE Kimes, R Ghai. 2013b. Expanding the marine virosphere using metagenomics. PLoS Genetics 9:e1003987.

Niu, B, Z Zhu, L Fu, S Wu, W Li. 2011. FR-HIT, a very fast program to recruit metagenomic reads to homologous reference genomes. Bioinformatics 27:1704-1705.

Olson, DM, E Dinerstein. 1998. The Global 200: A representation approach to conserving the Earth's most biologically valuable ecoregions. Conserv Biol 12:502-515.

Olson, DM, E Dinerstein, ED Wikramanayake, ND Burgess, GV Powell, EC Underwood, JA D'amico, I Itoua, HE Strand, JC Morrison. 2001. Terrestrial ecoregions of the world: A new map of life on Earth BioScience 51:933-938.

Symonds, EM, DW Griffin, M Breitbart. 2009. Eukaryotic viruses in wastewater samples from the United States. Appl Environ Microbiol 75:1402-1409.

---

[6] http://www.ncbi.nlm.nih.gov/books/NBK21097/

Appendix A5: Annotated Genome Maps

We prepared two versions of each phage genome: an overview that shows the genome as it would be 'seen' in the virion and a detailed genome map with each gene labeled with available information about its function, localization, and/or its relation to other phage genes (see POGs, below). Both renditions were created by consulting the primary literature and the annotated genome files in the NCBI GenBank repository. We do not claim that these GenBank annotations are the best available or that they will prove to be complete as more data are acquired. It is entirely possible that additional genes will be identified, gene boundaries will change, and gene functions will be discovered and clarified. We evaluated several annotation methods and sources and chose the GenBank annotations for several reasons: annotations for all of our featured phages were readily available; gene names familiar to the research community were usually included; researchers have had opportunities to update the annotations as research in phage biology progresses; and NCBI is a trusted source for cross-referenced 'omics data that can used by the keen reader seeking additional information.

To create a genome map, the genome was downloaded from the Viral Genomes database in GenBank format[1] (Table 1). Each genome file was then graphically manipulated in CLC Main Workbench[2] using custom Sequence Layout and Annotation Layout settings. UniProt[3] was searched for additional information regarding gene product function; if found, such information was incorporated by manually editing gene labels in CLC Main Workbench. Functional information was added to gene labels to indicate a putative function, a known function from research, and/or a predicted location in the virocell or virion (abbreviations in Table 2). Much of the functional information in the gene labels should be considered hypothetical.

Homologous genes shared between phage genomes were identified using the phage orthologous groups (POGs; [Kristensen et al. 2013]). When available, POG numbers were added to each gene label. A POG number is assigned if and only if at least two other distinct phages in the genome databases have a homolog of that gene (distinct phages being those that have less than 90% of their genes in common). Some genes were assigned multiple POGs due to an undetected domain fusion or paralogous relationship (Kristensen et al. 2013; D.M. Kristensen, personal communication). In the latter case the focal gene would most likely only belong to one POG; however we have included in the labels all associated POG numbers so as not to exclude any potentially useful information. This is particularly important because it cannot be determined which POG is the correct POG, or whether multiple POG assignments are due to domain fusion or unresolved paralogy. When these genome maps were created, genes from Enterobacteria phage χ, *Caulobacter* phage ϕCbK, and viruses of microbial eukaryotes were not yet incorporated into POGs.

Gene colors denote the life cycle stage when their protein (or RNA) product is likely most important for the phage. Several genes whose products are known to act in multiple life cycle stages were assigned to a separate color-keyed group. As expected, functional data were not available for many genes and these were categorized as *Unknown*. Some genes with similarity to genes that encode metabolic enzymes (e.g., NAD synthetase) were assumed to be necessary for maintaining the host until virion production is complete, and were thus classified as

---

[1] ftp.ncbi.nlm.nih.gov/genomes/Viruses/
[2] www.clcbio.com
[3] http://www.uniprot.org/

Replication-lytic. This category included genes whose products facilitate phage lytic replication in a broad sense (e.g., evading host restriction), rather than the more limited sense of genome replication. We acknowledge that many genes may be categorized erroneously and can only hope that future research into phage phunction will put these wayward genes back on track. For the large genome of Mimivirus, many genes were annotated based on similarity to a domain; however, we only included a function in the gene label if alignment similarities extended beyond a single domain.

Cited reference

Kristensen, DM, AS Waller, T Yamada, P Bork, AR Mushegian, EV Koonin. 2013. Orthologous gene clusters and taxon signature genes for viruses of prokaryotes. J Bact 195:941-950.

Table 1. Accession numbers for GenBank files used to create genome maps.

Phage	Accession number
Acanthamoeba polyphaga Mimivirus	NC_014649
Acholeplasma phage L2	NC_001447
Acidianus two-tailed virus	NC_007409
Bacillus phage φ29	NC_011048
Bacillus phage PZA	M11813 M13904 M13905
Bordetella phage BPP-1	NC_005357
Caulobacter phage φCbK	NC_019405
Enterobacteria phage χ	NC_021315
Enterobacteria phage f1	J02448
Enterobacteria phage HK97	NC_002167
Enterobacteria phage λ	NC_001416
Enterobacteria phage N15	NC_001901
Enterobacteria phage P4	NC_001609
Enterobacteria phage PRD1	NC_001421
Enterobacteria phage Qβ	NC_001890
Enterobacteria phage RB49	NC_005066
Enterobacteria phage RB51	NC_012635
Enterobacteria phage RB69	NC_004928
Enterobacteria phage T3	NC_003298
Enterobacteria phage T4	NC_000866
Enterobacteria phage T7	NC_001604
Mycobacteriophage Brujita	NC_011291
Pseudomonas phage φ6	M17461; M17462; M12921
Pseudomonas phage φ8	AF226851; AF226852; AF226853
Saccharomyces cerevisiae virus L-A	NC_003745; NC_001782
Salmonella phage Gifsy-2	NC_010393
Streptococcus phage 2972	NC_007019
Sulfolobus turreted icosahedral virus	NC_005892
Synechococcus phage S-SSM7	NC_015287
Yersinia phage φA1122	NC_004777

Table 2. Abbreviations used in gene labels.

Abbreviation	Definition
activ	activator
antirepress	antirepressor
BP	baseplate
connect	connector
DdRNAP	DNA-dependent RNA polymerase
dep	dependent
det	determinant
DHFR	dihydrofolate reductase
DHFR-TS	dihydrofolate reductase-thymidylate synthase
dsDNA	double-stranded DNA
endonuc	endonuclease
GALE	UDP-glucose 4-epimerase
imm	immunity
ind	inducible
LSU	large subunit
LTF	long tail fiber
m, meas	measure
mod	modifier
nt	nucleotide
nucleotidyltransfer	nucleotidyltransferase
PE	phosphatidylethanolamine
PG	peptidoglycan
PNK	polynucleotide kinase
PSII	photosystem II
PTOX	plastoquinol terminal oxidase
RdRNAP, RdRP	RNA-dependent RNA polymerase
recomb	recombination
reg	regulator
rep	replication
rev	reverse
ribo	ribosylase
RNAP	RNA polymerase
RNR	ribonucleoside diphosphate reductase
ssDNA	single-stranded DNA
SSU	small subunit
synth	synthetase
tc	transcription
TF	tail fiber
tl	translation
tp	transporter
util	utilization

Appendix B: Lexicon of Phage Behaviors
B1: Rationale

Why have a lexicon? Let's start with a simple question: Are phage and other viruses sexual organisms? Most biologists, including virologists, would say "No." Animals and plants have sex, microbes do not, and hence biology is split between sexual and asexual organisms. But let's look at this assumption in a little more detail.

It is an evolutionary necessity to eliminate deleterious mutations. All organisms, sexual and asexual, can accomplish this through homologous recombination, which often involves finding an outside source of DNA, i.e., a mate. Phage and other viruses actually have more potential for homologous recombination because 10 to >1,000 genomic copies are available in the same cell during phage replication; recombination can also occur between homologous phage and host genes (e.g., *psbA*).

But do phages find *mates*? The biological definition of sexual reproduction is the union of gametes, classified as mobile, male sperm (pollen in plants) and non-motile, female eggs (e.g., ovum, ovule). Virions are non-motile like ova. Many host cells are motile, making them more like sperm. From this perspective, it is apparent that the virocell is a gamete-producing organism that produces eggs (i.e., virions) and sperm (i.e., microbial cells). A phage's search for a host is a form of sexual selection in that it serves to bring two gametes together. As such it is subject to the same evolutionary dynamics, including co-evolution of receptors, exclusion of DNA from other species, and more. Thus a phage infecting a bacterial cell is another battleground in the classical war of the sexes.

The promiscuous phage also engage in what, from the Puritan macro-organismal point of view, is kinky sex. Instead of simply recombining between regions of homologous DNA (which requires ~200 bp of near identity), phages routinely recombine with total strangers. Non-homologous pieces of DNA (or RNA) recombine due to template-switching events that occur during the replication process. Here, before completely replicating one template strand, DNA polymerase can 'hop' to a new template strand that has only a short region of homology (<20 bp) with its newly synthesized strand. This can result in "hopeful monsters" (sensu Goldschmidt[1]) when the templates are from two unrelated phages that meet in the same host (e.g., a prophage and invader) or between phage and host genomes. With the advent of genomics, the genetic mosaicism produced by template switching became apparent, and showed that horizontal gene transfer (HGT) between phages was rampant (*see 5-48*). Only later was HGT documented in everything from Bacteria to plants.

That phage and other microbes engage in sexual practices highlights our point: that phage are biological entities that share many traits with the macrobes. Phage and other microbes are not just sexual, they have much more varied sex then the prudish macrobes. Phage are also predators that, as virions, hunt prey, sense their environment, and make choices. As a virocell they participate in the rich social life and diverse metabolic activities of their microbial host. However, a biologist's behavioral vocabulary, the product of millennia of observation of the macrobial

---

[1] Goldschmidt, R. 1933. Some aspects of evolution. Science 78:539-547.

world, has been regarded as applicable only to macrobes. People continue to argue whether or not viruses are alive! This often drowns out any mention of macrobe-microbe commonalities. A defined lexicon allows a biological synthesis that does not exclude the Earth's most numerous and most diverse creatures.

We started this process by creating a phage behavioral lexicon that combined our own knowledge of phage behavior with elements of an existing animal behavior ontology (ABOCore, kindly provided by Peter Midford). Lexicon terms were continually updated as we learned about new phage behaviors while researching the phage stories. This yielded a hierarchical lexicon structured first by life cycle stage, then by behavioral category, and lastly by discrete observed behaviors. We created this lexicon in OBOEdit[2] and then plotted it in Cytoscape[3] to create the network graphs shown in the following lexicon pages.

Studying behavior in phages offers a genetic advantage compared to using macro-organisms. For many lexicon terms we already know which phage gene(s) encodes that particular behavior. When such genes or gene products were featured in our action stories, they are shown next to their associated behavior term on the lexicon pages. We know that these single genes represent only a small sample of phage-encoded behavior. To provide a more complete picture, we also mapped gene families downloaded from ACLAME (for all phages in the ACLAME database, not just our featured phages). We assigned families with more than ten genes to the most relevant life cycle stage and gave them a collective functional assignment (e.g., recombination). This broad level of mapping was performed because each family may include multiple behaviors.

---

[2] http://oboedit.org/
[3] http://www.cytoscape.org/

Prowling for a Partner

RB49 wac (GI:33620663)

Extend tail fiber

Taste

Identify

Touch down

Sense

Assess host suitability
(stable interaction between
capsid and surface of host)

Interrogate

Recognize

RB51 hoc (GI:228861112)

Ambush

Waylay

Wait for host (molecular
interactions between capsid
and surrounding milieu)

Surprise

Lurk

Pounce

Watch

Hide

Perish

Die

Succumb

Decay

Resist

Respond to environmental
assaults

Dormancy

Defend genome

Prowling for a partner (free
virions searching for and
reversibly binding to host
cell surface)

Tumble

Walk

Forage

Stalk

ATV P800 (GI:75750445)
ATV P628 (GI:75750456)

Track

Enlarge or grow appendages

Float

Roam

Move to find host
(electrostatic interactions &
dipole moments)

Drift

Hop

Roll

Chase

PRD1 P2 (GI:9626353)
PRD1 P5 (GI:9626355)

Wander

Evaluate

Search

Ramble

Crawl

Walk-about

Collide

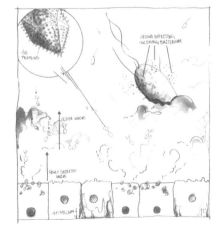

ACLAME protein families

Adsorption: 4 families

Attachment: 4 families

Neck: 3 families

Baseplate: 12 families

Recognize: 7 families

Capsid: 33 families

Tail fibers: 6 families

Decoration: 3 families

Tail: 53 families

Court Host

T3 gp14 (GI:17570830)
T3 gp15 (GI:17570831)
T3 gp16 (GI:17570832)
Extend appendage

Infiltrate

Φ6 P6 GI:215489
Fuse membranes

Mob

Kick T3 gp14 (GI:17570830)
T3 gp15 (GI:17570831)
T3 gp16 (GI:17570832)
Channel

Plug membrane hole

Digest
PRD1 P7 (GI:9626369)

Penetrate host

Gnaw
f1 g3p (GI:166208)

Hit

Secrete

Flush genome

Merge

Excavate

Burrow
Φ6 P5 (GI:215496)

Contract tail sheath
PRD1 P20 (GI:9626360)
PRD1 P22 (GI:9626363)
PRD1 P32 (GI:159192292)
PRD1 P11 (GI:159192287)
PRD1 P14 (GI:159192294)

Fertilize host (Deliver capsid
contents)

Remodel capsid to extend
tube

Court host; a virocell is born
(irreversibly bind to host cell
surface)

Reject host

**Translocate genome with
molecular motor**
T7 RNAP (GI:9627432)
T7 gp15 (GI:9627475)
T7 gp16 (GI:9627476)

Grapple   Attach

Grasp

Ride flagellum
ΦCbK gp263 (GI:414087789)

Probe

Accept host (Adsorption)   Hold

Hang

Climb   Land

Wrestle

Stick

Latch onto pilus
Φ6 P3 spike (GI:215490)
f1 g3p (GI:166208)

Bind

Lasso
χ gene 71 (GI:509139294)

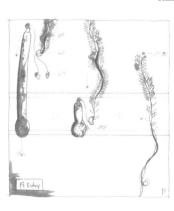

ACLAME protein families

Channel formation: 6 families

Diesterase: 1 family

DNA delivery: 6 families

Protein delivery: 3 families

Tail fibers: 2 families

Transglycosylase: 3 families

Translocation: 6 families

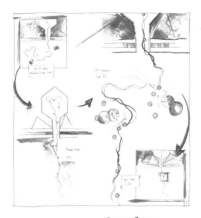

Ensure Virocell Viability

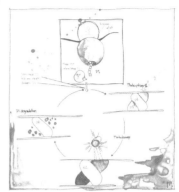

S-SSM7 DI (GI:326783933)   Prevent suicide

Photosynthesize

Resuscitate

Camouflage

RB69 ipI (GI:32453636)

Interfere   React   Recognize kin

Φ6 MCP P8 GI:215493

Hide   Allow super-infection

Maintain metabolism

Dispatch with host defenses

Avoid   Nurse virocell   Defend virocell   Prevent super-infection

Mimic

ΦA1122 Ocr GI:30387454   Disguise   Ensure virocell viability
                                            (takeover host cell's
                                            molecular machinery and
                                            metabolism)

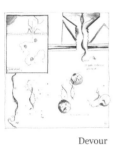   Injure

                                                                          CRISPRs
                                                                          Outwit

                                            Threat

                               Redirect   Coerce host   Outshine
                                                                                       S-SSM7 PDF (GI:326783953)
Devour

                               Hijack

Divert   Garner resources   Kill   Render helpless
                                                                        T4 alc GI:9632701

                                                                          Spar

                                                              Stun

Triumph   Commandeer

        Appropriate   Seize
        T4 asiA (GI:9632708)   Usurp   T4 alt (GI:9632702)   Pathogenicity: 5 families
        T4 motA (GI:9632776)   Phosphate metabolism: 1 family
                                                                        Restriction escape: 1 family
                                                                        Serotype conversion: 2 families
                                                                        Sigma factor: 1 family
ACLAME protein families   Super-infection exclusion: 5 families
Host killing: 2 families   Toxin production: 4 families
LPS biosynthesis: 1 family   Transcription: 1 family
Membrane protein: 2 families      Transporters: 1 family
Modulation of host functions: 6 families
Nucleoid-associated: 1 family

Hedge Life History Bet

<u>ACLAME protein families</u>

DNA repair: 2 families

Excision: 3 families

Homing endonuclease: 2 families

Integration: 2 families

Ligase: 4 families

Lysis/lysogeny switch: 6 families

Methyltransferase: 7 families

Nuclease: 9 families

Nucleotide metabolism: 17 families

Recombinase: 12 families

Recombination: 21 families

Replication: 42 families

Segregation: 2 families

Topoisomerase: 1 family

Transcription: 20 families

Translation: 1 family

Transposition: 7 families

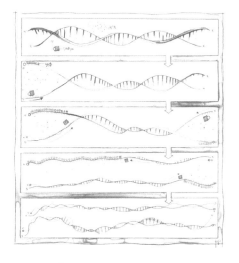

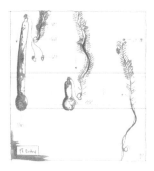

Multi-purpose proteins
T4 gp45 (GI:9632639)
Qβ A$_2$ (GI:9630318)
f1 g3p (GI:166208)
L-A Gag (GI:20428568)

Overlapping genes

Economize

Outcross (illegitimate
recombination)

PZA TP (GI:216050)

Innovate

Replicate genome

Protect single strands

Recombination/Sex

Inbreed (homologous
recombination)

Fix errors

Lytic replication

Manipulate nucleic acid
polymers

 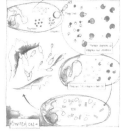

Mass produce genome copies

Initiate genome replication
PZA TP (GI:216050)

Open helix

Recycle

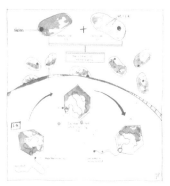

L-A Gag (GI:20428568)
Steal

Obtain materials

Cannibalize

Sponge

Reuse

Mooch

Scavenge   Thief

M1 gp1 (GI:9629211)
Produce toxin

Traitor   Betray

Enable colonization of new
niche

Defend host

Increase virocell fitness

Camouflage host   Inactivate

Φ29 OA-boxes
Φ29 parS sites
Hibernate

Jump around
Re-activate / Excise

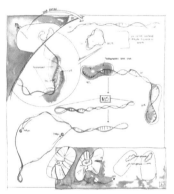

Temporarily collaborate   Propagate

Conspire

N15 SopA (GI:9630492)
N15 SopB (GI:9630491)
Self-sustaining

Mutate host

Take census

Integrate   Kill competitors

Hedge life history bet
(decide to replicate genome
now or later)
N15 gp29 (GI:9630493)

Linearize genome

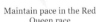

Express proteins
T4 gp33 (GI:9632625)
Circularize genome   *T4 gp55 (GI:9632660)*

Mutate

Maintain pace in the Red
Queen race

Capsid coat

Evolve
Go extinct
Diversify

Modularity

Modify nucleotides
RB69 p48 (GI:32453532)

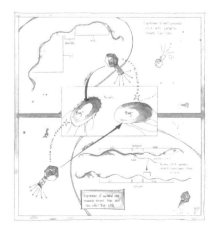

Genome

Tail fibers

Tropism switching

Morphogenesis of Progeny

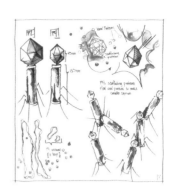

P4 Sid (GI:9627518)
P4 Psu (GI:9627520)

Steal

Coerce

Synthesize

Displace

Scavenge

Obtain components

Cheat

Quality control

Tail fiber assembly
Process DNA   Build shell

Symmetry
Measure tail length

Join head and tail   Morphogenesis of progeny
(assemble virions)   Manage production

Schedule

Offspring development and
care

Reinforce   Baseplate assembly   Mass produce
HK97 gp5 (GI:9634158)

Build virion factory

Gatekeeper of portal
Accessorize capsid   Incubate   Build nest

Tail assembly

Scaffold

Choreograph
Prepare tools   *Φ8 P4 (GI:7532972)*
Orchestrate   *Φ8 pac sequences*

Package DNA

Pump

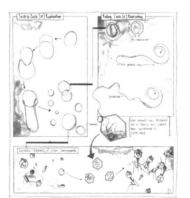

Coat with protein   Motor

Stuff prohead
Terminase

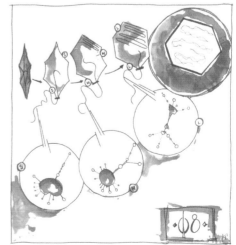

<u>*ACLAME protein families*</u>

Capsid-tail joining: 14 families   *Portal: 6 families*

DNA maturation: 15 families   *Procapsid: 1 family*

Genome packaging: 11 families   *Protease: 5 families*

Late transcription: 4 families   *Scaffold: 12 families*

Tape measure: 2 families

Wean Progeny

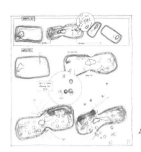

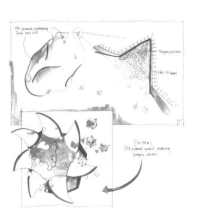

λ Rz (GI:9626310)
λ Rz1 (GI:160338810)
Annihilate outer membrane

Lyse

Detonate

STIV c92 (GI:48696998)
Build pyramids

Qβ A₂ (GI:9630318)
Sabotage

Blow up

Destroy virocell

End game

Degrade cell wall
λ R (GI:9626309)

Set timer
λ S anti-holin (GI:9626308)

Light fuse
λ S' holin (GI:160380505)

Wean progeny (free virions
from virocell)

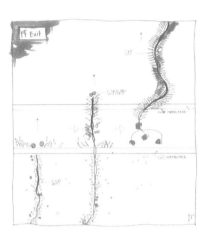

Quietly depart

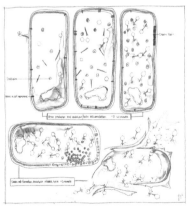

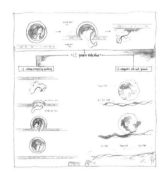

Abandon virocell

Re-appear
f1 g8p (GI:166207)

<u>*ACLAME protein families*</u>
Spanins: 4 families
Holins: 14 families
Endolysins: 10 families
Lysis: 6 families

Bud

Escape

Extrude

Glossary

Note: The first appearance of a glossary term within a story is highlighted in boldface. Some of these terms have different meanings depending on the context. The definitions provided here apply specifically to their usage in relation to phages.

Adsorption: the specific interaction between a phage and a **receptor** structure on the surface of a potential host cell; the first step in infection. Initial adsorption may be reversible; subsequent irreversible binding typically leads immediately to structural changes in the virion and transfer of the genome into the host cell.

Burst size: the number of progeny virions released when a phage terminates lytic replication and lyses the host cell.

Capsomer: a subunit of a viral capsid that is formed by the aggregation of protein monomers and that, in turn, self-assembles to construct a capsid shell.

Catenane: a molecular architecture composed of interlocked subunits, each subunit being a covalently-linked ring. To liberate any ring requires breaking ring covalent bonds.

Caudovirales: the tailed phages. In the **ICTV** taxonomy, *Caudovirales* is an order that includes three families: the *Myoviridae*, the *Siphoviridae*, and the *Podoviridae*.

Co-condensation: a method of genome **encapsidation** in which the capsid proteins assemble around the DNA or RNA genome.

Cohesive ends: an overhanging length of single-stranded DNA at both ends of some phage genomes. Either the 5′ or the 3′ end is extended at both ends of the genome. Upon arrival in the host cytoplasm, these complementary overhangs facilitate circularization of the genome, thereby forming the **cos site**.

Cos site: when the substrate for the packaging **terminase** is a circular double-stranded DNA molecule, the location in the DNA where the terminase makes two staggered, single-strand cuts to produce the linear form packaged in the virion. This cleavage leaves an overhang at each end ('sticky' or **cohesive ends**) that reanneal to circularize the genome after arrival in the host cytoplasm.

Direct repeat: a nucleotide sequence that is repeated in the same direction at two locations in a DNA or RNA strand.

Direct terminal repeat: (DTR) a direct repeat with the repeat sequences at the two strand termini.

Encapsidation: the process whereby a phage genome acquires its protective protein capsid or coat. *See also* **packaging** *and* **co-condensation**.

End replication problem: a difficulty that arises because DNA polymerase (DNApol) can synthesize DNA in only the 5′ → 3′ direction and, moreover, requires a primer, i.e., a hydroxyl group to which it can attach the initial 5′ nucleotide. If left unresolved, a few nucleotides at the 5′-terminus would not be replicated and the genome would shrink each generation.

Endolysin: a phage-encoded peptidoglycan-degrading enzyme used during host lysis.

Envelope: in Bacteria and Archaea, the cell membrane and associated outer layers, e.g., cell wall, S-layer, capsule, outer membrane. In phages, an external lipid layer.

Episome: In Bacteria and Archaea, a genetic element, usually circular, that can integrate into the cell's chromosome(s) or can replicate and partition into daughter cells independently.

Excisionase: an enzyme encoded by **temperate phages** that excises the **prophage** from the host's chromosome.

Flagellum: (plural, flagella) long filamentous appendages anchored to the membrane of some Bacteria and Archaea that confer motility. Each flagellum is rotated like a propeller by the 'motor' at its base. A cell may have one polar flagellum or numerous flagella scattered over the cell surface. In the latter case, the cell coordinates their rotation.

Foldon: a protein domain that folds quasi-independently.

Frameshift: (ribosomal frameshift) a mechanism that enables two proteins encoded by adjacent or overlapping genes to be consistently translated in specific relative numbers. Normal translation of a gene transcript yields one protein from one reading frame; a frameshift during translation yields a different protein, part of which is translated from a different reading frame. A typical ribosomal frameshift occurs at low frequency when the translating ribosome slips by one base in either the 5′(-1) or 3′(+1) direction, thus producing a -1 or +1 frameshift. Frameshifts are triggered by encoded factors such as the secondary structure of the RNA transcript and the sequence of codons. Where and how often they occur can be 'programmed' in the gene. Since frameshifts typically occur at low frequency during translation, many more copies of the normal protein will be made than of the frameshifted product.

Heptad repeats: a protein motif composed of a repeating pattern of seven amino acids. If the protein adopts an α-helical structure, the repeats enable two or more of these protein helices to align and entwine to form a coiled-coil protein.

Homolog: a gene that is related to another gene by virtue of having descended from a common ancestral gene. Proteins encoded by those genes are likewise considered to be homologous proteins. Homology is assessed by comparing nucleotide or amino acid sequences, or by comparing their three-dimensional structures ('folds').

Homologous recombination: **recombination** between genomic regions (DNA or RNA) that are similar in sequence.

Host tropism: the host range of a phage, i.e., which cells it can productively infect.

Hyperthermophile: a life form that grows optimally at extremely elevated temperatures, typically considered to be 80° C or above.

ICTV: (International Committee on Taxonomy of Viruses) an organization founded in the late 1960s and charged with responsibility for developing and maintaining a universal virus taxonomy.

Induction: the excision of a prophage from the host chromosome, usually followed by lytic replication and host lysis.

Integrase: an enzyme encoded by **temperate phages** that converts the phage DNA genome into a **prophage** by inserting it into the host's chromosome at a particular location by **site-specific recombination**.

Integration: *See* **integrase**.

Internal protein: a protein packaged inside a virion that, upon infection of a host, is delivered into the cell with the genome.

Inverted repeat: a nucleotide sequence whose reverse complementary sequence is located on the same DNA or RNA strand.

Inverted terminal repeat: (ITR) an inverted repeat in which the reverse complementary sequences are located at the strand termini.

Latent period: during lytic infection, the time between phage **adsorption** and host lysis.

Lipopolysaccharide: (LPS) a major component of the outer membrane of Gram-negative Bacteria. It is found in the outer leaflet of the bilayered cell membrane and is composed of a lipid and a polysaccharide core, extending outward from which are glycan chains (O-antigens). Numerous phages (e.g., T3, T4, T7, P2, P22, φX174) use LPS as their receptor.

Lysogen: a bacterium or archaeon that carries one or more phage genomes either integrated within its chromosome as a **prophage** or stably maintained extrachromosomally as a **plasmid** or in some other form. In this situation most, but not all, of the phage-encoded genes are not being expressed and progeny virions are not being released from the cell.

Lysogeny: the phage lifestyle in which phage-directed virion production is postponed indefinitely and the phage genome is stably maintained within the bacterial host (**lysogen**).

Methyltransferase: an enzyme that catalyzes the transfer of a methyl group from one molecule to another, typically from S-adenosyl methionine to a nucleic acid or protein.

MOI: See multiplicity of infection.

Mollicute: a class of Bacteria that lacks a cell wall, instead enclosing their cellular contents within only a sterol-containing membrane. Most members are small, less than half the size of *E. coli*.

Mucosal surface: regions on the surface of an animal covered by a protective mucus layer secreted by the underlying epithelium. These surfaces are involved in processes such as food adsorption and gas exchange, thus must be thin, permeable barriers. As a result they are particularly vulnerable to infection.

Multiplicity of infection: (MOI) at the start of an infection, the average number of phage genomes that have entered each host cell.

Muralytic enzyme: a protein that cleaves **peptidoglycan** (murein).

Negative-sense: for RNA, the strand that is complementary to the messenger RNA (mRNA). For DNA, the strand whose sequence (with T replaced by U) is complementary to the mRNA. Thus RNA or DNA strands can serve as the template for synthesis of mRNA. *See also* **positive-sense**.

Nuclease: an enzyme that cleaves a nucleic acid (RNA or DNA). An exonuclease progressively removes the terminal nucleotides, whereas an endonuclease cleaves the molecule at an interior location.

Nucleoid: the region within a bacterial or archaeal cell where the DNA is localized, thus also the site of DNA replication and transcription, as well as the translation of mRNA into protein.

Packaging: a method of encapsidation in which the genome is translocated into a preformed procapsid by an energy-requiring process.

Peptidoglycan: (or murein) in Bacteria, the cross-linked polymer of sugars and amino acids that forms the meshwork outside the cell membrane known as the cell wall.

Periplasm: in Gram-negative Bacteria, the cell compartment between the inner membrane (cell membrane) and the outer membrane. This zone contains a thin mesh of peptidoglycan and numerous enzymes, including **nucleases**.

Phagocytosis: ingestion of a particle, virus, or small cell by a eukaryotic cell. In the process, the cell membrane infolds around the object and eventually pinches off to form a vesicle (phagosome). This behavior is characteristic of amoeba and other amoeboid cells.

Phagosome: a vesicle inside a eukaryotic cell formed by **phagocytosis**.

Pilus: (plural, pili) hairlike appendages found on the surface of many Bacteria and Archaea that function in cell adherence, protein export, motility, or the transfer of DNA between cells. Pili are assembled from multiple copies of a single protein, pilin, that are added or removed at the pilus base.

Plasmid: typically a circular, double-stranded DNA replicon found in many Bacteria and Archaea that replicates independent of the host chromosome. Plasmids often carry specialization genes that are beneficial for the host in particular circumstances, e.g., antibiotic resistance factors. Small plasmids are frequently transferred between hosts, even between hosts in different, but closely-related, genera; large plasmids are not mobile and may be in the process of becoming secondary chromosomes. Some bacterial and phage plasmids are linear. Plasmids typically provide factors needed for their own replication; low copy number plasmids also encode a mechanism for their segregation to both daughter cells.

Polycistronic mRNA: a messenger RNA that contains more than one open reading frame (ORF) and thus encodes more than one polypeptide or protein.

Porins: a class of proteins that form channels through the outer membrane of Gram-negative Bacteria and that function as pores that allow the passage of small molecules by diffusion. As such, they are distinct from membrane transporter proteins that actively facilitate the movement of molecules through membranes.

Positive-sense: for RNA, the strand that can serve as messenger RNA (mRNA). For DNA, a strand whose sequence (with T replaced by U) is the same as the mRNA. *See also* **negative-sense**.

Procapsid: an assembled capsid before the genome has been packaged inside.

Prophage: a phage genome present in a **lysogen**. Most often it has integrated at a specific location into the chromosome of the host bacterium or archaeon and is replicated by the host as part of the host's chromosome. Many prophage genes are not transcribed (i.e., are repressed); those expressed include those that repress the lytic cycle and sometimes genes that benefit the host.

Protelomerase: the prokaryotic 'telomerase.' The protelomerase enzyme generates covalently-closed hairpin ends on linear prophage plasmids and subsequently assists in plasmid replication. Protelomerase is unrelated to eukaryotic telomerase, and likewise phage telomeres differ structurally from those on eukaryotic chromosomes.

Pseudolysogeny: the stalled development of a phage in a **virocell** in which its genome is neither replicated nor degraded, often associated with unfavorable growth conditions for the host. When conditions improve, the phage life cycle continues.

Receptor: a structure exposed on the surface of a bacterium or archaeon that is used by one or more different phages to **adsorb** to a potential host cell. Receptors may be components of the cell envelope (e.g., proteins, teichoic acids, **lipopolysaccharides**) or on a cell appendage (e.g., **pilus**, **flagellum**).

Recognition site: (restriction site) a short DNA sequence, typically 4-8 bp, that is recognized by a restriction-modification system and used to distinguish 'self' DNA from foreign DNA. Recognition as self requires the methylation of one or more specific bases within that sequence.

Recombination: (genetic recombination) in phages, the generation of a new nucleotide strand (DNA or RNA) from two or more 'parental' strands. The second parent can be a different region of the phage genome, another phage genome, a host genome, or some other mobile element. Because recombination can generate new genetic combinations, it is considered a form of sex. It has the potential to not only accelerate adaptation in phages but also possibly allow their escape from Muller's ratchet. Some of the machinery used in recombination also functions in genome replication and repair. *See also* **homologous recombination** *and* **site-specific recombination**.

Restriction: the site-specific cleavage of an invading phage genome by a host-encoded endonuclease.

Restriction endonuclease: See **restriction-modification system**.

Restriction-modification: a bacterial and archaeal defense against invading foreign DNA (phages, plasmids, etc.) that distinguishes 'self' DNA from 'non-self'. 'Self' is identified by being previously methylated at specific sites; DNA in which those sites are unmethylated is restricted, i.e., cleaved by a sequence-specific 'restriction' endonuclease. A few phages also encode such a system for cleavage of host DNA.

Secondary receptor: when phage **adsorption** or genome entry requires sequential interaction with two different **receptors**, the second of those receptors.

Segmented genome: a phage genome that is encoded in more than one molecule of DNA or RNA. Each molecule is termed a segment. When each segment is packaged in a separate virion, the segmented genome is said to also be multipartite.

Sigma (σ) factor: in Bacteria, a dissociable subunit of the RNA polymerase holoenzyme that is required for the recognition of promoter sequences and the initiation of transcription. When multiple genes have similar promoter sequences, their transcription can be regulated coordinately by the availability of the corresponding sigma factor.

Site-specific recombination: genetic recombination between specific short DNA sequences. For example, in **lysogeny** a phage-encoded **integrase** inserts the **prophage** into the host chromosome by **recombination** between the *attP* site on the phage genome and the *attB* site in the host chromosome. The phage **excisionase** carries out the reverse reaction to excise the phage genome.

Sporulation: differentiation of a bacterial cell into a metabolically dormant spore.

Stationary phase: in the laboratory, a period in the bacterial life cycle when cell numbers stop increasing due to the exhaustion of nutrients or accumulation of metabolic products in the medium. Cryptic growth may occur when viable cells feast on dead cells. In nature, Bacteria likely exist in a similar physiological state most of the time.

Superinfection exclusion: (superinfection immunity) protection afforded a lysogen by its resident prophage(s) against infection by the same or related phages.

Swarmer cells: motile cells of various Bacteria that have a complex life cycle that includes both sessile and motile phases (e.g., members of the genus *Caulobacter*).

Synteny: when comparing genomes, the conservation of gene order between genomes; an indication of shared ancestry.

Temperate phage: a phage that upon infection chooses between the lytic and **lysogenic** pathways.

Terminal protein: a protein bound to the 5′ ends of some linear dsDNA phage genomes. It primes DNA replication, thus is one strategy for overcoming the **end replication problem** faced by linear genomes. Terminal proteins may also assist with genome packaging.

Terminase: a phage enzyme that packages a linear dsDNA phage genome into a **procapsid**. The packaging substrate is often a concatenated chain of multiple genomes. The enzyme is composed of two subunits: TerL, the large subunit binds the procapsid, cleaves the DNA into monomeric units, and translocates the genome into the procapsid; TerS, the small subunit, recognizes and binds a specific sequence in the phage DNA, thus ensuring only that DNA is packaged.

Tropism: See **host tropism**.

Twitching motility: a mechanism of bacterial motility across a surface mediated by Type IV **pili**. The pilus tip adheres to the surface ahead and then the pilus is forcibly retracted, thereby moving the cell forward.

Virion: a phage genome enclosed within a capsid; the extracellular stage of the phage life cycle; an infectious viral particle.

Virocell: the intracellular stage of the phage life cycle that begins with adsorption and ends with release of progeny virions from the host cell.

Virospore: during sporulation in some Firmicute Bacteria, a spore that, in addition to the bacterial genome, contains a non-integrated phage genome.

Index

About Us

Forest Rohwer[1]**, PhD,** is a Professor of Biology at San Diego State University. Early in the 21st century he pioneered research in phage diversity and ecology, en route providing the world with new avenues for phage discovery through the development of innovative metagenomic tools. Currently this work continues in his lab at warp speed, in conjunction with research on coral and human ecosystems.

It was in the late 1990s that the phages captured Forest and diverted him from other anticipated research pursuits. Once he started down this path, there was no exit. The more he investigated them, the more he recognized their staggering importance to the health and functioning of every ecosystem. Looking around, he saw that much of biology was still overlooking or ignoring the Earth's most abundant and most diverse life forms. Something more was needed to secure for the phages their deserved attention: a global celebration. As 2012 drew to a close, he envisioned a book celebrating 2015, the centennial of their discovery, as the Year of the Phage. He infected others with this vision, and it came to pass.

Merry Youle, PhD, is a freelance microbiology editor and writer. Her writing for her beloved phages includes more than 30 stories contributed to the ASM-sponsored blog, Small Things Considered, a habit that she continues now at a slower pace as blogger emerita. She also co-authored the 2010 book *Coral Reefs in the Microbial Seas* with Forest.

As her contribution to the Year of the Phage, she edited all of the text in this book, authored most of the stories, encouraged her colleagues as they created the artwork and maps, then gave herself the title of Publishing Editor as she carried Forest's vision through to publication. The phages, however, are not yet finished with her. Next to bubble up from her lava tube home in Hawai'i will be a book to introduce a broad readership to the creativity of the phage multitude that supports all life on Earth.

Heather Maughan[2]**, PhD,** is a freelance writer and bacteriologist, and an independent scholar at The Ronin Institute. She spies on Bacteria as they evolve to make sense of their genomes and metagenomes. When not at her desk working for the microbes, she treks across the winter landscape on snowshoes or attends to the crops and critters on her small Ontario farm.

Despite their blatant cruelty to Bacteria, the phages succeeded in enticing Heather to contribute to their centennial celebration. In their service for this book she has written stories, coordinated art production, created detailed annotated maps of phage genomes, and established a lively lexicon from observed phage behaviors. But all this notwithstanding, has she truly become a phageophile? Or is she merely an undercover agent for the Bacteria spying on the enemy? We suspect the latter.

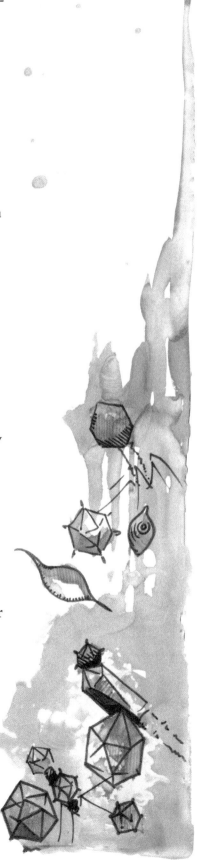

---

[1] phuckitphage.org/
[2] www.heathermaughan.ca

Nao Hisakawa, M.S., is a GIS Researcher in the Biology department at San Diego State University. She regularly creates GIS-based maps and uses remote sensing techniques to help track microbes on the surface of the Earth, e.g., to study the distribution of the Arctic algae that form red snow. As a geographer and an aspiring visualization scientist, she possessed the skills needed to wrestle hundreds of phage lineages and metagenomes into interpretable graphical displays.

Nao was admittedly shocked to encounter a global map replete with the familiar biomes, but now crawling with phages. Phages were everywhere anyone looked! Her assignment was to sort them out, determine who had been sighted at each location sampled and with what certainty. For her second mission, she was given 1220 genomes to be arranged in a visual schema where we could infer their relatedness. Missions completed.

Leah L Pantéa[3] is a classically trained artist who earned her BFA at Central Washington University and now specializes in drawing and painting. Her current mixed media abstract landscape work draws inspiration from the Abstract Expressionist movement. She lives and creates in San Diego.

From time to time, Forest entices Leah to briefly wander away from her fine art and transform into an illustrator charged with portraying entities that she cannot see. The phages knew Leah readily collaborates with scientists, and welcomed her role in this project as they knew words alone would not be sufficient to convey their crazy antics. Thus persuaded, she put ink on paper to tell their stories and infuse life into their nucleic acids. This task completed, the phage images still swirling in her head spilled over into one of her abstract landscapes and from this was born our book cover.

Benjamin Darby[4] is a vibrant artist who received his BFA from Cornish College of the Arts in Seattle. His paintings mix cynicism, social commentary, and humor using acrylic and unique texturing. His work has been exhibited in galleries throughout North America. He lives and creates in San Diego where Forest can readily nab him for challenging projects, a fate that he welcomes as he enjoys collaboration with scientists.

Being immodestly aware of their exquisite beauty, the phages sought Ben's skilled pen to picture the diversity of their detailed capsids, the grandeur of their incredible tails, and the weightlessness of their playful fibers. They knew that given a fuzzy electron micrograph, a T-number, and optimally a tomographic reconstruction or two, he would synthesize a mental picture of the virion in three dimensions, twirl it around in his mind, then draw it in his chosen orientation.

---

[3] www.llpfineart.com
[4] www.darbyarts.com/

Alexis Morrison[5] arrived in the phage world with an extensive background in graphic design and typography on board, eager to dig in on a major new project. We welcomed her with hundreds of files and asked her to make of them a beautiful book. She complied, solving innumerable problems en route much to the relief of the Publishing Editor. Multifaceted, she loves to paint and to sew, to work in papier mache, to create mysterious poetry. The ideal week is an artistic smorgasbord balanced by a few hours on the beach. Having strong interests in both fashion design and recycling, and delighting in miniatures, Alexis found she has much in common with the phages. After tasting both the East and West Coasts, she currently lives a quietly creative life on the Big Island of Hawai'i as one of the Zen Hens—a love-life-art support network forged with two imaginative friends. Her handmade treasures travel widely, leaving her contentedly creating at home.

Our Acknowledgements

We would like to thank all the people who in their individual ways supported and guided each of us in our efforts to assemble the myriad pieces that now fill the pages of this book. Any errors that may have slipped in are solely our responsibility. In particular, we are grateful to:

Ryan Bulger for writing the scripts that made our data handling and analysis more efficient.

Alex Burgin, Betty Kutter, Breeann Kirby, and members of the Rohwer lab for their helpful reading of our stories.

Ana Cobián for her persistence in determining the groups of the PPT.

Bas Dutilh for guidance in developing the phage behavior lexicon.

Rob Edwards for calculating the PPT and answering a myriad of questions.

David Kristensen for help with interpreting Phage Orthologous Groups.

Katelyn McNair for performing the fragment recruitment analysis used in our global maps.

Peter Midford for providing the ABOCore animal behavior ontology.

Anne Vidaver for ferreting out the currently accepted taxonomy of *Pseudomonas savastanoi* pv. phaseolicola.

---

[5] http://www.zenhensart.com/

CPSIA information can be obtained at www.ICGtesting.com
Printed in the USA
BVIW12n0934040318
509341BV00005B/32